ENDOCRINOLOGY OF EMBRYO–ENDOMETRIUM INTERACTIONS

REPRODUCTIVE BIOLOGY

Series Editor: Sheldon J. Segal

The Population Council
New York, New York

A Continuation Order Plan is available for this series. A continuation order will bring delivery of each new volume immediately upon publication. Volumes are billed only upon actual shipment. For further information please contact the publisher.

ENDOCRINOLOGY OF EMBRYO–ENDOMETRIUM INTERACTIONS

Edited by

Stanley R. Glasser and
Joy Mulholland
Baylor College of Medicine
Houston, Texas

and

Alexandre Psychoyos
Hôpital Bicêtre
Bâtiment INSERM Gregory Pincus
Bicêtre, France

SPRINGER SCIENCE+BUSINESS MEDIA, LLC

Library of Congress Cataloging-in-Publication Data

On file

Proceedings of a Satellite Symposium of the Ninth International Congress of Endocrinology on
Endocrinology of Embryo–Endometrial Interactions,
held September 6–10, 1992, in Bordeaux, France

ISBN 978-0-306-44809-6   ISBN 978-1-4615-1881-5 (eBook)
DOI 10.1007/978-1-4615-1881-5

© 1994 Springer Science+Business Media New York
Originally published by Plenum Press, New York in 1994

PREFACE

Early embryonic loss is a continuing social and economic global problem. In human populations the estimates of interruptions early in pregnancy range from 35-60%. In animal husbandry (swine, ruminants) fully 30% of pregnancies fail to survive early events of gestation. The futility associated with this persistant high risk is even more unsettling because of advances made in assisted reproductive technology which, although this very selective methodology has added to our knowledge of embryo-endometrial interactions, has resulted in a live birth rate of only 14%. These studies have instigated comparisons of the relative contributions of the embryo and the uterus to the outcome of pregnancy. These analyses have shown that we have learned significantly less about the role of the uterus in deciding the outcome of either natural or assisted pregnancies.

In 1979 a quotation by George Corner was used to set the tone of a meeting that was devoted to discussion of the cellular and molecular aspects of implantation. In spite of the proliferation in research activity which occurred in the following 15 years our real understanding of the embryo transfer process has fallen short of our expectations. We use the Corner quotation, once again, to preface this symposium so that we may recall that the fundamental nature of the process which regulates embryo-endometrial interactions still escapes us.

> "It is characteristic of eggs and early embryos of lower animals that they are prepared to develop without shelter and nutriment from the mother. When the mammals evolved the phenomenon of utero-gestation, the chosen place of shelter, the uterus, was developed from the part of the oviduct, a channel that had for its purpose the efficient transportation and discharge of eggs, not their retention and maintenance. To fit it for gestational functions, the endocrine mechanism of the corpus luteum was evolved. In the light of this thought it is not surprising that the uterine chamber is actually a less favorable place for early embryos than, say, the anterior chamber of the eye, except when the hormones of the ovary act upon it and change it to a place of superior efficiency for its new function."

George W. Corner
The Hormones in Human Reproduction,
1947

This meeting was designated, by the Ninth International Congress of Endocrinology, as the only satellite symposium to be charged with critically examining current knowledge of the regulatory biology of the peri-implantation period.

The conference was organized by an international committee of active investigators who share an interest in experimental and applied research related to embryo-endometrial interactions. Our organizing committee included J. Findlay (Australia), F. Leroy (Belgium), C. Tachi, T. Mori (Japan), G. Zeilmaker (The Netherlands), B.O. Nilsson (Sweden), S.C. Bell (United Kingdom) and S.K. Dey, G. Gibori, Z. Rosenwaks and J. Strauss (United States of America). The committee organized the symposium to provide a common ground that would facilitate the exchange of technology and ideas. Realization of these goals was optimized by inviting speakers and moderators from various disciplines who, independent of their seniority, were selected on the evidence of their innovative research and their commitment to constant redefinition of the state of their science. To this end the program included a special M.C. Chang Commemorative Lecture to recognize the long history of fundamental contributions that M.C. Chang, together with Gregory Pincus, made to reproductive science. These studies, initiated over a half-century ago, catalyzed and redirected the research that is the subject of this symposium. Interaction between all participants of the symposium provided the stimulus for free and critical evaluation of current research, assessment of theories and methods and would promote, if indicated, the development of new concepts and strategies.

The symposium focused on the biochemical and molecular mechanisms that program the generation of reciprocal maternal and embryonic signals. The transduction of these signals synchronizes the development of uterine receptivity to the embryo and coordinately initiates the processes that follow. The papers that appear in this volume critically discuss whether these signals (hormones, growth factors, cytokines, etc.) act directly on the uterine epithelial cell and/or indirectly via the uterine stromal cell/extracellular matrix compartment to transform the uterus sequentially from non-receptive to ovoreceptive to refractory. Interdisciplinary discussions of endometrial cell type responses to blastocyst signals serve to identify the cell-cell, cell-substratum processes by which the embryonic signals are translated and transduced in different species to establish and maintain gestation. Discussion of *in vitro* models for comparative studies of embryo (human, rodent, bovine)-endometrial relationships examined the validity of methods that can clarify the fundamental mechanisms common to the regulation of embryo-endometrial interactions in all species. After three days in Bordeaux this symposium concluded by celebrating the 100th anniversary of the description of lactational delayed implantation by Fernand Lataste in his home community of Gradignan.

We were fortunate that our concern about the issues addressed by the symposium were shared by others. The organizational resources required to plan and arrange a conference which addresses complex issues were made available by the Department of Cell Biology and the Center of Population Research and Studies in Reproductive Biology at Baylor College of Medicine (Dr.

Bert W. O'Malley, Chairman and Director). Enthusiastic intellectual and financial support was provided by members of the Population and Research Branch of the National Institute of Child Health and Human Development. The concept of interdisciplinary and international communication was so provocative to scientists who rarely have the opportunity to share so common a forum that it was not difficult to attract the support of people and organizations committed to the success of this congress. In addition to recognizing the support from NICHD the Organizing Committee acknowledges with gratitude the assistance provided by the following organizations and companies: Naturalia and Biologia, Ares Serono, Serono France, Organon, Syntex, Ipsen Biotech, Takeda, Besin-Iscovesco and CCD.

There are no words, in any of the languages represented at our meeting, that can properly acknowledge the debt owed by every participant at the Symposium on the Endocrinology of Embryo-Endometrium Interactions to the members of the Local Committee. The time, effort and enthusiasm of these friends and colleagues created a warm, a generous environment that gave rise to spirit of personal and intellectual ambience that made this meeting in Bordeaux (September 6-10, 1992) an indelible event in our lives. Our gratitude is due to R. Canivenc, the Honorary President of this Symposium, J.L. Latapie, G. Mayer, J.M. Meunier and, most notably, Alain Audebert, who chaired the local committee. Thanks are due, in no small measure, to Patricia Vann and her associates at Plenum Publishing Corporation, for the help, guidance and their great patience that transformed all these efforts into a published book.

Stanley R. Glasser
Joy Mulholland
Alexandre Psychoyos

CONTENTS

UTERINE RECEPTIVITY: EXPERIMENTAL STUDIES

UTERINE RECEPTIVITY: CLINICAL STUDIES

THE M.C. CHANG MEMORIAL LECTURE

PREGNANCY RECOGNITION

ENDOMETRIAL RESPONSES TO RECEPTIVITY

THE NEXT YEARS

ENDOCRINOLOGY OF EMBRYO-ENDOMETRIAL INTERACTIONS:

A HUNDRED YEARS OF FASCINATING DISCOVERIES

Alexandre Psychoyos

Hopital Bicetre, Batiment INSERM
94276 Kremlin-Bicetre, France

Only a hundred years ago, when Fernand Lataste(1891) discovered lactational delayed implantation, no one was able to suggest a rational explanation. Fifty years went by before Krehbiel(1941)and Weichert(1942)reported that this intriguing phenomenon could be due to low estrogen secretion.

At the time of Lataste's discovery, Mathias Duval, known for his studies on the human placenta, described egg-implantation and ovarian involvment in his book "Cours de Physiologie" published in 1892. He said, "The fertilized ovum arriving into the uterus induces by its presence hypertrophy of the uterine mucosa from which results the decidua: at the same time there occurs in the ovary, due to an act of sympathy(in French:act sympathique)or a reflex difficult to explain, the characteristic evolution of corpora lutea of pregnancy."

These few lines, summarize the background of knowledge from which the pioneers of our field started, at the beginning of this century, to add further information, answering two fundamental questions:

- Are the corpora lutea necessary for the normal evolution of pregnancy?

A positive answer was given by Ludwig Fraenkel who first demonstrated the importance of corpora lutea, by a series of elegant experiments.

- Is the modification of the uterine mucosa observed during pregnancy, dependent on the presence of a fertilized ovum?

The negative answer to this question was given by Pol Bouin, with his classical work with Ancel, in pseudopregnant rabbits; by Leo Loeb, with the experimental induction in the Guinea pig of a decidual transformation (deciduoma) of the uterine mucosa.

I will refer again to these main discoveries, while following the pathway which from Mathias Duval's "sympathy" led us to the "hormonal" concept.

We also owe to Fernand Lataste (1892), the first description of the cyclic changes occuring in the vaginal epithelium. However, it took some 25 years until Charles Stockard and George Papanicolaou at Cornell University, rediscovered this phenomenon in the Guinea pig, by a simple technique, the vaginal smear, the key to the discovery of the estrogenic hormone (Stockard and Papanicolaou, 1917). Joseph Long and Herbert Evans, at the University of California, inspired by Stockard and Papanicolaou's paper, then started to study vaginal cytology during the cycle in the rat, and published their findings in their classical monograph (Long and Evans, 1922).

It was now the turn of a young instructor in Anatomy at Washington University, Edgar Allen, to be inspired by the reading of Long and Evans' monograph. He collected follicle fluid from sow ovaries and injected it into ovariectomized mice. By using the vaginal smear method, characteristic of estrus, he could thus read his results within a day. It seems that Edgar Allen met Edward Doisy at St Louis on the occasion of a joint effort to form a faculty baseball team. In March 1923, Allen reporting his findings to Doisy, had speculated that the ovarian-vaginal relationship he observed could be of hormonal nature. The two young men decided to work together. Doisy started purifying the follicular fluid, observed that the estrogenic potency was found in the lipid fraction and obtained soon after the partial purification of the estrogenic hormone (Allen and Doisy, 1923; 1924).

At this point we have to give credit to a Doctor of Medicine and Maitre de Conferences at the Sorbonne in Paris: Henri Iscovesco. In 1912 he had already reported to the Société de Biologie that one of the lipid fractions he prepared from sow ovaries contained a substance which, injected into young rabbits, caused marked growth of their uteri. In a longer article two years later, Iscovesco made known that the active substance was soluble in alcohol, ether and petroleum ether, but not in acetone. He also reported that using this substance in his gynaecological practice, he obtained satisfactory results in dysmenorrhea, amenorrhea and hypogonadism (Iscovesco, 1912; 1914). Henri Iscovesco must also be considered as a pioneer and precursor of modern times from another point of view. It seems that he did a commercial use of his extract, and I guess that he was the founder of the Pharmaceutical company which bears his name today.

For English-speaking scientists an invaluable guide had appeared in 1910 in Cambridge: the book of Francis Marshall "The Physiology of Reproduction". Marshall and Jolly (1905), then in Edinburgh, had injected an anoestrus bitch with a saline extract of ovarian tissue from another bitch that was

in heat,and observed that this treatment induced estrus in the
recipient.However they did not further pursue this study.

     Already in 1900,Emil Knauer,a 29-year old Assistant in a
Gynaecological clinic in Vienna,had prevented the uterine
atrophy that would have followed ovariectomy of adult rabbits,
by regrafting fragments of the removed ovaries and had
suggested that these fragments produce "something" that is
transported to the uterus via the blood stream(Knauer,1900).
In the same year,another young Viennese gynaecologist Josef
Halban carried out a variation of these experiments.
Taking three infantile guinea pigs,he grafted fragments of
adult ovaries under their skin and found that their uteri grew
rapidly to adult size.He also concluded that a substance
produced by the ovary and taken up into the blood was able to
exert a specific effect upon the genital organs(Halban,1900).

     In Vienna at the turn of the century,Gustav Klimt was
painting his masterpieces,Gustav Mahler was composing his
Symphonies and Sigmund Freud was analysing dreams.
However,a widely applied "therapy" for dysmenorrhea and
various neuroses was the surgical removal of the ovaries.It
was thus well known that ovariectomy leads to uterine
atrophy,an effect described in animals already by Aristotle.
It is difficult for us to realize today that up to Knauer and
Halban's demonstration,the uterine atrophy which followed
ovariectomy was explained by a nervous disconnection between
the uterus and the ovary.

     At the same period,Louis-August Prenant in Paris and
Gustav Born in Breslau,based on histological criteria,were
suggesting an endocrine function for the corpus luteum.It
seems that Born was dying,and in his last moments,he asked the
young and brilliant gynaecologist Ludwig Fraenkel,to undertake
the experiments which he could not realize himself.He wanted
evidence for his hypothesis,namely,that the corpus luteum is
an organ of internal secretion,the function of which is related to the protection of the embryo.

     Fraenkel knew that the rabbit embryos implant 7-8 days
after coitus and decided to take out the corpora lutea before
this time,using either total ovariectomy or the selective
cauterisation of the corpora lutea.Pregnancy was arrested in
all cases.Fraenkel presented his results first in 1903,and
faced only criticism.It took him 7 years more to give the
definitive demonstration of his claims(Fraenkel,1903;1910).

     During these years and probably without knowing
Fraenkel's experiments,Pol Bouin and Paul Ancel in Nancy,were
following their own studies,also published in 1910.They
discovered that the rabbit uterine mucosa exhibits,in early
pregnancy,a specific proliferation which they named "dentelle
endometriale".Furthermore,they demonstrated by an ingenious
experiment that this modification was related to the presence
of the corpora lutea and not to the embryo.They had a
magnificent idea.to mate their females with males,which they
rendered sterile by ligation of the vas deferens.Ovulation
took place and corpora lutea were formed,but in the absence of
pregnancy.Nevertheless the uterus exhibited the typical

progestational proliferation,the "dentelle endométriale"(Bouin and Ancel,1910).

As the vaginal smear served as a key for the discovery of oestrogen,in the same way the "dentelle endométriale"of Bouin and Ancel,was the key to the discovery of the luteal hormone.At Rochester,some twenty years later,in 1929,George Corner and a medical student named Willard Allen were able to obtain this progestational change in ovariectomized rabbits by extracts of corpora lutea(Corner and Allen,1929).

At the same time as Fraenkel's,and Bouin and Ancel's experiments,the american Leo Loeb was demonstrating in non pregnant guinea pigs another relationship between the presence of corpora lutea and the uterus i.e.the possibility of the induction of a decidual reaction by the irritation of the uterine mucosa during a limited period of the normal cycle (Loeb,1908).I don't consider it necessary to emphasize to you the tremendous impact that this discovery had and continues to have for the understanding of the multiple aspects of the embryo-endometrial interactions and their hormonal control.

When speaking of the pioneers in our field and their fascinating discoveries,we have to give a prominent place to Walter Heape in Cambridge.Heape was profoundly influenced by the writings of Fernand Lataste on the genital cycle.As reported by Alan Parkes,he had annotated every one of the 676 pages of the book,on the breeding habits of rodents,offered to him by Lataste.Heape,in a paper published in 1900,redefined the French terms of Lataste,such as cycle genital etc,and introduced the terms" reproductive period","sexual season",and also the new terms "oestrus","pro-oestrus","dioestrus" etc.

Heape is considered to be the first one to use the transfer of fertilized ova as a technique,during his studies in the rabbit(Heape,1890).He is also the first to observe that ovulation in this species is dependant on,and happens 10 hours after mating.Heape publishing his results in 1905,suggested that ovarian activity was conditioned in this case by some extragonadal substance to which he gave the name of "generative ferment"(Heape,1905).

The first findings showing the origin and nature of Heape's "generative ferment" appeared almost simultaneously with the isolation of the ovarian hormones.In 1927,Philip Smith and Earle Engle at the Columbia University,following up their observations on hypophysectomized rats,tried the effect of placing small pieces of anterior pituitary gland under the skin of immature animals(Smith and Engle,1927). Zondek and Aschheim(1927)in Berlin,were searching for the source of extragonadal control of the gonads by carrying out identical experiments.Their results showed clearly that the anterior pituitary contained substances which would cause in the ovary,follicular maturation,ovulation and corpus luteum formation.However,the discovery of pituitary gonadotropins which followed,corresponded only partially to Heape's "generative ferment".

4

The release of pituitary hormones was found in the 50's by Geoffrey Harris(1955)in England and Jacques Benoit and Ivan Assenmacher in Paris at the College de France,to be dependant on some hypothalamic substance,which should reach the pituitary through the portal vessels.The experiments of the French investigators,published in 1953,were particularly demonstrative.They were done in the duck.The advantage was that in this species,lesions could be made in either the portal vessels or the nerve fibers of the stalk which connects the hypothalamus to the pituitary.Benoit and Assenmacher(1953)found that cutting the portal vessels resulted in atrophy of the testes,whereas section of the nerve fibers of the pituitary stalk (leaving the portal vessels intact) was compatible with normally developed gonads.

The first attempts to isolate the hypothalamic factor responsible for the release of pituitary gonadotropins,were also performed at the College de France by Robert Courrier, Roger Guillemin,Marian Jutisz,Edouard Sakiz,and Aschheim's son Pierre,in the early 60's(Courrier et al,1961).They led,as it is known,to the discovery of GnRH in the 70's and the award of the Nobel Prize to Roger Guillemin and to Andrew Schally.

The placenta,has been suspected to be another source of hormonal products since the early 1900's by the same Josef Halban of Vienna,who wondered "How the ovary knows that there is pregnancy".The experimental evidence for such a placental function takes us back to 1927.

Aschheim and Zondek,at this time still in Berlin,discovered that the urine of pregnant women contained a product which stimulated the ovaries of prepuberal mice and rats,and this observation was used for a long time as a biological test for the early detection of pregnancy(Aschheim and Zondek,1928).Aschheim and Zondek suggested initially that this product could be released from the pituitary of the pregnant woman.However,it was soon accepted that the source of this gonadotrophic factor was the foetal part of the placenta.

Hirose in Japan had already in 1920,prepared from human placentas an aqueous emulsion which induced the formation of corpora lutea in rabbits.Hirose's finding was confirmed and completed a few years later,also in Japan,by Murata and Adachi(1927).These investigators were able to induce,by such placental extracts,ovarian maturation,ovulation and luteinization in immature rabbits.Furthermore,they also prepared extracts from metastatic chorio-epithelioma and obtained the same result.

What an exciting decade of years,between 1925 and 35!

Within these years,estrogens,androgens,and progesterone were isolated and characterized,and their biological properties extensively investigated,while in addition the hypophyseal and placental gonadotropins were also discovered.

Most of the actors of this heroic age of reproductive Endocrinology met each other for the first time in Paris, at the Collège de France, for a "Colloque International sur les Hormones Sexuelles", held in June 1937 and sponsored by the Fondation Singer-Polignac. In that meeting, which can be considered as the first truly international Endocrinology meeting, Pol Bouin served as Chairman, and the participants were: E.C. Dodds, Ruth Deanesly, Robert Courrier, Alan Parkes, Edgar Allen, Carl Hartman, Solly Zuckerman, Frederic Hisaw, Marc Klein, Aura Severinghaus, Philip Smith, S. Aschheim, William Young, Francis Marshall, Hans Selye, Idwal Rowlands, among others.

By the end of the 40's, it was now clear, that estrogens and progesterone had the possibility to promote or to inhibit the various uterine functions, by their synergic or antagonistic interactions-a concept developed by Robert Courrier, a former student of Pol Bouin. The classical book of Robert Courrier "Endocrinologie de la Gestation" appeared in 1945. For French-reading investigators this was a precious source of inspiration and the stimulus for new discoveries, such as those of Alfred Jost, Raymond Kehl, Yves Chambon, Gaston Mayer, René Canivenc, Jean-Jacques Ailoiteau, among others.

Two major observations were of prime importance for the study of the functional relationships between progesterone and estrogen during early pregnancy, in the next two decades: a) the possibility of suspending the egg-implantation process in the rat and mouse by experimentally inducing a delay in implantation, b) the existence of a precise hormonal interplay controlling the timing of uterine receptivity for nidation.

We owe to Yves Chambon (1949) the first demonstration that delayed implantation in the rat can also occur after early ovariectomy and to Rene Canivenc and his collaborators (1956), to Bob Cohrane and R.K. Meyer (1957), the fundamental notion that progesterone alone cannot induce implantation in early ovariectomized rats.

During the 50's, another exciting finding came out, first by the classical paper of M.C. Chang (1950) in the rabbit and then that of Ann McLaren and D. Michie (1956) in the mouse, and of Bob Noyes and Zev Dickmann (1960) in the rat. Namely, the importance of a synchronisation in the development of the uterus and the embryo for successful implantation, after embryo transfer.

It was in those years that I had the great chance to be accepted by Robert Courrier, to join his "Laboratoire de Morphologie Expérimentale et Endocrinologie" at the Collège de France. Professor Courrier asked me to look into the possibility of transferring rat blastocysts in delay, to normal pseudopregnant rats at various moments of pseudopregnancy. The idea was that such blastocysts could be more resistant and could implant independantly of a synchronized endometrium.

6

I must confess that I lost two months before I found a rat blastocyst in uterine flushings performed on Day 5 of pregnancy. The rare papers available at this time, described their size and shape (a base-ball-like cellular mass, etc) with no photos. It also happened, in this period that I had to continue during the day my clinical specialization in the Hospital. I was thus able to work for this research project only during the evening.

It was on a holiday that, working during the morning in my lab, I had the great joy to immediately recognize, in uterine flushings performed before noon of Day 5, a lot of blastocysts, by the brilliancy of their as yet intact zona pellucida. Trying to induce delayed implantation by an ovariectomy on Day 4, as René Canivenc did, I also had no success at the beginning. All my rats were implanting normally. This, up to another holiday when I started my ovariectomies before noon of Day 4. In this case, all my animals exhibited delay, amply confirming Professor Canivenc's findings.

Having to work for my first research project late in the evening, I obtained thus two basic informations, not yet available at that time: a) the rat blastocyst loses its zona pellucida in the afternoon of Day 5 and b) by the evening of Day 4, exogenous estrogen is no longer necessary to induce egg-implantation in the rat. In other words, nidatory estrogen intervenes in this species late on Day 4. I succeeded finally, in realizing the initial project of transferring blastocysts in diapause to pseudopregnant hosts. The results were clear but disappointing. These blastocysts as well as the non diapausing ones, could implant if transferred to pseudopregnant uteri before Day 6 of pseudopregnancy. Transferred on Day 6 and beyond, they never implanted. I had to wait some years more to understand the reasons, i.e. the existence of a postreceptive uterine refractoriness for egg-implantation (Psychoyos, 1963).

By the end of the 50's, the study of embryo-endometrial interactions and their hormonal control entered a new era. I consider that this was mainly due to the efforts of Professor Shelesnyak and his collaborators at the Institute of Biodynamics at Rehovoth, in Israel. The new ideas advanced by him (Shelesnyak, 1957) stimulated several young investigators, including myself, to consider many new vistas. In less than ten years, during the 60's (see review: Psychoyos, 1973), several points concerning the process of egg-implantation and its hormonal correlates were clarified in various species, by the contribution of a small group of investigators: Ann McLaren, Vince De Feo, Koji Yoshinaga, Alan Enders, Colin Finn, Ove Nilsson, Gerard Zeilmacker, David Kirby.

David, one of the best, died young, following a car accident. During his brief scientific life, he had accomplished by his ectopic grafts of mouse embryos, many breakthroughs for our understanding of the embryo-maternal relationships.

Since then, the tremendous development of our knowledge on the mode of hormone action, redirected the study of embryo-endometrial interactions and their hormonal regulation to the cellular and molecular level. Particularly by the discovery of the intracellular receptors by Elwood Jensen for estrogens, by Étienne Baulieu and Edwin Milgrom, by Bert O'Malley, for progesterone.

The new achievements in cellular biochemistry stimulated several investigators by the early 70's, in particular, Stan Glasser in Houston, Fernand Leroy in Brussels, Hans Lindner in Rehovoth, Peter Heald and Stephen Bell in Glascow, who studied the molecular events involved in endometrial receptivity and decidualization. In addition, Harry Weitlauf, Azim Surani, Michael Harper, Tom Kennedy, S.K. Dey among others, tried to understand the process of blastocyst activation, the nature, synthesis and release of embryonic signals (see reviews: Weitlauf, 1988; Parr and Parr, 1989).

During the last decade, the discovery of several new uterine and embryonic factors involved in early pregnancy, was greatly advanced by the emergence of new multidisciplinary methods, such as those using monoclonal antibodies for immunocytochemistry, in situ hybridization or Polymerase Chain Reaction. It is with great pleasure that I can say that many of the actors of this new era, are participating in this meeting (Susan Kimber, Joy Mulholland amongst others).

I think that many of us working in animal models did not expect that our findings could have a such rapid impact on clinical practice. The drive was the urgent need for new contraceptive methods and also the success of Bob Edwards and Patrick Steptoe in obtaining in 1978, the first baby conceived by in vitro fertilization (Edwards et al., 1980).

I must mention here that both contraceptive technology and in vitro fertilization owe a lot to Gregory Pincus. Father of the contraceptive pill, he was also the pioneer of the In Vitro Fertilization and Transfer methodology. He had published already in 1934, his experiments showing that rabbit eggs fertilized in vitro could develop normally after being transferred to the genital tract (Pincus and Enzmann, 1934). At the present time, the existence of a limited period of endometrial receptivity for egg-implantation, the so-called Implantation Window, initially observed in rats (Psychoyos, 1973) appears to be also valid in the human (Navot et al., 1991). The chronological limits of this period appear also to obey a similar hormonal interplay.

The aim of my talk was to remind us of some crucial steps of the past. I had in mind that this could contribute to a better appreciation of the tremendous progress that has been achieved in our field. Our Symposium will give us the opportunity to know where we are at the present, but also, as we all hope, to identify new promising areas of study for the years to come, at least until the turn of the century.

REFERENCES

Allen,E.,and Doisy,E.A.,1923,An ovarian hormone,*J.Am.Med.Ass.*81:819-21.

Allen,E.,and Doisy,E.A.,1924,The extraction and some properties of an ovarian hormone,*J.Biol.Chem.*61:711-27.

Aschheim,S.,and Zondek,B.,1928,Die Schwangerschaftsdiagnose aus dem Harn durch Nachweis des Hypophysenvorderlappenhormons. *Klin.Wochensch.*7:1404-10,1453-57.

Benoit,J.,and Assenmacher,I.,1953,Rapport entre la stimulation sexuelle préhypophysaire et la neurosécretion chez l'oiseau. *Arch.Anat.microsc.*42:334-86.

Bouin,P.,and Ancel,P.,1910,Recherches sur les fonctions du corps jaune gestatif.I.Sur le déterminisme de la préparation de l'utérus à la fixation de l'oeuf.*J.Physiol.Pathol.génér.*12:1-16.

Canivenc,R.,Laffargue,M.,and Mayer,G.,1956,Nidations retardées chez la ratte castrée et injectée de progestérone,*C.R.Soc.Biol.*12:2208-12

Chambon,Y.,1949,Réalisation du retard de l'implantation par des faibles doses de progesterone chez la ratte.*C.R.Soc.Biol.*143:753-56.

Chang,M.C.,1950,Development and fate of transferred rabbit ova or blastocysts in relation to the ovulation time of recipients *J.Exptl.Zool.*114:197-225.

Cochrane,R.L.,and Meyer,R.K.,1957,Delayed implantation in the rat induced by progestérone.*Proc.Soc.Exptl.Biol.Med.*96:155-59.

Corner,G.,and Allen,W.,1929,Physiology of the corpus luteum.II.Production of a special uterine reaction(progestational proliferation)by extracts of the corpus luteum.*Am.J.Physiol.*88:326-39.

Courrier,R.,1945,"Endocrinologie de la gestation",Masson,Paris.

Courrier,R.,Guillemin,R.,Jutisz,M.,Sakiz,E.,and Aschheim,P.,1961, Présence dans un extrait d'hypothalamus d'une substance qui stimule la sécretion de l'hormone antéhypophysaire de lutéinisation (LH),*C.R.Acad.Sc.Paris*,253:922-27.

Edwards,R.G.,Steptoe,P.C.,and Purdy,J.M.,1980,Establishing full-term human pregnancies using cleaving embryos grown in vitro, *Br.J.Obstet.Gyn.*87:737-56.

Fraenkel,L.,1903,Die Function des Corpus luteum.*Arch.Gynäk.*68:438-45.

Fraenkel,L.,1910,Neue Experimente zur Function des Corpus luteum. *Arch.Gynäk.*91:705.

Halban,J.,1900,Ueber den Einfluss der Ovarien auf die Entwicklung des Genitales,*Mschr.Geburtsh.Gynäk.*12:496-505.

Harris,G.W.,1955,"Neural Control of the Pituitary Gland",Edward Arnold, London.

Heape,W.,1890,Preliminary note on the transplantation and growth of mammalian ova within a uterine foster-mother,*Proc.Roy.Soc.* 48:457-58.

Heape,W.,1905,Ovulation and degeneration of ova in the rabbit, *Proc.Roy.Soc.*,ser.B,76:260-68.

Iscovesco,H.,1912,Le lipoïde utérostimulant de l'ovaire.Propriétés physiologiques.*C.R.Soc.Biol.*63:104-6.

Iscovesco,H.,1914,Lipoïdes homo-stimulants de l'ovaire et du corps jaune,*Rev.Gynéc.*22:161-98.

Knauer,E.,1900,Die Ovarientransplantation.Experimentelle Studie, *Arch.Gynäk.*60:322-76.

Krehbiel,R.H.,1941,The effects of theelin on delayed implantation in the pregnant lactating rat,*Anat.Record*,81:381-92.

Lataste,F.,1891,Des variations de durée de la gestation chez les mammifères et des circonstances qui déterminent ces variations, *C.R.Soc.Biol.*43:21-31.

Lataste,F.,1892,Transformation periodique de l'épithelium du vagin des rongeurs,C.R.Soc.Biol.44:765

Loeb,L.,1908,The production of deciduomata and the relation between the ovaries and the formation of the decidua,J.Am.Med.Ass. 50:1897-1901

Long,J.A.,and Evans,H.M.,1922,"The Oestrus Cycle in the Rat and its associated phenomena",Mem.Univ.Calif.n°6,Univ.of California, Berkeley.

Marshall,F.H.A.,and Jolly,W.A.,1906,Contributions to the physiology of mammalian reproduction.I,the oestrus cycle of the dog.II,the ovary as an organ of internal secretion,Phil.Trans.B, 198:99-141.

McLaren,A.,and Michie,D.,1956,Studies on the transfer of fertilized mouse eggs to uterine foster-mothers.I.Factors affecting the implantation and survival of native and transferred eggs. J.Exptl.Biol.33:394-416.

Murata M.,and Adachi,K.,1927,Ueber die kunstliche Erzeugung des Corpus luteum durch Injection der Plazentarsubstanz aus frühen Schwangerschaftsmonaten,Zeitschr.fur Geburtshiefe 92:45-71.

Navot,D.,Scott,R.T.,Droesch,K.,Veeck,L.L.,Liu,H.,and Rosenwaks,Z.,1991, The window of embryo transfer and the efficiency of human conception in vitro,Fertil.Steril.55:114-18.

Noyes,R.W.,and Dickmann,Z.,1960,Relation of ovular age to endometrial development,J.Reprod.Fertil.1:186-96.

Parr,M.B.,and Parr,E.L.,1989,The implantation reaction,in:"Biology of the Uterus",R.M.Wynn,and Jollie,eds,Plenum Press,New York.

Pincus,G.,and Enzmann,E.V.,1934,Can mammalian eggs undergo normal development in vitro?,Proc.Nat.Acad.Sc.20:121-22.

Psychoyos,A.,1963,Précisions sur l'état de "non-réceptivite" de l'utérus.C.R.Acad.Sc.Paris,257:1153-56.

Psychoyos,A.,1973,Endocrine control of egg-implantation,in:"Handbook of Physiology",Section 7:Endocrinology,Vol.II,Part 2,R.O.Greep and E.B.Astwood,eds,Williams and Wilkins,Baltimore,Maryland.

Shelesnyak,M.C.,1957,Aspects of reproduction.Some experimental studies on the mechanism of ovo-implantation in the rat,Recent Progr. Hormone Res.13:269-322.

Smith,P.E.,and Engle,E.T.,1927,Experimental evidence regarding the role of the anterior pituitary in the devlopment and regulation of the genital system,Am.J.Anat.40:159-217.

Stockard,C.R.,and Papanicolaou,G.N.,1917,The existence of a typical oestrus cycle in the guinea-pig,with a study of its histological and physiological changes,Amer.J.Anat.22:225-84.

Weichert,C.K.,1942,The experimental control of prolonged pregnancy in the lactating rat by means of estrogen,Anat.Record 83:1-17.

Weitlauf,H.M.,1988,Biology of implantation,in:"The Physiology of Reproduction",E.Knobil and J.D.Neill,eds,Raven Press,New York.

Zondek,B.,and Aschheim,S.,1927,Das Hormon des Hypophysenvorderlappens; Testobject zum Nachweis des Hormons,Klin.Wochensch.6:248-52.

CONTRIBUTIONS OF COMPARATIVE STUDIES TO UNDERSTANDING MECHANISMS OF IMPLANTATION

Allen C. Enders
Department of Cell Biology & Human Anatomy
University of California School of Medicine
Davis CA 95616

INTRODUCTION

Studies comparing the different forms and mechanisms of implantation in different mammals are useful in understanding a number of factors in implantation, including the extent of various relationships between trophoblast and uterus and the relative durations of specific types of tissues, and in assessing maternal and fetal roles in implantation and in the establishment of the placenta. I would like to focus on a few areas only, in order to illustrate this concept.

INITIAL ADHESION AND PENETRATION OF UTERINE EPITHELIUM

Trophoblast-Epithelial Cell Adhesion

An interesting aspect of comparative implantation is the way in which trophoblast populations involved in initial adhesion of the blastocyst to the uterus are segregated. Adhesion, the result of molecular changes at the surface of trophoblast and uterus, is a necessary stage in implantation in all species. It is particularly interesting because it involves not only the adhesion of cells of two different origins but specifically the apical surfaces of these cells. The apical surfaces of simple epithelial cells are normally not adhesive, a necessary aspect of maintenance of the polarity of these cells and a quality that results in the production of a single-layered epithelium.

The importance of trophoblast-epithelial adhesion may well vary among species. Adhesiveness is clearly going to be more significant in a uterus with a small blastocyst and a relatively large luminal surface than in a uterus in which the implanting blastocyst swells and locally distorts the shape of the uterus, in which case the muscular clasping of the blastocyst may be more significant. When the uterus undergoes closure with luminal epithelial microvillar interdigitation followed by formation of a tubular implantation crypt such as in the rat and mouse (Nilsson, 1970; Weitlauf, 1988), minimal adhesive forces would appear to be necessary for retention of apposition of the blastocyst.

In a uterus such as that of the primate in which the orientation of the blastocyst is dependent on the adhesion of a few cells at one pole of the blastocyst, the role of adhesive changes in those cells may be very significant in determining the success or failure of the implantation. Because of the limited availability of primate blastocysts, these blastocysts will probably be studied largely by methods that localize adhesion molecules following leads obtained from other species. Considerable information on adhesion molecules has been obtained from the mouse (see for example Kimber, 1990; Carson et al., 1993), but less attention has been paid to species in which adhesion is limited to a single cell type, and which therefore has the advantage of an intrinsic control of nonadhesive cells, so that the differences in glycocalyx properties can be directly contrasted between adherent and nonadherent cells. A particularly fascinating example is found in the ruminants, in which the binucleate trophoblast cells become adhesive to the apical ends of luminal epithelial cells (with which they eventually fuse), while at the same time or shortly afterwards losing their association with adjacent trophoblast cells (Wooding, 1992).

Epithelial Invasion

It is clear both from comparative studies and from *a priori* reasoning that maintenance of the integrity of the maternal system is important. This can be achieved by trophoblast penetrating the uterine epithelium in such a manner as to preserve the integrity of that epithelium, as it does in the rhesus monkey, or by the uterus itself controlling the extent of isolation of the region of uterus involved in implantation, as the decidua does in the rat. One of the clearest lessons of comparative implantation is that trophoblast does not erode the endometrium. Indeed ulceration of this mucous membrane would be counter-productive to a successful implantation. Although we know that in many species trophoblast processes intrude between uterine luminal epithelial cells and share junctional complexes with these cells, at least partially maintaining the luminal integrity, there is no information on the exact nature of the apical portions of the junctions or on the relative effectiveness of the junction in blocking paracellular diffusion pathway.

RESTRICTION OF INVASIVE TISSUES

Trophoblast cells involved in epithelial penetration and stromal invasion are often specialized. There are many advantages to such a specialization. Only those specific cells that come into contact with maternal stromal constituents need to have specific surface features such as restricted histocompatability factors. Since these cells are generally postmitotic, the extent of trophoblast invasion into maternal tissue can be partially controlled such that any cells that might be displaced from the implantation site into the maternal system do not act as metastases. If these cells or tissues that do invade into the endometrium have a limited life span, this further tends to control the extent of invasion into the maternal system.

Penetration by Syncytial Trophoblast

The most common type of specialization of trophoblast at implantation is the use of a restricted region of syncytial trophoblast. A syncytium has the advantages of being able to adhere to the surfaces of several uterine epithelial cells simultaneously, and of having sufficient cytoplasm to flow into penetrated areas, and hence to exploit infiltration into the epithelium. The development of regions of syncytial trophoblast is widespread within many mammalian families.

In primates syncytial trophoblast is found adjacent to or overlying the inner cell mass of the periimplantation blastocyst. It is a normal developmental sequence that does not require the presence of endometrial tissue, since it can develop in vitro when blastocysts are cultured beyond the time of normal implantation (Pope et al., 1982; Enders et al.,1989), but is found in vivo only at the time of implantation. In some mammals (e.g. squirrels and caviomorphs), the syncytial trophoblast region may be restricted to a single abembryonic area (Mossman, 1987). In others (carnivores and rabbits), multiple regions of syncytium form a series of invasion regions over a broad area of the mural trophoblast within the implantation sites (Enders and Schlafke, 1979).

What is not as clearly understood is that, although syncytial trophoblast becomes widespread after early implantation stages, this more ubiquitous later syncytium has quite different characteristics than the initial invasive trophoblast. An example of this is seen in the rhesus monkey in which the first syncytial trophoblast consists of several multinucleate masses which invade the luminal epithelium and contribute to a flattened region of mixed syncytial and cellular trophoblast, then rapidly invade the underlying maternal vessels (Enders and King, 1991). Even before maternal vessels are tapped, some of the syncytial trophoblast begins to transform into a unilaminar microvillous absorptive layer which rapidly becomes the predominant trophoblast type as lacunae are formed (Enders, 1989). Subsequent invasion of the endometrium is accomplished slowly by cytotrophoblast of the anchoring villi, and invasion into the deeper endometrial vessels is also achieved by cytotrophoblast in this species (Blankenship et al., 1993).

Penetration by Cellular Trophoblast

Despite the apparent preference of mammals for the use of specialized syncytial trophoblast for initial invasion, there are many species in which epithelial penetration occurs using cytotrophoblast rather than syncytial trophoblast. However even in these cases the cells are considerably specialized. As far as we know they are generally postmitotic. Many of them are giant cells, such as those seen in the formation of the yolk sac placenta of the rat and mouse (Welsh and Enders, 1987, 1991). In other cases they are binucleate cells such as in ruminants in which these are the cells involved in fusion with the maternal epithelium (Wango et al., 1990; Wooding, 1992).

In the case of the vespertilionid bats, the layer of trophoblast present during epithelial penetration is cellular, but segregates into two populations of cells (Enders and Wimsatt, 1968). That layer of cells associated with the maternal endothelium is presumably postmitotic although this has not been documented. At any rate these cells rapidly fuse to form syncytial trophoblast, whereas the inner layer remains as progenitor cells.

BLASTOCYST AGGRESSION VS. MATERNAL COOPERATION

Comparative studies are also useful in assessing the relative contribution of trophoblast aggression and maternal cooperation. In delayed implantation it is obvious that the maternal organism controls implantation. In other instances the endometrial window of receptivity must intersect the appropriate developmental time of the blastocyst. Despite the widespread distribution of delayed implantation in some groups (rodents, carnivores, marsupials), other closely related species do not exhibit the phenomenon (Renfree and Calaby, 1981).

Perhaps the extreme of maternal control and involvement in the implantation process is seen in the rat, in which not only can implantation be delayed but, once implantation begins, the removal of the uterine epithelium in different regions of the implantation site proceeds without direct trophoblast participation; in the case of the primary implantation site, uterine epithelial cells undergo apoptosis which facilitates trophoblast removal of these cells (Parr and Parr, 1989; Welsh, 1993). In addition to participating in epithelial removal, decidual cells form an orienting implantation chamber, penetrate the basal lamina of the luminal epithelium, and reorganize the stroma surrounding the implantation site as well as participating in formation of blood channels between the maternal vasculature and the giant trophoblast cells of the yolk sac placenta.

It is generally thought that blastocysts that become interstitial are particularly aggressive, and that the human blastocyst is one of the most aggressive. Closer examination of some species makes this idea somewhat questionable, or at least suggests the necessity for modifying the statement; that is, aggressive with regard to epithelial penetration, stromal invasion, penetration of maternal vessels, depth of penetration, etc. The guinea pig blastocyst is considered to be aggressive since it rapidly penetrates the uterine luminal epithelium. However once it penetrates this epithelium decidualization of the endometrium occurs rapidly. Some of the trophoblast is lost during inversion of the yolk sac, and the subsequent development of trophoblast occurs within a decidual chamber which, other than being immediately beneath the luminal epithelium, is similar to that of the 'less aggressive' rat and mouse.

Similarly the human blastocyst, which becomes interstitial, is often thought of as being aggressive compared to the macaque and baboon blastocyst, which implant more superficially. However if we examine these species more closely we find that trophoblast of the baboon blastocyst rapidly invades both uterine epithelium and the endothelium of the underlying blood vessels (Enders and King, 1991). Subsequent development of the trophoblastic lacunae is extremely rapid, and growth of the placenta is by enlargement of these lacunar spaces rather than by invasion into the endometrial stroma. On the other hand syncytial trophoblast of the human blastocyst appears to surround endometrial capillaries when they are first encountered rather than immediately tapping them. This, coupled with the small size of the human blastocyst and the eventual conversion of all of the mural trophoblast into syncytial trophoblast overlying cytotrophoblast, leads to an interstitial implantation which is relatively slow in tapping maternal vessels. Thus the intervillous space in both species is similar shortly after tapping of maternal vessels, but in the human this has been achieved by involving more of the circumference of the blastocyst and deeper invasion of the endometrium, and in the baboon by more intraplacental growth. In the macaque, a similar area of lacunae is achieved by formation of a secondary placenta on the abembryonic uterine surface in most instances. Consequently the human blastocyst may be as aggressive as that of the baboon and macaque in penetrating the epithelium, but it is clearly less aggressive with regard to invasion of the endometrial vessels (Enders, 1993). Contrariwise, the trophoblast of the three primate species may have similar properties but human endometrial capillaries may be more resistant to disruption, whether due to differences in basal laminas or to endothelial cell adhesion. Unfortunately no human implantation sites have been examined by electron microscopic methods prior to the late lacunar stage (Knoth and Falck Larsen, 1972).

Although there are clearly advantages in trophoblast tapping maternal vascular systems with regard to positioning trophoblast for endocrine influence of the maternal system, and in placing trophoblast in a position that is advantageous for respiratory exchange, there are some possible disadvantages. For example, the pressure of blood entering the trophoblastic lacunae from the maternal vascular system could result in disruption of extraembryonic membranes, which are particularly fragile at this stage because of the lack of connective tissue supporting the epithelia in the first few days of implantation. Primates in general are rapid in formation of extraembryonic connective tissues, using parietal endoderm as the initial source of mesenchymal cells rather than primitive streak cells (Enders and King, 1988). In the baboon and macaque, trophoblast taps hypertrophied maternal vessels particularly early but trophoblast cells rapidly enter arte rioles, blocking or at least reducing blood flow from this potentially disruptive source. The combination of hypertrophy of the venules and reduction of arteriolar entry should result in a low pressure vascular bed with ample blood flow. In the human, which is further developed when maternal vessels are tapped, trophoblast does not invade the vascular system as rapidly. In all three species massive modification of larger arteries occurs later in gestation when the trophoblastic shell forms.

In the armadillo, another species having rapid trophoblast invasion of the maternal vascular system, trophoblast invades and expands a preexisting series of maternal vessels, using these as intervillous space rather than developing intervillous spaces within the placenta. Interestingly enough the fundic region of the uterus, the area where syncytial trophoblast of the armadillo blastocyst first invades, remains a more vulnerable region in which uterine rupture occasionally occurs in this species.

This brief overview of some comparative aspects of implantation is by no means comprehensive, and is intended only to demonstrate that there are still insights to be obtained from comparative studies as well as from in depth analysis of individual species and implantation sequences.

ACKNOWLEDGMENTS

These studies were supported by grant HD10342 from the National Institute of Child Health and Human Development, and NIH RR00169 to the California Regional Primate Research Center.

REFERENCES

Blankenship, T.N., Enders, A.C., and King, B.F., 1993, Trophoblastic invasion and the development of uteroplacental arteries in the macaque: an immuno-histochemical study, Cell Tissue Res. 272:227-236.

Carson, D.D., Tang, J.P., and Julian, J., 1993, Heparan sulfate proteoglycan (perlecan) expression by mouse embryos during acquisition of attachment competence, Devel. Biol. 155:97-106.

Enders, A.C., 1989, Trophoblast differentiation during the transition from trophoblastic plate to lacunar stage of implantation in the rhesus monkey and human, Am. J. Anat. 186:85-98.

Enders, A.C., 1993, Overview of the morphology of implantation in primates. In: "In Vitro Fertilization and Embryo Transfer in Primates," D.P. Wolf, R.L. Stouffer, and R.M. Brenner, eds., Springer-Verlag, New York, pp. 145-157.

Enders, A.C., Boatman, D., Morgan, P., and Bavister, B.D., 1989, Differentiation of blastocysts derived from in vitro-fertilized rhesus monkey ova, Biol. Reprod. 41:715-727.

Enders, A.C., and King, B.F., 1988, Formation and differentiation of extraembryonic mesoderm in the rhesus monkey, Am. J. Anat. 181:327-340.

Enders, A.C., and King, B.F., 1991, Early stages of trophoblastic invasion of the maternal vascular system during implantation in the macaque and baboon, Am. J. Anat. 192:329-365.

Enders, A.C., and Schlafke, S., 1979, Comparative aspects of blastocyst-endometrial interactions at implantation, In: "Maternal Recognition of Pregnancy," Ciba Foundation Symposium 64, North Holland, Amsterdam.

Enders, A.C., and Wimsatt, W.A., 1968, Formation and structure of the hemodichorial chorio-allantoic placenta of the bat (Myotis lucifugus), Am. J. Anat. 122:453-490.

Kimber, S.J., 1990, Glycoconjugates and cell surface interactions in pre- and peri-implantation mammalian embryonic development, Int. Rev. Cytol. 120:53-167.

Knoth, M., and Falck Larsen, J., 1972, Ultrastructure of a human implantation site, Acta Obstet. Gynecol. Scand. 51:385-393.

Mossman, H.W., 1987, "Vertebrate Fetal Membranes," Rutgers University Press, New Brunswick.

Nilsson, B.O., 1970, Some ultrastructural aspects of ovo-implantation. In: "Ovo-implantation, Human Gonadotropins and Prolactin", P.O. Hubinont, F. Leroy, C. Robyn, and P.Leleux, eds., S. Karger, New York, pp. 52-72.

Parr, M.B., and Parr, E.L., 1989, The implantation reaction. In: "Biology of the Uterus," R.M. Wynn and W.P. Jollie, eds., Plenum Medical Book Co., New York, pp.233-277.

Pope, V.Z., Pope, C.E., and Beck, L.R., 1982, Development of baboon preimplantation embryos to post-implantation stages in vitro, Biol. Reprod. 27:915-923.

Renfree, M.B., and Calaby, J.H., 1981, Background to delayed implantation and embryonic diapause, J. Reprod. Fert., Suppl. 29:1-9.

Wango, E.O., Wooding, F.B.P., and Heap, R.B., 1990, The role of trophoblast binucleate cells in implantation in the goat: a quantitative study, Placenta 11:381-394.

Weitlauf, H.M. , 1988, Biology of Implantation. In: "The Physiology of Reproduction," E. Knobil and J. Neil (eds) Raven Press, LTD., New York. pp. 231-262.

Welsh, A.O., 1993, Uterine cell death during implantation and early placentation, J. Micros. Res. Tech. 25:223-245.

Welsh, A.O., and Enders, A.C., 1987, Trophoblast-decidual cell interaction and establishment of maternal blood circulation in the parietal yolk sac placenta of the rat, Anat. Rec. 217:203-219.

Welsh, A.O., and Enders, A.C., 1991, Chorioallantoic placenta formation in the rat. I. Uterine epithelial cell death and extracellular matrix modifications in the mesometrial region of implantation chambers, Am. J. Anat.192:215-231.

Wooding, F.B.P., 1992, The synepitheliochorial placenta of ruminants: binucleate cell fusions and hormone production, Placenta 13:101-113.

CELL BIOLOGY OF ENDOMETRIAL RECEPTIVITY AND OF TROPHOBLAST-ENDOMETRIAL INTERACTIONS

Hans-Werner Denker

Institut für Anatomie
Universitätsklinikum Essen
Hufelandstr. 55
D-W-4300 Essen, Germany

SUMMARY

Implantation initiation is commonly thought to require that 1) the trophoblast or subpopulations of it have reached a state of "invasiveness" and, synchronously, 2) the endometrium a state of "receptivity" ("implantation window"). Many questions remain open, in particular for the situation in the human. The cell biological basis of "receptivity" as well as of "invasiveness" is still largely unknown, but recently it appears that the application of modern concepts of cell and developmental biology opens promising new views of it, concentrating on cell adhesion and cell polarity phenomena.

Implantation initiation involves that the trophoblast attaches with its apical plasma membrane to the apical plasma membrane of the uterine epithelium. Since apical plasma membranes of epithelia are normally non-adhesive, this has been called a cell biological paradox. In development, cells can attain two major phenotypes and can switch between these: 1) the mesenchymal/ fibroblastoid phenotype that is compatible with cells moving individually; 2) the epitheloid phenotype which is characterized by cells expressing apico-basal polarity and strong association with neighbouring cells via various junctions, so that they can migrate as sheets but not as individual cells. Application of this concept to embryo implantation allows to reconcile many perplexing observations about the receptive endometrium as well as the invasive trophoblast. Indeed it has been found that the uterine epithelium down-regulates a number of parameters of epithelial cell polarity in this phase. This applies in a somewhat similar way to the trophoblast of blastocysts which has to give up part of its typical epithelial organization when becoming invasive: It must express cell-cell adhesion molecules or matrix receptors non-typically at its apical plasma membrane and must change its motility apparatus. Interestingly, recent data show that, in both systems, a great number of differentiation parameters of cells change in addition. It appears that part of the epithelial differentiation program is down-regulated at this phase. This new concept appears to offer interesting aspects of the basis of steroid hormone action at the endometrium, as well as of trophoblast invasiveness, postulating that switches occur in the activity of regulatory "master" genes as also involved in decision making during development.

INTRODUCTION

During the intial phase of embryo implantation, the trophoblast of the blastocyst has to attach to the endometrium, and, in invasive types of implantation, it subsequently penetrates through the uterine epithelium into the endometrial stroma. As suggested by a number of experimental data, this process includes adhesive and invasive interactions between trophoblast and uterine epithelium which can be initiated only if both partners have entered a specific physiological state: the "**invasive state**" in case of the trophoblast, and the "**receptive (permissive) state**" in case of the endometrium. It is widely believed that implantation can indeed be initiated only when both partners enter these states in synchrony. Receptivity is maintained for only a limited period of time, which defines an "**implantation window**" (Psychoyos, 1973, 1986, 1988; Psychoyos and Casimiri, 1980). While receptivity of the endometrium is regulated by ovarian steroid hormones (notably by progesterone and changes in the estrogen/progesterone ratio), invasiveness of the trophoblast is attained when it has reached a certain state of differentiation, the regulation of which is unclear (see also below).

These concepts and the mentioned general terms describing them have been derived in the first place from experiments on asynchronous embryo transfer performed in laboratory rodents and the rabbit, and from investigations on the endocrine regulation of early pregnancy and implantation in these species. They have proven useful in interpreting the results obtained in such experiments, and recently it has been proposed that they are likewise applicable for the human in particular with respect to problems encountered with embryo transfer after in-vitro fertilization (Psychoyos, 1986, 1988; Martel et al., 1987; Psychoyos and Martel, 1990). On the other hand, these concepts do not directly help with defining the molecular processes going on in trophoblast and endometrial cells. However, recently new information became available from a number of experimental studies based on modern cell biological concepts on epithelial cell polarity, epithelial differentiation and epithelial-mesenchymal (E-M) transition in development, and on cell adhesion phenomena related to those processes. This review will concentrate on these new concepts.

IMPLANTATION INITIATION: A CELL BIOLOGICAL PARADOX

The morphology and general physiology of trophoblast-endometrial interactions at implantation have been reviewed before (Denker, 1990, 1993). For the first phase of this process, i. e., the interaction with the uterine epithelium, morphology has revealed three different modes realized in different species: the "**displacement type**" (rat and mouse), the "**fusion type**" (rabbit, binucleate cells in ruminants, perhaps the human) and the "**intrusion type**" (carnivores) (Schlafke and Enders, 1975; for additional literature see Denker, 1990, 1993).

All three modes have in common that the process always starts with attachment of the apical plasma membranes of trophoblast and uterine epithelial cells to each other. This attachment is characterized morphologically by membranes running parallel over longer stretches at a distance of about 200 Å, and by the development of a specialized submembranous filament network. Stability of attachment can be demonstrated experimentally because now blastocysts cannot be torn apart anymore from the uterine epithelium without breaking cells. It is thus necessary to define adhesion molecules involved in this attachment phenomenon, and to explain on a cell biological basis what mechanisms may cause the expression of these adhesive properties.

From a cell biological point of view, implantation must be regarded as an astonishing phenomenon and has been termed **a cell biological paradox** (Denker, 1986, 1988, 1990, 1993): The fact that, when implantation is initiated, the trophoblast of the blastocyst attaches with its apical plasma membrane to the apical plasma membrane of the uterine epithelium, is far from being trivial. A fundamental property of simple epithelia is to possess a polarized organization and, as one aspect of this, two distinct membrane domains: the apical and the basolateral plasma membrane domain (Hay, 1985a, b; Rodriguez-Boulan and Nelson, 1989; Simons and Fuller, 1985). In contrast to basolateral membranes which are rich in adhesion molecules so that they can mediate cell-to-cell and cell-to-matrix adhesion, apical plasma membranes normally lack most of these molecules and lack adhesive properties. In addition, they may contain bulky molecules that sterically hinder the interaction of potentially adhesive molecules. However, at implantation initiation we are confronted with the fact that trophoblast and uterine epithelium make their first contact exactly via their apical cell membranes, and this is what needs to be explained at a cellular and molecular level.

ENDOMETRIAL RECEPTIVITY

The **luminal epithelium** of the uterus appears to play a central role in mediating the properties of "receptivity" or "non-receptivity" of the endometrium, and it seems to be a unique property of this epithelium in contrast to other epithelia to be able to enter a state of "receptivity" under steroid hormone control. If the uterine epithelium is removed experimentally, blastocysts can "implant" completely independent of any hormonal control (Cowell, 1969). When the trophoblast is allowed to interact with various types of tissues without having to overcome an intervening epithelium, it can invade deeply regardless of the hormonal status of the host, even in males, e. g., when blastocyts are transplanted to ectopic sites (Kirby, 1965, 1967; Porter, 1967). Even the pig trophoblast, which never becomes invasive in the normal in-vivo situation, was reported to show adhesive and invasive interactions after ectopic transplantation or in in-vitro experiments (for literature, see Denker, 1993). The uniqueness of the changes in behaviour of the uterine epithelium as seen at "receptivity" is demonstrated by the fact that other epithelia (with the exception of mesothelia and endothelia) do not seem to allow the trophoblast to attach; these obviously include the tubal epithelium which the troboblast cannot penetrate in any hormonal state, at least not in animals (Tutton and Carr, 1984; Pauerstein et al., 1990).

It is well possible that changes seen in the physiology of the uterine epithelium at "receptivity" are secondary to changes which occur in the endometrial stroma (e. g., after a blastocyst-derived signal has before been transduced via the uterine epithelium, Denker, 1990). However, the exact sequence of events, be it as just described or any other possible variant, is of no major bearing for our arguments concerning the nature of the cell biological changes finally defining the "receptive state" of the uterine epithelium. The nature of that state will now be discussed on a molecular level.

In a number of investigations it has been tried to define molecular changes in the composition of the attaching apical plasma membranes, that of the uterine epithelium and the corresponding membrane of the trophoblast. Very consistently, a reduction in the thickness of the glycocalyx of uterine epithelial cells and in cell surface charge has been observed in various species (Anderson et al., 1986, 1990; Enders and Schlafke, 1977; Morris and Potter, 1984, 1990; Morris et al., 1988; Potter and Morris, 1990). On the other hand, the expression of new cell surface proteins has also been observed (Lampelo et al., 1985; Anderson et al., 1988; Hoffman et al., 1990). Knowledge about the identity of molecules involved, however, is still very limited. The most specific conclusions concerning the nature of the involved molecules have been drawn by Carson et al. (1990, 1993) who proposed that heparan sulfate proteoglycan (HSPG) receptors are expressed at the apical plasma membrane of the uterine epithelium specifically during the "receptive phase", and that the trophoblast attaches via its cell surface-bound HSPG molecules (perlecan, not syndecan being the core protein), in the mouse. (Somewhat conflicting observations concerning syndecan have been presented by Potter and Morris, 1992, so that many questions still appear to be open). On the other hand, there is indeed evidence for other carbohydrate recognition processes (Lindenberg et al., 1988) possibly including a galactosyltransferase-galactose mechanism (Chávez, 1990).

Of particular interest is that the changes seen in the uterine epithelium when entering the receptive phase are indeed surprisingly complex: they comprise many more characteristics than one would expect when focussing on changes specifically needed for allowing the trophoblast to attach to this apical plasma membrane. It was proposed, therefore, that an aspect critically involved in "receptivity" or "non-receptivity" of the uterine epithelium is the degree of expression of its polar organization (Denker, 1986, 1988, 1990, 1993; Glasser et al., 1990, 1991). Detailed investigations of the in-vivo situation have impressively shown that parameters related to the expression of general epithelial cell polarity change not only in the apical but also in the lateral and basal aspects of uterine epithelial cells at this phase. This has led to the proposition that receptivity involves a change in the expression of the general epithelial phenotype (Denker, 1986, 1988, 1990, 1993). As already mentioned, this phenotype is characterized by possessing, in simple epithelia, membrane domains (apical and basolateral) with strikingly differing composition, typical sets of adhesion molecules (like uvomorulin and certain integrins), a basal lamina at one pole, and cytokeratins.

The major relevant changes seen in the uterine luminal epithelium at receptivity can be listed as follows:

Plasma Membrane Domains

Apical Plasma Membrane

- Loss of marker enzymes of the brush border type (Classen-Linke et al., 1987).

- Reduced lectin binding properties.

    Literature is somewhat confusing insofar as it was proposed that the expression of glycoconjugates with terminal galactose is positively correlated with receptivity (Chávez and Anderson, 1985; Anderson et al., 1986; Anderson et al., 1990). However, in the implantation chamber there is an overall trend towards reduction of lectin binding including lectins that recognize galactose (Nalbach, 1985; Bükers et al., 1990). This is consistent with findings on reduction of the thickness of the glycocalyx and of cell surface charge at receptivity in various species as cited above.

- Increased density of intramembranous protein particles as seen in freeze fracture morphology, so that the resulting density of particles corresponds to that typical for basolateral membranes (Murphy at al., 1982a; Winterhager, 1985; Winterhager et al., 1990). Unfortunately, so far little is known about the interesting question to what extent these particles may represent adhesion-related molecules, e. g., cell-cell adhesion molecules, matrix receptors, glycosyl transferases, lectins and others. A re-distribution of HSPG "receptors" to the apical plasma membrane of mouse uterine epithelium in the receptive phase was proposed by Carson et al. (1990) (see also above). Alternatively, receptors may not be relocated but simply be made available for binding by release of previously bound HSPG (Morris and Potter, 1990; Morris et al., 1988; Potter and Morris, 1990).

- Formation of "reflexive" gap junctions (Murphy et al., 1982 d) and (under certain experimental conditions) hemidesmosome-like junctions (Denker, 1977). These observations demonstrate that, in addition to the HSPG receptors mentioned above, other types of adhesion molecules (such as those involved in formation of these junctions) obviously become expressed in the apical plasma membrane.

Lateral Plasma Membrane

- Translocation of the subapical band of intercellular junctions (an indicator of changes in functional polarity of epithelia, Chevalier at al., 1985; Kitajima et al., 1985).

    Tight junctions: strands proliferate towards the basal cell pole (Murphy et al., 1982 b; Murphy et al., 1982 c; Winterhager and Kühnel, 1982).

    Adhaerens junctions: Uvomorulin (E-Cadherin, cell-CAM 120/80) an integral membrane protein typically associated with the zonula adhaerens, is maximally concentrated in the subapical region of the lateral plasma membrane during pre-receptive phases; in those parts of the uterine epithelium that immediately surround a blastocyst in rabbit implantation chambers this adhesion protein is seen to lose its subapical maximum and to become more evenly distributed over the lateral plasma membrane, most obviously at the endometrial "placental folds" on days 8 and 9 post coitum immediately before the trophoblast attaches and fuses with this epithelium. In this part of the epithelium, relocation of uvomorulin locally reaches very impressive degrees so that it becomes maximally concentrated in parts of the basal plasma membrane where it cannot be shown (or only in traces) with the same methodology in pre-receptive phases. It is here located at small cytoplasmic processes of uterine epithelial cells that penetrate the basal lamina (see below) (Donner et al., 1991, 1992; Donner and Denker, unpublished; Denker, 1993).

The desmosome-associated protein desmoplakin equally shows a loss of maximal concentration in the subapical region like uvomorulin, although in this case a re-distribution to the basal plasma membrane was not seen (Classen-Linke and Denker, 1990; Donner et al., 1991).

Basal Plasma Membrane

- Reduced adhesion to the basal lamina in rodents (Bitton-Casimiri et al., 1977; Schlafke and Enders, 1975; Tachi et al., 1970).

- Formation of cytoplasmic processes of the uterine epithelium that penetrate through the basal lamina into the underlying stroma (Roberts et al., 1988; Marx et al., 1990).

Intracellular/Transcellular Transport Activities

- Stage-dependent changes in vectorial transport activities through the uterine epithelium are traditionally thought to serve the specific changes in secretory activity that provide a stage-specifically optimized milieu for blastocyst development (Beier, 1974; Parr, 1980, 1982, 1983; Parr and Parr, 1977, 1978; Marengo et al., 1986). However, in the context that we are discussing here they must be regarded as also potentially serving as a mechanism contributing to sorting and re-distribution of membrane precursors and differential transport of degradation products thus regulating differential composition of apical vs. basolateral plasma membranes. Problems of mimicking this in in-vitro systems have been addressed by Glasser et al. (1991).

Cytoskeleton

- Upregulation of vimentin and a re-distribution along the apico-basal axis of polarity in the implantation chamber in the rabbit (Hochfeld et al., 1990).

It was proposed that the changes seen in this large number of parameters can be interpreted as follows: All mentioned parameters are characteristics of the apico-basal polarity of epithelia. During the pre-receptive phases, these parameters are organized in a polarized fashion along the apico-basal axis, but during acquisition of "receptivity", there is an overall trend towards loss of this polar organization with many of these parameters, and with some of them polarity even appears to become inverted (e. g., uvomorulin and vimentin). It was proposed, therefore, that steroid hormone action may (directly or indirectly via the endometrial stroma) change the expressed genetic program of the uterine epithelium in such a way that **part of the epithelial type differentiation program is being downregulated at receptivity** (Denker, 1986, 1988, 1990, 1993). As a consequence, the receptive uterine epithelium shows changes in cell behaviour (cells detaching from their basal lamina in rodents, and behaving in a semi-invasive manner by sending projections through their basal lamina in the human and the rabbit), thus facilitating trophoblast invasion.

There is much evidence that locally acting signals derived from the blastocyst are contributing considerably to modifying the properties of the uterine epithelium. So, the behaviour of this epithelium would be determined in the first place by preconditioning through systemically acting maternal steroid hormones and would then be modulated in addition locally in the vicinity of the implanting blastocyst. Changes in polarity parameters are indeed particularly obvious in the **implantation chamber** in contrast to interblastocyst segments of the uterus. It appears that local signals provided by the blastocyst drive the switches even further than the maternal steroids (for literature, see the listing of changes, above). The nature of such local signals is at present a matter of discussion: They may include interferon-type molecules like oTP-1 and bTP-1 in ruminants (Roberts, 1989), or other cytokines, growth factors, steroids, prostaglandins, and others. As discussed previously (Denker, 1990), matrix-type molecules (including fragments retaining ligand properties) could also act as such short-range signals since it was shown in other cellular systems that they can very well promote changes in polar organization (Garbi et al., 1986; Greenburg and Hay, 1982, 1986, 1988; Hay, 1985a, b; Mauchamp et al., 1987).

TROPHOBLAST INVASIVENESS

Recent data on the regulation of trophoblast differentiation (e. g., on the actions of cytokines and growth factors as well as matrix molecules) have been reviewed by Aplin (1991), Graham and Lala (1991), Lala and Graham (1990), and Hohn et al. (1992). On the other hand, with respect to trophoblast invasiveness nearly as little is known as it is about the cellular basis of tumor cell invasion. However, recent progress in two fields appears promising: adhesion molecules and motility properties. It appears that in both systems regulation of the expression of adhesion molecules, matrix degrading hydrolases and motility factors as well as of their receptors seems to be of central importance (for reviews, see Mareel et al., 1990; Behrens et al., 1991; Birchmeier et al., 1991).

In the context of the concepts discussed in the present paper the following findings are of particular interest: As recent analysis (Aplin, 1991; Damsky et al., 1992; Korhonen et al., 1991) of trophoblast emigration out of the so-called cytotrophoblastic cell columns of anchoring villi suggests, acquisition of invasiveness of trophoblast cells is accompanied by:

- loss of polar organization with respect to integrin ($\alpha_6 \beta_4$, $\alpha_3 \beta_1$) distribution in relation to the basement membrane on which these cells sit originally,
- subsequent loss of certain integrins ($\alpha_6 \beta_4$),
- acquisition of new types of integrins ($\alpha_5 \beta_1$) that enable the emanating invasive cells to interact with interstitial matrix materials (such as fibronectin, type I collagen and fibrinogen/fibrin). Fibronectin was found to be the best substrate for adhesion of isolated normal (placental) and malignant trophoblast in vitro (Aplin and Charlton, 1990; Foidart et al., 1990).

Remarkably, therefore, a loss of polar organization of cells in addition to the expression of new types of adhesion molecules is found in invasive trophoblast as it is in receptive uterine epithelium. This will be discussed below with respect to genetic re-programming.

A UNIFYING CELL BIOLOGICAL VIEW OF ENDOMETRIAL RECEPTIVITY AND TROPHOBLAST INVASIVENESS

The described features of trophoblast invasiveness and of endo-metrial receptivity show certain fascinating similarities. The common denominator appears to be a decrease in expression of apico-basal polarity (Fig. 1). This is of particular interest when comparing it with a process in embryology that recently attracts much attention: **epithelial-mesenchymal (E-M) transformation**. During development, cells can switch (even various times subsequently) between two major phenotypes, the epitheloid and the mesenchymal or fibroblastoid phenotype (Hay, 1985a, b; Greenburg and Hay, 1986; Rodriguez-Boulan and Nelson, 1989; Ekblom, 1989). Characteristics of these two phenotypes include:

- epithelial phenotype: apico-basal polarity, cytokeratins, laminin, collagen type IV, the integrin $\alpha_6 \beta_4$, uvomorulin (E-cadherin).

- mesenchymal/fibroblastoid phenotype: front-rear polarity, vimentin, fibronectin, collagen type I, the integrin $\alpha_5 \beta_1$.

Switches between these two major phenotypes involve, in various experimental systems, many or all of the mentioned parameters. It is postulated, therefore, that certain master genes regulate these programs and switches. Recently there is great interest in these types of master regulatory genes, and attempts are being made to apply these views to the changes in cell behaviour seen in invasive tumors (Birchmeier et al., 1991; Mareel et al., 1990, 1991).

In the context of trophoblast-endometrial interactions it is of much interest to ask whether similar genes may be involved. As pointed out earlier (Denker, 1986), what appears as a paradox in implantation initiation, i. e., contact formation between two epithelia via their apical cell poles, is indeed found in many examples in embryology. Of greatest interest is that recent investigations increasingly show that those processes are typically combined with E-M transformations, and not primarily with cell death as thought traditionally. Examples include the following embryonic "fusion" processes:

Figure 1. *(see opposite page)*
Schematic sketch of changes in polar organization of trophoblast when acquiring attachment capability/invasiveness, and uterine epithelium when entering "receptivity", at implantation initiation. This scheme concentrates on distribution of adhesion molecules represented by the symbols. It is still highly speculative since only very limited data are available so far: it does not intend to be correct in detail but is meant to be thought-provocative, illustrating the principle behind the concept on partial loss of epithelial-type characteristics as discussed in the text. Whereas in the pre-invasive/pre-receptive state, apical plasma membranes of trophoblast (TR) and of uterine epithelium (U) are non-adhesive (A), they express adhesion molecules when acquiring attachment competence (invasiveness)/receptivity (B). (C) and (D) show attachment, fusion and beginning penetration into the endometrial stroma for the fusion type (rabbit), (E) and (F) for the intrusion type of epithelial penetration (carnivores). T: apical type integral membrane proteins (ectodomain non-adhesive); filled circles: homotypically binding cell-cell adhesion molecules; Y: heterotypically binding receptors (e. g., HSPG receptor); triangles: various ligands for Y receptors; stars: matrix receptors, e. g., mesenchymal type integrins. (Modified after Denker, 1990).

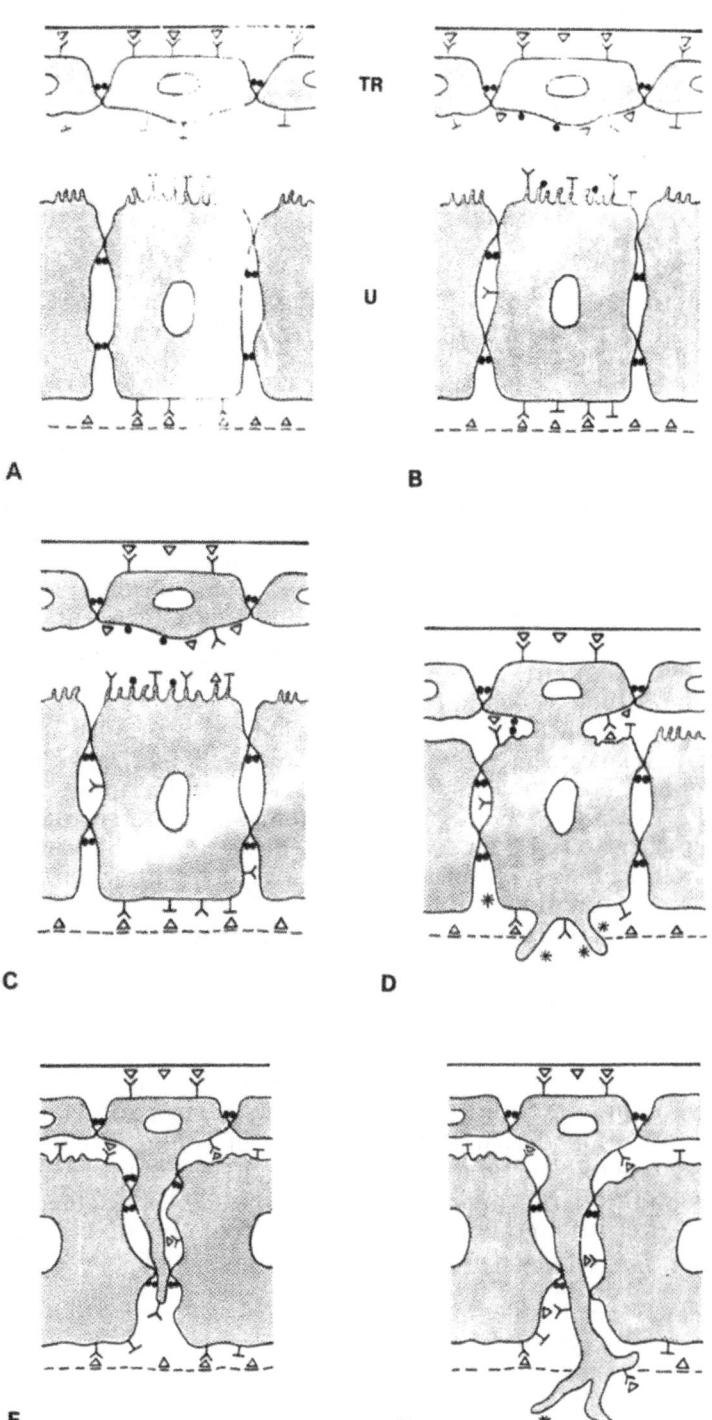

TR

U

A   B

C   D

E   F

- various epithelia:
  formation of the neural tube (combined with differentiation and
  emigration of neural crest cells), of the ear vesicle, the semicir-
  cular canals, the lens vesicle, the secondary palate and the naso-
  lacrimal duct as well as the fusion of nasal swellings;
- mesothelium:
  closure of the pleuroperitoneal canal at formation of the dia-
  phragm;
- endothelia:
  fusion of endocardial cushions at septation of the heart (see Denker, 1986, for literature).

Systematic studies of the cell biological changes that take place
during these processes in embryonic development are still largely
lacking. In particular, we do in most cases not yet have sufficient data
on the parameters listed above that would allow to monitor E-M
transformations, with the exception of neural crest formation and
formation of the secondary palate, the most widely studied examples.
In particular, the action of the postulated master regulatory genes
during embryonic development and their supposed de-regulation in
tumors are still little understood. If we are right with our supposition
that lines can be drawn between these processes and endometrial
receptivity, the endometrium could be a particularly valuable system
for continuing studies along these lines, since in this tissue the
"master" genes are obviously regulated by steriod hormones. This
may open excellent new experimental approaches to study their
identity and their regulation.

However, it must be pointed out that application of this concept
to endometrial receptivity and trophoblast invasiveness is still very
hypothetical. Loss of polar organization along the apico-basal axis is
certainly a common theme for all systems. Changes in molecular parameters appear to be far less consistent as far as data are available.
So the changes seen in the trophoblast at acquisition of invasiveness
and in uterine epithelium at receptivity do not seem to comprise the
complete set of parameters just mentioned for E-M transformation.
For example, loss of $\alpha_6 \beta_4$ integrin and acquisition of $\alpha_5 \beta_1$ integrin is
found in E-M transformation as well as in acquisition of the invasive
phenotype by the trophoblast (literature, see above), but has not been
described for the uterine epithelium. However, the latter does show
changes in expression of other integrin subunits (notably appearance
of the α_1, α_v and β_3 subunits in the secretory phase) and changes in the
polar distribution that may be indicative of such switches (Lessey et
al., 1992; Albers, Thie and Denker, unpublished data).

Upregulation of vimentin is found in E-M transformation and
receptive uterine epithelium but not in trophoblast. Uvomorulin (E-
cadherin) was reported to be down-regulated in E-M transformation
as well as in invasive tumor cells (Gumbiner et al., 1988; Mareel et
al., 1991; Behrens et al., 1991; Birchmeier et al., 1991). In the invasive trophoblast, this was reported for the mouse (Damjanov et al.,
1986) but may (Castellucci, personal communication) or may not
(Fisher et al., 1989) be seen in the human; in the latter, it was rather
described to occur in connection with fusion of cytotrophoblast to
syncytiotrophoblast (Coutifaris et al., 1991). Such down-regulation of
uvomorulin is also not seen in the uterine epithelium at receptivity
(Donner and Denker, unpublished). Data on other relevant parame-

ters (laminin vs. fibronectin, type IV vs. type I collagen) are still very incomplete for trophoblast and uterine epithelium, or as in case of syndecan and perlecan, partially contradictory (Carson et al., 1993; Potter and Morris, 1992).

Therefore, many questions remain open when we try to compare trophoblast invasiveness and endometrial receptivity with E-M transformation; this is in particular true with respect to the question what master gene-regulated switches in the general genetic programs transcribed may be involved. So far one can only say that there are reports on steroid-dependent changes in activity of regulatory genes coding for transcription factors in endometrium and that these findings are basically consistent with such a view (Webb et al., 1990; Jouvenot et al., 1990; Baker et al., 1992). Clearly data on genes that control major switches in differentiation pathways are urgently needed, for both the uterine epithelium and stroma cells at receptivity and for the trophoblast at acquisition of invasiveness. It can be expected that such data will become available during the next few years and they will clarify to what extent the hypothetical views presented in this paper will hold.

Acknowledgements
Results from our own laboratory discussed here are based on collaboration with a number of colleagues whose input is greatly appreciated: Dr. I. Classen-Linke, Dr. A. Donner, Dr. H.-P. Hohn, Dr. M. Marx, Dorothea Schünke, Dr. M. Thie, Prof. Dr. E. Winterhager. I like to thank Mrs. G. Freise for typing this manuscript, and Mrs. P. Dierkes for help with the list of references. These investigations were supported by Deutsche Forschungsgemeinschaft, the Minister für Wissenschaft und Forschung Nordrhein-Westfalen, and Mildred-Scheel-Stiftung für Krebsforschung.

REFERENCES

Anderson, T.L., Olson, G.E., and Hoffman, L.H.,1986, Stage-specific alterations in the apical membrane glycoproteins of endometrial epithelial cells related to implantation in rabbits, Biol. Reprod. 34:701-720.

Anderson, T.L., Sieg, S.M., and Hodgen, G.D., 1988, Membrane composition of the endometrial epithelium: Molecular markers of uterine receptivity to implantation, in: Human Reproduction, (Int'l. Congress Ser., No. 768), R. Iizuka and K. Semm, eds. Amsterdam-New York-Oxford, Excerpta Medica, pp. 513-516.

Anderson, T.L., Simon, J.A., and Hodgen, G.D., 1990, Histochemical characteristics of the endometrial surface related temporally to implantation in the non-human primate (Macaca fascicularis), in: Trophoblast Invasion and Endometrial Receptivity. Novel Aspects of the Cell Biology of Embryo Implantation, H.-W. Denker and J.D. Aplin, eds. (Trophoblast Research, Vol. 4). Plenum Medical Book Comp., New York and London, pp. 273-284.

Aplin, J.D., 1991, Implantation, trophoblast differentiation and haemochorial placentation: mechanistic evidence in vivo and in vitro, J. Cell Sci. 99:681-692.

Aplin, J.D., and Charlton, A.K., 1990, The role of matrix macromolecules in the invasion of decidua by trophoblast: Model studies using BeWo cells, in: Trophoblast Invasion and Endometrial Receptivity. Novel Aspects of the Cell Biology of Embryo Implantation, H.-W. Denker and J.D. Aplin, eds. (Trophoblast Research, Vol. 4). Plenum Medical Book Comp., New York and London, pp. 139-158.

Baker, D.J., Nagy, F., and Nieder, G.L., 1992, Localization of c-fos-like proteins in the mouse endometrium during the periimplantation period, Biology of Reproduction 47:492-501.

Behrens, J., Weidner, K.M., Frixen, U.H., Schipper, J.H., Sachs, M. , and Birchmeier, W., 1991, The role of E-cadherin and scatter factor in tumor invasion and cell motility, *Experientia* 59:109-126.

Beier, H.M., 1974, Oviducal and uterine fluids, *J. Reprod. Fert.* 37:221-237.

Birchmeier, W., Behrens, J., Weidner, K.M., Frixen, U.H., and Schipper, J., 1991, Dominant and recessive genes involved in tumor cell invasion, *Current Opinion in Cell Biology* 3:832-840.

Bitton-Casimiri, V., Rath, N.C., and Psychoyos, A., 1977, A simple method for separation and culture of rat uterine epithelial cells, *J. Endocr.* 73:537.

Bükers, A., Friedrich, J., Nalbach, B.P., and Denker, H.-W., 1990, Changes in lectin binding patterns in rabbit endometrium during pseudopregnancy, early pregnancy and implantation, *in:* Trophoblast Invasion and Endometrial Receptivity. Novel Aspects of the Cell Biology of Embryo Implantation, H.-W. Denker and J.D. Aplin, eds. (Trophoblast Research, Vol. 4). Plenum Medical Book Comp., New York and London, pp. 285-305.

Carson, D.D., Tang, J.-P., and Julian, J., 1993, Heparan sulfate proteoglycan (perlecan) expression by mouse embryos during acquisition of attachment competence, *Developmental Biology* 155:97-106.

Carson, D.D., Wilson, O.F., and Dutt, A., 1990, Glycoconjugate expression and interactions at the cell surface of mouse uterine epithelial cells and periimplantation-stage embryos, *in:* Trophoblast Invasion and Endometrial Receptivity. Novel Aspects of the Cell Biology of Embryo Implantation, H.-W. Denker and J.D. Aplin, eds. (Trophoblast Research, Vol. 4). Plenum Medical Book Comp., New York and London, pp. 211-241.

Chávez, D.J., 1990, Possible involvement of D-galactose in the implantation process, *in:* Trophoblast Invasion and Endometrial Receptivity. Novel Aspects of the Cell Biology of Embryo Implantation, H.-W. Denker and J.D. Aplin, eds. (Trophoblast Research, Vol. 4). Plenum Medical Book Comp., New York and London, pp. 259-272.

Chávez, D.J., and Anderson, T.L., 1985, The glycocalyx of the mouse uterine luminal epithelium during estrus, early pregnancy, the peri-implantation period, and delayed implantation. I. Acquisition of Ricinus communis I binding sites during pregnancy, *Biol. Reprod.* 32:1135-1142.

Chevalier, J., Bourguet, J., and Pinto da Silva, P., 1985, Osmotic gradient reversal induces massive assembly of tight junction strands at the basal pole of toad bladder epithelial cells, *J. Cell Biol.* 101:304a.

Classen-Linke, I., and Denker, H.-W., 1990, Preparation of rabbit uterine epithelium for trophoblast attachment: Histochemical changes in the apical and lateral membrane compartment, *in:* Trophoblast Invasion and Endometrial Receptivity. Novel Aspects of the Cell Biology of Embryo Implantation, H.-W. Denker and J.D. Aplin, eds. (Trophoblast Research, Vol. 4). Plenum Medical Book Comp., New York and London, pp. 307-322.

Classen-Linke, I., Denker, H.-W., and Winterhager, E., 1987, Apical plasma-membrane-bound enzymes of rabbit uterine epithelium: Pattern changes during the periimplantation phase, *Histochemistry* 87:517-529.

Coutifaris, C., Kao, L.-C., Sehdev, H.M., Chin, U., Babalola, G.O., Blaschuk, O.W., and Strauss, III, J.F., 1991, E-cadherin expression during the differentiation of human trophoblasts, *Development* 113:767-777.

Cowell, T.P., 1969, Implantation and development of mouse eggs transferred to the uteri of non-progestational mice, *J. Reprod. Fertil.* 19:239-245.

Damjanov, I., Damjanov, A., and Damsky, C.H., 1986, Developmentally regulated expression of the cell-cell adhesion glycoprotein cell-CAM 120/80 in peri-implantation mouse embryos and extraembryonic membranes, *Dev. Biol.* 116:194-202.

Damsky, C.H., Fitzgerald, M.L., and Fisher, S.J., 1992, Distribution patterns of extracellular matrix components and adhesion receptors are intricately modulated during first trimester cytotrophoblast differentiation along the invasive pathway, in vivo, *J. Clin. Invest.* 89:210-222.

Denker, H.-W., 1977, Implantation: The role of proteinases, and blockage of implantation by proteinase inhibitors, Adv. Anat. Embryol. Cell Biol., Vol. 53, Part 5, Springer-Verlag, Berlin-Heidelberg-New York.

Denker, H.-W., 1986, Epithel-Epithel-Interaktionen bei der Embryo-Implantation: Ansätze zur Lösung eines zellbiologischen Paradoxons, Verh. Anat. Ges., 80 (Prag 1985), *Anat. Anz.* Suppl. 160:93-114.

Denker, H.-W., 1988, Implantation: recent approaches to understand a cell biological paradox, in: Human Reproduction. Current Status / Future Prospect (Proceed. VIth World Congr. on Human Reproduction, Tokyo 1987) (International Congress Series No. 768), R. Iizuka and K. Semm, eds. Excerpta Medica, Elsevier Science Publ., Amsterdam, pp. 237-240.

Denker, H.-W., 1990, Trophoblast-endometrial interactions at embryo implantation: a cell biological paradox, in: Trophoblast Invasion and Endometrial Receptivity. Novel Aspects of the Cell Biology of Embryo Implantation, H.-W. Denker and J.D. Aplin, eds. (Trophoblast Research, Vol. 4). Plenum Medical Book Comp., New York and London, pp. 3-29.

Denker, H.-W., 1993, Implantation: a cell biological paradox, J. exp. Zool. (in press)

Denker, H.-W., Hohn, H.-P., and Bükers, A., 1989, Investigations on the cell biology of embryo implantation, in: Reproductive Biology and Medicine, A.F. Holstein, K.D. Voigt, D. Grässlin, eds, Diesbach Verlag, Berlin, pp. 224-238.

Donner, A., Classen-Linke, I., Frixen, U., Behrens, J., and Denker, H.-W., 1991, Änderungen im Verteilungsmuster von Markerproteinen für das basolaterale Membran-Kompartiment des Uterusepithels in der Periimplantationsphase: Desmoplakin und Uvomorulin. 8. Arbeitstagung d. Anat. Gesellschaft, Würzburg 1990, Anat. Anz. 172:33 (Abstract).

Donner, A., Behrens, J., Frixen, U., and Denker, H.-W., 1992, Uvomorulin, actin and the shift in uterine epithelial polarity at embryo implantation, Ann. Meet. Deutsche Ges. f. Zellbiologie, Konstanz 1992, Europ. J. Cell Biol., Suppl. 36, (Vol. 57):16 (Abstract).

Ekblom, P., 1989, Developmentally regulated conversion of mesenchyme to epithelium, The FASEB Journal 3:2141-2150.

Enders, A.C., and Schlafke S., 1977, Alteration in uterine luminal surface at the implantation site, J. Cell Biol. 75:70a.

Fisher, S.J., Cui, Tian-yi, Zhang, L., Hartman, L., Grahl, K., Guo-Yang, Z., Tarpey, J., and Damsky, C.H., 1989, Adhesive and degradative properties of human placental cytotrophoblast cells in vitro, J. Cell Biol. 109:891-902.

Fisher, S.J., Sutherland, A., Moss, L., Hartman, L., Crowley, E., Bernfield, M., Calarco, P., and Damsky, C., 1990, Adhesive interactions of murine and human trophoblast cells, in: Trophoblast Invasion and Endometrial Receptivity. Novel Aspects of the Cell Biology of Embryo Implantation, H.-W. Denker and J.D. Aplin, eds. (Trophoblast Research, Vol. 4). Plenum Medical Book Comp., New York and London, pp. 115-138.

Foidart, J.M., Christiane, Y., and Emonard, H., 1990, Interactions between the human trophoblast cells and the extracellular matrix of the endometrium. Specific expression of α-galactose residues by invasive human trophoblastic cells, in: Trophoblast Invasion and Endometrial Receptivity. Novel Aspects of the Cell Biology of Embryo Implantation, H.-W. Denker and J.D. Aplin, eds. (Trophoblast Research, Vol. 4). Plenum Medical Book Comp., New York and London, pp. 159-177.

Garbi, C., Tacchetti, C., and S.H. Wollman, S.H., 1986, Change of inverted thyroid follicle into a spheroid after embedding in a collagen gel. Exp. Cell Res. 163:63-77.

Glasser, S.R., Julian, J., Mulholland, J., Mani, S., Carson, D.D., and Jacobs, A.L., 1990, In vitro implantation on polarized uterine epithelia, in: Early Embryo Development and Paracrine Relationships. S. Heyner, and L.M. Wiley, eds., Alan R. Liss, Inc., New York, pp. 153-167.

Glasser, S.R., Mani, S.K., and Mulholland, J., 1991, In vitro models of implantation, in: Uterine and Embryonic Factors in Early Pregnancy, J.F. Strauss III, C.R. Lyttle, eds. Plenum Press, New York, pp. 33-50.

Graham, C.H., and Lala , P.K.,1991, Mechanism of control of trophoblast invasion in situ, J. Cell Physiol. 148:228-234.

Greenburg, G., and Hay, E.D., 1982, Epithelia suspended in collagen gels can lose polarity and express characteristics of migrating mesenchymal cells, J. Cell Biol. 95:333-339.

Greenburg, G., and Hay, E.D., 1986, Cytodifferentiation and tissue phenotype change during transformation of embryonic lens epithelium to mesenchyme-like cells in vitro, Dev. Biol. 115:363-379.

Greenburg, G., and Hay, E.D., 1988, Cytoskeleton and thyroglobulin expression change during transformation of thyroid epithelium to mesenchyme-like cells, Development 102:605-622.

Gumbiner, B., Stevenson, B., and Grimaldi, A., 1988, The role of the cell adhesion molecule uvomorulin in the formation and maintenance of epithelial junctional complexes, *J. Cell Biol.* 107:1575-1587.

Hay, E.D., 1985a, Matrix-cytoskeletal interactions in the developing eye. *J. Cell. Biochem.* 27:143-156.

Hay, E.D., 1985b, Extracellular matrix, cell polarity and epithelial-mesenchymal transformation, in: Molecular Determinants of Animal Form, G.M. Edelman, ed., Alan R. Liss, Inc., New York, pp. 293-318.

Hochfeld, A., Beier, H.M., and Denker, H.-W., 1990, Changes of intermediate filament protein localization in endometrial cells during early pregnancy of rabbits, in: Trophoblast Invasion and Endometrial Receptivity. Novel Aspects of the Cell Biology of Embryo Implantation, H.-W. Denker and J.D. Aplin, eds. (Trophoblast Research, Vol. 4). Plenum Medical Book Comp., New York and London, pp. 357-374.

Hohn, H.-P., Parker, Jr., C.R., Boots, L.R., Denker, H.-W., and Höök, M., 1992, Modulation of differentiation markers in human choriocarcinoma cells by extracellular matrix: on the role of a three-dimensional matrix structure, *Differentiation* 51:61-70.

Hoffman, L.H., Winfrey, V.P., Anderson, T.L., and Olson, G.E., 1990, Uterine receptivity to implantation in the rabbit: Evidence for a 42kDa glycoprotein as a marker of receptivity, in: Trophoblast Invasion and Endometrial Receptivity. Novel Aspects of the Cell Biology of Embryo Implantation, H.-W. Denker and J.D. Aplin, eds. (Trophoblast Research, Vol. 4). Plenum Medical Book Comp., New York and London, pp. 243-258.

Jouvenot, M., Pellerin, I., Alkhalaf, M., Marechal, G., Royez, M., and Adessi, G.L., 1990, Effects of 17β-estradiol and growth factors on c-fos gene expression in endometrial epithelial cells in primary culture, *Molecular and Cellular Endocrinology* 72:149-157.

Kirby, D.R.S., 1965, The role of the uterus in the early stages of mouse development, in: Ciba Foundation Symposium: Preimplantation Stages of Pregnancy, G.E.W. Wolstenholme and M. O'Connor, eds. Churchill, London, pp. 325-344.

Kirby, D.R.S. (1967) Ectopic autografts of blastocysts in mice maintained in delayed implantation. *J. Reprod. Fert.* 14:515-517.

Kitajima, K., Yamashita, K., and Fujita, H., 1985, Fine structural aspects of the shift of zonula occludens and cytoorganelles during inversion of cell polarity in cultured porcine thyroid follicles, *Cell Tiss. Res.* 242:221-224.

Korhonen, M., Ylänne, J., Laitinen, L., Cooper, H.M., Quaranta, V., and Virtanen, I., 1991, Distribution of the α1-α6 integrin subunits in human developing and term placenta, *Lab. Invest.* 65:347-356.

Lala, P.K., and Graham, C.H., 1990, Mechanisms of trophoblast invasiveness and their control: the role of proteases and protease inhibitors, *Cancer Metastasis Rev.* 9:369-379.

Lampelo, S.A., Ricketts, A.P., and Bullock, D.W., 1985, Purification of rabbit endometrial plasma membranes from receptive and non-receptive uteri, *J. Reprod. Fert.* 75:475-484.

Lessey, B.A., Damjanovich, L., Coutifaris, C., Castelbaum, A., Albelda, S.M., and Buck, C.A., 1992, Integrin adhesion molecules in the human endometrium. Correlation with the normal and abnormal menstrual cycle, *J. Clin. Invest.* 90:188-195.

Lindenberg, S., Sundberg, K., Kimber, S.J., and Lundblad, A., 1988, The milk oligosaccharide, lacto-N-fucopentaose I, inhibits attachment of mouse blastocysts on endometrial monolayers, *J. Reprod. Fert.* 83:149-158.

Mareel, M.M., Behrens, J., Birchmeier, W., De Bruyne, G.K., Vleminckx, K., Hoogewijs, A., Fiers, W.C., and Van Roy, F.M., 1991, Down-regulation of E-cadherin expression in Madin Darby canine kidney (MDCK) cells inside tumors of nude mice. *Int. J. Cancer* 47:922-928.

Mareel, M.M., Van Roy, F.M., and De Baetselier, P., 1990, The invasive phenotypes, *Cancer and Metastasis Reviews* 9:45-62.

Marengo, S.R., Bazer, F.W., Thatcher, W.W., Wilcox, C.J., and Wetteman, R.P., 1986, Prostaglandin F_2 as the luteolysin in swine: VI. Hormonal regulation of the movement of exogenous PGF_2 from the uterine lumen into the vasculature, *Biol. Reprod.* 34:284-292.

Martel, D., Frydman, R., Glissant, M., Maggioni, C., Roche, D., and Psychoyos, A., 1987, Scanning electron microscopy of postovulatory human endometrium in spontaneous cycles and cycles stimulated by hormone treatment, *J. Endocr.* 114:319-324.

Marx, M., Winterhager, E., and Denker, H.-W., 1990, Penetration of the basal lamina by processes of uterine epithelial cells during implantation in the rabbit, in: Trophoblast Invasion and Endometrial Receptivity. Novel Aspects of the Cell Biology of Embryo Implantation. H.-W. Denker and J.D. Aplin, eds. (Trophoblast Research, Vol. 4). Plenum Medical Book Comp., New York and London, pp. 417-430.

Mauchamp, J., Chambard, M., Verrier, B., Gabrion, J., Chabaud, O., Gerard, C., Penel, C., Pialat, B., and Anfosso, F., 1987, Epithelial cell polarization in culture: Orientation of cell polarity and expression of specific functions, studied with cultured thyroid cells, J. Cell Sci. Suppl. 8:345-358.

Morris, J.E., and Potter, S.W., 1984, A comparison of developmental changes in surface charge in mouse blastocysts and uterine epithelium using DEAE beads and dextran sulfate in vitro, Develop. Biol. 103:190-199.

Morris, J.E., and Potter, S.W., 1990, An in vitro model for studying interactions between mouse trophoblast and uterine epithelial cells. A brief review of in vitro systems and observations on cell-surface changes during blastocyst attachment, in: Trophoblast Invasion and Endometrial Receptivity. Novel Aspects of the Cell Biology of Embryo Implantation. H.-W. Denker and J.D. Aplin, eds. (Trophoblast Research, Vol. 4), Plenum Medical Book Comp., New York and London, pp. 51-69.

Morris, J.E., Potter, S.W., and Gaza-Bulseco, G., 1988, Estradiol-stimulated turnover of heparan sulfate proteoglycan in mouse uterine epithelium. J. Biol. Chem. 263:4712-4718.

Murphy, C.R., Swift, J.G., Mukherjee, T.M., and Rogers, A.W., 1982a, Changes in the fine structure of the apical plasma membrane of endometrial epithelial cells during implantation in the rat, J. Cell Sci. 55:1-12.

Murphy, C.R., Swift, J.G.,Mukherjee, T.M., and Rogers, A.W., 1982b, The structure of tight junctions between uterine luminal epithelial cells at different stages of pregnancy in the rat, Cell Tiss. Res. 223:281-286.

Murphy, C.R., Swift, J.G., Need, J.A., Mukherjee, T.M., and Rogers, A.W., 1982c, A freeze-fracture electron microscopic study of tight junctions of epithelial cells in the human uterus, Anat. Embryol. 163:367-370.

Murphy, C.R., Swift, J.G., Mukherjee, T.M., and Rogers, A.W., 1982d, Reflexive gap junctions on uterine luminal epithelial cells, Acta Anat. 112:92-96.

Nalbach, B.P., 1985, Lektinbindungsmuster in Uterus und Blastozyste des Kaninchens während der Präimplantationsphase und der frühen Implantation. Histochemie und Methodenkritik. Dissertation, Medizinische Fakultät der RWTH Aachen.

Parr, M.B., 1980, Endocytosis at the basal and lateral membranes of rat uterine epithelial cells during early pregnancy, J. Reprod. Fert. 60:95-99.

Parr, M.B., 1982, Apical vesicles in the rat uterine epithelium: A morphometric study, Anat. Rec. 202:145A-146A.

Parr, M.B., 1983, Relationship of uterine closure to ovarian hormones and endocytosis in the rat, J. Reprod. Fert. 68:185-188.

Parr, M.B., and Parr, E.L., 1977, Endocytosis in the uterine epithelium of the mouse, J. Reprod. Fert. 50:151-153.

Parr, M.B., and Parr, E.L., 1978, Uptake and fate of ferritin in the uterine epithelium of the rat during early pregnancy, J. Reprod. Fert. 52:183-188.

Pauerstein, C.J., Eddy, C.A., Kyoung Koong, M., and Moore, G.D., 1990, Rabbit endosalpinx suppresses ectopic implantation, Fertility and Sterility 54:522-526.

Porter, D.G., 1967, Observations on the development of mouse blastocysts transferred to the testis and kidney, Amer. J. Anat. 121:73-85.

Potter, S.W., and Morris, J.E., 1990, Immunolocalization of heparan sulfate proteoglycans in mouse uterine epithelium. Estrous cycle-dependent changes, J. Cell Biol. 111:266a.

Potter, S.W., and Morris, J.E., 1992, Changes in histochemical distribution of cell surface heparan sulfate proteoglycan in mouse uterus during the estrous cycle and early pregnancy, Anat. Rec. 234:383-390.

Psychoyos, A., 1973, Endocrine control of egg implantation, in: Handbook of Physiology, Sect. 7 (Endocrinology), Vol. II (Female Reproductive System) Part 2. R.O. Greep, ed., American Physiological Society, Washington, pp. 187-215.

Psychoyos, A., 1986, Uterine receptivity for nidation, Ann. NY Acad. Sci. 476:36-42.

Psychoyos, A., 1988, The "implantation window": Can it be enlarged or displaced? in: Human Reproduction. Current Status / Future Prospect. (International Congress Ser., No. 768), R. Iizuka and K. Semm, eds., Excerpta Medica, Amsterdam-New York-Oxford, pp. 231-232.

Psychoyos, A., and Casimiri, V., 1980, Factors involved in uterine receptivity and refractoriness, in: Blastocyst-Endometrium Relationships (Prog. Reprod. Biol., Vol. 7), F. Leroy, C.A. Finn, A. Psychoyos, and P.O. Hubinont, eds., Karger, S., Basel, München, Paris, London, New York, Sydney, pp. 143-157.

Psychoyos, A., and Martel, D, 1990, Réceptivité utérine pour l' ovo-implantation et microscopie électronique a balayage, Rech. Gynécol. 2:116-118.

Roberts, D.K., Walker, N.J., and Lavia, L.A., 1988, Ultrastructural evidence of stromal/epithelial interactions in the human endometrial cycle, Am. J. Obstet. Gynecol. 158:854-861.

Roberts, R.M., 1989, Conceptus interferons and maternal recognition of pregnancy, Biol. Reprod. 40:449-452.

Rodriguez-Boulan, E., and Nelson, W.J., 1989, Morphogenesis of the polarized epithelial cell phenotype, Science 245:718-725.

Schlafke, S., and Enders, A.C., 1975, Cellular basis of interaction between trophoblast and uterus at implantation, Biol. Reprod. 12:41-65.

Simons, K., and Fuller, S.D., 1985, Cell surface polarity in epithelia, Ann. Rev. Cell Biol. 1:243-288.

Tachi, S., Tachi, C., and Lindner, H.R., 1970, Ultrastructural features of blastocyst attachment and trophoblastic invasion in the rat, J. Reprod. Fert. 21:37-56.

Tutton, D.A., and Carr, D.H., 1984, The fate of trophoblast retained within the oviduct in the mouse, Gynecol. Obstet. Invest. 17:18-24.

Webb, D.K., Moulton, B.C., and Khan, S.A., 1990, Estrogen induced expression of the c-jun proto-oncogene in the immature and mature rat uterus, Biochem. Biophys. Res. Commun. 168:721-726.

Winterhager, E., 1985, Dynamik der Zellmembran: Modellstudien während der Implantationsreaktion beim Kaninchen, Habilitationsschrift, Med. Fak. der RWTH Aachen.

Winterhager, E., Classen-Linke, I., and Denker, H.-W., 1990, Strukturelle Veränderungen der apikalen Uterusepithelzellmembran in Vorbereitung auf die Embryoimplantation, Verh. Ant. Ges. 83, Anat. Anz., Suppl. 166:153-154.

Winterhager, E., and Kühnel, W., 1982, Alterations in intercellular junctions of the uterine epithelium during the preimplantation phase in the rabbit. Cell Tiss. Res. 224:517-526.

PROGESTERONE DIRECTED GENE EXPRESSION IN RAT UTERINE STROMAL CELLS

Joy Mulholland, Deana Roy, and Stanley Glasser

Department of Cell Biology
Baylor College of Medicine
Houston, TX 77030 U.S.A.

When embryos attach to the luminal epithelium in the rat uterus they somehow induce differentiation of the stromal cells underlying the attachment site, giving rise to an enormous increase in the size of the stromal cell compartment through both cell proliferation and cell growth. This reaction, decidualization, is localized to areas of embryo attachment (reviewed in DeFeo, 1967). Both attachment of the embryo to the luminal epithelium and differentiation of the stromal cells in response to embryo attachment require conditioning of the uterus with steroid hormones (Psychoyos, 1973). If the uterus is hormonally prepared, decidualization can be obtained using an intraluminal traumatic or chemical artificial stimulus. In response to these stimuli stromal cells along the entire length of the uterus will differentiate (DeFeo, 1967).

The growth of differentiating stromal cells following a decidual stimulus is dramatic. Undifferentiated cells contain scant cytoplasm, exhibit limited contact with neighboring cells, and are embedded in an extensive extracellular matrix (Figure 1). Following the induction of decidualization, cell cytoplasmic volume greatly increases (Figure 1), and cells actively synthesize DNA. Initially the stromal cells divide and subsequently DNA synthesis continues in the absence of cytokinesis leading to endopolyploidy of the decidual cells. Little extracellular matrix is present in decidualized tissue and numerous intercellular junctions connect closely apposed decidual cells.

Decidualization may be considered as a three step process. First stromal cells must be sensitized to respond to a decidualizing stimulus. Sensitization is dependent on progesterone treatment, and in ovariectomized rats, treatment with progesterone for 48h is required before stromal cells will differentiate in response to a decidual stimulus (Psychoyos 1973, Glasser and Clark, 1975). For the second step, induction of differentiation, an external stimulus is necessary. In the pregnant rat, embryo attachment provides this stimulus, however in ovariectomized, hormone-treated animals, either a traumatic (needle scratch) or chemical (sesame or peanut oil are most common) stimulus applied to the luminal epithelium will trigger the decidualization response. The stimulus is transmitted from the lumen to the stromal cells via the luminal epithelium (Lejeune et al., 1981). Although embryo attachment in the rat requires estrogen, stimulation of stromal cell differentiation is dependent only on progesterone. The third step is to maintain growth and development of the decidual cells by continued treatment with progesterone.

We were interested in the first step of decidualization, sensitization, and wished to address the question of how progesterone sensitizes stromal cells to respond to a decidual stimulus. We chose to try to answer this question by identifying genes which are expressed after progesterone treatment.

For these experiments we used mature, Sprague-Dawley rats which had been ovariectomized and left untreated for 10-14 days to clear residual steroids from the circulation. The animals were given 5mg/day progesterone for 4 days and killed within 4h of the last injection. Uteri were removed and the luminal epithelium isolated by the method of Bitton-Casimiri *et al.* (1977). Stromal cells were then separated from the myometrium with a scalpel blade and immersed in 4M guanidinium isothiocyanate for RNA extraction.

RNA isolated from both untreated (ovx) and progesterone treated (PX3) animals was used to prepare subtracted cDNA libraries. The flow chart shown in Figure 2 outlines the protocol used to construct the subtracted libraries. PolyA+ mRNA was selected by oligo-dT chromatography from both groups of rats and first-strand cDNA was prepared from the mRNA of the progesterone-treated animals. A dG-tail was added to the first-strand cDNA using terminal transferase and the g-tailed cDNA was hybridized for 24h with a 30-fold excess of mRNA from ovariectomized, untreated rats. The hybridization solution was passed over a hydroxylapatite column to separate single-stranded cDNA from cDNA-RNA hybrids. Unhybridized cDNA was eluted with 0.12M PO_4 buffer, pH 6.8. This cDNA was amplified by polymerase chain reaction, ligated into the lambdaGEM-4 vector (Promega) and propagated in E. coli LE 392. The subtracted library was not amplified and was directly screened with cDNA from both untreated and progesterone-treated rats.

Repeated screening yielded 6 clones which were found only in the stromal cell cDNA of the progesterone-treated animals. These are (1) a unique clone which has just been sequenced and does not match any Genbank entries, (2) mitochondrial cytochrome oxidase whose transcription is stimulated by androgens in the testis (B. Sanborn, personal communication), (3) and (4) two transcripts corresponding to repetitive DNA elements, LINE and B2, (5) heparin-binding epidermal growth factor-like growth factor (HB-EGF), a new member of the EGF family, (6) amyloid beta protein, which has been associated with Alzheimer's disease.

The first PX3 specific clone we isolated was comprised of two cDNAs, a LINE (long interspersed repetitive DNA) cDNA which was interrupted by the consensus B2 cDNA. LINE transcripts are abundant in the rat, mouse, and human genome, however, little is known about their function or regulation. It has been proposed that they may be pseudogenes or that they may be regulatory elements. In the rat, the LINE transcript

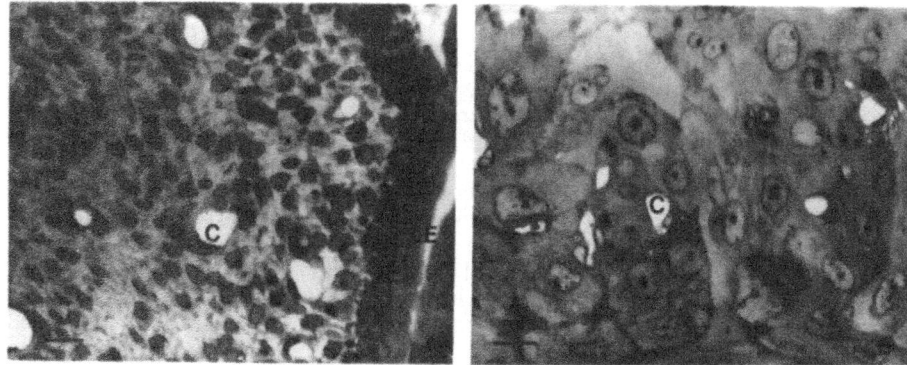

Figure 1. A. Undifferentiated rat uterine stromal cells from hormonally primed, unstimulated uterine horn. LE, luminal epithelium; C, capillary. B. Decidualized stromal cells in the contralateral uterine horn from the same rat, four days after application of a traumatic stimulus. C, capillary, bar = 50um.

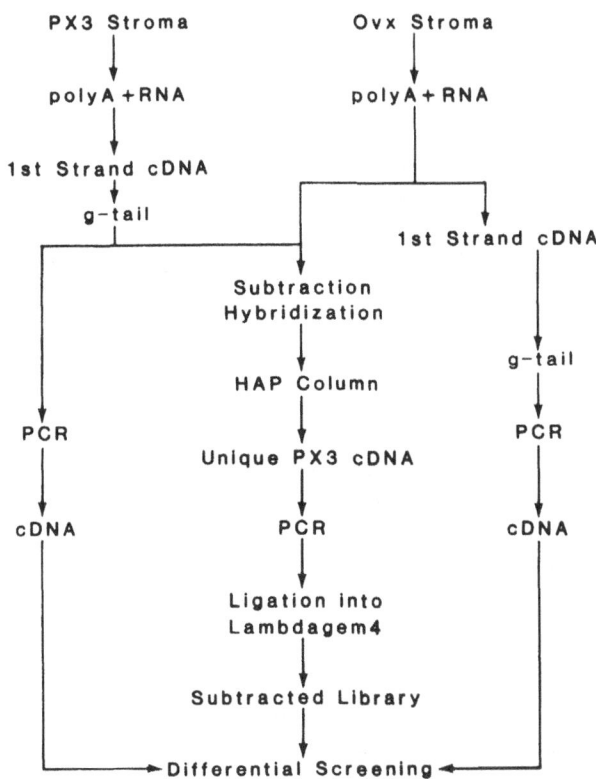

Figure 2. Protocol for construction of subtracted cDNA library from uterine stromal cell RNA of progesterone treated rats.

contains two open reading frames coding for a hypothetical protein (which has never been identified in any cell) with a transmembrane domain (Achten and Doerfler, 1986).

B2 is also a repetitive DNA element which is transcribed by RNA polymerase III. The consensus sequence is 180 base pairs in length and the transcript itself contains the A and B boxes of the polymerase III promoter, a polyadenylation signal, and a termination signal for the polymerase. B2 transcripts vary in size from 200-600 bases depending on the length of the poly A tail. B2 is abundant in the rat and mouse genomes which may contain as many as 10^5 copies of the gene. This gene is thought to be a retroposon, reverse transcribed from the RNA and inserted into adenosine rich regions of the genome (Rogers, 1985). Analysis of the B2 transcript revealed an open reading frame coding for a 40 amino acid protein, which has never been identified. The B2 gene is not expressed in quiescent or terminally differentiated cells, but expression is abundant in cells which have been stimulated to proliferate (Lania et al., 1987). Examples include the cells of rapidly cleaving mouse embryos (Taylor and Piko, 1987; Murphy et al., 1983); cultured cells which have been stimulated to divide by serum, heat shock, or transformation (Edwards et al., 1985, Kohnoe et al., 1987, Singh et al., 1985); rapidly dividing embryocarcinoma cells (Bladon et al., 1990). When cells differentiate, cell division ceases and the B2 transcript is no longer expressed. Induction of B2 transcription in dividing cells is specific for this gene and is not the result of a general increase in RNA polymerase III activity (Carey and Singh, 1988).

The expression of B2 in progesterone-treated uterine stromal cells may be associated with the proliferation of these cells. Both Martin and Finn (1968) and Huet-Hudson et al. (1989) have shown that administration of progesterone to ovariectomized mice induces stromal cell division within 24h of treatment. Although the function of B2 has not been demonstrated, a regulatory function has been proposed by Clemens (1987) based on the sequence of the transcript. Many transiently expressed genes such as *onc* genes and growth factors contain a destabilizing motif (AUUUAUUUA) in the 3' untranslated region. When this motif is joined to the globin gene and the transgene transfected into cultured cells, the half-life of the globin message is greatly reduced. Clemens noted that the B2 transcript contains a motif (UAAAUAAAU) which is complementary to this destabilizing sequence. He suggested that the B2 transcript may bind to this region and prevent degradation of these messages or regulate their rate of turnover. One of the genes whose message contains the destabilizing motif is c-*myc*, which provides some evidence for Clemen's hypothesis in the mouse uterine stroma. When ovariectomized mice are treated with progesterone, c-*myc* expression is detected in stromal cells within 3h and increases by 12h after treatment (Huet-Hudson et al., 1989). Expression of *myc* protein occurred 18h prior to the induction of DNA synthesis and cell division in the stromal cell compartment. In the pregnant mouse, *myc* protein appeared in the stromal cells of the primary decidual zone approximately 24h after embryo attachment. After 48h, *myc* expression had spread to the secondary decidual zone and was followed by proliferation of these cells.

We examined the expression of the B2 trancript in the rat endometrium following stimulation of decidualization by needle scratch using northern blotting. High levels of the transcript appeared 24h after stimulation and remained elevated until 120h (Figure 3), corresponding to the period of maximal DNA synthesis and cell division in the antimesometrial decidual tissue. After this time, DNA synthesis declines in both the antimesometrial and mesometrial decidual cells (Kleinfeld and O'Shea, 1983). Expression of B2 declined concomitantly and was nearly undetectable by day 9 when the antimesometrial decidua is regressing (Figure 3).

A second clone which appears to be specific to the PX3 library is homologous to a recently described member of the epidermal growth factor (EGF) family, heparin-binding EGF-like growth factor (HB-EGF). Like all members of the EGF family, HB-EGF is synthesized as a membrane-bound precursor protein with cytoplasmic, transmembrane, and extracellular domains (Higashiyama et al., 1991,1992). A signal peptide sequence targets the protein to the membrane and the mature, active protein is released into the extracellular matrix by proteolytic cleavage. Mature HB-EGF contains an EGF domain and an extended N-terminal region which is hydrophilic and contains the heparin-binding region of the protein.

HB-EGF protein was isolated from secretions of cultured human macrophages. Messenger RNA for human HB-EGF is 2.5kb and codes for 208 amino acids. There are four known forms which result from posttranslational modification, range in molecular weight from 19-23kD and pI 7.2-7.8, and are O-glycosylated. The mature (secreted) protein is comprised of 86 amino acids. Structurally it resembles amphiregulin, which also has an extended hydrophilic N-terminal region, more than other members of the EGF family (Higashiyama, 1991,1992).

HB-EGF is mitogenic for cultured fibroblasts and smooth muscle cells, but not endothelial cells. It has a ten-fold greater affinity for the EGF receptor than does EGF

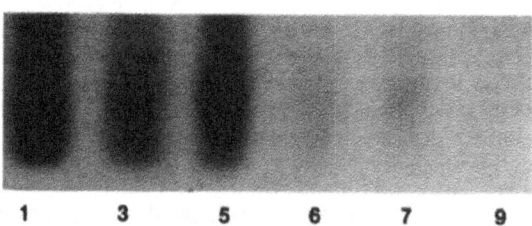

Figure 3. B2 mRNA expression in uterine stromal tissue following a decidual stimulus. RNA was isolated 1, 3, 5, 6, 7, and 9 days after a traumatic stimulus and hybridized with a cDNA probe of the B2 consensus sequence. Expression is elevated up to 5 days after stimulus and then declines.

and induces autophosphorylation of the EGF receptor. Transcription of HB-EGF and all members of the EGF family is stimulated in cultured cells by treatment with phorbol ester, but none of the genes for the EGF family growth factors contains an AP-1 transcription factor binding site. However, all members of the EGF family do express the AUUUAUUUA destabilizing motif in the 3' untranslated region of the transcripts. Like B2, expression of HB-EGF is associated with cell proliferation (Higashiyama 1991, 1992).

HB-EGF is distinguished from EGF by its heparin binding activity. Klagsbrun (1990) has suggested that heparin can act as a low affinity receptor for growth factors and has demonstrated that the mitogenic activity of fibroblast growth factor is enhanced by heparin binding. HB-EGF activity may also be stimulated by binding to heparin in the extracellular matrix. Heparin-binding growth factors may be particularly important in stimulating cell proliferation in the rodent uterus. The basal lamina of the uterine luminal epithelium is rich in heparan-sulfate proteoglycan (HSPG) and HB-EGF may initially bind here to stimulate proliferation of the subepithelial stromal cells destined to give rise to the primary decidual zone. One of the earliest biochemical changes associated with the differentiation of stromal cells in the mouse uterus is the secretion of HSPG around the decidualizing cell (Wewer et al., 1986). Secretion of heparin residues may be a mechanism to disseminate cell proliferation throughout the decidualizing stromal compartment. We are now examining the expression of HB-EGF mRNA in the rat uterine stroma during decidualization.

The last cDNA clone from the PX3 library to be discussed is homologous to the cDNA for amyloid beta precursor protein (APP). Like HB-EGF, APP is synthesized as a membrane-bound precursor protein which is cleaved to release a mature protein into the extracellular matrix. The APP gene has 18 exons and a number of variations of the protein result from both alternative RNA splicing and posttranslational modification of the protein (Yoshikai, et al., 1990). Four forms of the protein described as of this date are (1)APP695 - a "consensus" precursor protein which releases mature protein into the extracellular matrix giving rise to amyloid fibrils (De Strooper et al, 1991; Kang et al., 1987); (2,3) APP751, APP770 - both are comprised of APP695 but contain extra amino acids which give them proteinase inhibitor activity (Oltersdorf et al., 1989; De Strooper et al., 1991); (4) a form of APP which lacks a transmembrane domain and is directly secreted from the primary transcript rather than through proteolytic cleavage from a membrane-bound precursor (deSauvage and Octave, 1989). APP appears to be ubiquitous as it has been found in all tissues tested and some forms of the amyloid fibrils bind heparin (Schubert et al., 1989). The promoter region of the gene has an AP-1 transcription factor binding site and transcription can be induced in HeLa cells by phorbol ester, indicating that expression of this gene is also stimulated in proliferating cells (Yoshikai, 1990). In addition, the amyloid transcript also contains the destabilizing sequence characteristic of transiently expressed genes.

In summary, we have found in our subtracted library enriched for progesterone-stimulated transcripts the cDNAs for two transmembrane precursor proteins whose expression is stimulated by TPA and apparently by progesterone. Both HB-EGF and APP give rise to extracellular matrix proteins which can bind heparin, both contain the destabilizing motif characteristic of many transiently expressed genes, and both appear to be associated with DNA synthesis and/or cell division. A third cDNA clone codes for the repetitive DNA transcript, B2, which is also expressed during DNA synthesis and cell division and which may play a role in regulating the half-life of transiently expressed mRNA. The answer to our original question seems to be that progesterone treatment sensitizes stromal cells in part by inducing them to synthesize DNA and proliferate. This recalls a long-standing hypothesis that stromal cell mitosis is a prerequisite for decidualization. The evidence for this hypothesis has been thoroughly presented in the excellent review by O'Grady and Bell (1977). We speculate that HB-EGF, APP, and B2 play some role in inducing and/or maintaining DNA synthesis and mitosis in the rat uterine stroma and that they mediate the action of progesterone on stromal cell division.

Acknowledgements

Support for this research was provided by NIH, HD22785.

REFERENCES

Achten, S. and Doerfler, W., 1986, Characterization of a member of the highly repeated long interspersed rat DNA framily with long open reading frames, *J. Mol. Biol.* 192: 489-502.

Bitton-Casimiri, V., Rath, N.C., and Psychoyos, A., 1977, A simple method for separation and culture of rat uterine epithelial cells, *J. Endocr.* 73: 537-538.

Bladon, T.S., Fregeau, C.J., McBurney, M.W., 1990, Synthesis and processing of small B2 transcripts in mouse embryonal carcinoma cells, *Mol. Cell. Biol.* 10: 4058-4067.

Clemens, M.J., 1987, A potential role for RNA transcribed from B2 repeats in the regulation of mRNA stability, *Cell* 49: 157-158.

Carey, M.F. and Singh, K., 1988, Enhanced B2 transcription in simian virus 40-transformed cells is mediated through the formation of RNA polymerse II transcription complexes on previously inactive genes, *Biochemistry* 85: 7059-7063.

De Feo, V.J., 1967, Decidualization, in: "Cellular Biology of the Uterus," R. M. Wynn, ed., Appleton-Century-Crofts, N.Y., pp. 191-290.

De Strooper, B., Van Leuven, F., and Van Den Berghe, H., 1991, The amyloid beta protein precursor or proteinase nexin II from mouse is closer related to its human homolog than previously reported, *Biochim. Biophys. Acta* 1129: 141-143.

Edwards, D.R., Parfett, C.L.J., and Denhardt, D.T., 1985, Transcriptional regulation of two serum-induced RNAs in mouse fibroblasts: equivalence of one species to B2 repetitive elements, *Mol. Cell Biol.* 5: 3280-3288.

Glasser, S.R. and Clark, J.H., 1975, A determinant role for progesterone in the development of uterine sensitivity to decidualization and ovo-implantation, in "Developmental Biology of Reproduction," C. L. Markert, J. Papaconstantinou, Eds., Academic Press, N.Y., pp. 311-345.

Higashiyama, S., Abraham, J.A., Miller, J., Fiddes, J.C., and Klagsbrun, M., 1991, A heparin-binding growth factor secreted by macrophage-like cells that is related to EGF, *Science* 251: 936-939.

Higashiyama, S., Lau, K., Besner, G.E., Abraham, J.A., and Klagsbrun, M., 1992, Structure of Heparin-binding EGF-like Growth Factor, *J. Biol. Chem.* 267: 6205-6212.

Huet-Hudson, Y.M., Andrews, G.K., and Dey, S.K., 1989, Cell type-specific localization of c-Myc protein in the mouse uterus: modulation by steroid hormones and analysis of the periimplantation period, *Endocrinol.* 125: 1683-1690.

Kang, J., Lemaire, H-G., Unterbeck, A., Salbaum, M., Masters, C.L., Grzeschik, K-H., Multhaup, G., Beyreuther, K., and Muller-Hill, B., 1987, The precursor of Alzheimer's disease amyloid A4 protein resembles a cell-surface receptor, *Nature* 325: 733-736.

Klagsbrun, M., 1990, The affinity of fibroblast growth factors (FGFs) for heparin; FGF-heparan sulfate interactions in cells and extracellular matrix, *Curr. Opin. Cell Biol.* 2: 857-863.

Kleinfeld, R.G. and O'Shea, J.D., 1983, Spatial and temporal patterns of deoxyribonucleic acid synthesis and mitosis in the endometrial stroma during decidualization in the pseudopregnant rat, *Biol. Reprod* 28: 691-702.

Kohnoe, S., Maehara, Y., and Endo, H., 1987, A systematic survey of fepetitive sequences abundantly expressed in rat tumors, *Biochim. Biophys. Acta* 909: 107-114.

Lania, L., Pannuti, A., La Mantia, G., and Basilico, C., 1987, The transcription of B2 repeated sequences is regulated during the transition from quiescent to proliferative state in cultured rodent cells, *FEBS Lett.* 219: 400-404.

Lejeune, B., Van Hoeck, J., and Leroy, F., 1981, Transmitter role of the luminal uterine epithelium in the induction of decidualization in rats, *J. Reprod. Fert.* 61: 235-240.

Martin, L. and Finn, C.A., 1968, Hormonal regulation of cell division in epithelial and connective tissue of the mouse uterus, *J. Endocrinol.* 41: 363-371.

Murphy, D., Brickell, P.M., Latchman, D.S., Willison, K., and Rigby P.W.J., 1983, Transcripts regulated during normal embryonic development and oncogenic transformation share a repetitive element, *Cell* 35: 865-871.

O'Grady, J.E. and Bell, S.C., 1977, The role of the endometrium in blastocyst implantation, in "Development in Mammals, Vol. I," M.H. Johnson, Ed., pp. 165-243.

Oltersdorf, T., Fritz, L.C., Schenk, D.B., Lieberburg, I., Johnson-Wood, K.L., Beattie, E.C., Ward, P.J., Blacher, R.W., Dovey, H.F., and Sinha, S., 1989, The secreted form of the Alzheimer's amyloid precursor protein with te Kunitz domain is protease nexin-II, *Nature* 341: 144-147.

Psychoyos, A., 1973, Hormonal control of implantation, *Vitam. Horm.* 31: 201-256.

Rogers, J.H., 1985, The origin and evolution of retroposons, *Int. Rev. Cytol.* 91: 187-279.

de Sauvage, F. and Octave, J-N., 1989, A Novel mRNA of the A4 amyloid precursor gene coding for a possibly secreted protein, *Science* 245: 651-653.

Schubert, D., LaCorbiere, M., Saitoh, T., and Cole, G., 1989, Characterization of an amyloid beta precursor protein that binds heparin and contains tyrosine sulfate, *Proc. Natl. Acad. Sci. USA* 86: 2066-2069.

Singh, K., Carey, M., Saragosti, S., and Botchan, M., 1985, Expression of enhanced levels of small RNA polymerase III transcripts encoded by the B2 repeats in simian virus 40-transformed mouse cells, *Nature* 314: 553-556.

Taylor, K.D. and Piko, L., 1987, Patterns of mRNA prevalence and expression of B1 and B2 transcripts in early mouse embryos, *Development* 101: 877-892.

Wewer, U.M., Damjanov, A., Weiss, J., Liotta, L.A., and Damjanov, I., 1986, Mouse endometrial stromal cells produce basement-membrane components, *Differentiation* 32: 49-58.

Yoshikai, S., Sasaki, H., Doh-ura, K., Furuya, H., and Sakaki, Y., 1990, Genomic organization of the human amyloid beta-protein precursor gene, *Gene* 87: 257-263.

CELLULAR ASPECTS OF IMPLANTATION IN RUMINANTS

Michel Guillomot

I.N.R.A.
Station de Physiologie animale
78352 Jouy-en-Josas cédex, France

INTRODUCTION

Implantation implies major cell modifications of the trophoblast and the uterine endometrium. These cellular changes render the two tissues receptive and allow an intimate adhesion giving rise to a functional placenta. Classically the process of implantation is subdivided into 4 main steps: a pre-attachment period; apposition and adhesion of the blastocyst followed by invasion of the uterine endometrium (SCHLAFKE and ENDERS, 1975). In artyodactyls this division of implantation applies partly, with some specific characteristics. There is a long pre-attachment period (2-3 weeks), a diffuse and progressive attachment of the trophoblast along the uterine horns and a non-invasive phase of adhesion.

This paper reviews studies on the trophoblastic and uterine cellular changes which accompany the successive steps of implantation in the sheep and the cow. Since the major events are common to both species, the observations will refer to one or the other without differenciating between them.

PRE-ATTACHMENT PERIOD

The embryo enters the uterine cavity at the morula stage by day 4-5 p.i. and the blastocyst is formed two days later. Hatching from the zona pellucida occurs at days 9-10. By day 11 or 12 (in sheep and cows, respectively) the blastocyst begins a rapid and extensive elongation giving rise to a filamentous vesicle at day 14 (Figure 1) which fills the whole length of the uterine horns. If only one embryo is present one tip of the conceptus migrates into the opposite uterine horn (ROWSON and MOOR 1966; BINDON, 1971; BETTERIDGE et al., 1980).In places, the trophoblast and the uterine epithelium are juxtaposed but extracellular material prevent cell contacts between the two tissues (GUILLOMOT et al., 1981). The trophoblast exhibits structural characteristics of a polarized epithelium. The cuboidal cells present an apical surface covered by a dense network of microvilli (Figure 2). As observed in sections, the smooth lateral membranes form tight junctions, desmosomes and interdigitations from the apical to the basal poles of the cells. A basal lamina separates the trophoblastic cells from the endoderm which lines the inner face of the conceptus (Figure 3). The cytoplasmic distribution of cytoskeletal proteins (cytokeratins) confirms the epithelial nature of the trophoblast (GUILLOMOT and FLECHON, 1990). During the pre-attachment period the conceptus depends upon the uterine medium for its development. The trophoblastic cells are able to take up material from the uterine environment by endocytosis and phagocytocis of cell debris (WINTENBERGER-TORRES and FLECHON, 1974). This results in accumulation of cytoplasmic residual bodies and storage material such as lipid droplets (Figure 3) and proteinaceous crystalline inclusions (Figure 7).

Besides its nutritional function, the trophoblast also secretes specific regulatory proteins into the uterine lumen. Among these molecules are the so-called ovine and bovine Trophoblast Proteins (oTP and bTP, respectively). They are exclusively synthesized by the trophoblastic cells and are not detected in other embryonic tissues (Figure 4). These interferon-like proteins play a key role in maternal recognition of pregnancy (see review by ROBERTS, et al., 1992 and contributions of K. IMAKAWA and J. MARTAL in this Symposium).

The uterine mucosa of Ruminants presents two distinct tissues: the glandular endometrium and the aglandular caruncles (Figure 5). The uterine caruncles are constitutive endometrial differentiations which are dispersed along the uterine horns and are the maternal counterpart of the future placental cotyledons. In non-pregnant animals the caruncles have a dome-shaped appearance with a smooth surface. The monostratified epithelium is mainly composed of microvillous cells. Clusters of ciliated cells are also present around the gland openings. During the early stages of pregnancy the microvillous cells of the whole endometrium exhibit apical cytoplasmic protrusions (Figure 6). In rodents similar cellular features, named "pinopods", are involved in reabsorption of luminal fluid at time of

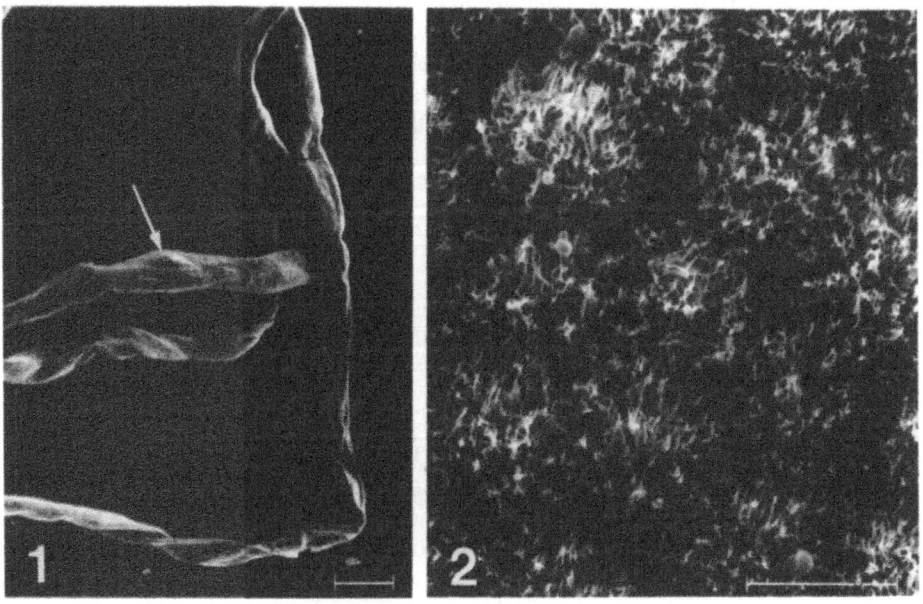

Figure 1. Day-14 ovine conceptus. The embryonic disc (arrow) protrudes from the filamentous trophoblast. SEM. Bar: 0.2 mm.

Figure 2. Cow, day 16. Cell surface of the trophoblast. SEM. Bar: 10 μm.

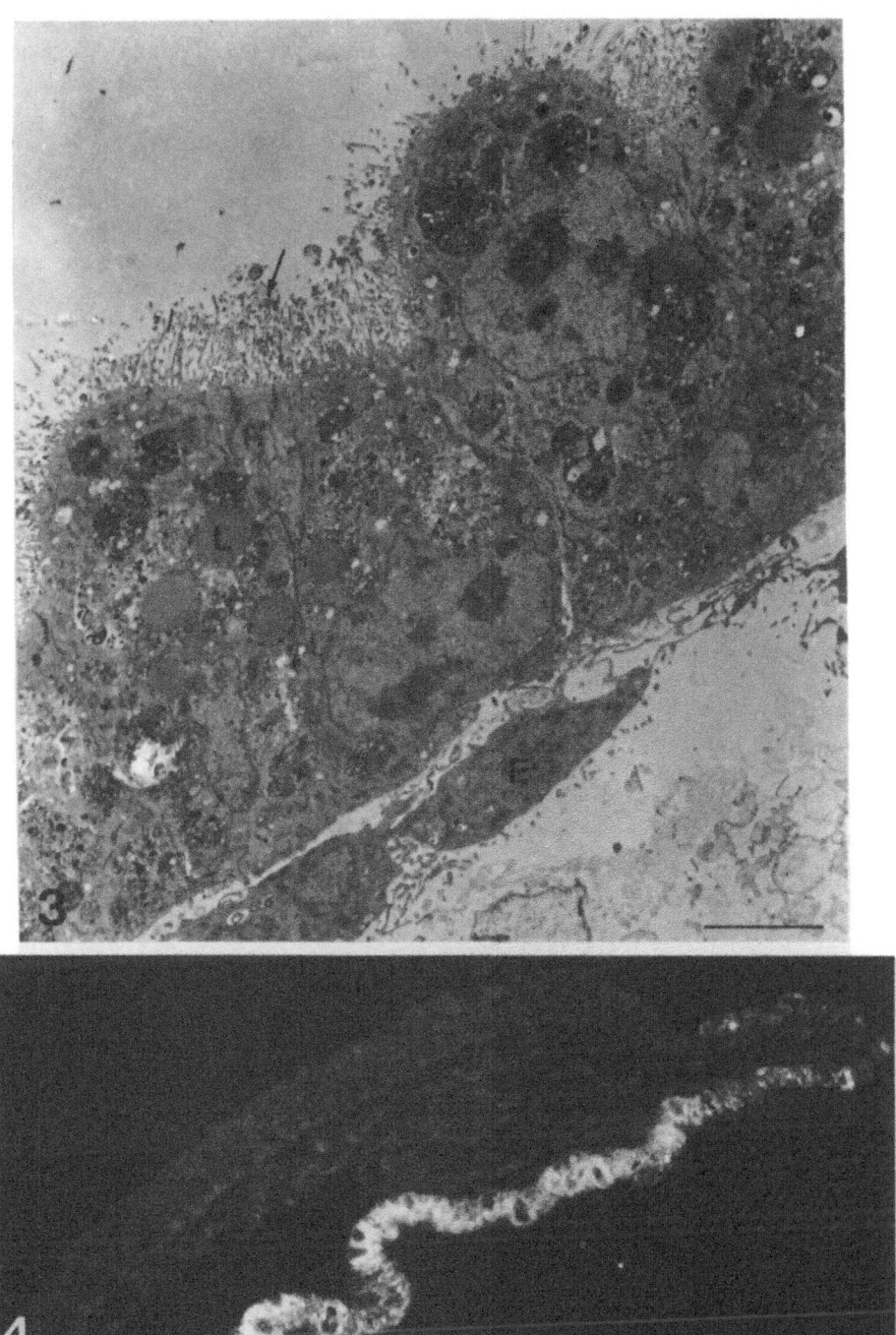

Figure 3. Bovine trophoblast, day 13. Residual bodies (asterisks) are scattered in the cytoplasm of the trophoblastic cells. Notice extracellular material (arrow) trapped by the apical microvilli. L: lipd droplets; E: endoderm cells.
TEM. Bar: 5 μm.

Figure 4. Day-14 ovine conceptus. Immunofluorescence localization of oTP in the extra-embryonic trophoblast. The embryonic disc is negative. Bar: 50 μm.

implantation (ENDERS and NELSON, 1973; PARR and PARR, 1977). However, no evidence has been demonstrated of specific endocytotic function for these cell protrusions, rather, they likely represent an apocrine mode of secretion in ruminants (GUILLOMOT et al., 1986). In cows and sheep, endocytosis of uterine intraluminal molecules is ensured by the microvillous cells; this endocytotic activity increases during the pre-attachment period and some of ingested proteins are transported to the uterine stroma via the baso-lateral intercellular spaces. Thus, undegraded macromolecules might be driven from the uterine lumen to the stroma without lysosomal denaturation (GUILLOMOT et al. 1986).

The pre-attachment period is ended by the arrest of elongation of the conceptus and its positioning at sites of implantation.

APPOSITION STAGE

This stage represents the first intimate cellular contacts between the trophoblast and the uterine epithelium. This occurs by day 15 in sheep (BOSHIER, 1969; GUILLOMOT et al., 1981) and by days 18-19 in cow (LEISER, 1975; KING et al., 1982). Trophoblast apposition first takes place close to the embryonic area, then spreads out along the whole conceptus during the following days. Apposed trophoblast is observed both on the caruncular and the intercaruncular endometrium. Early signs of endometrial changes associated with implantation are observed on the uterine caruncles. The caruncular surface becomes wrinkled and, in sheep, slightly depressed in its center (Figure 5). This central depression of the caruncles is the first indication of the subsequent shape of the ovine placental cotyledons. In the apposition areas, the trophoblast is tightly pressed against the endometrial epithelium so that bulging uterine cells imprint the apical membrane of the trophoblastic cells (Figure 7). As the epithelial cells restore a microvillous border, cell contacts are established between the tips of the uterine microvilli and the smooth trophoblastic cell membrane (Figure 8).
In the intercaruncular area, the trophoblast facing the uterine gland openings develops villous projections which invade the glandular lumen (Figures 9 and 10) (GUILLOMOT et al., 1981; GUILLOMOT and GUAY, 1982; WOODING and STAPLES, 1981). This villous trophoblast

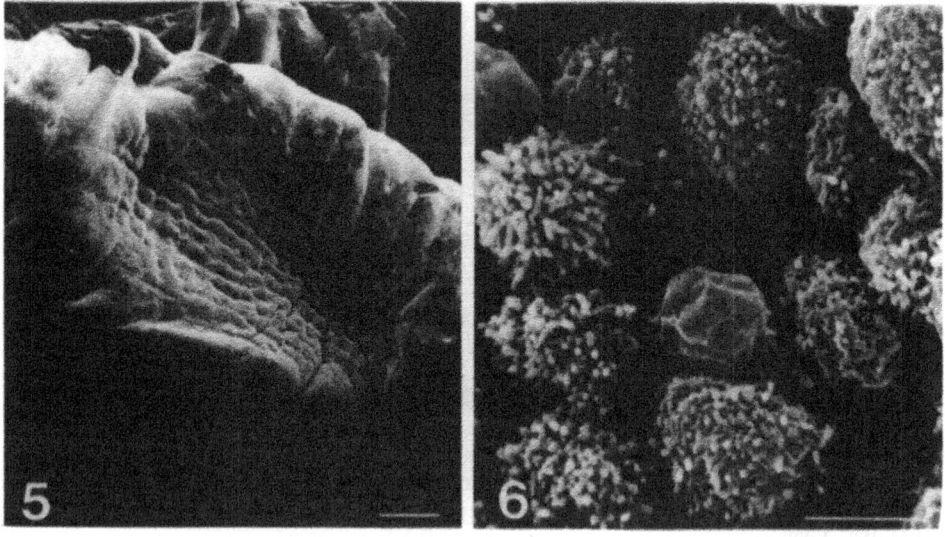

Figure 5. Ovine uterine caruncle from a day-15 pregnant animal. SEM. Bar: 0.2 mm

Figure 6. Sheep day 15. Uterine epithelial cells with bulbous secretory processes. SEM. Bar: 5 μm.

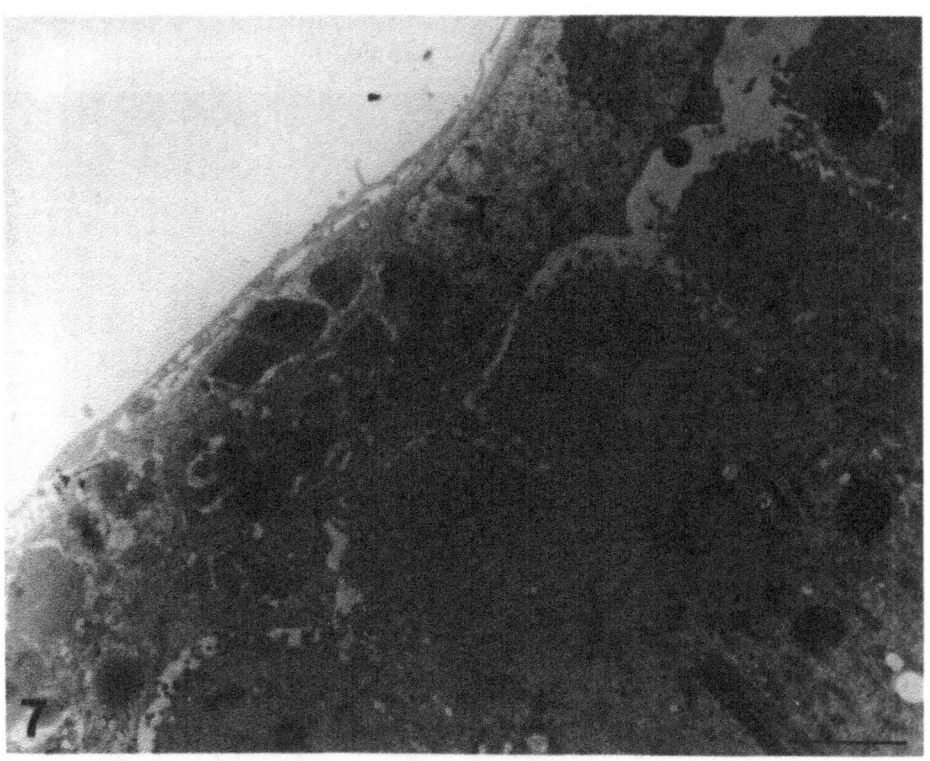

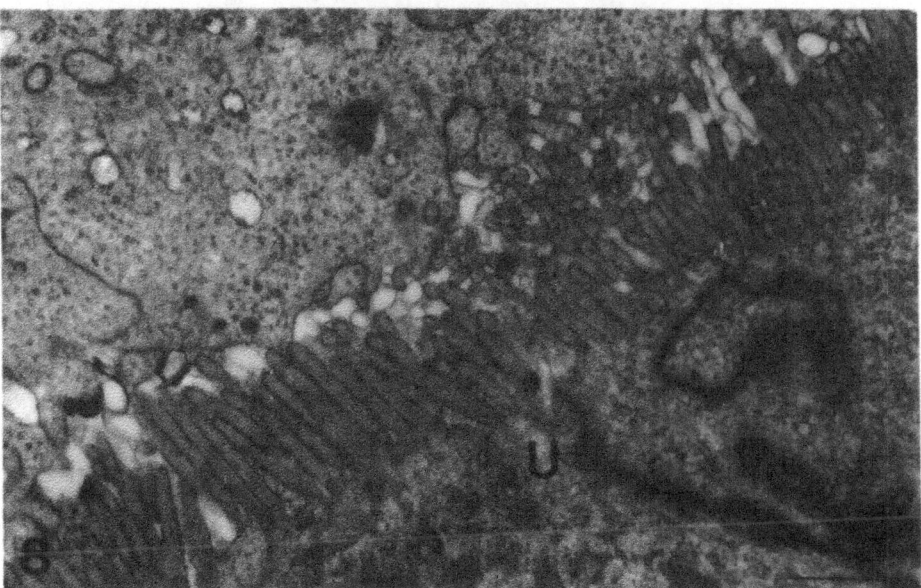

Figure 7. Apposition stage, sheep, day 15. Trophoblast (T) apposed on bulging uterine cells (U) which imprint the trophoblastic cell membrane. cr: crystalline inclusions. Bar: 3 μm.

Figure 8. Apposition stage, sheep day 15. Contact area between uterine microvillous cells (U) and the smooth membrane of trophoblastic cells. TEM. Bar: 1 μm.

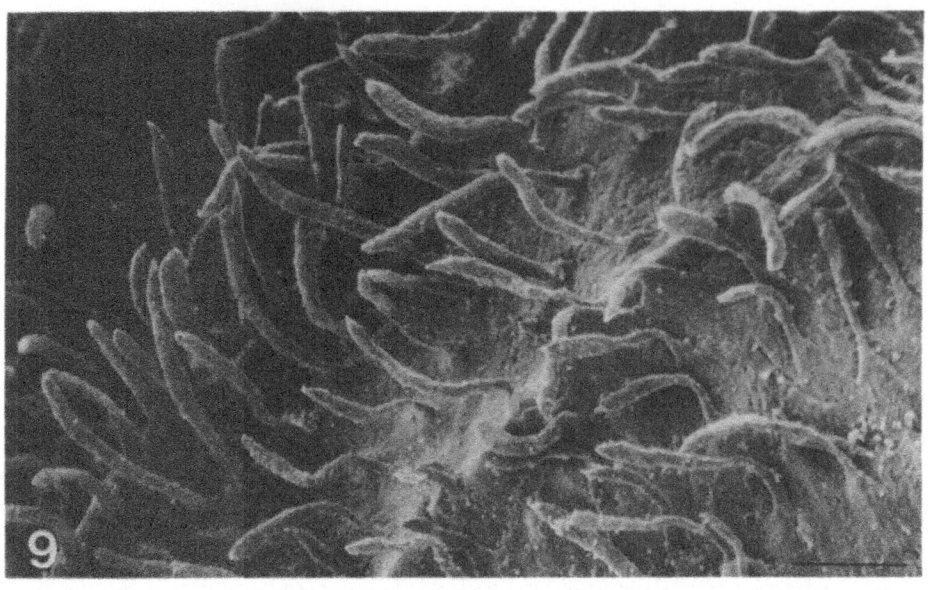

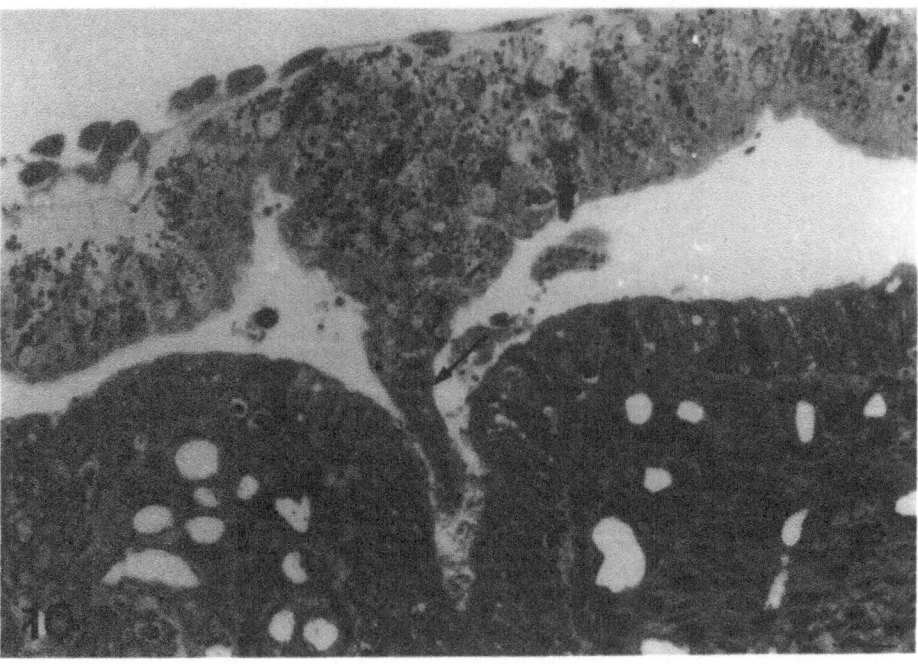

Figure 9. Sheep, day 18. Surface of the trophoblast with finger-like villi. SEM. Bar: 100 µm.

Figure 10. Sheep, day 18. Semithin section in a uterine glandular area. A trophoblastic villosity (arrow) is growing down into the glandular lumen. Bar: 100 µm.

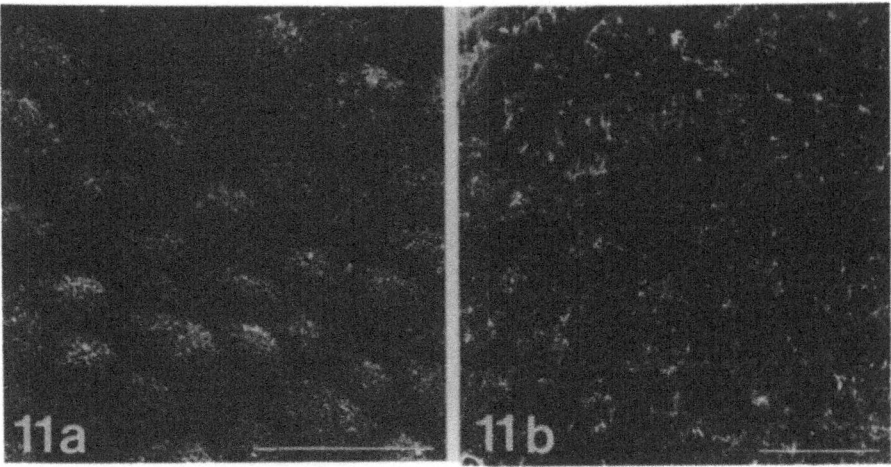

Figure 11. Apposition stage, cow, day 19. Surface of the trophoblastic cells of a) non-adherent and b) adherent trophoblast sampled from the same conceptus. Notice the reduction of microvilli in b. Bars: 10 µm.

constitutes a transient "para-placenta" which disappears thereafter. These villi might anchor the conceptus against the uterine epithelium, thus favouring cell contacts and absorption of the glandular secretions by the trophoblast.

During the apposition process the trophoblast undergoes cellular changes which affect both the ultrastructure of the cells and gene expression. One of the first evident modifications is the loss of apical microvilli on cells close to the embryonic area or on cells which have been in contact with the endometrium, whereas non-adherent portions of trophoblast show cells with a regular covering of microvilli (Figures 11a and b). In contact areas the smooth membrane of the trophoblastic cells rests on the tip of the uterine microvilli (Figure 8). The apical cell membrane presents biochemical changes which accompany the morphological modifications. The density of binding sites of Concanavalin A increases (Figures 12a and b), as do both cationic ferritin and ruthenium red staining , thus indicating changes in composition of the glycoprotein cell coat during apposition of the trophoblast on the uterine epithelium (GUILLOMOT et al., 1982). As shown by phosphotungstic acid (PTA) staining glycoproteins are still present at the interface between the trophoblast and the uterine epithelial cells (Figures 13 and 18a). These cell membrane glycoproteins probably increase the adhesiveness of both tissues. It will be of interest to determine the precise nature of these molecules and their role during implantation.

Protein synthesis by the trophoblast is also altered at time of implantation. Gene expression and production of oTP (GODKIN et al., 1982; CHARLIER et al., 1989) and bTP (GODKIN et al., 1988; CROSS and ROBERTS, 1991) decrease by this period of pregnancy. In sheep, results obtained by in situ hybridization and immunofluorescence techniques have given evidence that the end of oTP expression is an implantation-related phenomenom (GUILLOMOT et al., 1990). By day 15, arrest of oTP synthesis is first detected in the trophoblastic cells surrounding the embryo whereas its presence is still evident in the distal trophoblast (Figures 14a and c). Later during the implantation process oTP is still observed in the non-implanted trophoblast and has completely disappeared in the trophoblast at implantation sites (Figures 14d and e). In the same way, we have shown that expression of the c-fos proto-oncogene is strongly depressed in the trophoblast at implantation. Moreover the cellular localization of c-fos protein in non-implanted and implanted trophoblast parallels that of oTP (XAVIER et al., 1991). The proto-oncogenes control numerous of cell functions and are supposed to play a key role in embryonic development (ADAMSON, 1987). Whether these modifications of the genomic program are a cause or a consequence of restructuring of the trophoblastic cells at implantation, remains to be determined.

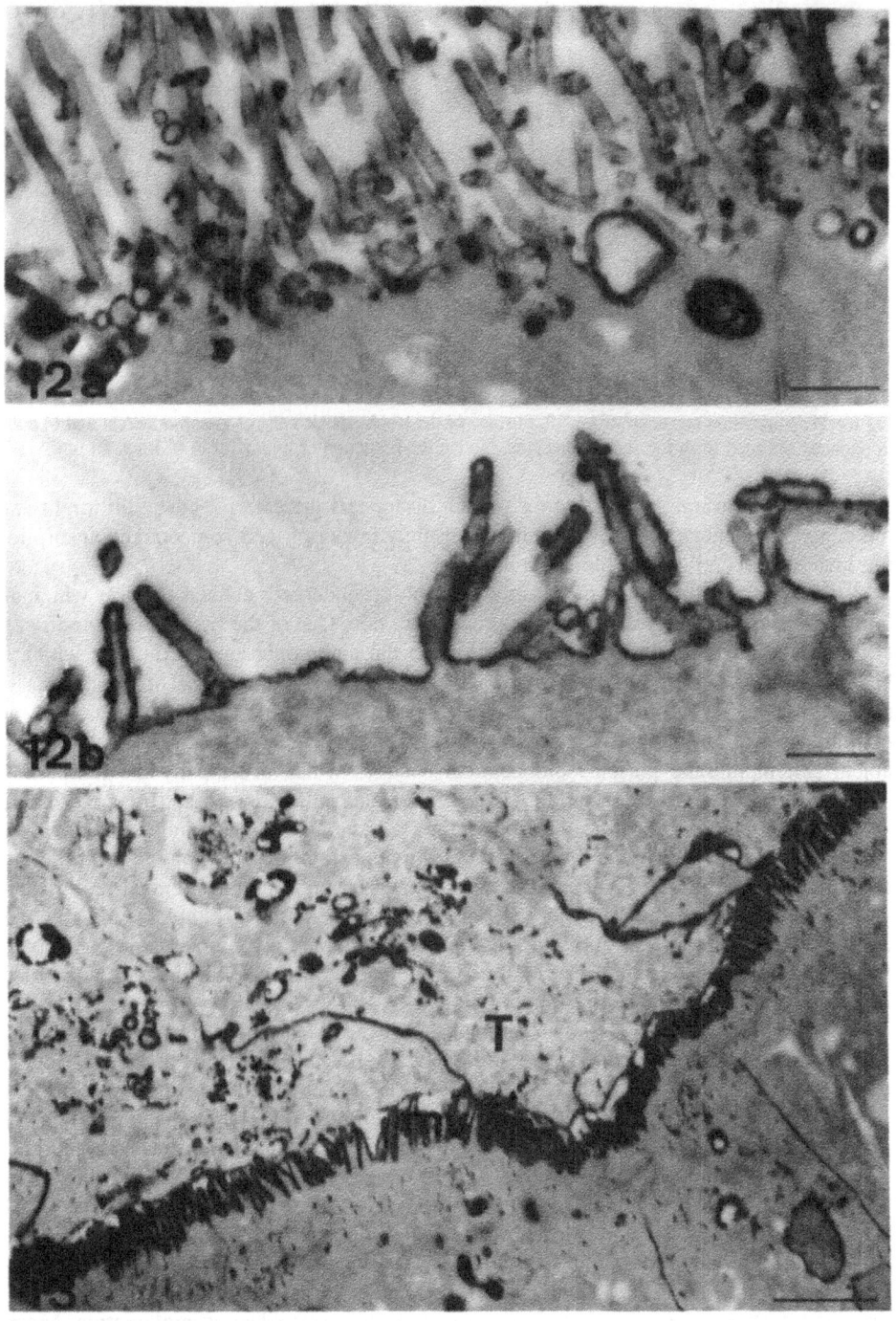

Figure 12. Ovine trophoblastic cells stained by the Concanavalin A - peroxidase technique at a) day 13 and b) day 15. Staining of the apical membrane is more intense at day 15. TEM. Bars: 5 μm.

Figure 13. Apposition area, sheep day 17. Phosphotungstic acid (PTA) staining of glycoproteins on the trophoblast (T) apical membrane and of the uterine cell microvilli (mv). TEM. Bar: 2 μm.

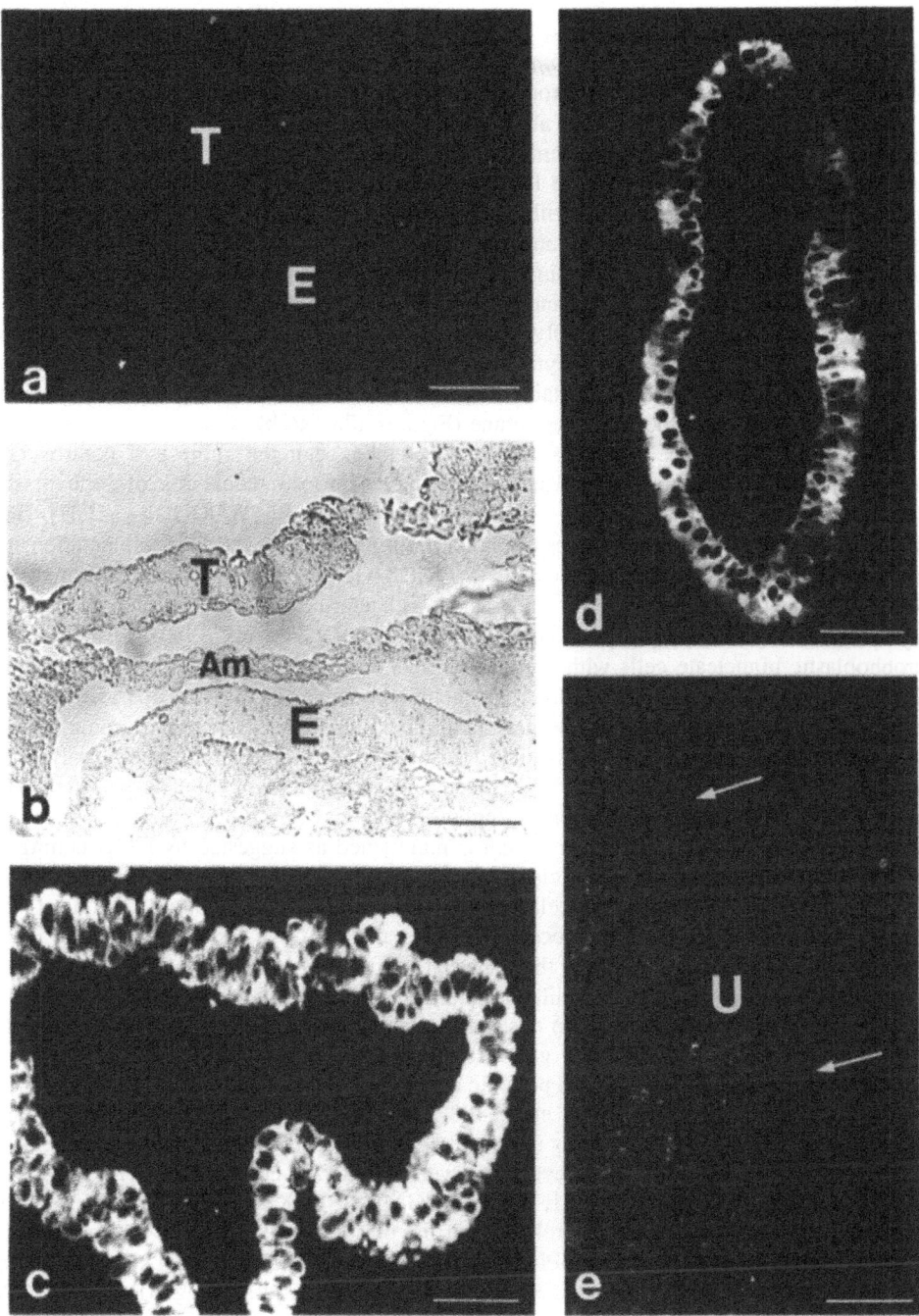

Figure 14. Immunofluorescence localization of oTP on ovine trophoblast at day 15 (a-c) and at day 17 (d,e). a) Section of the embryonic area showing non-stained trophoblast (T); b) same field observed by interference contrast, E: embryo; am: amnion. c) Distal trophoblast is still highly positive for oTP. d) Positively stained non-adherent trophoblast. e) Section in an adhesion area showing the negatively stained trophoblast (arrows) attached to the uterine endometrium (U). Bars: 50 μm.

ADHESION STAGE

By this stage the trophoblast is firmly adherent to the uterine epithelium, so the conceptus cannot be dislodged from the uterine lumen without damaging the tissues. Discrete areas of adhesion occurs around the embryo by day 16 in sheep and by day 19 in the cow (LEISER, 1975; KING et al., 1982). Adhesion between the trophoblast and the endometrium is strengthened by interdigitation of uterine microvilli and indentations of the trophoblastic cell membrane (Figure 15). This microvillar junction constitutes the placental barrier between the chorionic villi and endometrial crypts and characterizes the epitheliochorial placenta of the ruminants (BJORKMANN, 1965; 1968). During the adhesion process the uterine epithelium flattens and its height is reduced to few micrometers in places (Figure 16). In this manner the space between the trophoblast layer and the endometrial capillaries is considerably reduced. Another major transformation of the uterine epithelium is the appearance of syncytial masses scattered among the regular mononucleated cells. The syncytial cytoplasm contains three or more nuclei and electron-dense granules which accumulate near the basal membrane (Figures 17a and b).

The origin of the uterine syncytium has long been the subject of controversy. However results from ultrastructural studies have now demonstrated the role of trophoblastic binucleate cells in the syncytium formation (see review by WOODING, 1982). The binucleate cells differentiate from the trophoblast during the early stages of implantation and account for 20% of the trophoblastic cells (WOODING, 1982). Migration of the binucleate cells through the trophoblast layer and fusion with uterine cells give rise to the uterine symplasm (WOODING and WATHES, 1980; WOODING et al., 1980). Fusion of the trophoblastic binucleate cells with the uterine cells represents the most invasive phase of implantation in ruminants. The uterine basal membrane retains its integrity apart from places where syncytial cytoplasmic projections pass through it and reach the uterine stroma (Figure 17b). In the cow, the syncytial masses are transient and are replaced after cellular lysis by subsequent cell migrations; whereas in sheep and goats the syncytium is persistent and grows by continuous cell migration throughout pregnancy (WOODING, 1982). In ewes, protein synthesis by the uterine syncytium is maintained as suggested by PTA staining of glycoprotein material in the Golgi cisternae (Figure 18b).

Migration of binucleate cells might contribute to the immobilization of the trophoblast during the adhesion process (KING et al.,1982). An additional function is the delivery of embryonic products to the maternal tissues and blood circulation. Arguments supporting such a role are given by ultrastructural and immunocytochemical studies. Similar types of electron-dense granules are observed in the cytoplasm of both the binucleate cells and the uterine syncytium. These granules are easily identifiable by their glycoprotein content which stains positively with periodic acid-Schiff (PAS) (BOSHIER, 1969) and PTA (WOODING, 1980; GUILLOMOT et al., 1982) (Figures 18a and b). Moreover, ovine and bovine placental lactogen hormones (oPL and bPL) produced by the binucleate cells, have been localized in the maternal syncytium as well (MARTAL et al., 1977; WATKINS and REDDY, 1980; WOODING, 1981; WOODING and BECKERS, 1987). However, the placental lactogen hormones are probably not the only proteins produced and transported by the binucleate cells. In sheep, oPL is not a glycosylated molecule (CHAN et al., 1976) thus, the cytoplasmic granules stained by glycoprotein-specific stainings (PAS and PTA) must contain other secretory products. Another group of placental proteins, named Pregnancy Specific Proteins (PSP), have been detected in the maternal serum at increasing levels throughout gestation and are used as specific antigens in pregnancy diagnosis in ruminants (BUTLER et al., 1982; RUDER et al., 1988; SASSER et al., 1986). The PSP are a family of immunologically related glycoproteins with molecular weights ranging from 40 to 78 kD (SASSER et al., 1989). One of these proteins has been isolated from bovine cotyledons and has been named Pregnancy Serum Protein 60 (PSP-60) because of its

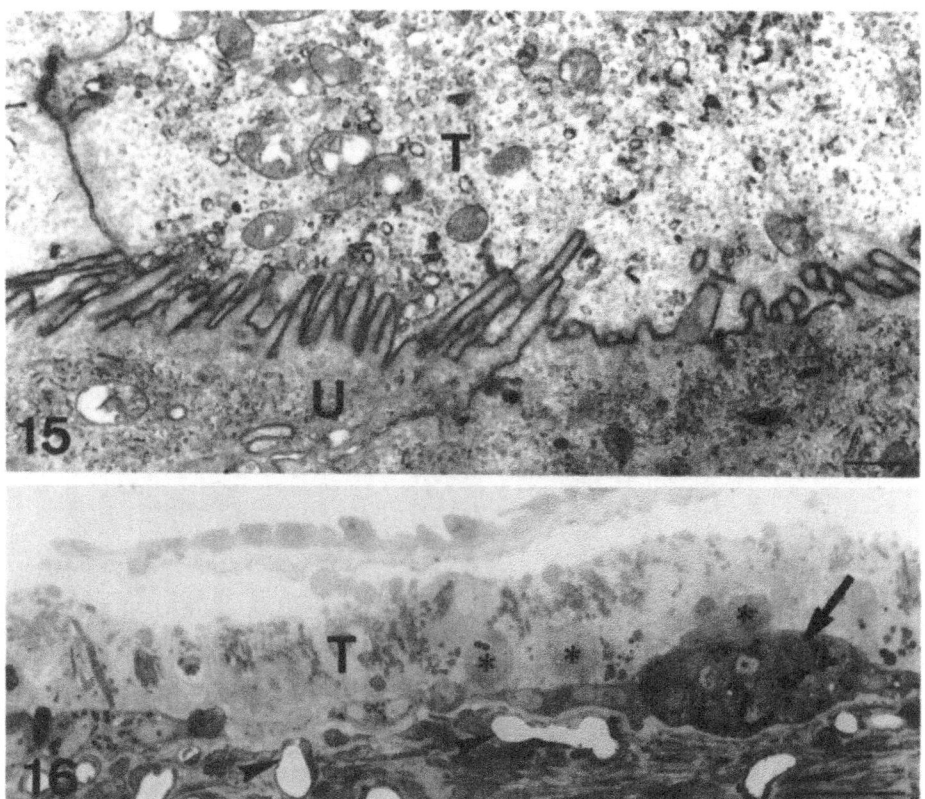

Figure 15. Adhesion stage, sheep day 20. Aspect of the microvillar junction interlocking the uterine (U) and trophoblastic cells (T). Bar: 1 µm.

Figure 16. Adhesion stage, sheep day 18. Semithin section showing the adherent trophoblast (T) on the uterine epithelium which presents syncytial transformation (arrow). Binucleate cells (asterisks) are scattered among the trophoblastic cells. Uterine capillaries (arrowheads) are close to the uterine basal membrane. Bar: 50 µm.

molecular weight: 60 kD (CAMOUS et al., 1988). Antibodies against PSP-60 recognize specific antigens in bovine, ovine and caprine maternal serum during pregnancy (CAMOUS, personal communication). By indirect immunofluorescence PSP-60 was localized specifically in the trophoblastic binucleate cells and in the uterine syncytium in the early stages of adhesion (Figure 19). The binucleate cell origin of other PSP-related proteins and their co-localization in the uterine syncytium have been reported (ECKBLAD et al., 1985; ZOLI et al., 1992). Thus, it is likely that PSP are transported to the maternal circulation by the migrating binucleate cells via the uterine syncytium. Analogous transport occurs in the equine placenta. In horses, binucleate cells originating from the chorionic girdle, migrate through the uterine stroma where they secrete equine chorionic gonadotrophin (ecG) (HAMILTON et al. 1973). In ruminants, the role of the PSP during gestation is still unclear. XIE et al. (1991) have reported that PSP-related proteins, named pregnancy-associated glycoproteins (PAG), belong to the aspartic proteinase family but were enzymatically inactive, although the molecules might still bind polypeptide ligands. It is intriguing that proteolytic enzymes are produced by the unique "invasive" cells of the ruminant trophoblast. Further studies are necessary to precise the functions of these trophoblastic proteinases during pregnancy in ruminants.

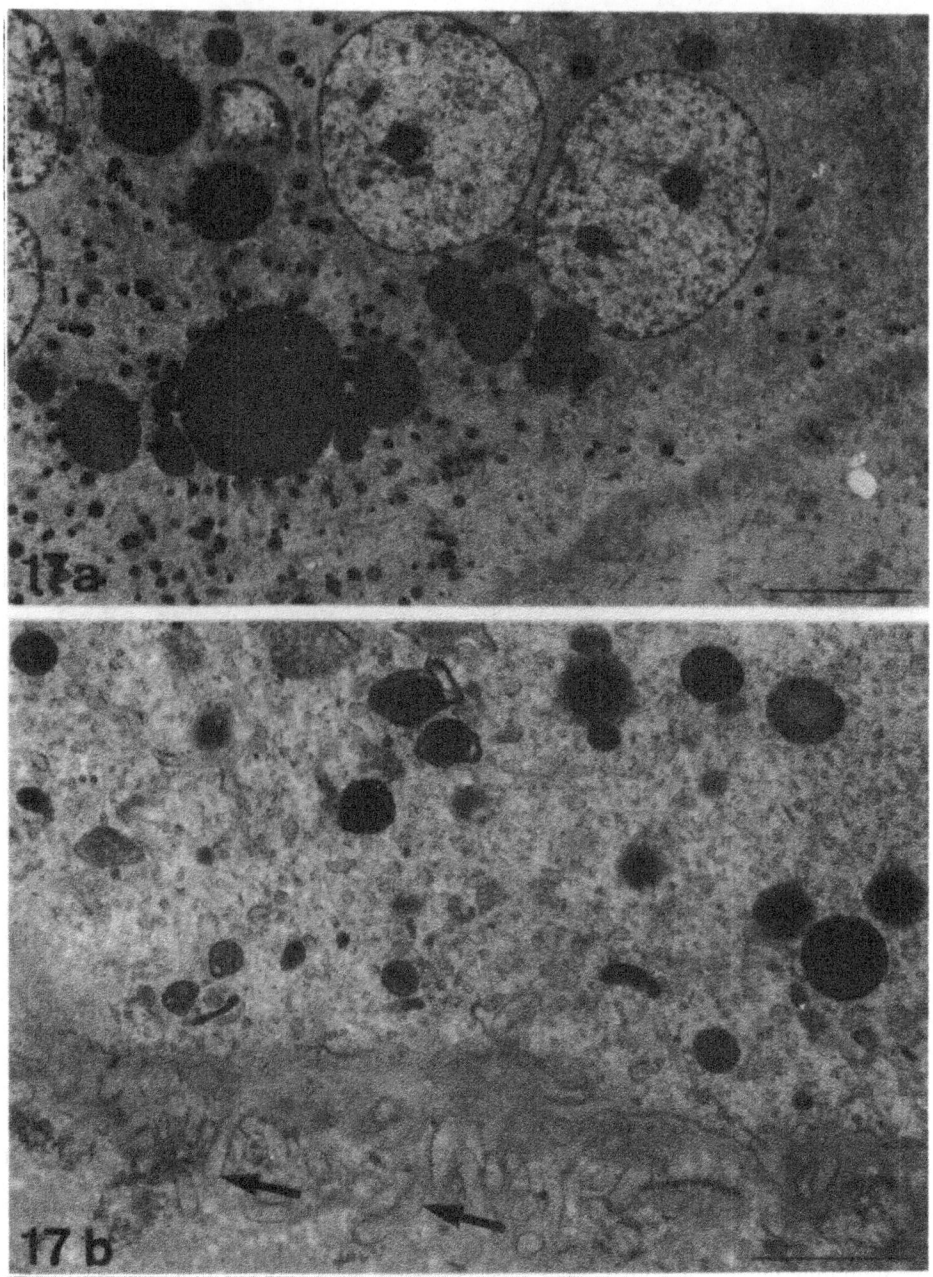

Figure 17. Adhesion stage, sheep day 18. a) Uterine syncytium with numerous electron dense granules scattered at the basal pole of the cytoplasm. TEM. Bar: 5 µm. b) Basal pole of a uterine symplasm with cytoplasmic processes (arrows) crossing the basal lamina. TEM. Bar: 1 µm.

Figure 18. Adhesion stage, day 18. Phosphotungstic acid staining (PTA). a) Section of an adhesion area between the trophoblast (T) and the uterine syncytium (Sy). PTA selectively stains granules in both the binucleate cells (bn) and the uterine syncytium. Glycoproteins of the microvillar junction are also stained. Bar: 5 µm. b) Basal cytoplasm of the uterine syncytium showing positively PTA-stained granules and Golgi cisternae (Go). Cytoplasmic processes (arrow) crossing the basal lamina. Bar: 1 µm.

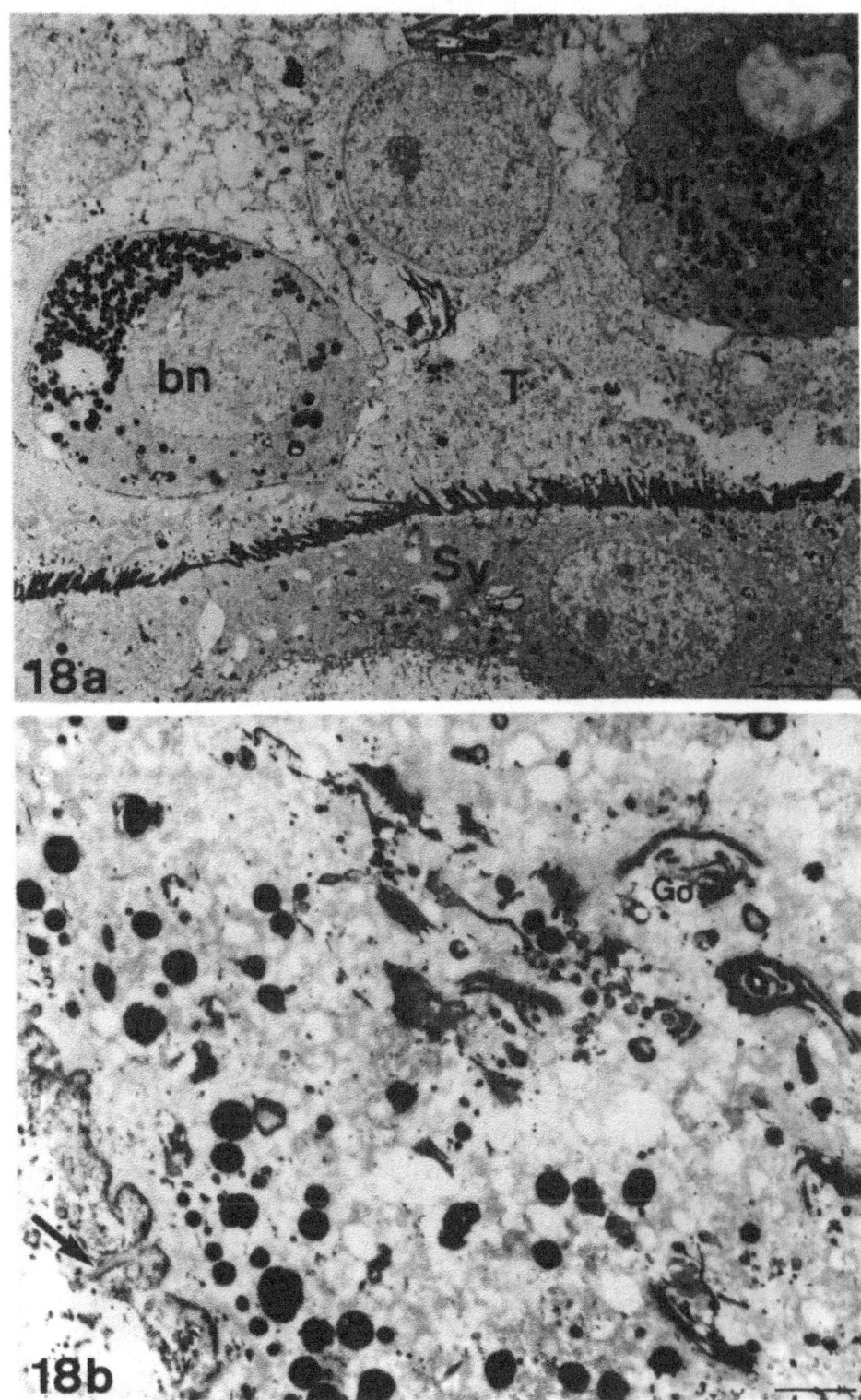

18a

bn

bn

T

Sv

18b

Go

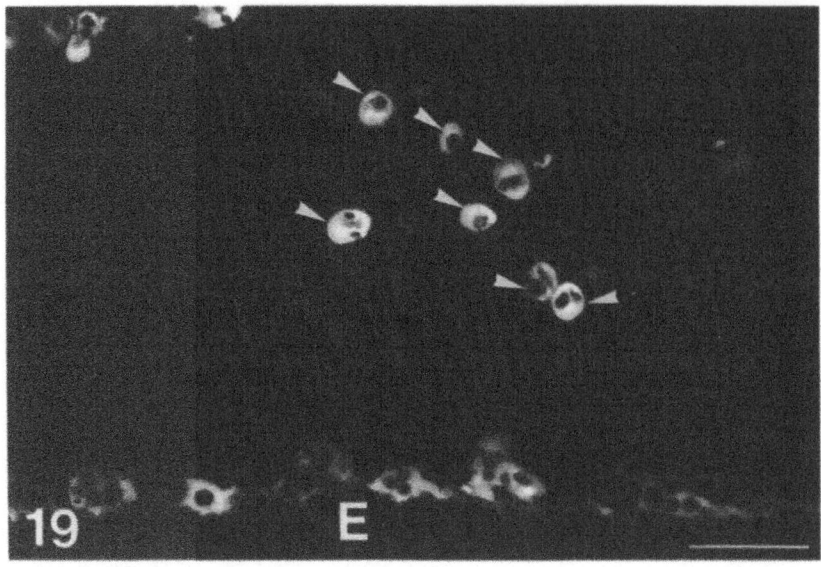

Figure 19. Sheep, day 17. Cellular localization by immunofluorescence of a Pregnancy Serum Protein (PSP-60) in the binucleate cells (arrowheads) of the trophectoderm and in the uterine syncytia. The trophoblast layer has been artefactually detached from the endometrium (E) during tissue processing. Bar: 50 μm.

CONCLUSIONS

In ruminants, as in other orders, the establishment of close adhesion between the trophoblast and the uterine epithelium induces dramatic changes of the cellular structures and activity of both tissues. As shown here the passage from non-adhesive to adhesive trophoblastic cells implies cell remodellings and shifts in gene expression. The regulation of such cellular modifications have to be determined. In sheep, the ultrastructural changes and the arrest of the expression of two genes (oTP and c-fos) are first restricted to the implanting trophoblastic cells close to the embryo. This suggests that local factors might control the ability of the trophoblast to implant. The nature and the origin of these putative products has to be determined. Cytokines which are known to act as paracrine and/or autocrine cell regulators might be involved in this process. Recently, it has been shown that expression of uterine LIF (Leukemia inhibiting factor) is essential for blastocyst implantation in mouse (STEWART et al., 1992). It will be of interest to find out whether such cytokines could play a role in conceptus attachment in other species.

Acknowledgements

I am very grateful to Dr. K. J. Betteridge for useful revision of the manuscrpit and to my colleagues at Jouy-en-Josas and at St-Hyacinthe (Que. Canada) for their help during these studies.

REFERENCES

Adamson, E.D., 1987, Oncogenes in development, *Development* 99: 449-471.
Betteridge, K.J., Eaglesome, M.D., Randall, G.C.B. and Mitchell, D., 1980, Collection, description and transfer of embryos from cattle 10-16 days after oestrus, *J. Reprod. fert.* 59: 205-216.
Bindon, B.M., 1971, Systematic study of preimplantation stage of pregnancy in the sheep, *Austr. J. Biol. Sci.* 24: 131-147.

Bjorkman, N., 1965, Fine structure of the ovine placentome, *J. Anat.* 99: 283-297.

Bjorkman, N., 1968, Fine structure of the cryptal and trophoblastic giant cells in the bovine placentome, *J. Ultrastruct. Res.* 24: 249-258.

Boshier, D.P., 1969, A histological and histochemical examination of implantation and early placentome formation in sheep, *J. Reprod. Fert.* 19: 51-61.

Butler, J.E., Hamilton, W.C., Sasser, R.G., Ruder, C.A., Hass, G.M., and Williams, R.J., 1982, Detection and partial characterization of two bovine pregnancy-specific proteins, *Biol. Reprod.* 26: 925-933.

Camous,S., Charpigny, G., Guillomot, M., Martal, J., and Sasser, R.G., 1988, Purification of one bovine pregnancy specific protein by high performance liquid chromatography (HPLC), Proc. BARD workshop on maternal recogniton of pregnancy and maintenance of the corpus luteum, Jerusalem, March, 20-24, Abstr. 2.

Chan, J.S.D., Robertson, H., and Friesen, H.G., 1976, The purification and characterization of ovine placental lactogen, Endocrinology 98: 65-76.

Charlier, M., Hue, D., Martal, J., and Gaye, P., 1989, Cloning and expression of cDNA encoding ovine trophoblastin: its identity with a class II alpha interferon, *Gene* 77: 341-348.

Cross, J.C., and Roberts, R.M., 1991, Constitutive and trophoblastic-specific expression of a class of bovine interferon genes, *Proc. Natl. Acad. Sci.* 88: 3817-3821.

Eckblad, W.P., Sasser, R.G., Ruder C.A., Panlasigui, P., and Kucznski, T., 1985, Localization of pregnancy-specific B (PSPB) in bovine placental cells using a glucose oxidase-anti-glucose oxidase immunohistochemical stain, *J. Anim. Sci.* 61 (Suppl. 1), Abstr. 149.

Enders, A.C., and Nelson, M., 1973, Pinocytotic activity of the uterus of the rat, *Am. J. Anat.* 138: 277-300.

Godkin, J.D., Bazer, F.W., Moffat, J., Sessions, F., and Roberts, R.M., 1982, Purification and properties of a major, low molecular weight protein released by the trophoblast of sheep blastocyst at days 13-21, *J. Reprod. Fert.* 65: 141-150.

Godkin, J.D., Lifsey, B.J., and Gillepsie, B.E., 1988, Characterization of bovine conceptus proteins produced during the peri-and post-attachment periods of early pregnancy, *Biol. Reprod.* 38: 703-711.

Guillomot, M., Betteridge, K.J., Harvey, D., and Goff A.K., 1986, Endocytotic activity in the endometrium during conceptus attachment in the cow, *J. Reprod. Fert.* 78: 27-36.

Guillomot, M., and Fléchon, J.E., 1990, Place de la microscopie dans l'étude des conditions de l'implantation chez les Ruminants, *Contr. fert. sexual.* 18: 875-885.

Guillomot, M., Fléchon, J.E., and Wintenberger-Torres, S., 1981, Conceptus attachment in the ewe: an ultrastructural study, *Placenta* 2: 169-182.

Guillomot, M., Fléchon, J.E., and Wintenberger-Torres, S., 1982, Cytochemical studies of uterine and trophoblastic surface coats during blastocyst attachment in the ewe, *J. Reprod. Fert.* 65: 1-8.

Guillomot, M., and Guay, P., 1982, Ultrastructural features of the cell surfaces of the uterine and trophoblastic epithelia during embryo attachment in the cow, *Anat. Rec.* 204, 315-322.

Guillomot, M., Michel, C., Gaye, P., Charlier, M., Trojan, J., and Martal, J., 1990, Cellular localization of an embryonic interferon, ovine trophoblastin and its mRNA in sheep embryos during early pregnancy, *Biol. Cell* 68: 205-211.

Hamilton, D.W., Allen, W.R., and Moor, R.M., 1973, Origin of endometrial cups. III. Light and electron microscopic study of fully developed equine endometrial cups, *Anat. Rec.* 177: 503-518.

King, G.J., Atkinson, B.A., and Robertson, H.A., 1982, Implantation and early placentation in domestic ungulates, *J. Reprod. Fert., Suppl.* 31: 17-30.

Leiser, R., 1975, Kontaktaufnahme zwischen Trophoblast und Uterusepithel wärhend der frühen Implantation beim Rind, *Anat. Histol. Embryol.* 4: 63-86.

Martal, J., Djiane, J., and Dubois, M.P., 1977, Immunofluorescent localisation of ovine placental lactogen, *Cell Tiss. Res.* 184: 427-433.

Parr, M.B., and Parr, E.L., 1977, Endocytosis in the uterine epithelium of the mouse, *J. Reprod. Fert.* 50: 151-153.

Roberts, R.M., Cross, J.C., and Leaman, D.W., 1992, Interferons as hormones of pregnancy, *Endocr. Rev.* 13: 432-452.

Rowson, L.E., and Moor, R.M., 1966, Development of the sheep conceptus during the first fourteen days, *J. Anat.* 100: 177-185.

Ruder, C.A., Sasser, R.G., Dahmen, J.J., and Stellflug, J.N., 1988, Detection of pregnancy in sheep by radioimmunoassay of sera for pregnancy-specific protein B, *Theriogenology* 29: 905-912.

Sasser, R.G., Ruder, C.A., Ivani, K.A., Butler, J.E., and Hamilton, W.C., 1986, Detection of pregnancy by radioimmunoassay of a novel pregnancy-specific protein in serum of cows and a profile of serum concentration during gestation, *Biol. Reprod.* 35: 936-942.

Sasser, R.G., Crock, J. Ruder-Montgomery, C.A., 1989, Characteristics of pregnancy-specific protein B in cattle, *J.Reprod.Fert. Suppl.* 37: 109-113.

Schlafke, S., and Enders, A.C., 1975, Cellular basis of interaction between trophoblast and uterus at implantation, *Biol. Reprod.* 12: 41-65.

Stewart, C.L., Kaspar, P., Brunet, L.J., Bhatt, H., Gadi, I., Köntgen, F., and Abbondanzo, S.J., 1992, Blastocyst implantation depends on maternal expression of leukemia inhibitory factor, *Nature* 359: 76-79.

Watkins, W.B., and Reddy, S., 1980, Ovine placental lactogen in the cotyledonary and intercotyledonary placenta of the ewe, *J. Reprod. Fert.* 58: 411-414.

Wintenberger-Torres, S., and Fléchon, J.E., 1974, Ultrastructural evolution of the trophoblast cells of the pre-implantation sheep blastocyst from day 8 to 18, *J. Anat.* 118: 143-153.

Wooding, F.B.P., 1980, Electron microscopic localisation of binucleate cells in the sheep placenta using phosphotungstic acid, *Biol. Reprod.* 22: 357-365.

Wooding, F.B.P., 1981, Localization of ovine placental lactogen in sheep placentomes by electron microscope immunocytochemistry, *J. Reprod. Fert.* 62: 15-19.

Wooding, F.B.P., 1982, The role of the binucleate cell in ruminant placental structure, *J. Reprod. Fert., Suppl.* 31:31-39.

Wooding, F.B.P., and Beckers, J.F., 1987, Trinucleate cells and ultrastructural localisation of bovine placental lactogen, *Cell. Tis. Res.* 247: 667-673.

Wooding, F.B.P., Chambers, S.G., Perry, J.S., George, M., and Heap, R.B., 1980, Migration of the binucleate cells in the sheep placenta during normal pregnancy, *Anat. Embryol.* 158: 361-370.

Wooding, F.B.P., and Wathes, D.C., 1980, Binucleate cell migration in the bovine placentome, *J. Reprod. Fert.* 59: 425-430.

Wooding, F.B.P., and Staples, L.D., 1981, Functions of the trophoblast papillae and binucleate cells in implantation inthe sheep, *J. Anat.* 133: 110-112.

Xavier, F., Guillomot, M., Charlier, M., Martal, J., and Gaye, P., 1991, Co-expression of the proto-oncogene FOS (c-fos) and an embryonic interferon (ovine trophoblastin) by sheep conceptus during implantation, *Biol. Cell* 73: 27-33.

Xie, S., Low, B.G., Nagel, R.J., Kramer, K.K., Anthony, R.V., Zoli, A.P., Beckers, J.F., and Roberts, R.M., 1991, Identification of the major pregnancy-specific antigens of cattle and sheep as inactive members of the aspartic proteinase family, *Proc. Natl. Acad. Sci.* 88: 10247-10251.

Zoli, A.P., Demez, P., Beckers, J.F., Reznik, M., and Beckers, A., 1992, Light and electron microscopic immunolocalization of bovine pregnancy-associated glycoprotein in the bovine placentome, *Biol. Reprod.* 46: 623-629.

EXPRESSION AND REGULATION OF INSULIN-LIKE GROWTH FACTOR BINDING PROTEIN-1 AND RETINOL BINDING PROTEIN IN THE BABOON (Papio anubis) UTERUS

Asgerally T. Fazleabas and Harold G. Verhage

Department of Obstetrics and Gynecology
University of Illinois College of Medicine
Chicago, IL 60680

INTRODUCTION

There has been increasing appreciation in recent years for the importance of fetal and maternal tissue interactions in pregnancy. Numerous interactions are associated with this mutual relationship, the most important of which is the achievement of implantation and maintenance of pregnancy (Enders and Schlafke 1979). In a broad sense, the establishment of pregnancy in all mammalian species could be divided into the maternal and fetal components, whose synergistic interaction is essential for a viable pregnancy. The maternal contribution is to provide an appropriate milieu within the oviduct to optimize fertilization, secrete or sequester substances within the uterus to sustain conceptus development and limit the degree of blastocyst invasiveness. Coincidentally, the embryonic component secretes luteotrophic or anti-luteolytic substances to prevent corpus luteal regression, in addition to maintaining uterine secretory activity and blood flow, and achieving immunologic privilege as a fetal allograft.

In a number of domestic species, antiluteolytic factors secreted by the conceptus alters endometrial secretory activity (Bazer, 1992). For example estrogen secreted by the elongating pig conceptus causes the release of progesterone induced secretory proteins from the uterine epithelial cells directly into the uterine lumen (Geisert et al., 1982a,b). Ovine trophoblast protein-1, which is secreted during the period of maternal recognition of pregnancy in the sheep, also induces the release of specific endometrial polypeptides (Godkin et al., 1984, Vallet et al., 1987, Ashworth and Bazer, 1989). In our studies on the non-human primate endometrium, we have also demonstrated the synthesis of specific endometrial proteins synthesized during the mid-luteal stage of the menstrual cycle (see Bell et al., 1989 and Fazleabas et al., 1989a for reviews), which are also upregulated during pregnancy (Fazleabas et al., 1993). Whether the conceptus/placental factor responsible for the upregulation of endometrial protein synthesis is the primate luteotrophin, chorionic gonadotrophin, or another as yet unidentified product is unclear at present. However, it may be more than coincidental that the highest period of synthetic activity by both the endometrium and placenta occurs at the same time as the peak of measurable CG in the peripheral circulation (Fortman et al., 1993) and is also associated with the time point when progesterone synthesis undergoes a luteal-placental shift in the baboon (Castracrane and Goldzieker, 1986).

Of the major progesterone-regulated proteins synthesized by the human endometrium, prolactin, insulin-like growth factor binding protein 1 (IGFBP-1) and a β-lactoglobulin homologue (PP_{14}) have been the most extensively characterized during the menstrual cycle and early pregnancy (see Bell, 1986 and Fay & Grudzinskas, 1991 for reviews). Comparative studies on the baboon endometrium have demonstrated that IGFBP-1 but not the β-lactoglobulin homologue are synthesized during the luteal phase and/or during periods of progesterone dominance (see Bell et al., 1989 and Fazleabas et al., 1991 for reviews). Interestingly, however, the baboon uterus produces retinol binding protein (RBP) in a manner that parallels β-lactoglobulin synthesis by the human endometrium during the menstrual cycle and early pregnancy (Donnelly et al., 1991, 1992; Fazleabas et al., manuscript in preparation).

Insulin-like Growth Factor Binding Protein-1

In the past decade there has been a proliferation of information relating to the IGF autocrine/paracrine system which is comprised of the IGF I and II peptides, the Type I and Type II receptors, and six distinct IGF binding proteins (IGFBP's; see Sara and Hall, 1990; Rosenfeld et al., 1990; Rotwein, 1991; Lamson et al., 1991; Nissley and Lopaczynski, 1991 for reviews). We have focused our studies on the regulation of IGFBP-1 and the Type-I receptor in the baboon uterus during the menstrual cycle and early pregnancy. The IGFBP's may function to prolong the half-life of IGF's, serve as a reservoir and control the delivery of IGF's to target cells (Binoux and Hossenlopp, 1988; Blum et al., 1989), or either inhibit or potentiate the mitogenic action of these peptides (Sara and Hall, 1990; Rosenfeld et al., 1990; Lamson et al., 1991). In both the baboon (Fazleabas et al., 1992; Tarantino et al., 1992) and human (Giudice et al., 1991) endometrium, protein and messenger RNA for IGFBP-1, 2, 3 and 4 are differentially regulated. IGFBP-2, 3 and 4 synthesis increases between the proliferative and secretory stages of the menstrual cycle, while IGFBP-1 synthesis is only observed during the secretory stage and in the baboon requires estrogen-priming followed by progesterone for its induction (Julkunen et al., 1988; Fazleabas et al., 1989b; Giudice et al., 1991; Tarantino et al., 1992).

Our initial studies (Fazleabas et al., 1989b,c) demonstrated that although IGFBP-1 in the baboon and human were immunologically and biochemically similar, their site of synthesis in the non-pregnant uterus was distinctly different. In the human, the major site of IGFBP-1 localization is the decidual cell both during the cycle and pregnancy, whereas in the baboon IGFBP-1 synthesis is confined to the epithelial cells of the deep basal glands during the late luteal stage. Synthesis of IGFBP-1 within the glandular epithelium begins to increase at day 18 of pregnancy. However, it is only when decidual synthesis of IGFBP-1 is induced at the implantation site that it becomes a significant labelled product on fluorographs (Fazleabas et al., 1993; Tarantino et al., 1992). As decidualization continues, IGFBP-1 synthesis increases and this protein eventually becomes the major secretory product of the primate decidualized endometrium (Fazleabas et al., 1989c).

Retinol Binding Protein

Retinol binding protein (RBP) is the circulating carrier protein for retinol (Vitamin A), which is a lipid soluble vitamin. In certain target tissues, uptake of retinol is dependent upon a membrane bound receptor that recognizes RBP (Ottonello et al., 1987; Sivaprasadarao and Findlay, 1988a,b). Intracellular transport is facilitated via distinct cellular RBP's (Chytil and Ong, 1984) and gene expression mediated by a family of nuclear retinoic acid receptors, that are structurally homologous to the steroid receptor superfamily (Petkovich et al., 1987, Giguere et al., 1987; Brand et al., 1988; Krust et al., 1989). Retinol plays a critical role in normal embryonic development (Goodman, 1984) and both deficiency and excess of this vitamin results in congenital defects in the embryo (Warkany and Roth, 1948; Cohlan, 1953).

In addition, Vitamin A is associated with normal differentiation of reproductive tract epithelia (Underwood, 1984) and is also capable of inducing retinoic acid receptor (Prentice et al., 1992). Thus, as has been proposed for the IGF autocrine/paracrine system, retinol, RBP and retinoic acid receptors may also play a role in cyclic proliferation and differentiation of the primate endometrium during the menstrual cycle and pregnancy.

Endometrial protein 15, also known as a α_2 pregnancy-associated endometrial globulin (α_2-PEG), is the major secretory product of the human glandular epithelium during the late luteal phase and early pregnancy (Bell and Bohn, 1986; Bell and Smith, 1988). This protein is identical to placental protein 14 (Bell and Bohn, 1986), and has sequence homology with β-lactoglobulin and RBP although its ability to bind retinol has not as yet been demonstrated. α_2-PEG may be involved in retinol transport during implantation and early embryonic development (Bell, 1988). Interestingly, although α_2-PEG constitutes the major endometrial luteal phase secretory product in the human, it is absent in the baboon (Fazleabas and Verhage, 1987; Fazleabas et al., 1989b). Retinol binding protein is a major secretory product of both the porcine (Adams et al., 1981; Harney et al., 1990) and bovine (Liu and Godkin, 1992) progestational endometrium. Since α_2-PEG (β-lactoglobulin) has potential for binding retinol (Bell et al., 1988), and due to its apparent absence in the baboon endometrium, we hypothesized that, as in the pig and sheep, the baboon endometrium synthesizes RBP instead. Our data showed that the synthesis and localization of RBP in the non-pregnant baboon endometrium is similar to that observed for IGFBP-1 (Donnelly et al., 1991), i.e., induced by progesterone following estrogen priming. Synthesis of RBP is up-regulated in the epithelium of glands in the mid-functionalis and basalis during pregnancy and reaches a maximum at day 32, similar to IGFBP-1. However, unlike IGFBP-1, RBP synthesis is confined to the glandular epithelium and decreases as pregnancy proceeds (Donnelly et al., 1991; Fazleabas et al., 1992). The synthetic profile of RBP in the baboon endometrium is similar to that of the glycosylated β-lactoglobulin homologue in the human (Bell et al., 1989) and is perhaps the evolutionary counterpart of this human protein (Fazleabas et al., 1989a, & 1992). One major difference between IGFBP-1 and RBP synthesis, compared to the overall endometrial protein expression in the pregnant uterus, is regionalization of cell specific expression. When the total protein synthetic profile of the endometrium in contact with the placenta (implantation site) was compared with the remainder of the endometrium no major differences were apparent. This is in contrast to IGFBP-1 and RBP synthesis. In the pregnant endometrium, IGFBP-1 and RBP expression is higher in the implantation site endometrium compared to the endometrium not in direct contact with the placenta (non-implantation sites; Donnelly et al., 1991, 1992; Tarantino et al., 1992). With IGFBP-1 in particular, this increased synthesis at the implantation site is maintained throughout the entire first third of pregnancy and appears to be regulated by the extent of placental contact with the endometrium (Fazleabas et al., 1992).

Postulated Functions of IGFBP-1 and RBP in the Baboon Uterus

Our in vivo studies on the modulation of the uterine environment during early pregnancy in the baboon have indicated there is an overall increase in protein biosynthesis (Fazleabas et al., 1993), cell-specific changes in IGFBP-1 expression (Fazleabas et al., 1989b,c; Tarantino et al., 1992), upregulation of RBP synthesis (Donnelly et al., 1991; Fazleabas et al., 1992) and alterations in estrogen and progesterone receptor localization and IGF Type I receptor expression (Hild-Petito et al., 1992a,b). Of all these changes, only IGFBP-1, RBP and IGF Type I receptor localization and expression are markedly different between the implantation site and non-implantation site, suggesting that the baboon conceptus/placenta modulates these proteins within the uterus. The fact that IGFBP-1, which binds the mitogenic peptides IGF I and II with an affinity equal to its receptor (see Rosenfeld et al., 1990; Lamson et al., 1991 for reviews), and RBP, which carries the cell differentiating agent retinol (Sporn et al., 1984), are specifically regulated at the implantation site implies that these two proteins may play an important role in the establishment and maintenance of pregnancy in this primate.

IGFBP's in general are thought to inhibit the mitogenic actions of the IGF's. However, IGFBP-1 and IGFBP-3 have been shown to potentiate the mitotic activity of IGF's (Sara & Hall, 1990, Rosenfeld et al., 1990; Lamson et al., 1991). Although the precise mechanisms by which IGFBP's enhance the actions of IGF's have not been delineated, it has been suggested that IGFBP-3 could act as a reservoir and regulate the release rate of IGF into the local environment (Blum et al., 1989; Conover et al., 1990; Conover and Powell, 1991). Alternatively, IGFBP-1, which contains the tripeptide recognition marker RGD for integrins near its carboxy terminal end, may mediate its action by binding to cell membranes (Sara & Hall 1990; Rosenfeld et al., 1990; Lamson et al., 1991; Shimasaki & Ling, 1992). We have postulated that in the baboon, IGFBP-1 initially has a stimulatory role in enhancing stromal cell and trophoblast proliferation in either an autocrine and/or paracrine manner and once placentation is established and decidual formation complete, the IGFBP-1 then becomes an inhibitory protein (Fazleabas et al., 1991, 1992).

The baboon endometrium undergoes dramatic growth and differentiation during the first third of pregnancy, and if IGF's are required for these processes, the local production of IGFBP-1 in its stimulatory form could provide a mechanism to locally enhance the growth promoting effects of IGF's. The prerequisite to the hypothesis that IGF's have a direct autocrine effect on decidual cell proliferation is that IGF I receptors have to be present on the same cell. Studies in our laboratory (Hild-Petito et al., 1992b) have shown the presence of IGF I receptors on decidualized cells at the implantation site on day 18 of pregnancy, and a continued increase in receptor expression as more stromal cells continue to decidualize. Alternatively, a potential paracrine function for IGFBP-1 is suggested when the cellular localization of synthesis and secretion is considered with reference to the behavior of the embryonic trophoblast during implantation and placental development in the primate. It is possible that the induction of IGFBP-1 synthesis in decidual cells at the placental-endometrial interface is a mechanism by which IGF's can be locally concentrated. IGF I receptors are present in the basal plasma membrane of cytotrophoblast cells at day 18 in the developing placenta (Hild-Petito et al., 1992b) coincident with IGFBP-1 expression in the luminal glandular epithelial cells and decidual cells. These two correlated events may enhance trophoblast proliferation. IGFBP-1 complexed with IGF's and localized in the decidual cells at the placental/endometrial interface and luminal glandular epithelium may act as a mitogen on conceptus tissues and facilitate trophoblastic penetration and contact with the maternal vasculature. The mechanism by which IGF's are released from IGFBP to enable it to bind to its receptor may be regulated locally by plasmin. Campbell et al. (1992) have demonstrated that activation of plasminogen to plasmin by plasminogen activators results in the dissociation of bioactive IGF I from IGFBP-1. The trophoblast produces plasminogen activators (Strickland et al., 1976; Fisher et al., 1989). Therefore, the potential exists for the penetrating and proliferating trophoblast to locally activate plasminogen to plasmin, which in turn releases IGF's from decidual IGFBP-1 thereby making this mitogenic peptide available for binding to its decidual and placental receptor. Following the establishment of placental contact with the endometrium, IGFBP-1 synthesis by the decidua increases (Fazleabas et al., 1989c). This increase in IGFBP-1 may be a mechanism to inhibit IGF binding to its receptor following implantation, and thereby control additional trophoblastic proliferation and invasion. This may be the process by which the baboon, a superficial implanter, is able to rapidly establish contact with the maternal vasculature and yet control trophoblast invasion.

Retinol primarily exerts its effects by inducing cellular differentiation and inhibiting cellular proliferation (Sporn et al., 1984). It is required for a wide variety of biological processes including vision, reproduction, maintenance of epithelial tissues, bone development and linear growth (Ross, 1993). Two principal models have been proposed by which RBP delivers retinol to its site of action. The first, which is an aqueous diffusion model, suggests that like IGFBP-3, retinol slowly dissociates from RBP in the extracellular environment enabling free retinol to be easily taken up by the cells (Ross, 1993). The second suggests that there is a specific association between RBP and the plasma membrane which results in the saturable uptake of retinol by these cells. Specific binding of RBP has been reported for the testis and human placenta (Blomhoff et al., 1991).

We hypothesize that in the baboon uterus during early pregnancy both potential delivery mechanisms for retinol are operational via the upregulation of glandular RBP synthesis. Retinol, in vitro, induces the upregulation of retinoic acid receptor-β transcripts in human endometrial stromal cells (Prentice et al., 1992) and increases IGFBP synthesis in MCF-7 cells (Fontana et al., 1991). Thus, it is conceivable that in the local uterine environment, retinol complexed to RBP may initially up-regulate glandular epithelial IGFBP-1 synthesis (Tarantino et al., 1992) and induce stromal cell differentiation and IGFBP-1 biosynthesis via a receptor-mediated process. The second potential function of RBP may be to carry retinol to the developing baboon conceptus. Unlike the pig (Adams et al., 1981; Harney et al., 1990), sheep (Liu et al., 1992) or bovine (Liu and Godkin 1992), the baboon placenta does not synthesize RBP (Donnelly et al., 1991, 1992). Since RBP has been reported to bind to human placental membranes (Blomhoff et al., 1991), we suggest that in the baboon, endometrial RBP, which is synthesized primarily during fetal organogenesis (Hendrickx 1971; Fazleabas et al., 1992), is necessary for the delivery of retinol to the fetal compartment during a critical phase of embryonic development.

In summary, our studies have clearly demonstrated that in addition to the general changes that the primate uterine endometrium undergoes during the establishment of pregnancy, the baboon conceptus appears to modulate the site and cell-specific synthesis of two major binding proteins, IGFBP-1 and RBP, both of which carry peptides that are essential for the growth and differentiation of both fetal and maternal tissues. The challenge for the future will be to attempt to delineate the mode of action by which uterine proteins initiate and sustain a successful pregnancy.

Acknowledgements

The studies described in this manuscript were supported by a NIH grant HD 21991. The authors wish to acknowledge the contributions of Ms. Kathy Donnelly, Dr. Sheri Hild-Petito, Ms. Patty Mavrogianis and Ms. Esther Vergara to these studies and we also thank Ms. Margarita Guerrero for her secretarial skills.

REFERENCES

Adams, K.L., Bazer, F.W., and Roberts, R.M., 1981, Progesterone induced secretion of a retinol-binding protein in the pig uterus. J. Reprod. Fertil. 62:39-47.

Ashworth, C.J., and Bazer, F.W. 1989, Changes in ovine conceptus and endometrial function following asynchronous embryo transfer or administration of progesterone, Biol. Reprod. 40:425-433.

Bazer, F.W., 1992, Mediators of maternal recognition of pregnancy in mammals, Proc. Soc. Expt. Biol. Med. 199:373-384.

Bell, S.C., 1986, Secretory endometrial and decidual proteins: studies and clinical significance of a maternally derived group of pregnancy-associated serum proteins, Hum. Reprod. 1:129-143.

Bell, S.C., 1988, Secretory endometrial/decidual proteins and their function in early pregnancy, J. Reprod. Fertil. Suppl. 36:1-17.

Bell, S.C., Bohn, H, 1986, Immunochemical and biochemical relationship between human pregnancy-associated secreted α_1 and α_2 endometrial globulins and the soluble placental proteins 12 and 14, Placenta. 7:283-294.

Bell, S.C., Fazleabas, A.T., and Verhage, H.G., 1989, Comparative aspects of secretory proteins of the endometrium and decidua in the human and non-human primates, in: "Blastocyst Implantation", K. Yoshinaga, ed., Adams Publishing Group, Boston. pp 151-162.

Bell, S.C., and Smith, S., 1988, The endometrium as a paracrine organ, in: "Contemporary Obstetrics and Gynecology, G.J.P. Chamberland, ed., Butterworks Scientific Ltd., London. pp 273-297.

Binoux, M., and Hossenlopp, P., 1988, Insulin-like growth factor (IGF) and IGF-binding proteins: comparison of human serum and lymph, J. Clin. Endocrinol. Metab. 67:509-514.

Blomhoff, R., Green, M.H., Green, J.B., Berg, T., and Novum, K.R., 1991, Vitamin A metabolism: New perspectives on absorption, transport and storage, Physiol. Rev. 71:951-990.

Blum, W.F., Jenne, E.W., Reppin, F., Kietzmann, K., Ranke, M.B., Bierich, J.R., 1989, Insulin-like growth factor-I (IGF-I)-binding protein complex is a better mitogen than free IGF-I. Endocrinology. 125:766-772.

Brand, M., Petkovich, M., Krust, A., Chambon, P., de The, H., Marchio, A., Tiollais, P., and Dejean, A., 1988, Identification of a second human retinoic acid receptor, Nature. 332:850-853.

Campbell, P.G., Novak, J.F., Yasonich, T.B., McMaster, J.H., 1992, Involvement of the plasmin system in dissociation of the insulin-like growth factor binding protein complex. Endocrinology. 130:1401-1412.

Castracane, V.D., Goldzieker, J.W., 1986, Timing of the luteal-placental shift in the baboon (Papio cynocephalus), Endocrinology. 118:506-511.

Chytil, F., Ong, D.E., 1984, Cellular retinoid-binding proteins, in, "The Retinoids", Sporn, M.B., Roberts, A.B., Goodman, eds., Academic Press, New York. pp 89-123.

Cohlan, S.Q., 1953, Excessive intake of vitamin A as a cause of congenital anomalies in rat, Science. 117:535-536.

Conover, C.A., Powell, D.R., 1991, Insulin-like growth factor (IGF)-binding protein-3 blocks IGF I induced receptor down-regulation and cell desensitization in cultured bovine fibroblasts, Endocrinology. 129:710-716.

Conover, C.A., Ronk, M., Lombana, F., and Powell, D.R., 1990, Structural and biological characterization of bovine insulin-like growth factor binding protein-3, Endocrinology. 127:2796-2803.

Donnelly, K.M., Vergara, E.F., Mavrogianis, P.A., and Fazleabas, A.T., 1991, Endometrial synthesis of retinol binding protein (RBP) in the non-pregnant and pregnant baboon (Papio anubis), Biol. Reprod. 44 Suppl. 1:57, Abstract #20.

Donnelly, K.M., Vergara E.F., Mavrogianis, P.A., and Fazleabas, A.T., (1992), Insulin-like growth factor binding proteins (IGFBPs) and retinol binding protein (RBP) expression in the baboon (Papio anubis) uterus during the first trimester of pregnancy, Biol. Reprod. 46 Suppl. 1:142 Abstract #365.

Enders, A.C., and Schlafke, S., 1979, Comparative aspects of blastocyst endometrial interactions at implantation, in, "Maternal Recognition of Pregnancy," Ciba Foundation Series 64 (new series), Excerpta Medica. pp 3-22.

Fay, T.N., and Grudzinskas, J.G., 1991, Human endometrial peptides: a review of their potential role in implantation and placentation, Hum. Reprod. 6:1311-1326.

Fazleabas, A.T., Bell, S.C., Verhage, H.G., 1991, Insulin-like growth factor binding proteins: a paradigm for conceptus-maternal interactions in the primate, in, "Uterine and Embryonic Factors in Early Pregnancy," J.F. Strauss, III and C.R. Lyttle, eds., Plenum Press, New York. pp 157-165.

Fazleabas, A.T., Donnelly, K.M., Mavrogianis, P.A., and Verhage, H.G., 1993, Secretory and morphological changes in the baboon (Papio anubis) uterus and placenta during early pregnancy, Biol. Reprod. (In Press).

Fazleabas, A.T., Hild-Petito, S., Donnelly, K.M., Mavrogianis, P.A., and Verhage, H.G., 1992, Interactions between the embryo and uterine endometrium during implantation and early pregnancy in the baboon (Papio anubis), in, "In Vitro Fertilization and Embryo Transfer in Primates," R.M. Brenner, D.R. Wolf, and R.L. Stouffer, eds., Springer-Verlag, New York (In Press).

Fazleabas, A.T., Verhage, H.G., and Bell, S.C., 1989a, Steroid-induced proteins of the primate oviduct and uterus: potential regulators of reproductive function, in, "Autocrine and Paracrine Mechanisms in Reproductive Endocrinology," L.C. Krey, B.J. Guylas, and J.A. McCraken, eds., Plenum Publishing Corp., New York. pp 115-136.

Fazleabas, A.T., Jaffe, R.C., Verhage, H.G., Waites, G., and Bell, S.C., 1989b, An insulin-like growth factor binding protein (IGFBP) in the baboon (Papio anubis) endometrium: synthesis, immunochemical localization and hormonal regulation, Endocrinology. 124:2321-2329.

Fazleabas, A.T., Verhage, H.G., Waites, G., and Bell, S.C., 1989c, Characterization of an insulin-like growth factor binding protein analogous to human pregnancy-associated secreted endometrial α_1 globulin, in decidua of the baboon (Papio anubis) placenta, Biol. Reprod. 40:873-885.

Fisher, S.J., Cui, T-Y., Zhang, L., Hartman, L., Grahl, K., Zhang, G.V., Tarpey, J., and Damskey, C.H., 1989, Adhesive and degradative properties of human placental cytotrophoblast cells in vitro, J. Cell Biol. 109:891-902.

Fontana, J.A., Mezu-Burrows, A., Clemmons, D.R., and LeRoith, D., 1991, Retinoid modulation of insulin-like growth factor binding proteins and inhibition of breast carcinoma proliferation, Endocrinology. 128:1115-2002.

Fortman, J.D., Herring, J.M., Miller, J.B., Hess, D.L., Verhage, H.G., and Fazleabas, A.T., 1993, Chorionic gonadotrophin estradiol and progesterone levels in baboons (Papio anubis) during early pregnancy and spontaneous abortion. Biol. Reprod. (In Press).

Geisert, R.D., Renegar, R.H., Thatcher, W.W., Roberts, R.M., and Bazer, F.W., 1982a, Establishment of pregnancy in the pig: I. Interrelationships between preimplantation development of the pig blastocyst and uterine endometrial secretions, Biol. Reprod. 27:925-939.

Geisert, R.D., Thatcher, W.W., Roberts, R.M., Bazer, F.W., 1982b, Establishment of pregnancy in the pig: III. Endometrial secretory response to estradiol valerate administered on day 11 of the estrous cycle, Biol. Reprod. 27:957-965.

Giguere, V., Ong, E.S., Segui, P., and Evans, R.M., 1987, Identification of a receptor for the morphogen retinoic acid, Nature. 330:624-629.

Giudice, L.C., Milkowski, D.A., Lamson, G., Rosenfeld, R.G., and Irwin, J.C., 1991, Insulin-like growth factor binding proteins in human endometrium: steroid-dependent messenger ribonucleic acid expression and protein synthesis, J. Clin. Endocrinol. Metab. 72:779-787.

Godkin, J.D., Bazer, F.W., and Roberts, R.M., 1984, Ovine trophoblast protein, 1, an early blastocyst protein, binds specifically to uterine endometrium and affects protein synthesis, Endocrinology. 114:120-130.

Goodman, D.S., 1984, Vitamin A and retinoids in health and disease, N. Engl. J. Med. 310:1023-1031.

Harney, J.P., Mirando, M.A., Smith, L.C., and Bazer, F.W., 1990, Retinol-binding protein: a major secretory product of the pig conceptus, Biol. Reprod. 42:523-532.

Hendrickx, A.G., 1971, Embryology of the baboon, The University of Chicago Press, Chicago, IL.

Hild-Petito, S., Verhage, H.G., and Fazleabas, A.T., 1992a, Immunocytochemical localization of estrogen and progestin receptors in the baboon (Papio anubis) uterus during implantation and early pregnancy, Endocrinology. 130:2343-2353.

Hild-Petito, S., and Fazleabas, A.T., 1992b, Characterization of receptors for insulin-like growth factor I in the baboon uterus during the cycle and pregnancy, Biol. Reprod. 46 Suppl 1:140 Abstract #360.

Julkunen, M., Koistinen, R., Aalto-Setalak, K., Seppala, M., Janne, O.A., and Kontulak, K., 1988, Primary structure of human insulin-like growth factor binding protein/placental protein 12 and tissue specific expression of its mRNA, FEBS Lett. 236:295-300.

Krust, A., Kastner, P., Petkovich, M., Zelent, A., and Chambon, P., 1989, A third human retinoic acid receptor, hRAR-γ, Proc. Natl. Acad. Sci. USA. 86:5310-5314.

Lamson, G., Giudice L.C., and Rosenfeld, R.G., 1991, Insulin-like growth factor binding proteins: structural and molecular relationships, Growth Factors. 5:19-28.

Liu, K.H., Gao, K., Baumbach, G.A., and Godkin, J.D., 1992, Purification and immunolocalization of ovine placental retinol-binding protein, Biol. Reprod. 46:23-29.

Liu, K.H., and Godkin, J.D., 1992, Characterization and localization of bovine uterine retinol-binding protein, Biol. Reprod. 47:1099-1104.

Nissley, P., and Lopaczynski, W., 1991, Insulin-like growth factor receptors, Growth Factors. 5:29-43.

Ottonello, S., Petrucci, S., and Maraini, G., 1987, Vitamin A uptake from retinol-binding protein in a cell-free system from pigment epithelial cells of bovine retina, J. Biol. Chem. 262:3975-3981.

Petkovich, M., Brand, N., Krust, A., and Chambon, P., 1987, A human retinoic acid receptor which belongs to the family of nuclear receptors, Nature. 330:444-450.

Prentice, A., Mathews, C.J., Thormas, E.J., and Redforn, C.P.F., 1992, The expression of retinoic acid receptors in cultured human endometrial stromal cells and effects of retinoic acid, Hum. Reprod. 7:692-700.

Rosenfeld, R.G., Lamson, G., Pham, H., Oh, Y.M., Conover, C., DeLeon, D., Donovan, S.M., Ocrant, I., and Giudice, L.C., 1990, Insulin-like growth factor binding proteins, Recent Prog. Horm. Res. 46:99-163.

Ross, A.C., 1993, Cellular metabolism and activation of retinoids: roles of cellular retinoid-binding proteins, FASEB J. 7:317-327.

Rotwein, P., 1991, Structure, evolution, expression, and regulation of insulin-like growth factors I and II, Growth Factors 5:3-18.

Sara, V.R., and Hall, K., 1990, Insulin-like growth factors and their binding proteins, Physiol. Rev. 70:591-614.

Shimasaki, S., Ling, N., 1992, Identification and molecular characterization of insulin-like growth factor binding proteins (IGFBP-1, 2, 3, 4, 5, and 6), Prog. Growth Factor Res.3:243-266.

Sivaprasadarao, A., and Findlay, J.B.C., 1988a, The interaction of retinol-binding protein with its plasma membrane receptor, Biochem. J. 255:561-569.

Sivaprasadarao, A., and Findlay, J.B.C., 1988b, The mechanism of uptake of retinol by plasma-membrane vesicles, Biochem. J. 255:571-578.

Sporn, M.B., Roberts, A.B., Goodman, D.S., 1984, The Retinoids, Academic Press, New York.

Strickland, S., Reich, E., and Sherman, M.I., 1976, Plasminogen activator in early embryogenesis: enzyme production by trophoblast and parietal endoderm, Cell. 9:231-240.

Tarantino, S., Verhage, H.G., and Fazleabas, A.T., 1992, Regulation of insulin-like growth factor binding proteins in the baboon (Papio anubis) uterus during early pregnancy, Endocrinology. 130:2354-2362.

Underwood, B.A., 1984, Vitamin A in animal and human nutrition, in, "The Retinoids," M.B. Sporn, A.B. Roberts, and D.S. Goodman, eds., Academic Press, London. 281-392.

Vallet, J.L., Bazer, F.W., and Roberts, R.M., 1987, The effect of ovine trophoblast protein-one on endometrial protein secretion and cyclic nucleotides, Biol. Reprod. 37:1307-1316.

Warkany, J., and Roth, C.P., 1948, Congenital malformations induced in rats by maternal vitamin A deficiency. II. Effects of varying the preparatory diet in the yield of the normal young, J. Nutr. 35:1-11.

THE EXPRESSION OF PEPTIDASE ANTIGENS, CD10/NEUTRAL ENDOPEPTIDASE, CD13/AMINOPEPTIDASE N, AND CD26/DIPEPTIDYL PEPTIDASE IV IN HUMAN ENDOMETRIUM

Hideharu Kanzaki,[1] Kimitoshi Imai,[1] Hiroshi Fujiwara,[1] Michiyuki Maeda,[2] and Takahide Mori[1]

[1] Department of Gynecology and Obstetrics, Faculty of Medicine and
[2] Chest Disease Research Institute, Kyoto University
Shogoin Kawahara-cho, Sakyo-ku, Kyoto 606-01, Japan

INTRODUCTION

Peptidase enzymes are known to be widely distributed in plasma and tissues. As such, they are considered to play important roles in the local regulation of biologically active peptides, including peptide hormones, growth factors, and cytokines. In the uterine endometrium, it has been suggested that some peptidase enzymes in the endometrial epithelium play an important role in blastocyst implantation. In rabbits, the uterine epithelium undergoes extensive morphological change during the peri-implantation phase, and dynamic changes in aminopeptidase and dipeptidyl peptidase IV activities on the surface of endometrial epithelial cells have been reported in histochemical experiments (Classen-Linke et al., 1987). Furthermore, proteinase inhibitors have been shown to block implantation in rabbits (Denker, 1977). In humans, we have reported that the cluster of differentiation (CD) antigens, CD10 and CD13, were expressed in the endometrial stromal cells, and that CD26 antigens were localized in the glandular epithelial cells (Imai et al., 1992a, 1992b). CD10, CD13, and CD26 antigens were shown to be identical to neutral endopeptidase (NEP) (Letarte et al, 1988), aminopeptidase N (APN) (Look et al.1989), and dipeptidyl peptidase IV (DPP IV) (Mattern et al., 1989; Stein et al., 1989;Ulmer et al., 1990) respectively, all of which are cell surface peptidases. Since a variety of peptide factors, including peptide hormones, growth factors, and cytokines, have been suggested to play important roles in regulating endometrial cell proliferation/differentiation, these peptidase enzymes are thought to be involved in the regulation of human endometrial function and the implantation process.

IMMUNOHISTOCHEMICAL STAINING FOR PEPTIDASE ANTIGENS

Cryostat sections were prepared from human endometria at various phases of development and from the decidua of early pregnancy. Indirect immunofluorescence staining was carried out using monoclonal antibodies for CD2, CD10, CD11b, CD13, CD14, CD20, and

CD26. It was necessary to examine the lymphohematopoietic cell markers CD2 (T lymphocytes), CD11b (monocytes/granulocytes), CD14 (monocytes/granulocytes), and CD20 (B lymphocytes) in order to evaluate antigen expression in endometrial cells. Peptidase antigens of CD10/NEP, CD13/APN, and CD26/DPP IV have been identified as surface markers for lymphoid progenitor cells, granulocytes/monocytes, and activated T lymphocytes, respectively. In the proliferative phase, a majority of cells in the endometrial stroma demonstrated weak expression of CD10 and CD13 antigens (Fig. 1A, 1B), while endometrial glandular epithelial cells showed weak CD26 antigen expression (Fig. 1C). In the secretory phase, the expression of CD10 and CD13 antigens on endometrial stromal cells increased significantly (Fig. 2A, 2B). While endometrial glandular cells expressed CD26 antigen weakly to moderately in the early secretory phase. CD26 was expressed moderately in the mid-secretory phase, and strongly in late secretory phase (Fig. 2C). Furthermore, the endometrial surface epithelium expressed CD26 antigen at the same intensity as the glandular epithelium. No cells in the glandular epithelium expressed CD10 or CD13 antigens, and only a few cells in the stroma expressed CD26 antigen. Vascular endothelial cells in the uterine myometrium expressed CD26 antigen moderately throughout the menstrual cycle. In early pregnancy decidua, stromal cells with decidual transformation expressed CD13 antigen more strongly than secretory endometrial cells (Fig. 3B). CD10 antigen expression in decidual cells, however, was much weaker than in secretory endometrium (Fig. 3A). Moderate or strong expression of CD26 antigens was observed in the glandular epithelial cells of early pregnancy decidua (Fig. 3C). It appeared that CD26 antigen expression in the apical membrane of glandular cells and surface epithelium was stronger than in the basolateral membranes (Fig. 2C, 3C). In serial sections prepared during these experiments, CD2, CD11b, CD14 or CD20 antigen-bearing cells were sparsely distributed in the endometrium throughout the menstrual cycle and during the first

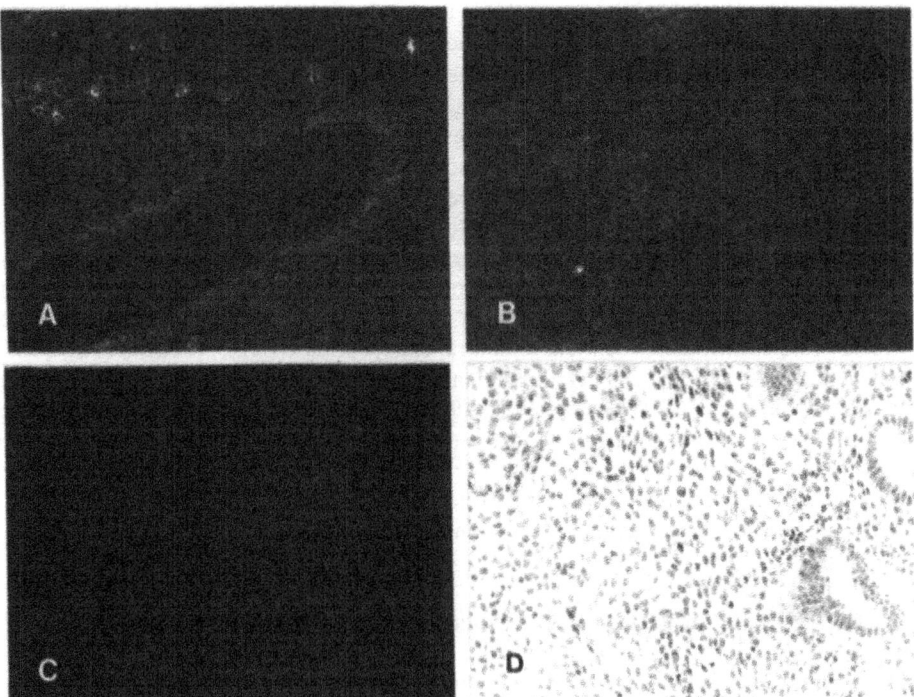

Fig. 1. Indirect immunofluorescence staining of human endometrium in proliferative phase. Cryostat sections were stained by using following monoclonal antibodies: A, CD10/Nu-N1; B, CD13/MCS2; C, CD26/Ta1; D, hematoxylin and eosin.

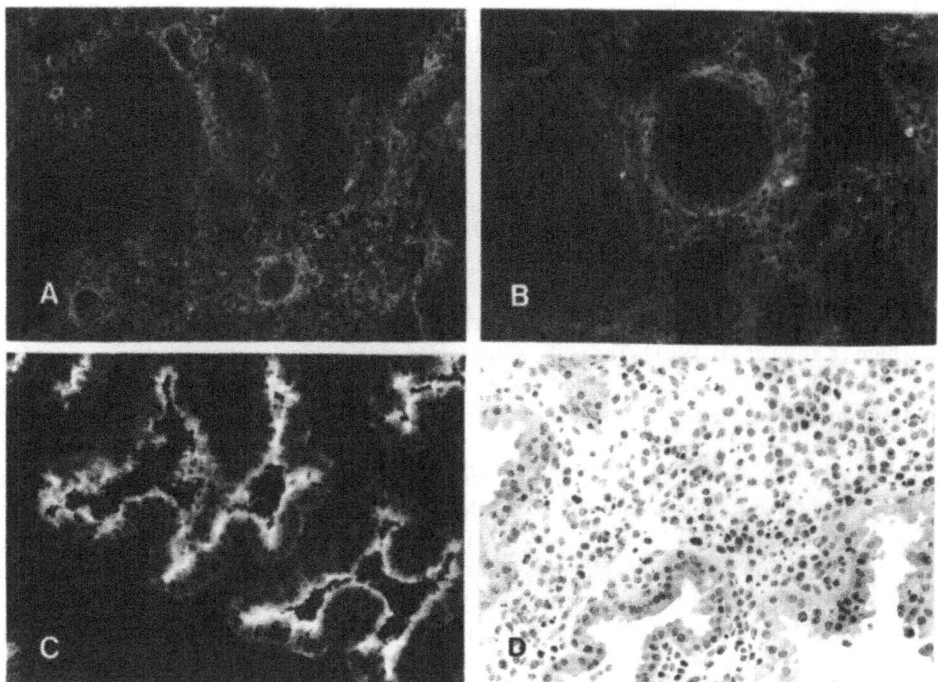

Fig. 2. Indirect immunofluorescence staining of human endometrium in secretory phase. Cryostat sections were stained by using following monoclonal antibodies: A, CD10/Nu-N1; B, CD13/MCS2; C, CD26/Ta1; D, hematoxylin and eosin.

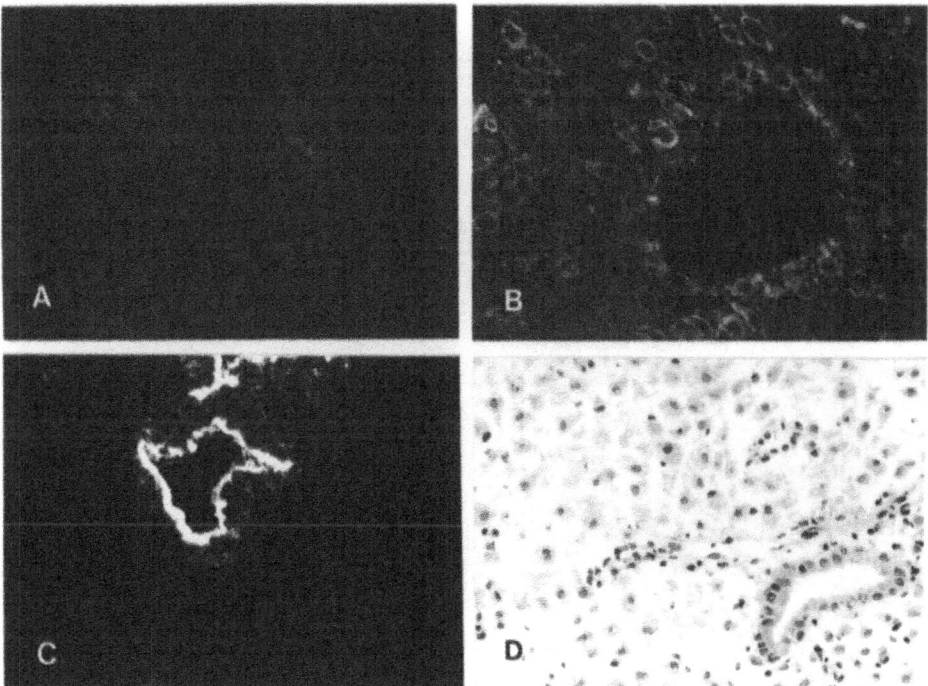

Fig. 3. Indirect immunofluorescence staining of human decidua of 7 weeks gestation. Cryostat sections were stained by using following monoclonal antibodies: A, CD10/Nu-N1; B, CD13/MCS2; C, CD26/Ta1; D, hematoxylin and eosin.

trimester of pregnancy. Therefore, it was apparent that CD10/NEP and CD13/APN antigens were expressed in stromal cells/decidual cells while CD26/DPP IV antigen was expressed in glandular cells. The intensity of CD10 and CD26 antigen expression was maximal during the secretory phase (peri-implantation periods), whereas the intensity of CD13 expression increased steadily following the establishment of pregnancy.

FLOW CYTOMETRIC ANALYSIS OF ENDOMETRIAL STROMAL CELL-ENRICHED PREPARATIONS

Human endometrial stromal cell-enriched preparations were obtained as described elsewhere (Kariya et al., 1991). The cells were stained with monoclonal antibodies against CD2, CD10, CD11b, CD13, CD14, and CD20 antigens, then examined using a flow cytometer. More than 80% of the cells examined were positive for CD10/NEP and CD13/APN antigens in every experiment, and the fluorescence intensity was relatively strong. CD2, CD11b, CD14 or CD20 antigen-bearing lymphohematopoietic cells usually accounted for less than 10% of observed fluorescence (Fig. 4). These indings confirmed the results of previous immunohistochemical experiments indicating that the stromal cells themselves express CD10/NEP and CD13/APN antigens in human endometrium. Dynamic changes in peptidase antigen expression during the menstrual cycle imply the involvement of ovarian gonadal steroids. Accordingly, the effect of estrogen, progesterone, or androgen on the expression of CD10/ NEP and CD13/APN antigens in cultured stromal cells was subsequently examined. Cells were cultured for 7 days with or without estradiol (10^{-8}M), progesterone (10^{-7}M), or testosterone (10^{-8}M), and CD10/NEP and CD13/APN antigen expressions on these cells were assessed by mean fluorescence intensity. During the culture period, CD10 antigen

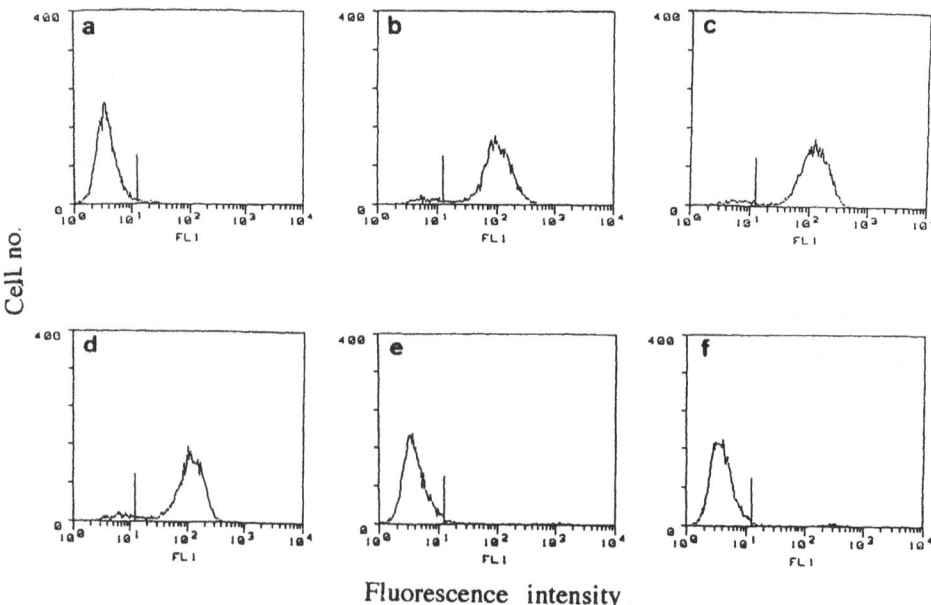

Fluorescence intensity

Fig. 4. Flow cytometric analysis of human ESC-enriched preparation from late-proliferative endometrium. Cells were stained as follows : a, a second antibody only (negative control); b, CD13/My7; c, CD13/MCS2; d, CD10/Nu-N1; e, CD14/LeuM3; f, CD11b/Bear-1. Percent positivity was as follows: a, 2.1; b, 92.8; c, 93.1; d, 92.2; e, 3.9; f, 3.3. Over 90% of the cells were positive for CD10 and CD13 antigens.

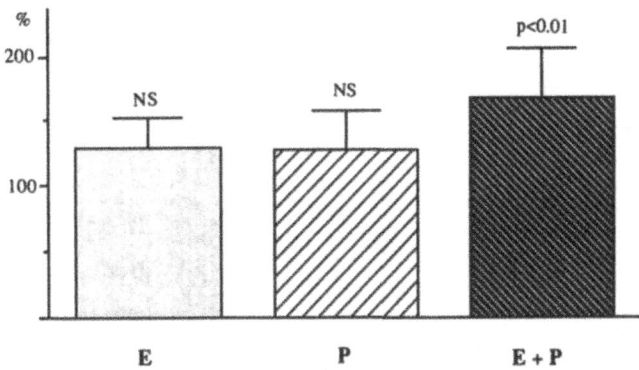

Fig. 5. CD13/APN antigen expression on endometrial stromal cells after 7 days culture with ovarian steroids. Mean fluorescence intensity was assessed by flow cytometry. Results were expressed as % of control cultures without ovarian steroids. (N=7)

expression on the stromal cells rapidly decreased, and the addition of gonadal steroids had no apparent effect on this expression. This observation suggests that some indispensable factor(s) are present *in vivo* for the maintenance of expression of NEP molecules on the cell surface. In contrast, CD13/APN antigen expression on stromal cells did not decrease even after 14 days' culture, and the expression of CD13/APN antigen was similarly unaffected by the addition of estrogen, progesterone, or androgen alone. A significant enhancement of CD13/APN antigen expression was observed, however, when both estradiol and progesterone were added to the cultures (Fig. 5). This *in vitro* experiment strongly suggests that the increase in APN expression in stromal cells during the proliferative phase and during decidualization is regulated by the effects of both estrogen and progesterone.

DETECTION OF AMINOPEPTIDASE AND DIPEPTIDYL PEPTIDASE IV ACTIVITIES

Aminopeptidase activity of the endometrial stromal cell-enriched preparation was examined using a method based on the hydrolysis of alanine-*p*-nitroanilide, to yield *p*-nitroaniline and alanine (Amoscato et al., 1989). The endometrial stromal cell-enriched preparation was suspended in PBS containing 0.4mM alanine-*p*-nitroanilide and incubated at 37°C with continuous stirring. The reaction was stopped by the addition of cold sodium acetate-acetic acid buffer. The optical density of the supernatant at 385nm was examined with a spectrophotometer, and the amount of *p*-nitroaniline formed was determined from a standard curve. Endometrial stromal cell-enriched fractions showed distinct aminopeptidase activity, and the amount of *p*-nitoaniline formed was linearly related to the number of the stromal cells (Fig. 6) and to the duration of the incubation period. This experiment demonstrated biochemically that human endometrial stromal cells have aminopeptidase activity. Interestingly, a significant increase in the *in vitro* peptidase activity was observed in the cultured endometrial stromal cells in the presence of both estrogen and progesterone, whereas estrogen, progesterone or androgen alone had no apparent effect (Fig. 7).

Histochemical detection of DPP IV activity was performed according to Lodja's method (1979). Briefly, 4mg of glycyl-prolyl-4-methoxy-β–naphthylamide was dissolved in 0.5ml N,N-dimethylformamide and mixed with 10mg Fast Blue B salt dissolved in 10ml 0.1M phosphate buffer. The mixture was filtered and overlaid on frozen sections of human endometrium and early pregnancy decidua. After 6-7 minutes of incubation, the slides were

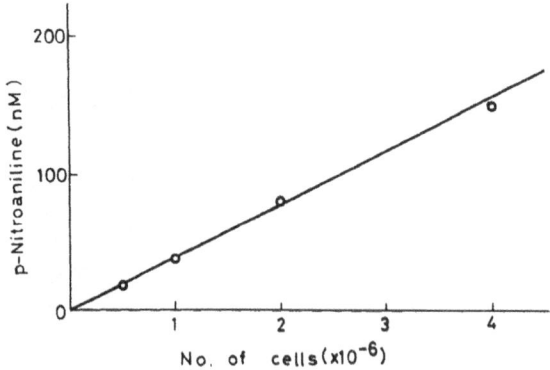

Fig. 6. Assay of peptidase activity of endometrial stromal cell-enriched preparation from the endometrium in mid-secretory phase. The amount of p-nitroaniline hydrolyzed from alanine-p-nitroanilide is linearly related to the number of cells. Incubation time: 20 min.

washed and red reaction products were observed with a light microscope. During the proliferative to early secretory phases, no DPP IV activity was observed in the endometrium (Fig. 8A). In the mid-secretory phase, most endometrial glandular cells showed weak to moderate DPP IV activity, and in the late secretory phase, glandular cells showed moderate to strong DPP IV activity (Fig. 8B). In the early pregnancy decidua, glandular cells showed moderate DPP IV activity (Fig. 8C), while vascular endothelium in the uterine myometrium had weak DPP IV activity. Lojda (1979) reported that, histochemically, DPP IV activity in the endothelium of blood vessels was heterogenous. In our study, most of the endothelium in the muscle layer was DPP IV positive, but the endothelium in the endometrial layer did not show DPP IV, either histochemically or immunohistologically.

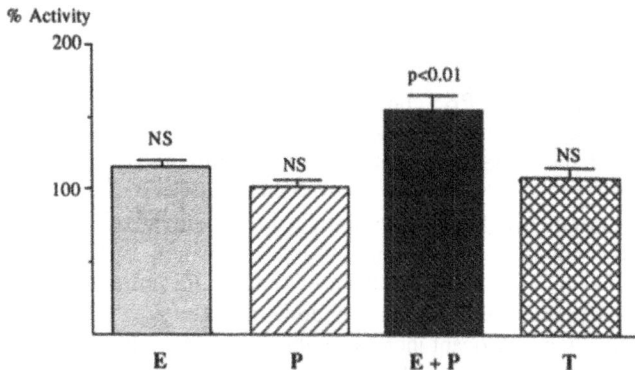

Fig. 7. Aminopeptidase activity of cultured endometrial stromal cells. Endometrial stromal cells were cultured for 7 days with ovarian steroids: E, 10^{-8}M; P, 10^{-7}M; T, 10^{-8}M. Results were expressed as % of control cultures without ovarian steroids. (N=6)

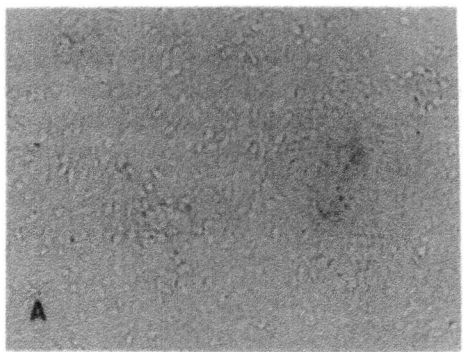
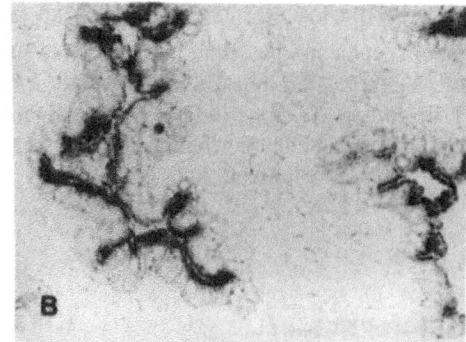
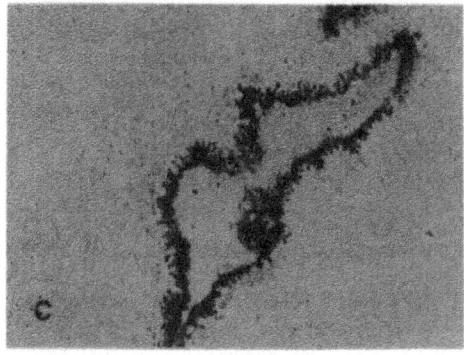

Fig. 8. Enzyme histochemistry of DPP IV activity in the human endometrium. A, proliferative phase; B, mid-secretory phase: C, first trimester of pregnancy (7 weeks).

PEPTIDASES AS POSSIBLE LOCAL REGULATORS OF PEPTIDE FACTORS

We have shown that human endometrial stromal cells express CD10/NEP and CD13/ APN antigens, and that endometrial glandular cells express CD26/DPP IV antigen; the intensity of expression changes with endometrial differentiation. All three enzymes have their active domain exposed on the extracellular surface. NEP cleaves peptides at the amino side of hydrophobic residues, such as enkephalins, bradykinin, oxytocin, substance P, neurotensin, chemotactic peptides (fMet-Leu-Phe), gastrin, atrial natriuretic peptide, interleukin 1 (Erdös et al., 1989) etc. APN catalyzes the removal of N-terminal amino acids from peptides, such as met-lys-bradykinin and met-enkephalin, somatostatin (Sidorowicz et al., 1981), (Asn[1]) angiotensin II (Ward et al., 1990) etc. DPP IV removes dipeptides from the N-termini of polypeptides. Its substrates include human gastrin-releasing peptide, human pancreatic polypeptide, α chain of human chorionic gonadotropin (Nausch et al., 1990), and substance P (Püschel et al., 1982). These peptidases are well known to have roles in the final steps of digestion in the intestinal canal. Recently, these enzymes have been proven to have a wide distribution on the surfaces of various cell types, and it has been suggested that they might have specific roles in the control of growth and differentiation in both hemopoietic and epithelial cell systems. In addition to having proteolytic activity, APN was reported to be a major receptor molecule for human coronavirus, and its significance in viral infection has been discussed (Delmas et al., 1992; Yeager et al., 1992). DPP IV has been shown to be involved in cellular adhesion in extracellular matrix proteins (Hanski et al., 1985, 1988; Piazza et al., 1989; Dang et al., 1990). Since DPP IV is expressed on the apical membrane of endometrial epithelium during the peri-implantation period, this antigen may have a role in the attachment of fertilized eggs to the endometrial surface.

It has become apparent that peptide factors including peptide hormones, growth factors, and cytokines play important roles during the peri-implantation period, and many of these peptides are possibly inactivated or functionally modulated by peptidase enzymes. Therefore, these enzymes may have a key role in the control of growth and differentiation of cells in human endometrium through the activation or inactivation of peptide factors and the regulation of their access to target cells.

REFERENCES

Amoscato, A.A., Alexander, J.R. and Babcock, G.F., 1989, Surface aminopeptidase activity of human lymphocytes. I. Biochemical and biologic properties of intact cells, *J. Immunol.* 142:1245.

Classen-Linke, I., Denker, H.W. and Winterhager, E., 1987, Apical plasma membrane-bound enzymes of rabbit uterine epithelium; pattern changes during the periimplantation phase, *Histochemistry* 87:517.

Dang, N.H., Torimoto, Y., Schlossman, S.F. and Morimoto, C., 1990, Human CD4 helper T cell activation: functional involvement of two distinct collagen receptors 1F7 and VLA integrin family, *J. Exp. Med.* 172:649.

Delmas, B., Gelfi J., L'Haridon R., Vogel L.K., Sjöström H., Nóren, O. and Laude, H., 1992, Aminopeptidase N is a major receptor for the enteropathogenic coronavirus TGEV, *Nature* 357:417.

Denker, H.W., 1977, Implantation. The role of proteinases, and blockage of implantation by proteinase inhibitors, *Adv. Anat. Embryol. Cell Biol.* 53 Fasc 5.

Erdös, E.G. and Skidgel, R.A., 1989, Neutral endopeptidase 24-11 (enkephalinase) and related regulators of peptide hormones, *FASEB J.* 3:145.

Hanski, C., Huhle, T. and Reutter, W., 1985, Involvement of plasma membrane dipeptidyl peptidase IV in fibronectin-mediated adhesion of cells on collagen, *Biol. Biochem. Hoppe-Seyler* 366:1169.

Hanski, C., Huhle, T., Gossrau, R. and Reutter, W., 1988, Direct evidence for the binding of rat liver DPP IV to collagen in vitro, *Exp. Cell Res.* 176:64.

Imai, K., Maeda, M., Fujiwara, H., Okamoto, N., Kariya, M., Emi, N., Takakura, K., Kanzaki, H. and Mori, T., 1992a, Human endometrial stromal cells and decidual cells express cluster of differentiation (CD)13 antigen/aminopeptidase N and CD10 antigen/neutral endopeptidase, *Biol. Reprod.* 46:328.

Imai, K., Maeda, M., Fujiwara, H., Kariya, M., Takakura, K., Kanzaki, H. and Mori, T., 1992b, Dipeptidyl peptidase IV as a differentiation marker of the human endometrial glandular cells, *Hum. Reprod.* 7:1189.

Kariya, M., Kanzaki, H., Takakura, K., Imai, K., Okamoto, N., Emi, N., Kariya, Y. and Mori, T., 1991, Interleukin-1 inhibits in vitro decidualization of human endometrial stromal cells, *J. Clin. Endocrinol. Metab.* 73:1170.

Letarte, M., Vera, S., Tran, R., Addis, J.B.L., Onizuka, R.J. and Quackenbush, E.J., 1988, Common acute lymphocytic leukemia antigen is identical to neutral endopeptidase, *J. Exp. Med.* 168:1247.

Lodja, Z., 1979, Studies on dipeptidyl(amino)peptidase IV (glycyl-proline naphthylamidase). II. Blood vessels, *Histochemistry* 54:153.

Look, A.T., Ashmun, R.A., Shapiro, L.H. and Peiper, S.C., 1989, Human myeloid plasma membrane glycoprotein CD13 (gp150) is identical to aminopeptidase N, *J. Clin. Invest.* 83:1299.

Mattern, T., Fled, H.D., Feller, A.C., Heymann, E. and Ulmer, A.J., 1989, Anti-DPP-IV and anti-Ta1 but not IOT 15 should be clustered in CDw26, in: "Leukocyte Typing IV. White Cell Differentiation Antigens. Oxford," Knapp, W., Dörken, B., Gilks, W.R., Rieber, E.P., Schmidt, R.E., Stein, H. and von den Borne, A.E.G.Kr., eds. University Press, Oxford, New York, Tokyo.

Nausch, I., Mentlein, R. and Heymann, E., 1990, The degradation of bioactive peptides and proteins by dipeptidyl peptidase IV from human placenta, *Biol. Chem. Hoppe-Seyler* 371:1113.

Piazza, G.A., Callanan, H.M., Mowery, J. and Hixson, D.C., 1989, Evidence for a role of dipeptidyl peptidase IV in fibronectin-mediated interactions of hepatocytes with extracellular matrix, *Bioch. J.* 262:327.

Püschel, G., Mentlein, R. and Heymann, E., 1982, Isolation and characterization of dipeptidyl peptidase IV from human placenta, *Eur. J. Biochem.* 126:359.

Sidorowicz, W., Zownir, O. and Behahl, F.J., 1981, Action of human pancreas alanine aminopeptidase on biologically active peptides:kinin converting activity, *Clin. Chem. Acta.* 111:69.

Stein, H., Schwarting, R. and Niedobitek, G., 1989, Cluster Report:CD26, in: "Leukocyte Typing IV. White Cell Differentiation Antigens. Oxford," Knapp, W., Dörken, B., Gilks, W.R., Rieber, E.P., Schmidt, R.E., Stein, H. and von den Borne, A.E.G.Kr., eds. University Press, Oxford, New York, Tokyo.

Ulmer, A.J., Mattern, T., Feller, A.C., Heymann, E. and Flad, H.D., 1990, CD26 antigen is a surface dipeptidyl peptidase IV (DPP IV) as characterized by monoclonal antibodies clone TII-19-4-7 and 4EL1C7, *Scand. J. Immunol.* 31: 429.

Ward, P.E., Benter, I.F., Dick, L. and Wilk, S., 1990, Metabolism of vasoactive peptides by plasma and purified renal aminopeptidase M, *Biochem. Pharmacol.* 40:1725.

Yeager, C.L., Ashmun, R.A., Williams, R.K., Cardellichio, C.B., Shapiro, L.H., Look, A.T. and Holmes, K.V., 1992, Human aminopeptidase N is a receptor for human coronavirus 229E, *Nature* 357:420.

ANGIOGENESIS IN THE RAT UTERUS DURING PREGNANCY

Rolf H. B. Christofferson, B. Erik Wassberg, and B. Ove Nilsson

Department of Human Anatomy
University of Uppsala
Box 571
S-751 23 Uppsala, Sweden

INTRODUCTION

The endometrium responds to blastocyst signals by decidualization and placentation. Angiogenesis — new blood vessel formation — is considered to be a prerequisite for the development of the vascular placenta (Folkman and Shing, 1992). Pharmacological modulation of angiogenesis in the uterus would be of interest for e.g. (1) stimulation of placentation in infertile women during an in vitro-fertilization programme, (2) stimulation of placentation in pregnacies with an intrauterine growth-retarded baby, (3) inhibition of placentation as a contraceptive, and (4) inhibition of angiogenesis in endometrial cancers. In order to stimulate or inhibit endometrial angiogenesis, it is important that the formation of the vascular part of the placenta during normal pregnancy be investigated.

The rat is often used in placental research since it exhibits a discoid, hemochorial, chorioallantoic placenta as does man. But the rat chorioallantoic placenta is hemotrichorial and labyrinthine, while that of man is hemomonochorial and villous. The rat embryo also develops a yolk sac placenta (choriovitelline placentation) prior to chorioallantoic placentation, which does not take place in man (see Mossman, 1987, for a review of placentation in vertebrates).

Formation of the maternal part of the vascular placenta in the rat has been studied by light microscopy of histological sections (e.g. Duval, 1891; Bridgman 1948a,b), ink-injections (Holmes and Davies, 1948), microangiography (Young, 1956), and corrosion casts (Bøe, 1950). These investigators are unfortunately often at variance, probably due incomplete vascular injections and the three-dimensional distribution of the microcirculation.

Microvascular corrosion casting with analysis in the scanning electron microscope (SEM) is a morphological method designed for visualization of the three-dimensional microangioarchitecture of tissues and organs (Murakami, 1971; Lametschwandtner et al., 1990). It is considered to be a sensitive and specific method for demonstrating angiogenesis by sprouting from pre-existing vessels (Christofferson and Nilsson, 1992).

The aim of this study is to demonstrate the location and extent of angiogenesis by sprouting in the rat uterus during of pregnancy by SEM analysis of microvascular corrosion casts and light microscopical analysis of histological sections.

MATERIALS AND METHODS

Experimental delay of implantation

Implantation in rodents can be demonstrated by the blue dye test (Psychoyos, 1971). Unfortunately, there is a considerable variation in the timing of the appearance of blue bands following intravenous injection of a dye, ranging from early day 4 to no visible bands late day 5 in normal pregnancies (Rogers et al., 1982). Blastocyst implantation can, however, be synchronized by means of an experimental delay of implantation, involving bilateral ovariectomy and exogenous progesterone administration, followed by activation of implantation by estrogen administration (Bergström, 1978). In the rat the blue dye test becomes positive 26 hours after estrogen activation (Lundkvist and Ljungkvist, 1977). The pregnancy following such a procedure can successfully be maintained by administration of exogenous hormones (Nutting and Meyer, 1963). For details on the procedures used in this study, please refer to Christofferson and Nilsson (1988a).

Vascular casting

Forty-one sexually mature female Sprague-Dawley rats (Alab, Sollentuna, Sweden), with a body weight of 203-447 g at the time of killing, were used. The rats were caged 1-2 overnight with males until pregnant. Vascular casts were made on days 5, 6, 7, 8, 9, 10, 12, 14, 16, 18, 20, and 22 of pregnancy, day 1 of pregnancy being the day on which sperm was found in vaginal smears. Casts were also made the day after delivery (day 22 or 23), and 26 and 50 hours after estrogen activation from an experimental delay of implantation. At least two animals were cast at each stage. Briefly, the animal was given an overdose sodium pentothal and perfused through the thoracic aorta with warm, heparinized, phosphate-buffered saline with papaverine and a colloid, followed by infusion of methyl methacrylate resin with a viscosity similar to rat´s blood. The infusion pressure did not exceed 100 mm Hg, as measured in the left femoral artery. The density of the polymerized resin was determined by weighing a given volume. For details on the casting procedure and resin preparation, please refer to Christofferson and Nilsson (1992).

Preparation of specimens

For SEM of vascular casts, implantation sites (as identified by the blue dye test or as a macroscopically visible node) were cut out with razor blades and subjected to differential corrosion (Christofferson and Nilsson, 1988b). After corrosion, the dry weights of 5 unsectioned implantation site casts from each stage were recorded. The casts were then sectioned with a razor blade while frozen in distilled water on a chiller plate, dried in air, mounted on aluminum stubs with colloidal silver, sputter coated with gold, and observed in a Philips 525 SEM at an acceleration voltage of 4-10 kV.

For light microscopy, implantation sites were cut out and immersion-fixed in 2.5% glutaraldehyde or 3.5% formaldehyde in phosphate-buffered saline for at least 24 h. The specimens were then dehydrated, embedded in paraffin, sectioned at 8-10 μm, and stained with hematoxylin and eosin.

RESULTS

The blue dye test was positive on days 5-10 of pregnancy. The uterine vascular pattern at the time of implantation is presented in Figs. 1-5. Arterial and venous vessels were identified by their branching pattern, their luminal diameter, and their imprint pattern of endothelial cell nuclei (Fig. 6).Vascular sprouts were defined as sharp-pointed, blindly ending extensions from capillaries and venules. In order to discriminate vascular sprouts from incompletely cast capillaries, sprouts had to exhibit imprints of endothelial cell nuclei. Extravasation of resin was frequently observed in areas exhibiting vascular sprouts.

The density of the polymerized resin was 1.16 g/ml. The maternal blood volume at implantation sites at different stages of pregnancy could be calculated by dividing the implantation site cast weight with the known resin density (Fig. 7).

Implantation (Fig. 8) induced a vascular shut-down in the primary decidual zone, observed as an avascular area in casts (Figs. 9-11) 26 and 50 hours after activation from an experimental delay of implantation. The area remained avascular until day 10 of pregnancy.

Two areas were identified where angiogenesis by sprouting from pre-existing maternal venules and capillaries occurred: mesometrially on days 8-10 and antimesometrially on days 8-12 of pregnancy (Fig. 12). Neither sprouting from arterioles nor transcapillary tissue pillars (Patan et al., 1993) were observed in the decidual vessels.

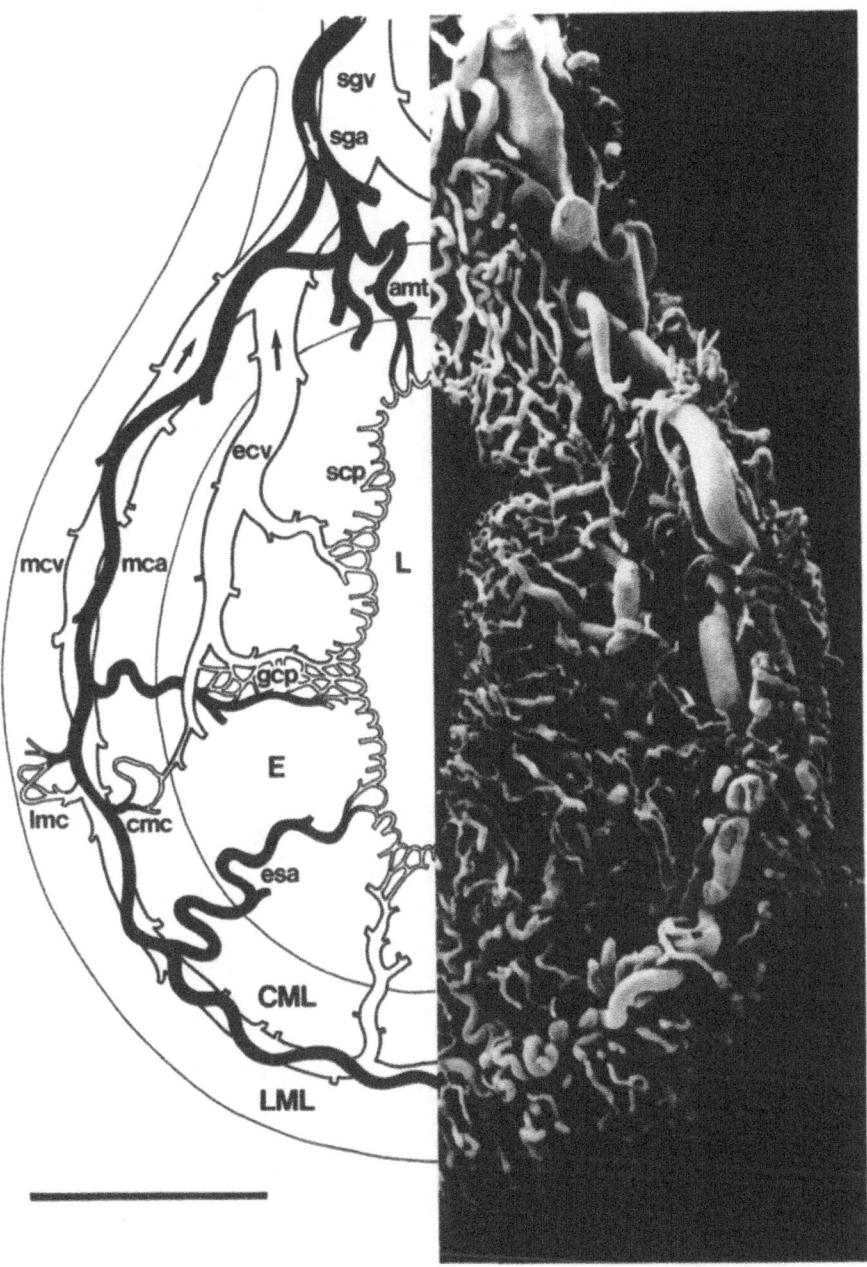

Figure 1. Vascular patterns of the rat uterus at the time of implantation. Arterial vessels black, mesometrium top. Arrows indicate the direction of flow. Abbreviations: amt — arterioles of the mesometrial triangle, cmc— circular muscle capillary plexus, CML — circular muscle layer, E — endometrium, ecv — endometrial circumferential venule, esa — endometrial spiral arteriole, gcp — glandular capillary plexus, L — lumen, lmc— longitudinal muscle capillary plexus, LML — longitudinal muscle layer, mca — myometrial circumferential arteriole, mcv — myometrial circumferential venule, scp — subepithelial capillary plexus, sga — segmental arteriole, sgv — segmental venule. Modified from Rogers and Gannon (1981). Right, SEM of vascular cast from day 4 of pregnancy. Bar = 1 mm.

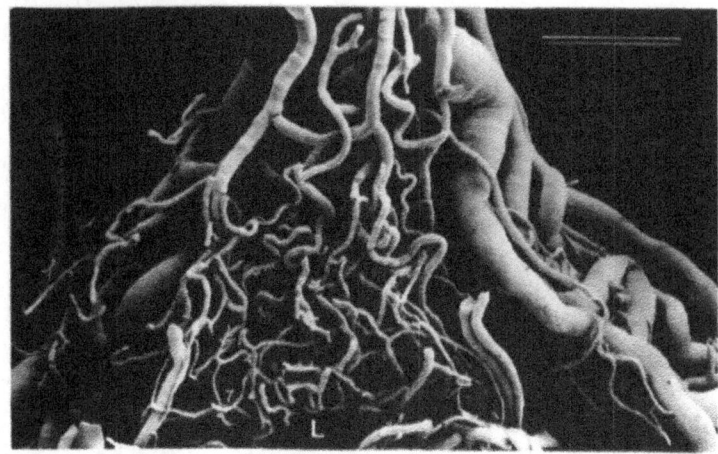

Figure 2. The mesometrial triangle in cross-section. The mesometrial triangle is bounded by the circular muscle layer and the two mesometrial extensions of the longitudinal muscle layer. Note the tortuous arterioles with abundant anastomoses, the complete absence of venules, and the comparatively few capillaries present in this area. L — lumen. Vascular cast, 50 hours after activation from an experimental delay of implantation. Bar = 300 μm.

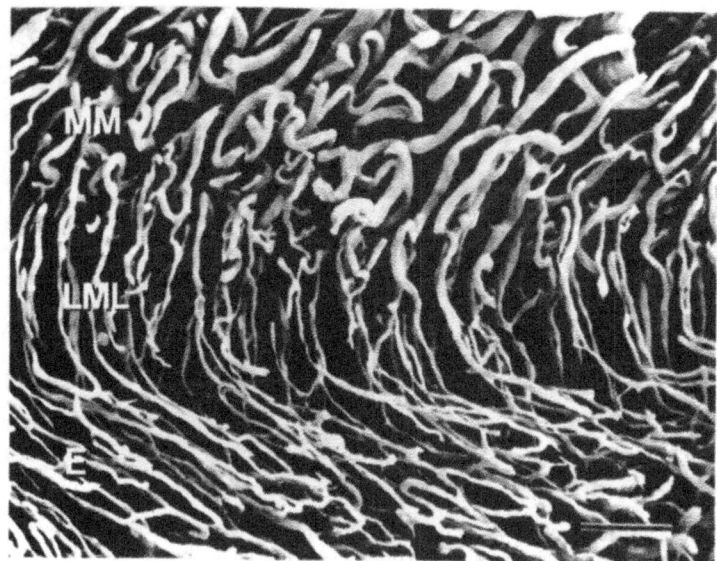

Figure 3. The mesometrial triangle in longitudinal section. The chorioallantoic placenta is formed in this area. MM — mesometrium, LML — longitudinal muscle layer, E — endometrium. Vascular cast, 26 hours after activation from an experimental delay of implantation. Bar = 100 μm.

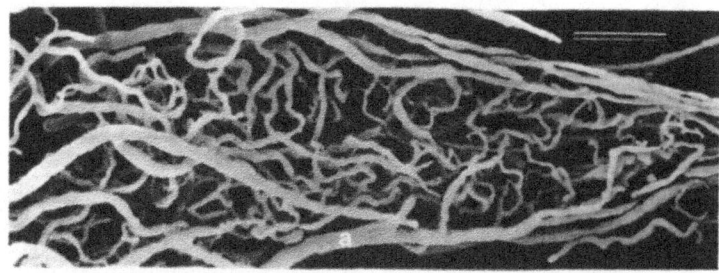

Figure 4. Uterine gland. The uterine gland capillary plexus is complex, but is essentially fed by a straight arteriole (a) with several draining venules. Vascular cast, dissected specimen, 26 hours after activation from an experimental delay of implantation. Bar = 100 µm.

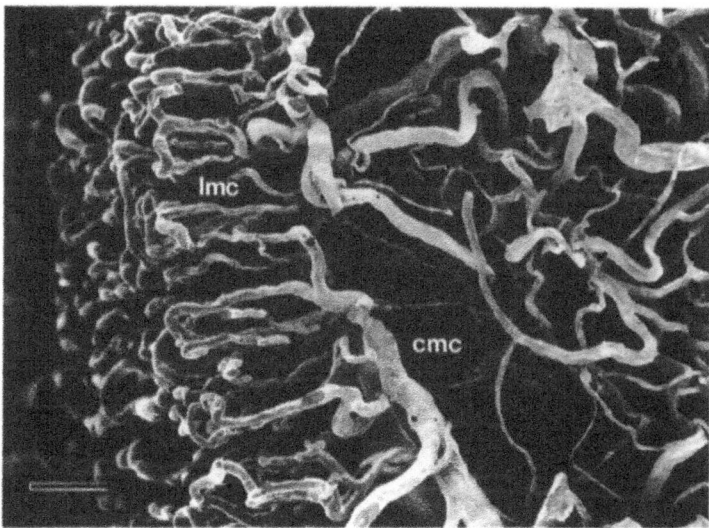

Figure 5. Capillary plexuses of the longitudinal and circular muscle layers. The myometrial circumferential vessels give rise to two capillary plexuses; the plexus of the longitudinal muscle layer (lmc) and that of the circular muscle layer (cmc), the latter consisting of few and small (4-6 µm) capillaries. Vascular cast, 26 hours after activation from an experimental delay of implantation. Bar = 100 µm.

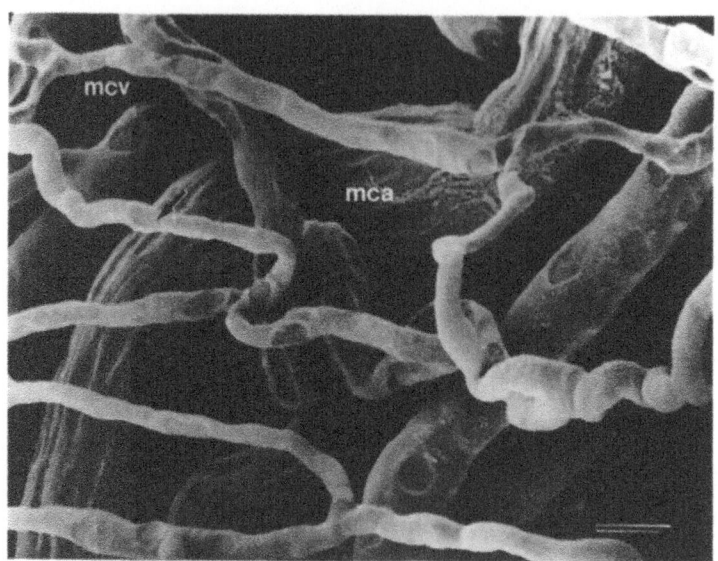

Figure 6. Quality of replication in vascular casts. The low-viscosity resin used in this study permitted filling of the microcirculation at an intra-arterial pressure not exceeding the physiological, and it also replicated endothelial cell nuclear impressions (arrow) in all vessels including capillaries. mca — myometrial circumferential arteriole, mcv — myometrial circumferential venule. Vascular cast, 26 hours after activation from an experimental delay of implantation. Bar = 20 μm.

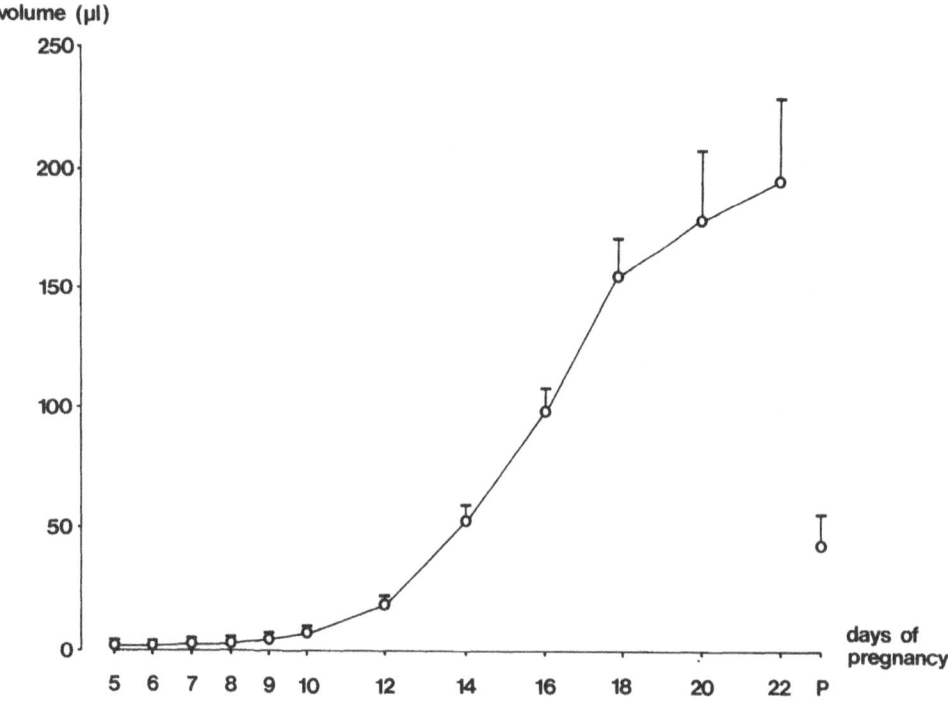

Figure 7. Estimated implantation site maternal blood volume. Values obtained by weighing 5 unsectioned casts from each day and dividing the weight with the resin density.

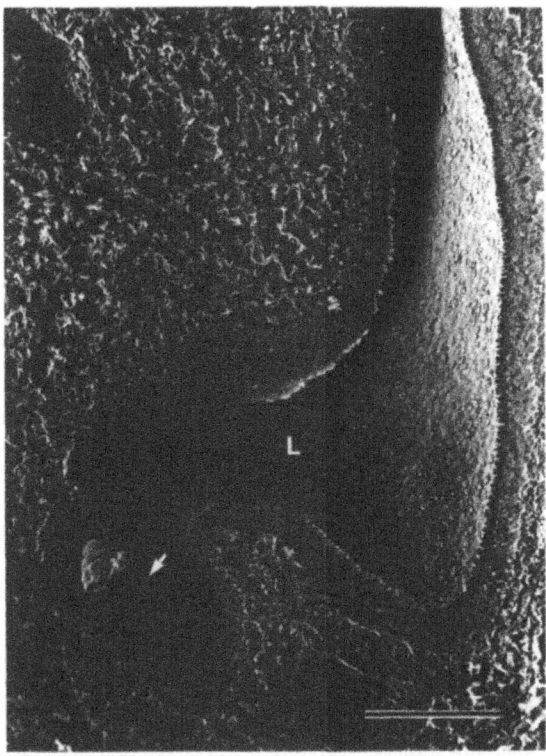

Figure 8. Implantation site, 26 hours after activation from an experimental delay of implantation. The blastocyst (arrow) has been touched in section and is slightly displaced, but still located in the antimesometrial extension of the uterine lumen (L). SEM of perfusion-fixed and vibratome-sectioned tissue. Bar = 100 μm.

Mesometrial sprouting

The mesometrial angiogenesis was observed as vascular sprouts arising from the dilated venules ("sinusoids") in the mesometrial decidua lateral to the remaining uterine lumen and the mesometrial chamber (Figs. 13, 14). Sprouts were occasionally found day 7 and were abundant day 8. The vascular sprouts were mainly pointing towards the mesometrial chamber. The newly formed vessels did not fuse to form a capillary network, but became continuous with the blood spaces within the Träger trophoblast on day 10 (Figs.15, 16). It was not possible to determine how trophoblast tapped the sinusoids and the newly formed vessels in casts or in light microscopical sections (Fig. 15). The maternal blood spaces appeared as highly irregular, anastomosing "blebs" of varying size (6-100 μm; Fig. 16), consistently connected to the dilated subepithelial capillary plexus (supplying arterialized blood) and to the mesometrial sinusoids and their vascular sprouts (draining venous blood). The "blebs" did not exhibit typical endothelial cell nuclear imprints, but larger (12-18 μm) nuclear imprints distributed irregularly over their surface. From day 12 — when the allantoic stalk made contact with the ectoplacental cone — and onwards, these blood spaces constituted the maternal part of the zona intima. The remaining venous vessels mesometrially formed the vessels of trophospongium and basal decidua, and were drained by the endometrial circumferential venules.

Antimesometrial sprouting

The antimesometrial angiogenesis was observed in the whole antimesometrial decidua, apart from the avascular primary decidual zone, on days 8-12 of pregnancy (Figs. 17, 18). Vascular sprouts emanated from capillaries and small venules, and were oriented at random. Tips of sprouts facing each other were frequently observed (Fig.19). The parental vessels were irregular in outline and exhibited deep imprints of endothelial cell nuclei. During formation of the antimesometrial lumen, vascular sprouts were observed on both sides of the cleavage line, i.e. both in the capsular decidua and in the thin endometrium lining the uterine wall. When the antimesometrial lumen was continuous between inter-implantation sites on day 14, no vascular sprouts could be detected in the condensed capsular decidua or in the attenuated uterine wall.

DISCUSSION

Microvascular corrosion casts of the maternal part of the rat placenta has previously been described by Takemori et al. (1984; 1985) and Christofferson and Nilsson (1989), but these investigations do not focus on angiogenesis by sprouting. With the technique applied in this study, most casts were successful in terms of retainment of the gross anatomical outline of the uterus, the well-preserved morphology of endothelial cell nuclear imprints, and the apparently normal morphology of the uterus and conceptus in histological sections. Such histological sections were necessary when correlating the casts with the morphology of the uterus and the conceptus (Fig. 20). It is concluded that the applied methods were relevant for the question at issue, reliable and reproducible.

Calculations of the implantation site maternal blood volume revealed a 100-fold increase in volume from implantation to day 22 of pregnancy. This approach has several confounding factors, however, e.g. What is an implantation site? How much of this increase in blood volume represents a sluggish blood flow in large vessels and how much represents an increase in the number of exchange vessels? Despite such uncertainties, the data imply that there is a considerable angiogenesis during pregnancy in the rat.

Mesometrial angiogenesis by sprouting was observed 2-3 days after implantation and ceased when the first maternal blood spaces appeared in the trophoblast. The sprouts did not join to form a new vascular bed, but seemed to become incorporated into the maternal blood spaces. Since the vessels in this area exhibit migratory endothelial cells and an attenuated basal membrane (Welsh and Enders, 1991), it is suggested that the mesometrial angiogenesis is a process intended to facilitate trophoblast access to the maternal circulation after the opening-up of sinusoids into the uterine lumen.

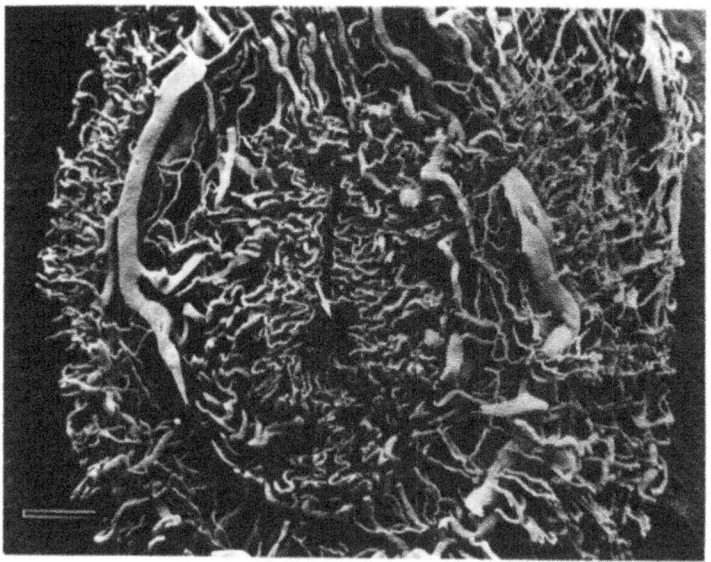

Figure 9. Implantation triggers a paradoxical shut-down of the vessels in the primary decidual zone, leading to an avascular area (arrow) surrounding the blastocyst. Implantation site, 26 hours after activation from an experimental delay of implantation. Vascular cast. Bar = 300 μm. Reprinted by permission of Springer-Verlag from Christofferson and Nilsson (1988a).

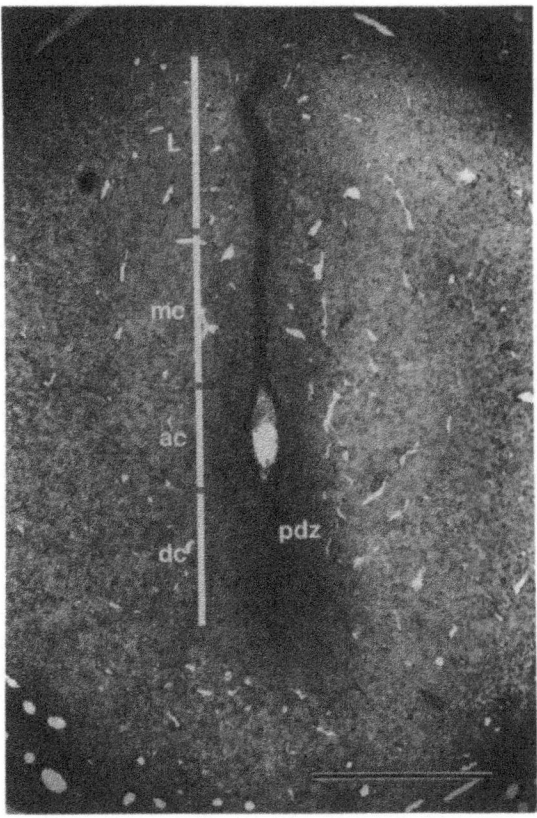

Figure 10. The implantation chamber is the antimesometrial extension of the uterine lumen (L) and can be divided into three parts, a mesometrial chamber (mc), an antimesometrial chamber (ac) with the blastocyst, and a decidual crypt (dc). Note the absence of blood vessels in the primary decidual zone (pdz). Light micrograph, 50 hours after activation from an experimental delay of implantation. Bar = 500 μm.

Antimesometrial angiogenesis by sprouting was also observed 2-3 days after implantation and ceased when the antimesometrial lumen had formed. The sprouts exhibited signs of fusion, and the microvascular bed of the endometrium lining the uterine wall was restored. The antimesometrial angiogenesis can hence in part be seen as a restorative process.

The biological reason for the observed sprouting in the developing capsular decidua is more difficult to explain, since the capsular decidua degenerated and had ruptured on day 16. One interpretation would be that all the sprouting observed in this study represents the action of a diffusible angiogenic substance (Torry and Rongish, 1992) affecting all capillaries and venules in the decidua. The origin of such a substance is difficult to determine, since the decidua exhibits several different cell types. One or several angiogenic substances may hence — primarily or secondarily — be secreted by the trophoblast, the uterine epithelium, the decidual cells, macrophages, mast cells, large granular lymphocytes, or by intravascular lymphocytes, granulocytes, platelets, or even the endothelial cells themselves.

"Angiogenesis" has become synonymous with angiogenesis by sprouting from pre-existing vessels (Folkman and Shing, 1992), but can also occur by at least two other processes; in situ-formation from angioblastic mesenchyme (so-called vasculogenesis; Risau, 1991), and by intussusception (Patan et al., 1993). In this study, neither process gave rise to the maternal blood spaces of the placenta. Thus, it is likely that the maternal blood spaces of the zona intima (which constitute the largest part of the maternal placenta at term) are formed by yet another process. In vascular casts, the maternal blood spaces appeared as irregular, anastomosing "blebs" of varying size interconnected between the subepithelial capillary

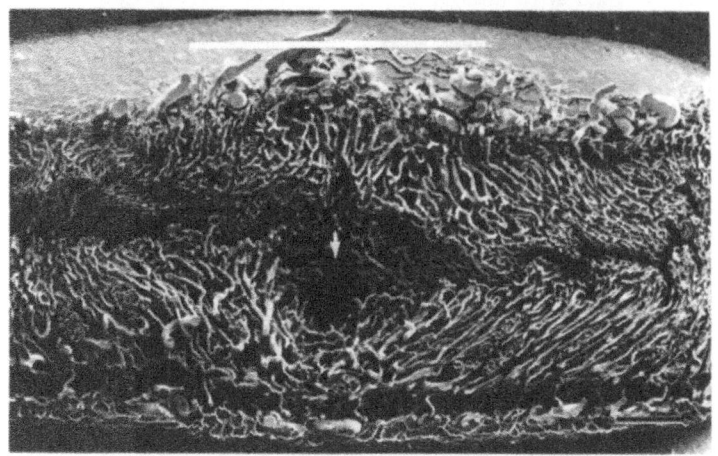

Figure 11. The first two vascular reactions to implantation, the increase in vascular permeability and the vascular shut-down, were observed in casts as spacing of vessels at the implantation site (white bar) with concomitant bunching at inter-sites, and an avascular area (arrow) corresponding to the primary decidal zone surrounding the implanting blastocyst. Vascular cast, longitudinal section, 50 hours after activation from an experimental delay of implantation. Bar = 500 μm. Reprinted by permission of Springer-Verlag from Christofferson and Nilsson (1988a).

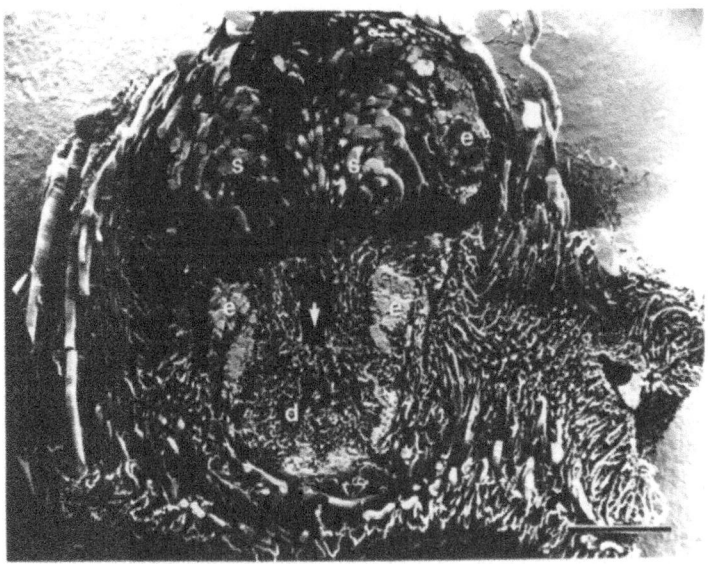

Figure 12. After implantation, angiogenesis occurs by sprouting from the mesometrial sinusoids (s), and from venules and capillaries in the antimesometrial decidua (d). Note extravasations of resin (e) and the avascular area (touched in section; arrow). Vascular cast, longitudinal section, day 8 of pregnancy. Bar = 1 mm.

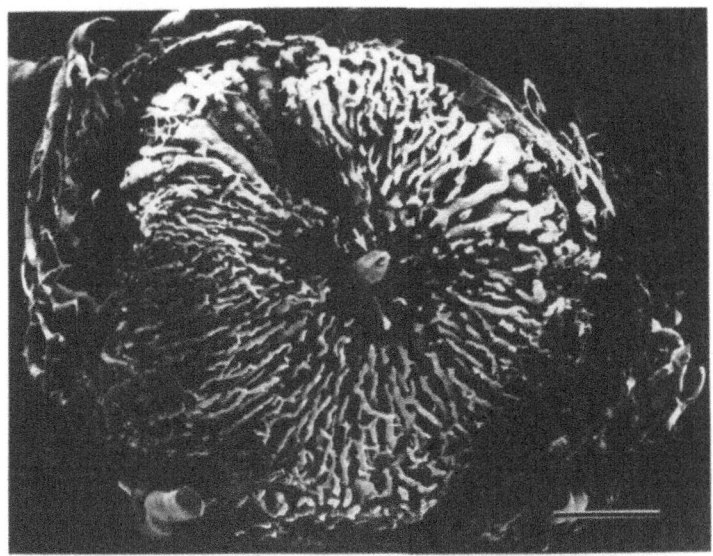

Figure 13. Mesometrial angiogenesis. Numerous vascular sprouts develop from the sinusoids and elongate towards the mesometrial chamber, here represented by extravasated resin (arrow). The uterine lumen and the mesometrial chamber serve as a transient blood conductor, since the sinusoids and the mesometrial part of the subepithelial capillary plexus open up into the lumen, and expose the Träger trophoblast and mural trophoblast to maternal blod from day 8-9 of pregnancy. Vascular cast, polar section seen from antimesometrially, day 8 of pregnancy. Bar = 1 mm.

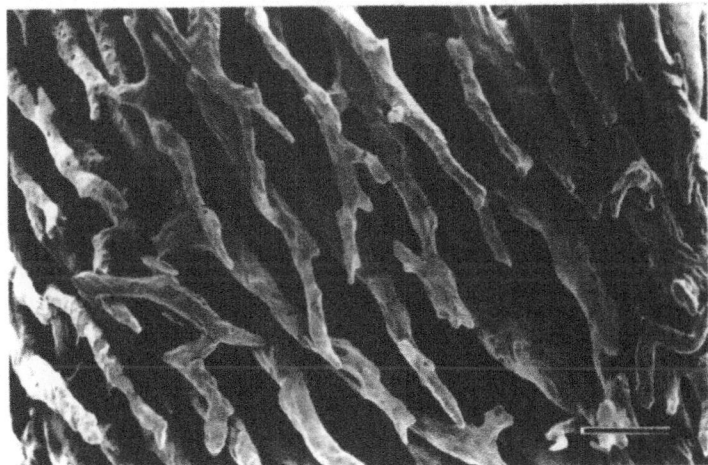

Figure 14. Mesometrial angiogenesis. The vascular sprouts appear as sharp-pointed, blindly ending extensions from the mesometrial sinusoids and point towards the mesometrial chamber. Vascular cast, day 8 of pregnancy. Bar = 100 μm.

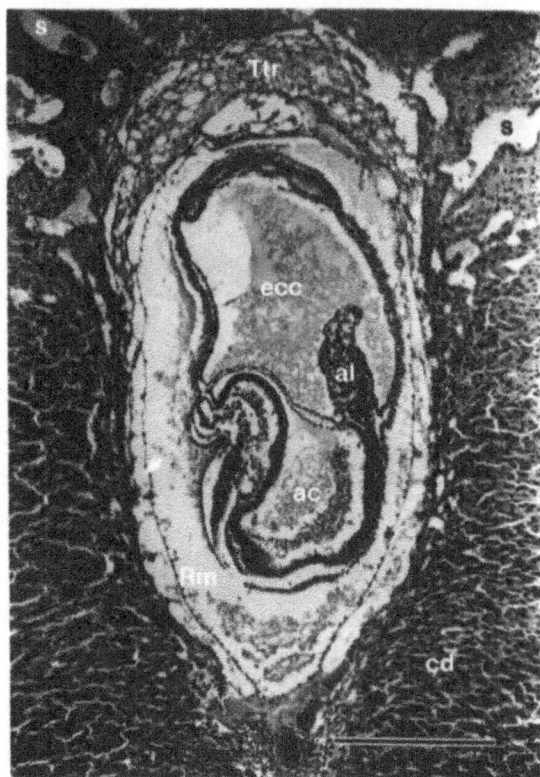

Figure 15. Formation of maternal blood spaces. Maternal blood is circulating in the yolk sac placenta, located in the mural trophoblast (arrow), and in the blood spaces of the Träger trophoblast (Ttr) from day 8-9 of pregnancy. Arterialized maternal blood enters the remnant uterine lumen — which at this stage is occupied by the Träger — and reaches the maternal blood spaces and yolk sac placenta, which are drained by the mesometrial sinusoids (s). ac — amniotic cavity, al — allantois, cd — capsular decidua, ecc — exocelomic cavity, Rm — Reichert's membrane. Light micrograph of immersion-fixed, cast tissue, day 10 of pregnancy. Bar = 500 µm. Reprinted by permission (Copyright © John Wiley & Sons, Inc., 1989) from Christofferson and Nilsson (1989).

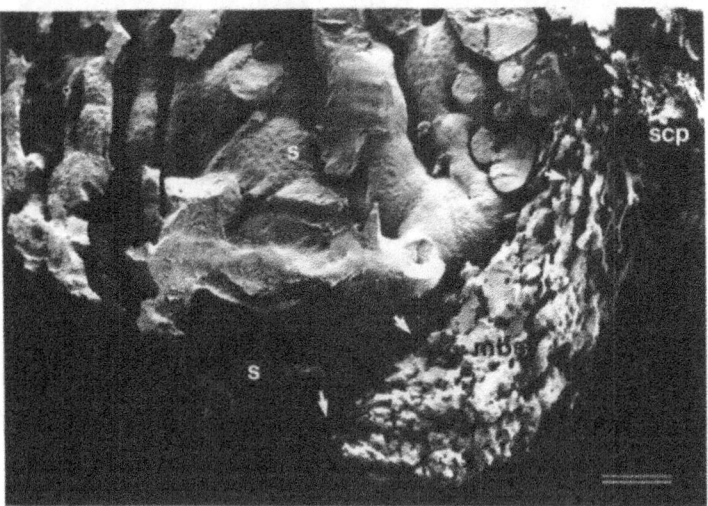

Figure 16. Formation of maternal blood spaces. The maternal blood spaces (mbs) were observed as irregular blebs of resin within the Träger trophoblast. The blebs were connected between the dilated subepithelial capillary plexus (scp), and the sinusoids (s) and their vascular sprouts (arrows). Vascular cast, day 12 of pregnancy. Bar = 200 µm.

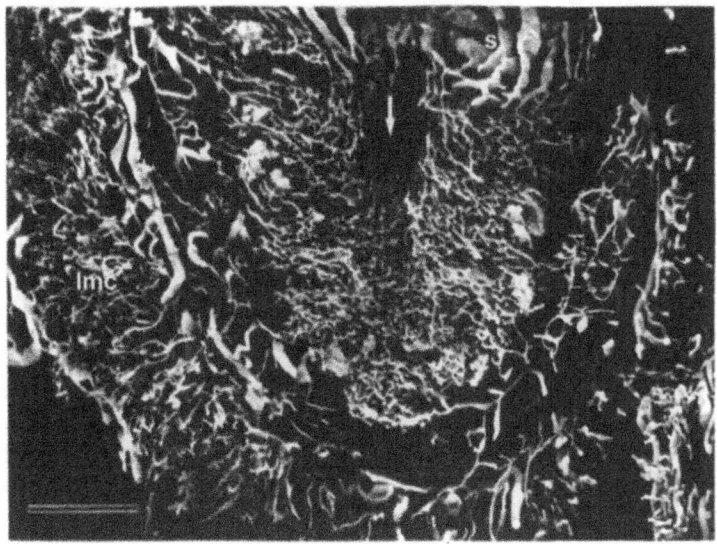

Figure 17. Antimesometrial angiogenesis. The vessels of the antimesometrial decidua are irregular in outline and exhibit numerous vascular sprouts. Compare endometrial vascular density with Fig. 11. lmc — longitudinal muscle capillary plexus, s — mesometrial sinusoid, arrow — avascular area. Vascular cast, cross-section, day 8 of pregnancy. Bar = 300 µm.

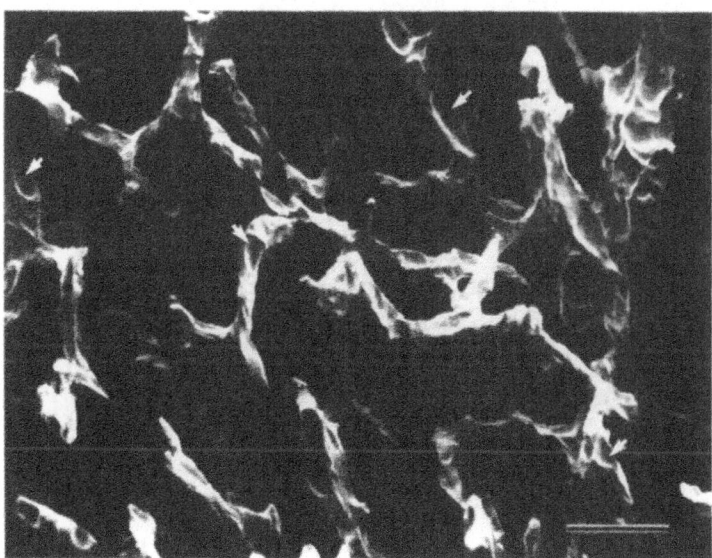

Figure 18. Antimesometrial angiogenesis. Note the sharp-pointed, blindly ending extensions from decidual capillaries and venules. Sprouts are oriented at random. The parental vessels are irregular in outline and exhibit prominent imprints of endothelial cell nuclei (some of which indicated by arrows), suggesting an increased metabolic activity of endothelial cells. Vascular cast, day 8 of pregnancy. Bar = 30 µm.

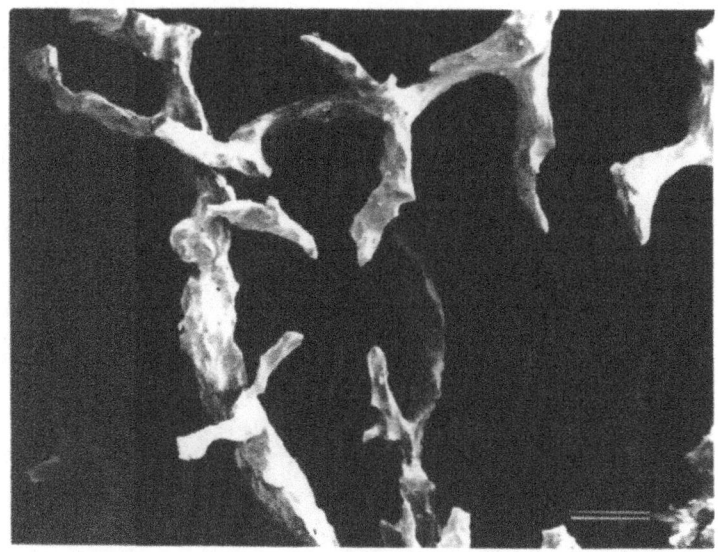

Figure 19. Antimesometrial angiogenesis. Vascular sprouts were frequently seen facing each other, which was interpreted as a sign of a future fusion between them. Vascular cast, day 8 of pregnancy. Bar = 30 μm.

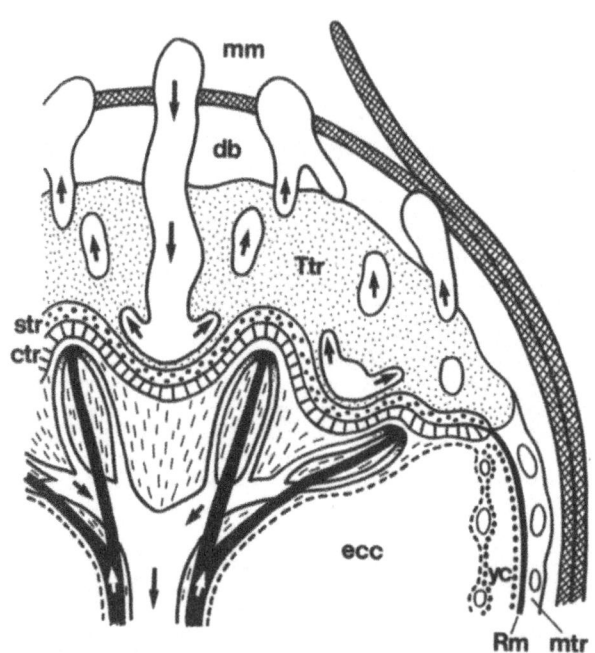

Figure 20. Schematic representation of the rat placenta on day 14 of pregnancy. Arrows indicate the direction of blood flow. Stippled: Träger trophoblast (Ttr), cross-hatched: myometrium, hatches: fetal mesoderm. db — decidua basalis, ctr — cytotrophoblast, ecc — exocelomic cavity, mm — mesometrium, mtr — mural trophoblast with maternal blood spaces of the parietal yolk sac, Rm — Reichert's membrane, str — syncytiotrophoblast, and yc — yolk sac cavity. Modified from Mossman and Fischer (1969).

plexus and the mesometrial sinusoids and their vascular sprouts, and appeared in the area where Träger trophoblast were identified in light microscopical sections. It is concluded that a transmission electron microscopic study focussed on the interface between trophoblast and the vessels of the mesometrial decidua on days 10-14 of pregnancy is required to determine how the maternal blood spaces are formed.

ACKNOWLEDGMENTS

Excellent technical assistance was provided by Barbro Einarsson, Leif Ljung, Marianne Ljungkvist, and Lena Kårud. This work was supported by the Swedish Medical Research Council (project No. 00070), the Swedish Society of Medicine, and grants from the Faculty of Medicine at Uppsala University.

REFERENCES

Bergström, S., 1978, Experimentally delayed implantation, *in:* "Methods in Mammalian Reproduction", J.C. Daniels ed., Academic Press, New York, pp. 419-435.

Bøe, F., 1950, Studies on placental circulation in rats. II. Vascular pattern illustrated by corrosion preparations, *Acta Endocrinol (Copenh)*. 5:369-375.

Bridgman, J., 1948a, A morphological study of the development of the placenta in the rat. I. An outline of the development of the placenta of the white rat, *J Morphol*. 83:61-87.

Bridgman, J., 1948b, A morphological study of the development of the placenta in the rat. II. An histological and cytological study of the development of the chorioallantoic placenta of the white rat, *J Morphol*. 83:195-225.

Christofferson, R.H. and Nilsson, B.O., 1988a, Morphology of the endometrial microvasculature during early placentation in the rat, *Cell Tissue Res*. 253:209-220.

Christofferson, R.H. and Nilsson, B.O., 1988b, Microvascular corrosion casting with analysis in the scanning electron microscope, *Scanning* 10:43-63.

Christofferson, R.H. and Nilsson, B.O, 1989, Placentation in the rat: a SEM study of microvascular casts, *in:* "Developments in Ultrastructure of Reproduction", P.M. Motta, ed., Alan R. Liss Inc., New York, pp. 435-442.

Christofferson, R.H. and Nilsson, B.O., 1992, Microvascular corrosion casting in angiogenesis research, *in:* "Scanning Electron Microscopy of Vascular Casts: Methods and Applications", P.M. Motta, T. Murakami, and H. Fujita, eds., Kluwer Academic Publishers, Boston, pp. 27-37.

Duval, M., 1891, Le placenta des rongeurs, *J Anat. (Paris)* , 27:24-73, 334-395, 515-612.

Enders, A.C., 1991, Current topic: Structural responses of the primate endometrium to implantation, *Placenta* 12:309-325.

Folkman, J. and Shing, Y., 1992, Angiogenesis, *J Biol Chem*. 267:10931-10934.

Holmes, R.P. and Davies, D.V., 1948, The vascular pattern of the placenta and its development in the rat, *J Obstet Gynaecol Brit Emp*. 55:583-607.

Lametschwandtner, A., Lametschwandtner, U., and Weiger, T., 1990, Scanning electron microscopy of vascular corrosion casts — Technique and applications: Updated review, *Scanning Microsc*. 4:889-941.

Lundkvist, Ö. and Ljungkvist, I., 1977, Morphology of the rat endometrial stroma at the appearance of the Pontamine blue reaction during implantation after an experimental delay, *Cell Tissue Res*. 184:453-466.

Mossman, H.W., 1987, "Vertebrate Fetal Membranes", Macmillan, London.

Mossman, H. W. and Fischer T.V., 1969, The preplacenta of pedetes, the träger, and the maternal circulatory pattern in rodent placentae, *J Reprod Fertil, Suppl*. 6:175-184.

Murakami, T., 1971, Application of the scanning electron microscope to the study of the fine distribution of the blood vessels, *Arch Histol Jap*. 32:445-454.

Nutting, E.F. and Meyer, R.K., 1963, Implantation delay, nidation and embryonal survival in rats treated with ovarian hormones, *in:* "Delayed Implantation", A.C. Enders, ed., University of Chicago Press, Chicago, pp. 233-252.

Patan, S., Haenni, B., and Burri, P.H., 1993, Evidence for intussusceptive capillary growth in the chicken chorio-allantoic membrane (CAM), *Anat Embryol*. 187:121-130.

Psychoyos, A., 1971, Methods for studying changes in capillary permeability of the rat endometrium, *in:* "Methods in Mammalian Embryology", J.C. Daniel, ed., W.H. Freeman and Co, San Francisco, pp. 334-338.

Risau, W., 1991, Embryonic angiogenesis factors, *Pharmacol Ther*. 51:371-376.

Rogers, P.A.W. and Gannon, B.J., 1981, The vascular and microvascular anatomy of the rat uterus during the oestrus cycle, *Aust J Biol Sci.* 59:667-679.

Rogers, P.A.W., Murphy, C.R., and Gannon, B.J., 1982, Changes in spatial organization of the uterine vasculature during implantation in the rat, *J Reprod Fertil.* 65:211-214.

Takemori, K., Okamura, H., Kanzaki, H., Koshida, M., and Konishi, I., 1984, Scanning electron microscopy study on corrosion cast of rat uterine vasculature during the first half of pregnancy, *J Anat.* 138:163-173.

Takemori, K., Okamura, H., Kanzaki, H., Koshida, M., Konishi, I., and Mori, T., 1985, Scanning electron microscopy study on corrosion casts of rat uterine vasculature during the second half of pregnancy and post partum, *J Anat.* 142:21-31.

Torry, R.J. and Rongish, B.J., 1992, Angiogenesis in the uterus: potential regulation and relation to tumor angiogenesis, *Am J Reprod Immunol.* 27:171-179.

Welsh, A.O. and Enders A.C., 1991, Chorioallantoic placenta formation in the rat: II. Angiogenesis and maternal blood circulation in the mesometrial region of the implantation chamber prior to placenta formation, *Am J Anat.* 192:347-365.

Young, A., 1956, The vascular architecture of the rat uterus during pregnancy, *Trans R Soc Edinb.* 63:167-184.

RECEPTIVE AND REFRACTORY PERIOD IN HUMAN IMPLANTATION

J. Mandelbaum[1], M. Plachot[1], A.M. Junca[1],
J. Cohen[2], J. Salat-Baroux[3]

[1] U 173 Hôpital Necker,
149, Rue de Sèvres 75743 Paris Cédex 15 - France

[2] CHI Jean Rostand - Sèvres (92)

[3] Hôpital Tenon - 75020 Paris

INTRODUCTION

The successful establishment of pregnancy after embryo transfer requires a healthy blastocyst and a uterus that accepts it.

In rodents, it has been well established that the synchrony between embryo development to the blastocyst stage and the maturation of the uterine endometrium is very important for a successful implantation. The timing and hormonal conditions of nidation are well defined. (Psychoyos, 1976).

First, priming of the endometrium by progesterone for about 48 hours is required to establish a prereceptive neutral phase. During this phase, the uterus allows a transferred blastocyst to survive and even wait for the receptive phase also called "window of implantation". The receptivity period is triggered in rodents by estrogens and is a brief event that after 36 hours automatically leads to the refractory phase during which the endometrium is no longer capable of decidual response and during which the uterine environment becomes insensitive and even hostile to unimplanted eggs. This transition tooks place regardless of the occurrence of implantation during the receptive phase.

In all other mammals studied, the results of asynchronous transfers of embryos into the uterus of surrogate mothers underline the importance of a chronological relationship between the age of the embryo and the developmental stage of the endometrium.

I. THE NEUTRAL PRERECEPTIVE PHASE IN HUMANS

In humans, the embryo issued from in vitro fertilization should be transferred in utero at the moment of its physiological entry in the uterine cavity in vivo. This moment was investigated on short series and seems to occur around the fourth day following fertilization. The youngest normal embryos recovered in fertile women, after uterine flushing, were a 12 cell (Hertig and Rock, 1949) and a 16 cell- morulae (Diaz et al, 1980), the latter collected 120 hours after the LH surge, whereas 8 cell-embryos were the most advanced embryonic stages observed in vivo in the tubes.

In fact, it is possible to transfer in utero mature oocytes inseminated 2 to 4 hours earlier as was reported by Rizk et al (1990) and obtain 20% of pregnancy per transfer. Similar results were also reported by Ajuha et al in 1985 when replacing pronuclear zygotes the day after insemination. Cleaved embryos at day 2 post insemination led to 18% of pregnancy per transfer according to the pooled french data of FIVNAT 1993 and, at day 3 post-insemination, embryos at the second or third cleavage stage gave identical results as 2 days old embryos on an homogenous series reported by Belaisch-Allart et al (1990), 32 and 29% of pregnancies per transfer respectively.

These early transfers are widely used in all IVF centers in the world, and are mainly performed at the second day following insemination and fertilization with embryos at the 2 to 4 cell stage.

Some attempts were carried out to transfer late embryos in order to better mimic the in vivo chronology of embryo-endometrium interactions.

Zygotes are cultured on cell monolayers, monkey kidney epithelial cells (called Vero cells), human tubal cells and even bovine oviductal epithelial cells, in order to improve embryo viability in vitro and obtain 50% or more of morulae and blastocysts at day 5 post insemination. These techniques of cocultures significantly improve the percentage of zygotes reaching the blastocyst stage in vitro (Menezo et al, 1992 ; Bongso et al, 1992 ; Wiemer et al, 1993) when compared to conventional cultures (Bolton et al, 1991).

However, at the moment, the results of such late transfers do not show any statistical improvement in the pregnancy rate when compared to earlier transfers, at least in randomized series (Guerin et al, 1991). Thus, the uterine milieu is tolerant to the young developing embryo in humans as in monkeys (Marston et al, 1977) from the day of fertilization on, and if the synchrony between embryo growth and endometrium maturation is maintained, the blastocyst will hatch during the receptive phase of the uterus and be able to implant.

One can postulate that in humans the **neutral prereceptive phase** of the uterus do extend to the period of the embryo tubal transport, but this does not prove that

Table 1. Uterine transfer of embryos in humans at various times following oocyte recovery

Day of transfer	Embryos	Pregnancy per transfer (%) D0	D1	D2	D3	D5
Day 0	oocytes + Zona bound sperms	20% (Rizk et al, 1990)				
Day + 1	zygotes	(Ahuja et al, 1985)	21%			
Day + 2	1e and 2e cleavage	(FIVNAT, 1993)		18%		
Day + 3	2e and 3e cleavage	(Belaïsch-Allart et al, 1990)			29% ...	32%
Day + 5	Morulae and Blastocysts	(Guérin et al, 1991)			17.8%	20%

Day 0 = the day of oocyte recovery and insemination.

Table 2 . Comparative results of assisted procreations in France : embryo uterine transfer (FIV) or zygote tubal transfer (ZIFT). Data collected on 108 IVF centers (FIVNAT, 1993).

	FIV (1990)	ZIFT (1990)
oocyte recovery (**or**)	23883	253
clinical pregnancy per transfer	24.8%	23.7%
clinical pregnancy per **or**	19.7%	20.2%
delivery **per or**	15%	20.2%

Such results again point out the existence of a long prereceptive uterine phase in the humans.

transferring human embryos in the uterine cavity at the time they should have been in the tubes represents the best moment and the best place.

To avoid these "inadequate" transfers, many teams tried to transfer in the tubes zygotes (ZIFT : zygote intra-fallopian transfer) or early embryos (TEST, TET : tubal embryo stage transfer). These procedures are applied to patients with patent tubes.

Nevertheless, only few groups showed an advantage to these tubal transfers (Yovich, 1990) and in the survey of french IVF results no difference appeared in the birth rate per oocyte recovery whether IVF or ZIFT were performed.

II) THE UTERINE RECEPTIVE PHASE : IMPLANTATION WINDOW

1) Asynchronous transfers in mammals

The existence and duration of the receptive phase of the endometrium to nidation were studied in several species by means of asynchronous transfers.

In most species, this kind of studies were performed with late embryos, morulae and blastocysts just before the moment of their implantation and therefore, the degree of asynchrony between embryo and endometrium ages that keeps tolerable and does not impair the pregnancy rate may define the time limits of the implantation window.

When asynchronous transfers are performed with early embryos, such an extrapolation becomes less accurate as interactions between the embryo and the endometrium may occur. The embryo can modify by its secretion the uterine receptivity, whereas the uterus may have an effect on embryo development.

In rodents, a high synchrony between the embryo and the recipient is required (Chang, 1950 ; Dickmann and Noyes, 1960 ; Doyle et al, 1963 ; Fischer et al, 1989). When asynchrony between endometrium and embryonic development exceeds 2 days, embryo survival is drastically impaired. During a short period, the blastocyst adjusts its growth to the maternal environment : when transferred in an older uterus, cell proliferation and embryo growth is enhanced whereas a slowing down appears in the development of blastocysts transferred in a "younger" uterus. Nevertheless, important alterations soon occur in the morphology and growth of the asynchronously transferred embryos.

In the pig, Polge (1982) showed that it is possible to transfer blastocysts in a 2 day-younger uterus whereas when transferred in a 2 day-older recipient, embryos degenerate rapidly and the pregnancy rate decreased from 77% to 9%.

In sheep, embryos can respond to a small degree of asynchrony by an adjustement of their developmental rate in order to recover a state of growth corresponding to the age of the endometrium (Wilmut et al, 1986). If the asynchrony exceeds 2 days, embryos display anomalies, do not implant and degenerate.

In cattle, Geisert et al (1991) obtained 61% of pregnancies per synchronous transfers of blastocysts recovered in vivo at Day 8. If the transfer was performed in recipients which estrus took place 72 hours after the donor's one, the pregnancy rate decreased to 4.8%. There too, better results were obtained with synchronous transfers.

In monkeys, Hodgen (1983) by surrogate tubal transfers of embryos at various developmental stages in castrated and treated females, could estimate the duration of the implantion window to 3 days.

It is more difficult to define the limits and duration of the receptive phase in humans.

2) The implantation window in humans

Though it is easy to link the neutral phase to the tubal transport and intrauterine embryo development, it is more difficult to define accurately the limits and duration of the receptive phase in humans as asynchronous transfers are impossible in conventional IVF where embryos are transferred in their own mothers. A dissociation between embryo and endometrium ages could be realized in two circumstances : first, in frozen-thawed embryo transfers and second in oocyte donation programs.

In theses 2 cases, it became possible, in humans, to perform asynchronous transfers and therefore study the window of transfer, which is supposed to reflect the window of implantation.

- asynchronous frozen-thawed embryo transfers

In a search of the implantation window limits, we carried out synchronized and desynchronized frozen-thawed embryo transfers (Mandelbaum et al, 1988, 1990).

As other groups, we routinely used, since 1986, embryo freezing for excess embryos resulting from IVF.

The protocol, widely described, comprised propanediol and sucrose as cryoprotectants, diluted in B2 medium supplemented with 20% maternal serum.

Seeding was performed manually in a programmable freezer and the procedure was applied on cleaved embryos 40 hours post-insemination.

During a 6 year period (1986-1992), 69% of frozen embryos displayed after thawing 50% at least of their blastomeres intact, which represents their survival rate.

Eighty eight per cent of the patients having an attempt of embryo thawing could be replaced with a mean number of 2.2 embryos. And on a large series of 2489 transfers, the pregnancy rate reached 14.4% resulting in 359 clinical pregnancies with an implantation rate by transferred embryo of 7% and per thawed embryo of 5%.

Transfers were performed either in spontaneous, stimulated or artificial cycles.

In spontaneous cycles, to determine the day of transfer, patients were monitored from day 10 by morning daily plasma LH and estradiol assays, sometimes progesterone

Table 3 . Transfers of frozen-thawed human embryos (spontaneous cycles)

Embryo-endometrium synchrony*	Transfers (T)	Embryo/T	Clinical pregnancies (%)
In synchrony	565	2.2	95 (17%)[a]
Out of synchrony	287	2.1	32 (10%)[b]

X2 test : a-b : p < 0.05

* according to the LH surge initiating rise (LH SIR).

Table 4 . Transfers of frozen-thawed human embryos (spontaneous cycles)

Embryo-endometrium synchrony *	Transfers (T)	Embryo/T	Clinical pregnancies (%)
0	565	2.2	95 (17%)
- 1	83	2.3	10 (12%)
+ 1	186	2.0	22 (12%)
+ 2	18	2.2	----------

X2 TEST : NS

* according to the LH surge initiating rise (LH SIR).

assays and also with the help of follicular ultra-sound. During a period of time, transfers were performed either in synchrony, when 2 days-old embryos were transferred on the fourth day following the LH surge initiating rise (LH SIR), or with a relative asynchrony. Asynchronous transfers could be carried out in a uterus one day "younger" than the embryo (-1) or one or two days older (+1 and + 2).

When comparing 565 synchronous and 287 asynchronous transfers, the uterus age being calculated according to the LH SIR, it appeared that, when ovulation could be accurately determined and for an identical number of embryos transferred, transfers in synchrony led to 17% of clinical pregnancies instead of 10% for transfers out of synchrony, the difference being statistically significant. (Table 3).

However, differences lost their signification when considering transfers in strict synchrony or with a one day shift only. On the contrary, no pregnancy arose from 18 transfers performed in a uterus 2 days older than the embryo, a procedure that was rapidly forsaken (Table 4).

Despite strict synchrony appears as the most favourable condition for embryo transfer in humans, implantation is still possible when the transferred embryo is one day older or younger than the uterus suggesting either a 3 day period of uterine receptivity or an adaptation of the embryo developmental speed to counteract the asynchrony as was demonstrated in sheep, pig and rabbit.

Implantation of frozen-thawed embryos and timing of the luteal phase

Animal experiments widely stated that endometrium receptivity to nidation depends on steroid environment and especially the timing of the sequence of estradiol and progesterone secretion. By exogenous administration of progesterone, it is possible to create a pharmacological asynchrony between a young embryo and an overmature treated endometrium.

This was done in rodents : Schacht and Foote (1978) reported that in the rabbit the preovulatory administration of progesterone decreased to 5% the implantation rate of embryos transferred to the uterus of treated recipients and in the rat Dickmann (1970) also obtained a dramatic decrease in the percentage of transferred morulae devoloping in fetuses when the administration of progesterone began in the preovulatory period. So, an overmature endometrium may be refractory to implantation .

The first day of the luteal phase can be determined by daily plasma progesterone assays and occurs when progesterone rises up above 1ng/ml and goes on increasing. Therefore, transfers of thawed embryos that have been frozen 2 days after insemination should occur on the first or second day following the progesterone rise.

In the course of 376 transfers of frozen-thawed embryos performed in spontaneous cycles, progesterone was assessed daily in plasma.

Table 5 . Transfers of frozen-thawed human embryos (spontaneous cycles)

Day of transfer	Transfers (T)	Embryo/T	Clinical pregnancies (%)
T0*	17	2.8	4 (24%)
T -1	140	2.2	24 (17%)a
T -2	122	2.1	19 (16%)b
T -3	84	2.4	10 (12%)c
T -4	13	2.5	3 (23%)

TO : Transfer performed the first day of the rise in plasma progesterone ($\geq$ 1 ng/ml)
$X2$ = a, b, c NS.

The progesterone rise could occur the day of the transfer to or 1 (To) 4 days earlier T-1, -2, -3, -4 (Table 5).

No significative difference emerged in the clinical pregnancy rate during a 3-day period corresponding to the second to fourth days of elevated progesterone concentrations in plasma. Moreover, and despite the short series, no reduction in the pregnancy rate was noted when transfers were performed the day of progesterone rise or 4 days later. The relative increase in the pregnancy rate could perhaps be related to the higher mean number of embryos transferred in these 2 groups. These results, as the data obtained with asynchronous transfers, again suggest either a large receptive phase in humans lasting at least 3 days or a remarkable ability of the embryo to adapt its growth to the uterine environment.

- transfers of embryos resulting from oocyte donation

In the same way, embryo transfers following oocyte donation in women deprived of endogenous ovarian function led to successful implantations even when 2 to 8 cell embryos were transferred 2, 3, 4 or 5 days after the beginning of the progesterone supply representing the artificial luteal phase : 30% of pregnancy per transfer when 2 days-old embryos were transferred on day 2 of the progestative phase as reported in our collaborative group by Salat-Baroux et al (1988) ; seventeen to thirty per cent on the data that Scott and Rosenwaks could collect in a multicentric study (1990).

Nevertheless, even if the receptive period lasts at least 3 days in humans, it ends and Navot et al (1986) and Rosenwaks (1987) reported that no evolutive pregnancy arose from transfers performed after the fifth day of progesterone substitution which suggests that the replaced embryos reached their own maturity after the entry of the uterus into the refractory period and could have been prevented from implanting.

III THE POST-RECEPTIVE REFRACTORY PERIOD

The uterus that has already experienced the receptive phase, automatically enters in the refractory phase regardless of the occurence of implantation (Yoshinaga, 1988).

It has been reported that the uterus in the refractory phase is not only indifferent but also, in some species and especially in rodents, even hostile to the blastocyst. Consequently, this embryotoxicity could explain the failure of asynchronous transfers.

In the rabbit, as related by Adams (1971), day 6 blastocysts transferred into the uterus of a pseudopregnant female at day 9, 10 or 11, degenerate. A 24 hour exposition of 6 days-old blastocysts into an older uterus leads to an arrest in embryo development even after a further transfer in a synchronous recipient.

In the rat too, as reported by Psychoyos and Casimiri (1981) day 5 blastocysts degenerate and are expelled in the vagina when transferred in a one day older uterus. Moreover, cultured with the flushings of day 6 uterus for 24 hours, they degenerate.

Is the embryotoxicity of the uterine environment in the refractory period observed in rodents applicable to humans ?

Psychoyos et al, (1989), collecting flushings from human uteri in the late luteal phase (day 22 to day 25) showed a strong embryotoxic effect in vitro on rat blastocysts. We observed the same effect on a model of human triploid embryos cultured in straws in aspirates of human fluids from the late luteal phase (Mandelbaum et al, 1990).

In 1989, Psychoyos and colleagues reported that flushings from human uteri contained a dialysable component of low molecular weight responsible for this embryotoxic effect.

Analysis in HPLC showed two main peaks with the same retention time as taurocholic acid and cholic acid. Taurine and taurocholic acid displayed no embryotoxic effect in vitro.

On the contrary, cholic acid induced in vitro, at a concentration of 0.2 M, the degeneration of 100% of rat blastocysts within 24 hours.

The cholic acid content per human uterine sample reached 100 mcg for an average volume of fluid of 40 μl in the late luteal phase, representing 6 mM ; these concentrations 30 fold higher than those displaying an embryotoxicity in rats in vitro, also contrasted with the low levels of this acid in the blood serum which excluded the possibility of a contamination of the uterine samples by cholic acid originating from blood.

In an attempt to determine if cholic acid exerts the same embryotoxicity in humans as in rats, we used, as model, the human triploid embryos.

Human triploid embryos exhibiting 3 pronuclei the day after fertilization, are obtained in our routine IVF program as they represent 5% of all fertilized eggs ; they are obviously discarded from transfer.

Two days after insemination, cleaved triploid embryos at 2 to 4 cell stage, without cytoplasmic fragmentation were transferred in B2 medium supplemented by 15% maternal serum and containing or not cholic acid at concentrations of 0.2, 2 and 6 mM. The addition of cholic acid did not modify medium's pH and osmolarity.

Embryo development was checked daily until day 5 to 6 post-insemination all embryos were endly examined by direct Tarkowski's (1986) technic to assess the number of nuclei and differentiate embryo development from embryo regular fragmentation. Almost half of the control triploid embryos (48%), cultured in B2 enriched with serum were able to cleave twice or more at least and reach the eight cell to morula stage. Only 19% were blocked and 33% experienced only one segmentation before their developmental arrest. No more than 3% were able to develop into a blastocyst in the absence of feeders. When cultured in the presence of cholic acid, human triploid embryos were rapidly arrested in their growth and this effect increased with acid cholic concentrations : 32% for 0.2 mM, 44% with 2 mMol, 79% with 6 mM. At 6 mMol, concentrations similar to what Psychoyos et al (1989) found in the late luteal human uterus, 100% of triploid embryos did not go beyond one mitotic cycle and 79% of such embryos were highly degenerative when compared to control embryos. It was a particular degenerative aspect as the embryos displayed an intact zona pellucida containing clear blastomeres with weakened plasma membranes. Cholic acid thus appears as an embryotoxic factor in humans as well as in rodents. Many questions, however remain unsolved : what is the origine of uterine taurocholic and cholic acids which are bile acids, presumed to be only produced by the liver ? If cholic acid is really present at high concentrations during the post receptive period, does it exert an embryotoxic effect in vivo as in vitro ? And therefore, can it be involved in transfer failures or responsible for a particular kind of sterility ? The answer to those questions could add to our knowledge of the implantation window in humans.

References

Adams C.E., 1971, The fate of fertilized eggs transferred to the uterus or oviduct during advancing pseudopregnancy in the rabbit, *J. Reprod. Fertil.*, 26 : 99-111.

Ahuja K.K., Smith W., Tucker M., Craft I., 1985, Successful pregnancies from the transfer of pronucleate embryos in an outpatient in vitro fertilization program, *Fertil. Steril.*, 44, 181-184.

Belaisch-Allart J., Briot P., Allart J.P., Dufetre C., Mussy M.A., Adle F., Soria B., 1990, The value of embryo transfer three days after oocyte retrieval, *Contracept. Fertil. Sex.*, 18, 612-613.

Bolton V.N., Wren M.E., Parsons J.H., 1991, Pregnancies after in vitro fertilization and transfer of human blastocysts, *Fertil. Steril.*, 55, 830-832.

Bongso A., Ng S.C., Fong C.Y., Anandakumar C., Marshall B., Edirisinghe R., Ratnam S., 1992, Improved pregnancy rate after transfer of embryos grown in human fallopian tubal cell coculture, Fertil. Steril, 58 : 569-574.

Chang M.C., 1950, Development and fate of transferred rabbit ova or blastocyst in relation to the ovulation time in recipients, *J. Exp., Zool.*, 114 : 197-216.

Diaz S., Ortiz M.E., Croxatto H.B., 1980, Studies on the duration of ovum transport by the human oviduct III Time interval between the luteinizing hormone peak and recovery of ova by transcervical flushing of the uterus in normal women, *Am. J.Obstet. Gynecol.*, 137 : 116-121.

Dickmann Z., Noyes R.W., 1960, The fate of ova transferred into the uterus of the rat, *J. Reprod. Fertil.*, 1 : 197-212.

Dickmann Z., 1970, Effects of progesterone on the development of the rat morula, *Fertil. Steril.*, 21 : 541-548.

Doyle L.L., Gates A.H., Noyes R.W., 1963, Asynchronous transfer of mouse ova, *Fertil., Steril.*, 14 : 215-225.

Fischer B., 1989, Effects of asynchrony on rabbit blastocyst development, *J. Reprod. Fertil.*, 86 : 479-491.

FIVNAT, 1993, French national IVF registry : analysis of 1986 to 1990 data, *Fertil. Steril.*, 59 : 587-595.

Geisert R.D., Fox T.C., Morgan G. L.,Wells M.E., Wettemann R.P., Zavy M.T., 1991, Survival of bovine embryos transferred to progesterone treated asynchronous recipients, *J. Reprod., Fertil.*, 92 : 475-482.

Guérin J.F., Mathieu C., Pinatel M.C., Régnier-Vigouroux G., Lornage J., Boulieu D., Roudon T., Nachury L, Cognat M., Menezo Y., 1991, Coculture of human embryos with monkey kidney epithelial cells : clinical data concerning transfers delayed at D3 and D5, *Contracept. Fertil. Sex.*, 19 : 635-638.

Hertig A.T., Rock J., 1949, A series of potentially abortive ova recovered from fertile women prior to the first missed menstrual period, *Am. J. Obstet. Gynecol*, 58 : 968.

Hodgen G.D., 1983, Surrogate embryo transfer combined with estrogen-progesterone therapy in monkeys. *J. Am. Med. Assoc.*, 250 : 2167-2171.

Mandelbaum J., Junca A.M., Plachot M., Cohen J., Salat-Baroux J., 1988, Timing of embryo transfer and pregnancy rate in humans, *Reprod., Nutr. Develop.*, 28 : (6B), 1763-1772.

Mandelbaum J., Junca A.M., Plachot M., Cohen J., Alvarez S., Cornet D., Alnot M.O, Salat-Baroux J., 1990, The implantation window in human after fresh or frozen-thawed embryo transfers *in* "Advances in Assisted Reproductive

Technologies", eds Mashiach S., Ben-Rafael Z., Laufer N., Schenker J.G., Plenum Press., 729-735.

Marston J.M., Penn R., Sivelle P.C., 1977, Successful auto-transfer of tubal eggs in the rhesus monkey (Macaca mulatta), J. Reprod. Fertil., 49 : 175-176.

Menezo Y., Hazout, Dumont M., Herbaut N., Nicollet B., 1992, Coculture of embryos on Vero cells and transfer of blastocysts in humans, *Human. Reprod.*, 7 : 101-106.

Navot D., Laufer N., Kopolovic J., Rabinowitz R., Birkenfeld A., Lewin A., Granat M., Margalioth E.J., Shenker J.G., 1986, Artificially induced endometrial cycles and establishement of pregnancies in the absence of ovaries, *New Engl. J. Med*, 314 : 806-811.

Polge C., 1982, Embryo transplantation in the pig. *In* Merieux Ch., Bonneau M. Embryo transfer in Mammals, 235-242.

Psychoyos A., 1976, Hormonal control of uterine receptivity for nidation, J. Reprod., Fertil. Suppl. 25 : 17-18.

Psychoyos A., Casimiri V., 1981, Uterine blastotoxic factors. *In* Cellular and Molecular aspects of implantation. SR Glasser and DW Bullock (eds), Plenum Press New York pp. 327-334.

Psychoyos A., Roche D., Gravanis A., 1989, Is cholic acid responsible for embryotoxicity of the post-receptive uterine environnment ? *Human Reprod*, 4 : 832-834.

Rizk B., Bekir J.S., Avery S., Smith S., Edwards R.G., 1990, Intrauterine replacement of oocytes with bound sperms : a pilot study, *Human. Reprod.*, Esco-Eshre Meeting, Milan, Abstract 18 p. 5.

Rozenwaks Z., 1987, Donor eggs : their application in modern reproductive technologies, *Fertil. Steril.*, 47 : 895-909.

Salat-Baroux J., Cornet D., Alvarez S., Antoine J.M., Tibi Ch., Mandelbaum J., Plachot M., 1988, Pregnancies after replacement of frozen-thawed embryos in a donation program, *Fertil., Steril.*, 49 : 817-821.

Schacht C. J., Foote R.H., 1978, Progesterone induced asynchrony and embryo mortality in rabbits, *Biol. Reprod.*, 19 : 534-539.

Scott R.T., Rosenwaks Z., 1990, Oocyte donation. State of the art 1989 *in* : "Advances in assisted reproductive technologies" eds, Maschiach S., Ben-Rafel Z., Laufer N., Schenker J.G., Plenum Press, 729-735.

Tarkowski A.K., 1986, An Air-Drying method for chromosome preparation from mouse eggs, *Cytogenetics*, 5 : 394-400.

Wiemer K.E., Hoffman D.I., Maxson W.S., Eager S., Muhlberger B., Fiore I., Cuervo M., 1993, Embryonic morphology and rate of implantation of human embryos following co-culture on bovine oviductal epithelial cells, *Human. Reprod*, 8 : 97-101.

Wilmut I., Sales D. I., Ashworth C.J., 1986, Maternal and embryonic factors associated with prenatal loss in mammals, *J. Reprod. Fertil.*, 76 : 851-864.

Yoshinaga K., 1988, Uterine receptivity for blastocyst implantation, *Ann. N.Y. Acad. Sci.*, 541 : 424-431.

Yovich J., 1990, Tubal transfers : Prost and Test *in* "Gamete Physiology" Eds Asch., R. M., Balmaceda J.P., Johnston I., Serono Symposia USA, 305-317.

IMPLANTATION OF HUMAN BLASTOCYSTS FOLLOWING IN VITRO FERTILISATION

Virginia N. Bolton

Assisted Conception Unit
Department of Obstetrics & Gynaecology
King's College School of Medicine & Dentistry
Denmark Hill
London SE5 8RX, UK

INTRODUCTION

There have been a number of advances in therapeutic *in vitro* fertilisation (IVF) since the first live birth resulted from this form of treatment in 1978 (Steptoe and Edwards 1978). The most notable have been clinical rather than scientific, and include the introduction of the use of LHRH analogues (Fleming et al., 1982; Porter et al., 1984), and the use of vaginal ultrasound both for monitoring of ovarian response to superovulation (Popp et al., 1985), and oocyte retrieval (Wikland et al., 1985). However, these technical advances have resulted in only moderate improvements in the rate of successful pregnancy following treatment (Figure 1). Thus, data from all the centres practising IVF in the UK shows that while there has been a slight increase in the live birth rate per treatment cycle between 1985 and 1991, from 9% to 13%, this can be accounted for almost entirely by the reduction in cancelled cycles due to the use of LHRH analogues (Tan et al., 1993). Indeed, data from our unit illustrate clearly that only 6% of cycles are abandoned before oocyte retrieval, and that the majority (72%) proceed to embryo transfer (ET). It is after ET that the failure of IVF is manifested, with only 14% of cycles in our centre resulting in successful pregnancy (Figure 2).

These figures are even more disappointing when viewed in terms of implantation rate per embryo transferred. Thus, during 1991 in our unit, 1485 embryos were transferred in 663 ET procedures. While the overall pregnancy rate per ET was 27%, the implantation rate per embryo transferred was only 14%; after failed pregnancies were taken into consideration this figure fell to 13% per embryo.

This observation that up to 87% of human embryos fail after ET needs to be addressed. Factors that may be responsible include uterine receptivity, sub-optimal culture conditions, and differential potential of the oocytes and embryos. It is the last of these that will be considered below.

Endocrinology of Embryo-Endometrium Interactions
Edited by S.R. Glasser *et al.*, Plenum Press, New York, 1994

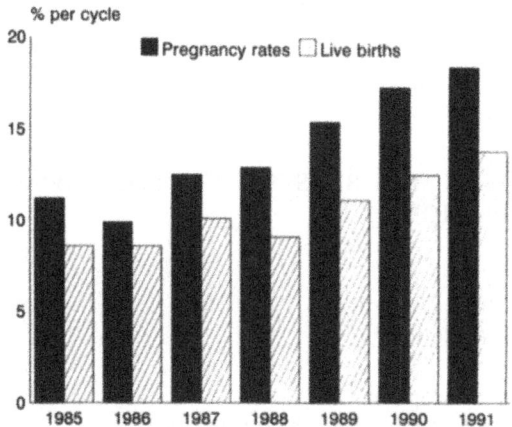

Figure 1. IVF pregnancy and live birth rates per treatment cycle in the UK, 1985-1990 (data compiled by the Human Fertilisation and Embryology Authority).

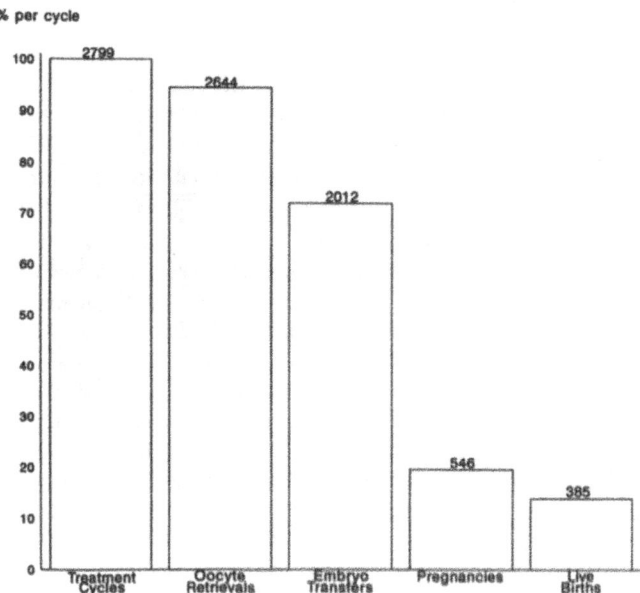

Figure 2. Results of IVF treatments commenced at King's College Hospital between January 1988 and December 1991.

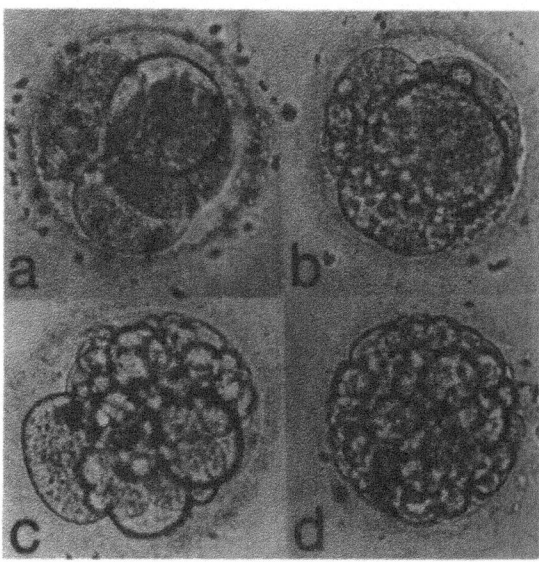

Figure 3. Photomicrographs of human 4-cell embryos of different morphological grades. (a) Grade 4: regular, spherical blastomeres with no extracellular fragmentation; (b) Grade 3: regular, spherical blastomeres with some extracellular fragmentation; (c) Grade 2: blastomeres slightly irregular in size and shape with considerable extracellular fragmentation; (d) Grade 1: barely defined blastomeres with considerable extracellular fragmentation.

EMBRYO TRANSFER ON DAY 2

Following IVF, ET in the human is performed traditionally at the two- to four-cell stage (day 2 after oocyte retrieval). This is primarily to minimise the duration of exposure to sub-optimal culture conditions, while at the same time allowing sufficient time in culture to ensure that normal fertilisation has taken place, and that cleavage has begun.

On day 2, the selection of those embryos from a cohort that are most suitable for transfer remains largely subjective, being based primarily on two criteria of development, namely cleavage rate and gross morphology (Figure 3). It is widely held that those embryos which divide more rapidly and which have regular, spherical blastomeres are more likely to lead to a pregnancy. However, a study of the *in vitro* development of spare human embryos derived by IVF has shown that even if embryos show little or no extra-cellular fragmentation, only 19%, 25% and 33% of embryos that have developed to the 2-cell, 3-cell and 4-cell stages respectively by day 2 will develop into fully expanded blastocysts by day 5 (Bolton et al., 1989).

Clearly, the criteria of cleavage rate and gross morphology are inadequate in the accurate prediction of the potential of human embryos either to develop into blastocysts *in vitro*, or to implant and lead to a viable pregnancy *in vivo* following ET. Some workers have attempted to derive more objective criteria for assessment of embryo viability, for example by measuring the thickness of the zona pellucida together with accurate quantification of extracellular fragmentation using videomorphometry (Cohen et al., 1989), or by the measurement of uptake and production of certain metabolites (Gott et al., 1990; Conaghan et al., 1993). However, definitive criteria for the assessment of human embryo potential on day 2 after oocyte retrieval have yet to be described.

EMBRYO TRANSFER ON DAY 3

Although the tradition of day 2 ET has been adhered to by most IVF units in the majority of treatment cycles, the desire to avoid working outside the 5-day week means that some, including our own, have introduced modification. Thus, for oocyte retrievals performed on Fridays, routine day 3 ETs avoid the necessity for work on Sundays.

In a prospective, randomised study of 398 ETs performed either on day 2 or day 3 after oocyte retrieval, there appeared to be a marginal advantage in terms of clinical pregnancies per ET in the latter group, although this did not reach significance (van Os et al., 1989). Therefore, delaying ET by one day until day 3 has been shown to have no disadvantages, and may, in a larger study, prove to have significant advantages.

ACTIVATION OF THE EMBRYONIC GENOME

There are sound scientific arguments in favour of day 3, or even later ETs. In a study of 317 spare human embryos cultured for 5 days following IVF, the majority (85%) of embryos survived to the 4-cell stage, but only 17% developed into fully expanded blastocysts (Bolton et al., 1989). The stage at which most embryos arrested was at the third cleavage division (i.e. between the 4- and 8-cell stages).

A series of experiments using the transcriptional inhibitor α-amanitin have demonstrated that in the human, embryonic gene activation occurs between the 4- and 8-cell stages (Braude et al., 1988). Thus, in the presence of α-amanitin, human embryos cultured *in vitro* were found to undergo cleavage to the 4-cell stage but no further, while control embryos proceeded to the 8-cell stage and beyond. The polypeptide synthetic profiles of both control and drug-treated embryos, examined by SDS polyacrylamide gel electrophoresis after incubation in ^{35}S-methionine, demonstrated that changes in the patterns of polypeptide synthesis of human embryos that normally occur between the 4-cell and blastocyst stages did not take place in embryos in which transcription had been blocked by α-amanitin (Figure 4).

A number of mammalian embryos undergo developmental arrest *in vitro*, and the stage of arrest tends to coincide with the stage at which activation of the embryonic genome occurs. It is tempting to speculate that there may be a causal relationship between susceptibility to developmental arrest and activation of transcription in the embryo. However, if such a relationship exists, it cannot be straightforward, since it has been shown in both the mouse and the human embryo that even those embryos which arrest undergo gene activation (Goddard and Pratt, 1983; Artley et al., 1992).

Whatever the explanation for developmental arrest in the human, there is no doubt that there is a temporal relationship between the onset of this elevated rate of embryonic attrition and the cessation of the exclusive control of embryogenesis by the maternal inheritance, which is coincident with the commencement of embryonic control. In terms of therapeutic IVF, clearly there should be an advantage in delaying ET until embryos

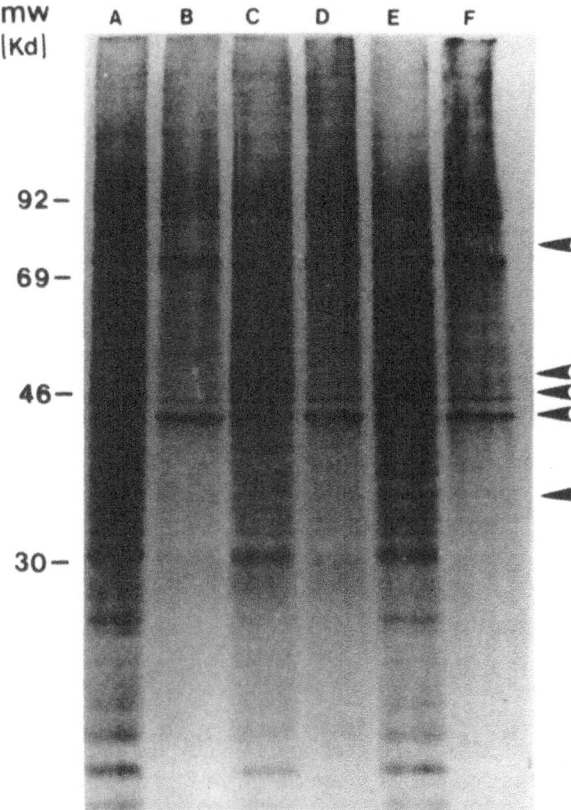

Figure 4. Fluorograph of ^{35}S-methionine labelled polypeptides synthesised by human preimplantation embryos between 70 hours and 73 hours post insemination and separated by one dimensional polyacrylamide gel electrophoresis. Embryos were cultured for 30 hours from the late 2-cell stage (lane A) or the early 4-cell stage (lanes C and E) in control medium, and from the late 2-cell stage (lanes B and D) or the early 4-cell stage (lane F) in the presence of α-amanitin. Arrowheads indicate polypeptide bands whose pattern of synthesis changes between the 2- to 4-cell and the blastocyst stages.

have cleaved to the 8-cell stage, so that only those embryos that at least have the potential to develop beyond this apparent developmental hurdle are included in the cohort from which embryos are selected for transfer.

EMBRYO TRANSFER ON DAY 5

Background

In light of the considerations outlined above, it follows that there are a number of reasons to re-evaluate the efficacy of performing ETs on day 5 after oocyte retrieval. First, ET is performed routinely at the blastocyst stage in domestic animals, leading to high pregnancy rates (Adams, 1982). Second, it is physiologically more appropriate for blastocysts to be transferred to the uterus than cleavage stage embryos. Third, as for day 3 transfers, it may be that delaying ET until embryos have successfully developed into fully expanded blastocysts will reduce the element of subjectivity in selecting embryos from a cohort for transfer. Fourth, a number of studies have demonstrated encouraging pregnancy rates following the transfer of human blastocysts that have developed *in vitro* using co-culture techniques (Menezo et al., 1992a, b; Bongso et al., 1992). Fifth, the preimplantation stage diagnosis of genetic disease may be facilitated by performing embryo biopsy at the blastocyst stage, since not only might it be possible to biopsy more cells from the blastocyst for genetic diagnosis, but also, only extra-embryonic cells need be biopsied, both of which would reduce the chances of the biopsy procedure compromising the viability of the embryo (Tarin et al., 1992; Tarin and Handyside, 1993). Finally, since it is the blastocyst that interacts with the endometrium at implantation, it is more likely that secretions by blastocyst stage, rather than cleavage stage embryos, may be identified, quantified and used to distinguish between embryos that do, and those that do not have the potential to lead to a pregnancy.

Prospective Randomised Study

In an attempt to examine the above points, a prospective randomised study was undertaken, with patients undergoing IVF followed by ET either on day 2 or day 5 after oocyte retrieval. Superovulation, ultrasound-directed follicle aspiration, IVF and ET were carried out exactly as described previously (Waterstone and Parsons, 1992; Bolton et al., 1989), except that follicles were aspirated and flushed carefully in order both to obtain an accurate measure of follicle volume, as well as an uncontaminated sample of follicular fluid.

A total of 72 couples, all of whom were suffering from tubal damage as the only identified cause of their infertility, underwent 128 cycles of IVF; in 63 cycles, ET was performed on day 2 after oocyte retrieval, and in 65 cycles, on day 5. The median age of the women in both groups was 31 years (range 24-37 years), and a median of 14 and 13 (ranges 2-31, and 4-39) oocytes were collected in the day 2 and day 5 groups respectively. After insemination *in vitro*, a median of 10 and 9 oocytes (ranges 2-22 and 2-24) became fertilised in the two groups respectively.

All the cycles resulted in ET, with a maximum of two embryos transferred. The pregnancy rates, and the implantation rates per embryo transferred in the day 2 and day 5 groups are shown in Table 1.

It can be seen that both the overall and ongoing pregnancy rates were similar in the two groups, as were the implantation rates and the rate of development of a fetal heart per embryo transferred. Clearly, these data provide no evidence to suggest that day 5 ET reduces the chance of pregnancy, but neither do they provide any evidence to support the contention that day 5 ETs confer an advantage. However, it should be noted that in the day 5 ET group, there were eight cases where embryos were transferred despite the fact that they had not developed beyond the 16-cell stage; none of these ETs resulted in pregnancy (data not shown). Indeed, while 57 of the 63 (90%) cases in the day 2 ET group had at least two "good quality" (i.e. Grade 4 or Grade 3; Bolton et al., 1989) embryos available for ET, only 52 of 65 (80%) cases in the day 5 ET group had at least 2 morulae or blastocysts available; had ET been performed on day 2 in the latter group, 61 (91%) would have had two "good quality" embryos transferred.

Despite this observation, the pregnancy rate in the day 5 group was not compromised, providing circumstantial evidence to suggest that those embryos that do not survive the additional period *in vitro* do not have the potential to develop into blastocysts and implant, even if they are returned to the *in vivo* environment at the 2- to 4-cell stage. If this proves to be correct, then it follows that as long as it remains the

Table 1. Results of prospective randomised trial comparing outcome of IVF followed by ET either on day 2 or day 5 after oocyte retrieval

| | ET Performed on:- | |
	Day 2	Day 5
No. (%) of:-		
Cycles	63	65
ETs	63	65
Pregnancies (+ve βhCG)	14 (22)	17 (26)
Ongoing Pregnancies (FH on ultrasound)	9 (14)	12 (18)
Embryos transferred	126	124
Implantations	20 (16)*	20 (16)*
Fetal Hearts	14 (11)*	16 (13)*

*% per embryo transferred

convention to perform ET on day 2, in many cases embryos will be transferred that do not have the potential to develop into a viable pregnancy; a proportion of these cases would be identified prior to ET if this procedure was delayed until day 5, through delayed cleavage or developmental arrest of non-viable embryos. Indeed, among those cases in the day 5 ET group where only morula or blastocyst stage embryos were transferred, the pregnancy rate per ET was 17/52 (33%), the implantation rate per embryo transferred was 20/108 (19%), and the fetal heart (FH) rate per embryo transferred was 16/108 (15%).

It can be concluded that ET on day 5 results in at least an equivalent pregnancy rate to that achieved following ET on day 2; a larger series of day 5 tranfers will demonstrate whether or not the trend of slightly higher pregnancy and implantation rate per embryo on day 5 illustrated in Table 1 is significant. Furthermore, it appears that those embryos that have not undergone morula or blastocyst formation by day 5 *in vitro* will not implant following ET, and that by delaying ET until day 5, it may be possible to select those embryos that are more likely to implant following transfer. Finally, and perhaps most significantly, it is clear that not all blastocysts have the potential to implant following ET.

FAILURE OF BLASTOCYSTS TO IMPLANT

Two possible reasons for the failure of *in vitro*-derived blastocysts to implant following ET on day 5 after insemination are (i) the embryos are intrinsically non-viable, despite overt morphological evidence of cavitation and blastocyst formation, and (ii) the blastocysts themselves are viable, but they are unable to hatch from the zona pellucida.

Blastocyst Viability Assessed by Cell Numbers

There are many intrinsic factors that may affect the viability of blastocysts, including chromosomal abnormalities which are reported to occur in nearly 40% of human embryos derived by IVF (Plachot, 1991), and impaired metabolic activity (Gott et al., 1990). Another factor is delayed or abnormal cytokinesis and/or karyokinesis, resulting in blastocysts with reduced cell numbers and/or multinucleate cells (Hardy et al., 1989a; Winston et al., 1991).

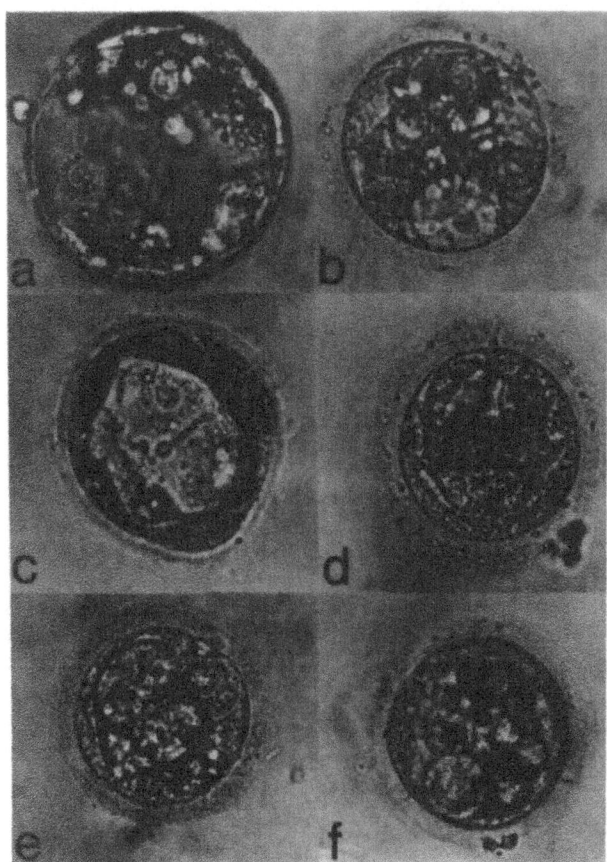

Figure 5. Photomicrographs of human embryos on day 5 after insemination *in vitro*. (a) fully expanded viable blastocyst which implanted following embryo transfer and led to the live birth of a healthy male infant; (b) embryo displaying multiple cavities or intracellular vacuoles; (c) embryo with a "blastocyst" appearance, but with too few cells; (d) embryo with outer layer of cells surrounding a "blastocoelic cavity", but with unincorporated blastomeres and too few cells; (e) embryo with a small number of outer cells, many unincorporated blastomeres and multiple cavities; (f) embryo displaying unincorporated blastomeres.

Table 2. Total nuclear numbers in *in vitro*-derived human blastocysts on days 5, 6 and 7 post insemination [1]

"Blastocyst" morphology & age (days post fertilisation)	No. Analysed	No. of Nuclei	Range
"Normal" Day 5	9	58.3±8.1	24-90
"Abnormal" Day 5	1	18	
"Normal" Day 6	25	84.4±5.7	27-136
"Abnormal" Day 6	15	38.3±6.2	3-86
"Normal" Day 7	11	125.5±19	60-283

[1]Data from Hardy et al., 1989a

There are few studies of cell numbers in human blastocysts that have developed *in vivo*. Two blastocysts, recovered at hysterectomy and estimated to be 5 days old, were found, by serial sectioning, to consist of 158 and 107 cells respectively (Hertig et al., 1954). A single blastocyst of undefined age, but thought to be more than 5 days old, was obtained by uterine flushing and estimated by serial sectioning to contain 186 cells (Croxatto et al., 1972). In comparison, two *in vitro*-derived blastocysts at 7 days post insemination were found to contain 112 and 110 nuclei respectively (Steptoe et al., 1971).

Two recent studies have examined in detail relatively large numbers of human blastocysts developed *in vitro* following IVF, both of which have identified abnormalities in a large proportion of such embryos. The first investigated cell number and allocation to either trophectoderm (TE) or inner cell mass (ICM) in blastocysts that developed on days 5, 6 and 7 after insemination *in vitro*, by nuclear staining (Hardy et al., 1989a). Blastocysts were classified as morphologically abnormal if they displayed one or more of the following characteristics: unincorporated blastomeres; multiple cavities (possibly intracellular vacuoles); no discrete ICM visible; low numbers of mural TE cells (Figure 5). Of 149 normally fertilised embryos, 48 (32%) developed into apparently normal blastocysts, and 15 (10%) developed into blastocysts displaying morphological abnormalities; 61 of these embryos were analysed successfully for total nuclear number. Results of the analysis are summarised in Table 2, and demonstrate that morphologically abnormal blastocysts had lower total cell numbers.

The implications of these results for the potential of blastocysts to develop into a viable pregnancy are considerable. First, it seems likely that a number of "abnormal" blastocysts display morphological evidence of degeneration, which is confirmed by the low cell numbers, indicating that cytokinesis is retarded or arrested. Second, some blastocysts with abnormally low cell numbers may remain viable, but with limited potential, since blastocysts with too few cells will be unable to allocate sufficient cells to form both functioning TE and viable ICM. Allocation of cells to the TE and ICM in the blastocyst is of fundamental importance to subsequent development; following implantation, the TE gives rise only to placenta and extra-embryonic membranes, while the ICM forms all three germ layers of the fetus and contributes to the extra-embryonic membranes (Gardner & Papaioannou, 1975). It follows that blastocysts with too few ICM cells may be responsible for biochemical pregnancies where there is initial production of hCG by TE cells, but no fetal development.

The second detailed study of in vitro-derived human blastocysts on day 5 investigated the number of cells present in the embryos by serial sectioning (Winston et al., 1991). Of 19 blastocysts, only 4 (21%) were found to contain more than 60 cells. Moreover, the high incidence of multinucleate cells found in this as well as earlier studies (Lopata et al., 1983; Tesarik et al., 1987) suggests that the majority of human blastocysts that develop in vitro contain multinucleated cells. Indeed, any estimate of cell numbers in such embryos are likely to be overestimates. From these data it was concluded that of those embryos that fertilise normally in vitro, the majority fail to complete sufficient cell cycles to produce viable blastocysts with sufficient cell numbers to differentiate a normal ICM.

Both these detailed studies lead to the conclusion that the majority of in vitro-derived human blastocysts are unlikely to be developmentally competent even if implantation should occur; an implanted blastocoelic vesicle, for example, could produce symptoms and signs of a pregnancy which fails early (a biochemical pregnancy) or could even progress further to produce an anembryonic pregnancy or abortion.

Inability of Blastocysts to Hatch

It has been hypothesised that some blastocysts are unable to hatch from the zona pellucida, and are therefore unable to implant despite the fact that they may be intrinsically viable. The phenomenon of zona hardening (Drobnis et al., 1988), which may be a function of prolonged culture, may also be related to zona thickness, which is itself related to embryonic stage and rate of development (Wright et al., 1990; Keenan et al., 1991) as well as maternal age (Cohen et al., 1992). The suggestion that there may be a direct relationship between a thickened and/or hardened zona and failure to implant has been investigated using microsurgical techniques.

In a preliminary study in which the zonae pellucida of 49 2- to 8-cell stage embryos were split by partial zona dissection (Cohen et al., 1991), the implantation rate following transfer was 25% (12/49), compared with only 6% (2/35) in control embryos, suggesting that implantation was enhanced by "assisted hatching" (Cohen et al., 1990). This study was followed by a more extensive series (Cohen et al., 1992). In one study, in which ET was performed on day 3 after insemination in vitro, 137 patients were randomised prospectively into two groups; in one group (68 patients) 229 embryos were transferred after zona drilling (Gordon and Talansky, 1986), and in the other group (69 patients), 239 embryos were transferred with their zonae intact. The rates of development of a fetal heart per embryo transferred were 28% and 21% in the two groups respectively (P<0.023; Cohen et al., 1992).

In a further study, embryos were identified which would theoretically benefit most from "assisted hatching" (i.e. those whose zonae were >15um thick, and/or where the cell number was <5 on the morning of day 3, and/or where >20% of the perivitelline space was filled with extracellular fragments). A total of 172 patients whose embryos displayed these properties were randomly allocated for zona drilling prior to ET. Of the 285 control, zona-intact embryos that were transferred in 92 patients, 51 (18%) implanted, all off which developed a fetal heart. Of the 278 embryos in the experimental, selective assisted hatching group, 70 (25%) implanted following ET, all of which developed a fetal heart (P<0.05).

These data support the suggestion that there is a relationship between zona thickness, hatching and implantation. While the precise mechanism by which assisted hatching promotes implantation remains unclear, it would seem likely that it is mechanical, in that the embryo has a greater chance of escaping from the split zona. However, it is possible that a thickened zona impairs the transport of metabolites and growth factors to the embryo, thereby inhibiting its development.

IDENTIFICATION OF EMBRYOS WITH THE POTENTIAL TO IMPLANT

It is clear that irrespective of the stage at which ET is performed, human embryos manifest a differential potential to implant and progress to a viable pregnancy. In order to improve the chances of pregnancy resulting from IVF, it is essential to develop methods by which those embryos with the potential to continue development can be identified. Currently, the criteria used most commonly for the selection of embryos for ET are cleavage rate and gross morphology (Bolton et al., 1989), neither of which are accurate.

The study in which ET was performed on day 5 after insemination *in vitro* has enabled the detailed examination of the follicular environment of oocytes that fertilise normally, develop into blastocysts and implant following ET, as well as the secretions of those blastocysts immediately prior to transfer.

The Follicular Environment

The ovarian follicles which develop under the influence of superovulatory stimulation regimens vary in size, and the oocytes derived from them are developmentally asynchronous (Veeck, 1989; Hammitt et al., 1993), despite the fact that each follicle is exposed to the same duration and levels of gonadotrophins. Clearly, the potential of each follicle to respond to gonadotrophins must vary, and this may be linked to oocyte quality. Thus, it is possible that examination of the follicular environment, which reflects its response to stimulation, may provide information regarding the potential of each oocyte.

Follicle Volume. During ultrasound-directed follicle aspiration for patients randomised to day 5 ET, the volume of the follicle from which each oocyte was collected was measured carefully. The fate of each oocyte, in terms of normal fertilisation *in vitro* and development to the blastocyst stage, related to the volume of the follicle from which it was derived is illustrated in Figure 6. It can be seen that there is no relationship between the size of the follicle and whether or not its oocyte achieves normal fertilisation *in vitro*, or develops into a blastocyst.

Follicular biochemistry. It is likely that there are measurable factors within each follicle that may give an indication of the developmental potential of the oocyte within it. This is not a novel idea, and a number of studies have investigated the relationship between follicular biochemistry and the potential of the oocyte to which each follicle gives rise, without identifying any influential factor (Nayudu et al., 1989; De Sutter et al., 1991; Gonzales et al., 1992; Phocas et al., 1992). However, no study has examined the relationship between the levels of different components of follicular fluids and the development of human oocytes to the blastocyst stage following fertilisation *in vitro*. In the study in which embryos were transferred on day 5 after insemination *in vitro*, follicular fluids from individual follicles were analysed for steroid (oestradiol, progesterone, testosterone and androstenedione) content, as well as for the levels of insulin-like growth factor-1 (IGF-1). Although the levels of the individual factors showed no correlation with the potential of the oocyte from a given follicle to undergo normal fertilisation, blastocyst formation, or to implant following ET on day 5, preliminary evidence was obtained suggesting that measurement of a combination of follicular factors may provide accurate information regarding oocyte quality (data not shown). Whether or not a relationship exists between levels of PAF-acether (Amiel et al., 1991), tumour necrosis factor (Roby and Terranova, 1988; Zolti et al., 1990), colony stimulating factor or interleukin-6 (Zolti et al., 1991, 1992), all of which have been suggested to influence oocyte quality, remains to be elucidated.

Non-Invasive Assays of Embryo Potential

Assays of embryo potential that will be useful in selecting embryos for ET must, by definition, be non-invasive. Microassays for uptake and incorporation of metabolites have been developed (Leese et al., 1986; Hardy et al., 1989b), but these have proved of little use in predicting embryo potential if ET is to be performed on day 2 (Conaghan et al., 1993). In contrast, studies of glucose and pyruvate uptake, and lactate production by human embryos *in vitro* have demonstrated significant differences on days 5 and 6 after insemination between those embryos that undergo developmental arrest and those that form blastocysts (Gott et al., 1990).

It seems unlikely that embryos secrete factors signalling their presence to the mother as early as the 2- to 4-cell stage, since they are free-floating and autonomous; *in vivo* they are in transit in the fallopian tube, and clearly, they can continue development in simple media *in vitro*. In contrast, by day 5, the blastocyst is approaching, or has entered the uterine cavity, and is poised for implantation; the endometrium is primed for its arrival, and it seems likely that the embryo secretes factors that signal its presence to which the endometrium (and corpus luteum) respond.

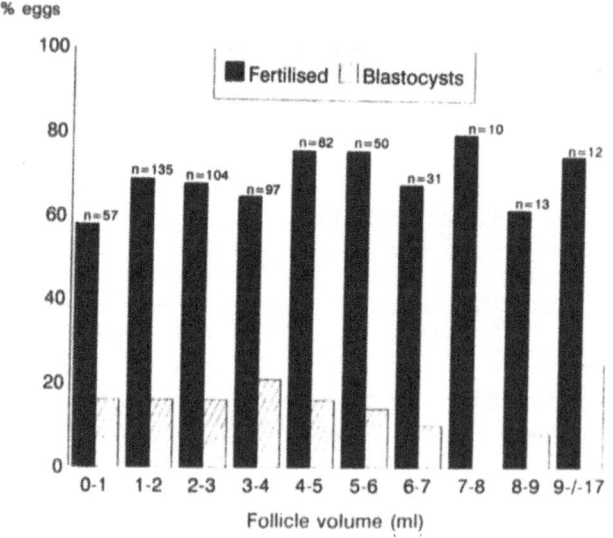

Figure 6. Histogram illustrating the percentage of eggs collected from follicles of different volumes that fertilise, and that develop into blastocysts by day 5 after fertilisation.

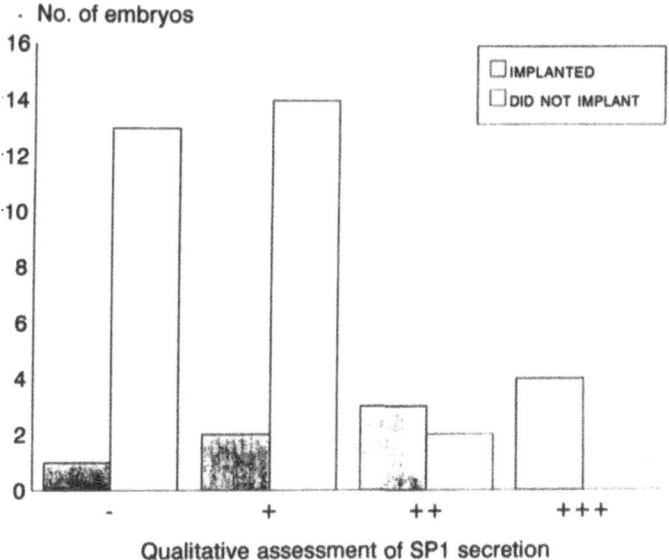

Figure 7. Histogram illustrating qualitative assessments of levels of SP$_1$ secreted by human embryos on day 5 after fertilisation *in vitro* prior to transfer.

Human chorionic gonadotrophin (hCG) is secreted by trophectoderm cells, and has been detected by immunoradiometric assay in the medium in which human embryos have been cultured for 8 days (Lopata and Hay, 1989) or for 7 days (Dokras et al., 1991; Lopata and Oliva, 1993). Recently, using a mouse Leydig cell bioassay, we have detected hCG scretion by day 6 blastocysts (Jones et al., 1992). Whether or not this assay will be of practical use in predicting which day 5 embryos should be selected for transfer during therapeutic IVF remains to be established.

A further hormone that may prove predictive of embryo potential is pregnancy specific β$_1$-glycoprotein (SP$_1$), which is known to be synthesised by the syncytiotrophoblast (Horne et al., 1976). Although a recent report, in which SP$_1$ secretion by human embryos between days 2 and 14 after fertilisation *in vitro*, suggested that nanograms of the protein are secreted by individual embryos from day 4 (Dimitriadou et al., 1992), we were able to detect only picogram levels secreted by day 5 embryos (Bolton and Bersinger, in preparation). The levels of SP$_1$ in the medium in which embryos had been cultured for 5 days prior to transfer were measured, and related to whether or not the embryo subsequently implanted and led to a clinical pregnancy. The preliminary results of this study are shown in Figure 7; because the numbers are as yet still small, the data are not shown quantitatively. Nonetheless, they show that most of the embryos that fail following ET secrete relatively low levels of SP$_1$, while of the embryos that develop into clinical pregnancies, most secrete higher levels of the hormone.

Among the cytokines, there are a number of potential candidates for a putative "embryonic signal", including platelet-activating factor (O'Neill et al., 1987; Nakatsuka et al., 1992), platelet-derived growth factor (Svalander et al., 1991), and interferon-α (Chard, 1991). We have measured interferon-α levels in medium in which human embryos had been cultured for 5 days. We were unable to detect this cytokine in any of the samples assayed; this included the culture medium from blastocysts that implanted, and those that failed to implant, following ET. This result was somewhat surprising, since substantial quantities of a molecule of the interferon-α family, which exhibits antiluteolytic activity and has been designated trophoblast interferon, are secreted by sheep, cow and pig embryos (reviewed by Chard, 1991). However, the role of trophoblast interferon is by no means universal, since it appears that it is not secreted by the mouse blastocyst.

CONCLUSIONS

Our understanding of human preimplantation embryogenesis has increased considerably since the development of successful techniques for human IVF. However, therapeutic IVF remains relatively unsuccessful, and any improvement will rely upon a better understanding of the processes leading to normal blastocyst formation and implantation. This objective may be achieved by performing ET on day 5 after fertilisation *in vitro*. During the additional period of culture it should be possible to identify and subsequently apply criteria which will enable the more accurate selection for transfer of those embryos with the greatest chance of developing into a viable pregnancy. In addition, it will be possible to investigate the nature of the interactions between the blastocyst and the endometrium during the peri-implantation period, and all without reducing the chance of achieving a pregnancy.

ACKNOWLEDGEMENTS

The author would like to thank Dr. B. Thiliganathan of the Harris Birthright Centre for Fetal Medicine, King's College School of Medicine and Dentistry, for his assistance with the assays for interferon-α.

REFERENCES

Adams, C.E., 1982, "Mammalian Egg Transfer", C.R.C. Press, Boca Raton, Florida.

Amiel, M., Testart, J., and Benveniste, J., 1991, Platelet-activating factor-acether is a component of human follicular fluid, 1991, *Fertil. Steril.* 56:62-64.

Artley, J.K., Braude, P.R., and Johnson, M.H., 1992, Gene activity and cleavage arrest in human pre-embryos, *Human Reprod.* 7:1014-1021.

Bolton, V.N., Hawes, S.M., Taylor, C.T, and Parsons, J.H., 1989, Development of spare human preimplantation embryos *in vitro*: an analysis of the correlations among gross morphology, cleavage rates and development to the blastocyst, *J. In Vitro Fert. Embryo Transfer* 6:30-35.

Bongso, A., Ng, S.C., Fong, C.F., Anandakumar,C., Marshall, B., Edirisinghe, R., and Ratnam, S., 1992, Improved pregnancy rate after transfer of embryos grown in human fallopian tubal cell coculture, *Fertil. Steril.* 58:569-574.

Braude, P., Bolton, V., and Moore, S., 1988, Human gene expression first occurs between the four- and eight-cell stages of preimplantation development, *Nature* 332:459-461.

Chard, T., 1991, Interferon-α is a reproductive hormone, *J. Endocrinol.* 131:337-338.

Cohen, J., Inge, K.L., Suzman, M., Wiker, S.R., and Wright, G., 1989, Videocinematography of fresh and cryopreserved embryos: a retrospective analysis of embryonic morphology and implantation, *Fertil. Steril.* 51:820-827.

Cohen, J., Elsner, C., Kort, H., Malter, H., Massey, J., Mayer, M.P., and Wiemer, K., 1990, Impairment of the hatching process following IVF in the human and improvement of implantation by assisted hatching using micromanipulation, *Human Reprod.* 5:7-13.

Cohen, J., Alikani, M., Malter, H.E., Adler, A., Talansky, B.E., and Rosenwaks, Z., 1991, Partial zona dissection or subzonal sperm insertion: microsurgical fertilization alternatives based on evaluation of sperm and embryo morphology, *Fertil. Steril.* 56:696-706.

Cohen, J., Alikani, M., Trowbridge, J., and Rosenwaks, Z.., 1992, Implantation enhancement by selective assisted hatching using zona drilling of human embryos with poor prognosis, *Human Reprod.* 7:6850691.

Conaghan, J., Hardy, K., Handyside, A.H., Winston, R.M.L., and Leese, H.J., 1993, Selection criteria for human embryo transfer: a comparison of pyruvate uptake and morphology, *J. Assisted Reprod. Genet.* 10:21-30.

Croxatto, H.B., Diaz, S., Fuentealba, B., Croxatto, H.,Carrillo, D., and Fabres, C., 1972, Studies on the duration of egg transport in the human oviduct. I. The time interval between ovulation and egg recovery from the uterus in normal women, *Fertil. Steril.* 23:447-458.

De Sutter, P., Dhont, M., Vanluchene, E., Vandekerckhove, D., 1991, Correlations between follicular fluid steroid analysis and maturity and cytogenetic analysis of human oocytes that remained unfertilized after in vitro fertilization, *Fertil. Steril.* 55:958-963.

Dimitriadou, F., Phocas, I., Mantzavinos, T., Sarandakou, A., Rizos, D., and Zourlas, P.A., 1992, Discordant secretion of pregnancy specific ß₁-glycoprotein and human chorionic gonadotropin by human pre-embryos cultured in vitro, *Fertil. Steril.* 57:631-636.

Dokras, A., Sargent, I.L., Gardner, R.L., and Barlow, D.H., 1991, Human trophectoderm biopsy and secretion of chorionic gonadotrophin, *Human Reprod.* 6:1453-1459.

Drobnis, E.Z., Andrew, J.B., and Katz, D.F., 1988, Biophysical properties of the zona pellucida measured by capillary suction: is zona hardening a mechanical phenomenon? *J. Exp. Zool.* 245:206-219.

Fleming, R., Adam, A.H., Barlow, D.H., Black, W.P., MacNaughton, M.C. and Coutts, J.R.T., 1982, A new systematic treatment for infertile women with abnormal hormone profiles, *Br. J. Obs & Gynae.* 89:80-83.

Gardner, R.L., and Papaioannou, V.E., 1975, Differentiation in the trophectoderm and inner cell mass, *in*: "The Early Development of Mammals," M. Balls and A.E. Wild eds., Cambridge University Press, London, pp107-132.

Goddard, M.J., and Pratt. H.P.M., 1983, Control of events during early cleavage of the mouse embryo: an analysis of the "2-cell block", *J. Embryol. exp. Morph.* 73:111-133.

Gonzales, J., Lesourd, S., Van Dreden, P., Richard, P., Lefebvre, G., and Vauthier Brouzes, D., 1992, Protein composition of follicular fluid and oocyte cleavage occurrence in *in vitro* fertilization (IVF), *J. Ass. Reprod. Genet.* 9:211-216.

Gott, A.L., Hardy, K., Winston, R.M.L., and Leese, H.J., 1990, Non-invasive measurement of pyruvate and glucose uptake and lactate production by single human preimplantation embryos, *Human Reprod.* 5:104-108.

Gordon, J.W., and Talansky, B.E., 1986, Assisted fertilization by zona-drilling: a mouse model for correction of oligospermy, *J. Exp. Zool.* 239:347-354.

Hammitt, D.G., Syrop, C.H., Van Voordis, B.J., Walker, D.L., Miller, T.M., and Barud, K.M., 1993, Maturational asynchrony between oocyte cumulus-coronal morphology and nuclear maturity in gonadotropin-releasing hormone agonist stimulations, *Fertil. Steril.* 59:375-381.

Hardy, K., Handyside, A.H., and Winston, R.M., 1989a, The human blastocyst: cell number, death and allocation during late preimplantation development *in vitro*, *Development* 107:597-604.

Hardy, K., Hooper, M.A.K., Handyside, A.H., Rutherford, A.J., Winston, R.M.L., and Leese, H.J., 1989b, Non-invasive measurement of glucose and pyruvate uptake by individual human oocytes and preimplantation embryos, *Human Reprod.* 4:188-191.

Hertig, A.T., Rock, J., Adams, E.C., and Mulligan, W.J., 1954, On the preimplantation stages of the human ovum: a description of four normal and four abnormal specimens ranging from the second to fifth day of development, *Contrib. Embryol.* 35:199-220.

Horne, C.H.W., Towler, C.M., Pugh-Humphrey, R.G.P., Thomson, A.Q., and Bohn, H., 1976, Pregnancy specific ß₁-glycoprotein, a product of the syncytiotrophoblast, *Experientia* 32:1197-1199.

Jones, T., Ellison, Z., Bolton V., Webley, G., and Milligan, S., 1992, Bioassayable human chorionic gonadotrophin production by human preimplantation embryos, *J. Reprod. Fert.* Abst. Series 9, p58.

Keenan, D., Cohen, J., Suzman, M., Wright, G., Kort, H., and Massey, J., 1991, Stimulation cycles suppressed with gonadotrophin-releasing hormone analog yield accelerated embryos, *Fertil. Steril.* 55:792-796.

Leese, H.J., Hooper, M.A.K., Edwards, R.G., and Ashwood-Smith, 1986, Uptake of pyruvate by early human embryos determined by a non-invasive technique, *Human Reprod.* 1:181-182.

Lopata, A., and Hay, D.L., 1989, The potential of early human embryos to form blastocysts, hatch from their zona and secrete HCG in culture, *Human Reprod.* 4 (Suppl.):97-94.

Lopata, A., and Oliva, K., 1993, Chorionic gonadotrophin secretion by human blastocysts, *Human Reprod.* 8:932-938.

Lopata, A., Kohlman, D., and Johnston, I., 1983, The fine structure of normal and abnormal human embryos developed in culture, *in*: "Fertilization of the Human Egg In Vitro," H.M. Beier and H.R. Lindner, eds, Springer Verlag, Heidelberg, pp189-210.

Menezo, Y., Hazout, A., Dumont, M., Herbaut, N., and Nicollet, B., 1992a, Coculture of embryos on Vero cells and transfer of blastocysts in humans, *Human Reprod.* 7:101-106.

Menezo, Y., Nicollet, B., Herbaut, N., and Andre, D., 1992b, Freezing cocultured human blastocysts, *Fertil. Steril.* 58:977-980.

Nakatsuka, M., Yoshida, N., and Kudo, T. (1992), Platelet activating factor in culture media as an indicator of human embryonic development after in-vitro fertilization, *Human Reprod.* 7:1435-1439.

Nayudu, P.L., Lopata, A., Jones, G.M., Gook, D.A., Bourne, H.M., Sheather, S.J., Brown, T.C., and Johnston, W.I.H., 1989, An analysis of human oocytes and follicles from stimulated cycles: oocyte morphology and associated follicular fluid characteristics, *Human Reprod.* 4:558-567.

O'Neill, C., Gidley-Baird, A.A., Pike, I.L., and Saunders, D.M., 1987, A bio-assay for embryo-derived platelet-activating factor as a means of assessing quality and pregnancy potential of human embryos, *Fertil. Steril.* 47:969-975.

Phocas, I., Mantza Vinos, T., Rizos, D., Dimitriadou, F., Arvaniti, K., and Zourlas, P.A., 1992, Hormone levels of follicular fluids with and without oocytes in patients who received gonadotropin-releasing hormone analogues and gonadotropins in an *in vitro* fertilization programme, *J. Ass. Reprod. Genet.* 9:233-237.

Plachot, M., 1991, Chromosome analysis of oocytes and embryos, *in*: "Preimplantation Genetics," Y. Verlinsky and A. Kuriev, eds., Plenum Press, New York, pp103-112.

Popp, L.W., Lemester, S., Hinrichs, S., Heesen, D., Muller-Holve, W., Martin K., 1985, Intravaginale Ultraschalldiagnostik (Vaginosonographie) - erste Erfahrungen mit dem Panoramesektor, *in*: "Ultraschalldiagnostik 84," G. Judmaier, H. Frommhold, and A. Kratochwil, eds., Thieme, Stittgart, p320.

Porter, R.N., Smith, W., Craft, I.L., Abdulwahid, N.A., and Jacobs, H.S., 1984, Induction of ovulation for in-vitro fertilisation using buserelin and gonadotropins, *Lancet* ii:1284-1285.

Roby, K.F., and Terranova, P.F., 1988, Tumor necrosis factor-α alters follicular steroidogenesis in vitro, *Endocrinology* 123:2952-2954.

Steptoe, P.C., Edwards, R.G., and Purdy J.M., 1971, Human blastocysts grown in culture, *Nature* 229:132-133.

Steptoe, P.C., and Edwards, R.G., 1978, Birth after the reimplantation of a human embryo, *Lancet* ii:366.

Svalander, P.C., Holmes, P.V., Olovsson, M., Wikland, M., Gemzell-Danielsson, K., and Bygdeman, M., 1991, Platelet-derived growth factor is detected in human blastocyst culture medium but not in human follicular fluid - a preliminarly report, *Fertil. Steril.* 56:367-369.

Tan, S.L., Maconochie, N., Doyle, P., Balen, A., and Brinsden, P., 1993, Cumulative live birth rate after *in vitro* fertilisation with and without pituitary desensitisation, *Gynae. Endocrinol.* 7:Supp 2, p50

Tarin, J.J., Conaghan, J., Winston, R.M.L., and Handyside, A.H., 1992, Human embryo biopsy on the 2nd day after insemination for preimplantation diagnosis: removal of a quarter of embryo retards cleavage, *Fertil. Steril.* 58:970-976.

Tarin, J.J., and Handyside, A.H., 1993, Embryo biopsy strategies for preimplantation diagnosis, *Fertil. Steril.* 59:943-952.

Tesarik, J., Kopecny, V., Plachot, M., and Mandelbaum., J., 1987, Ultrastructural and autoradiographic observations on multinucleated blastomeres of human cleaving embryos obtained by *in vitro* fertilisation, *Human Reprod.* 2:127-136.

Van Os, H.C., Alberda, A.T., Jansen-Caspers, H.A.B., Leerentveld, R.A., Scholtes, M.C.W., and Zeilmaker, G.H., 1989, The influence of the interval between *in vitro* fertilization and embryo transfer and some other variables on treatment outcome, *Fertil. Steril.* 51:360-362.

Veeck, L.L., 1989, Oocyte assessment and biological importance, Ann. N.Y. Acad Sci. 442:259-274.

Waterstone, J.J., and Parsons, J.H., 1992, A prospective study to investigate the value of flushing follicles during ultrasound-directed follicle aspiration, *Fertil. Steril.* 57:221-223.

Wikland, M., Lennart E., and Hamberger, L., 1985, Transvesical and transvaginal approaches for the aspiration of follicles by the use of ultrasound, *Ann. N.Y. Acad. Sci.* 442:184-192.

Winston., N.J., Braude, P.R., Pickering, S.J., George, M.A., Cant, A., Currie, J., and Johnson, M.H., 1991, The incidence of abnormal morphology and nucleocytoplasmic ratios in 2-, 3- and 5-day human pre-embryos, *Human Reprod.* 6:17-24.

Wright, G., Wiker, S., Elsner, C., Kort, H., Massey, J., Mitchell, D., Toledo, A., and Cohen, J., 1990, Observations on the morphology of human zygotes, pronuclei and nucleoli and implications for cryopreservation, *Human Reprod.* 5:109-115.

Zolti, M., Meirom, R., Shemesh, M., Wallach, D., Mashiach, S., and Shore, L., 1990, Granulosa cells as source and target organ for tumor necrosis factor-α, *FEBS Lett.* 261:253-255.

Zolti, M., Ben Rafael, Z., Meirom, R., Shemesh, M., Bider, D., and Maschiach, S., 1991, Cytokine involvement in oocytes and early embryos, *Fertil. Steril.* 56:265-272.

Zolti, M., Bider, D., Seidman, D.S., Mashiach, S., and Ben Rafael, Z., 1992, Cytokine levels in follicular fluid of polycystic ovaries in patients treated with dexamethasone, *Fertil. Steril.* 57:501-504.

CHARACTERIZATION OF THE HUMAN ENDOMETRIUM IN RELATION TO IMPLANTATION

Peter C. Svalander,[1] Paul V. Holmes[1], Kristina Gemzell-Danielsson, [2] Marja-Liisa Swahn,[2] Matts Wikland,[1] and Marc Bygdeman[2]

[1]Fertility Center Scandinavia, Carlanderska Hospital, Göteborg, Sweden
[2]Dept. of Obstetrics and Gynec., Karolinska Institute, Stockholm, Sweden

INTRODUCTION

Endometrial receptivity for implantation is highly relevant in the clinical practice of *in vitro* fertilization and embryo transfer (IVF-ET). By the advancement of assisted reproductive technology, it has become increasingly evident that human implantation is indeed a complex biological process, having many characteristics in common with an inflammatory reaction and a tumour invasion. Since medico-legal contraints limit the possibility of investigating and revealing the mechanisms of human implantation, most research is confined to studies of the separate tissues involved, i.e. the human blastocyst grown *in vitro*, the endometrium *in situ*, endometrial biopsy material, and culture of the separate cell types. Without reviewing the entire scientific literature on implantation in animals and man, we would like to present some fundamental views on human implantation with emphasis on studies of the mid-luteal phase endometrium of normal, fertile women. In a recent series of experiments, we obtained endometrial biopsies from the ovulation-timed cycles of normal women, both in control cycles and in treatment cycles, using the progesterone antagonist RU 486. Endometrial biopsies were obtained in the mid-luteal phase at the time corresponding to implantation and these were analysed by immunohistochemistry with a panel of monoclonal antibodies. The aim of these studies was to gather basic data that would aid in the manipulation of implantation to improve the success rate of human IVF-ET treatment.

Defining the human "Window of implantation"

Before scientific data can be correctly gathered and interpreted regarding human implantation, it is essential to attempt to define the correct time coordinates of the human implantation process. A valuable model to define is the time of implantation that is provided by one particular infertility treatment using oocyte donation in hormonal replacement cycles. In order to induce pregnancy, which implies that the endometrium should be receptive for the blastocyst, Navot and coworkers (1991) found it neccessary to time the embryo transfer in relation to the steroid hormone replacement regimen used. Specifically, they found that implantation only occurred when 2-day-old human conceptuses were transferred to the uterus at days 17-19 of the cycle (day 15 was considered the first day of progesterone administration). This period of the endometrial cycle can be referred to as the "window of transfer", which has to be distinguished from the term "window of implantation". The developmental age of the tranferred conceptuses must be taken into account in order to calculate the plausible time of implantation. The human blastocyst stage is reached at day-5 *in vitro* (Svalander et al., 1991). Given that human preimplantation development may be somewhat retarded *in vitro*, as is usually the case in mouse embryo cultures and, furthermore, that blastocyst activation, hatching and trophectodermal proliferation commences within 24 hours after the embryo has reached the blastocyst stage, one may assume that the earliest day of blastocyst-uterine attachment would be 3 days after the "period of transfer", as reported by Navot and co-workers (1991). This time period would be a fairly good estimation of the beginning of the "window of implantation", that is, during the optimum receptive conditions of the endometrium in the human female, days 20-22 in a normal cycle or, in other words, 7 days after the luteinizing hormone (LH) surge.

Clinical IVF and implantation success

Implantation failure after IVF-ET is often unexplained, typically when good quality embryos have been transferred to a patient in a "normal" treatment cycle. In a well-functioning IVF-program, the "live birth rate" per treatment cycle may be in the range of 25% but, with respect to implantation, the efficacy of human IVF-ET is low. Human implantation is defined as a positive serum hCG and the transvaginal ultrasound detection of a chorio-amniotic sac. Moreover, usually 2-3 embryos are replaced in each cycle, in order to compensate for the poor implantation rate, i.e., the number of viable fetuses per embryo transferred. We analyzed this in 472 consecutive IVF-treatment cycles performed at the Fertility Center Scandinavia in Göteborg during the period August 1991 to May 1992 and found that the total yield of "good quality" embryos (frozen and transferred embryos) was about 44% per retrieved oocyte (Fig. 1). In this early study material, the implantation rate was about 11% with all patient infertility indications included (Fig. 2). These data clearly show that, although great efforts are being put into the several steps of the IVF-ET procedure (superovulation, oocyte retrieval, sperm preparation, culture, transfer and luteal-phase support), there is still much to be learned about the diagnosis and treatment of the factors controlling implantation. In the remainder of the introduction, blastocyst signals, the uterine environment, and endometrial maturation will be discussed.

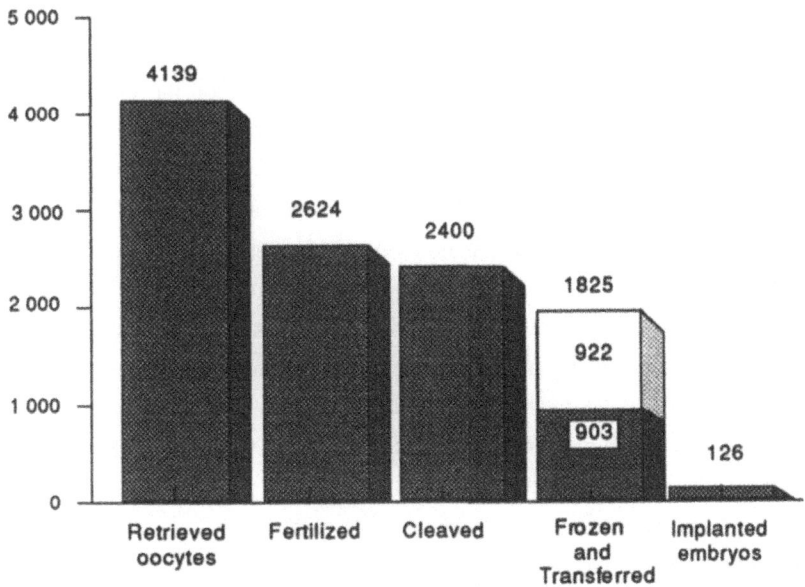

Figure 1. 472 consecutive human IVF-ET treatment cycles (all patient infertility indications) during the period August 1991 to May 1992 at the Fertility Center Scandinavia in Göteborg, Sweden.

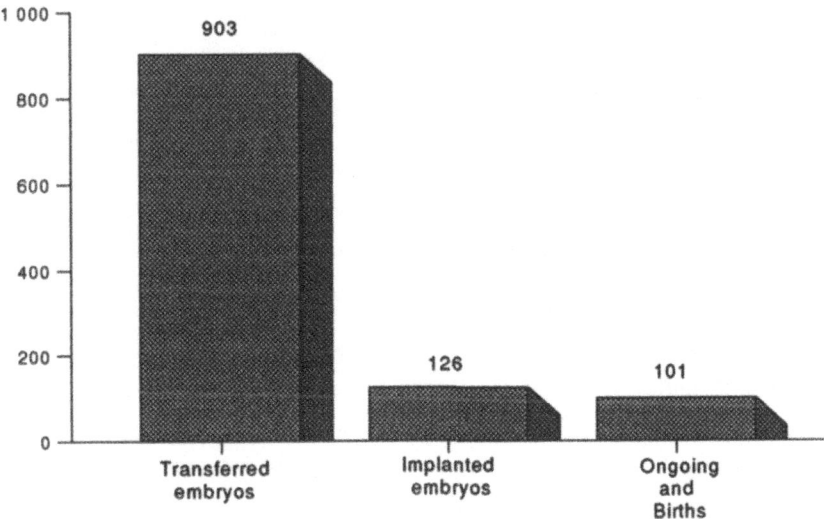

Figure 2. Human IVF-ET treatment efficiency, with the end-point ongoing pregnancies and births, in 472 consecutive cycles representing a clinical implantation rate of 11.2%. For background data, see figure 1.

Human blastocyst signals

Data from human and animal implantation models show that substances produced by blastocysts, e.g., Prostaglandin E2 (PGE2) (Holmes et al., 1989), Colony Stimulating Factor-1 (CSF-1) (Robertson et al., 1992), and Leukemia Inhibitory Factor (LIF) (see chapter by Stewart in this book) can induce decidualization of the endometrium when injected intraluminally. In addition, monoclonal antibodies to either Platelet-derived growth factor (PDGF), CSF-1, Transforming growth factor-β (TGFβ), Epidermal growth factor (EGF) or the EGF-receptor, when microinjected intraluminally into the mouse uterus, significantly reduces the rate of implantation when compared to the denatured and vehicle controls (Holmes and Svalander, unpublished). All of these experimental studies have lead to the assumption that blastocyst signals are also essential for successful human implantation. That blastocyst signals exist in the human implantation process is supported by evidence from some considerable time ago. In the work of Hertig and Rock (1941), the gross examination of an early human implantation site found on day-25 of the menstrual cycle (specimen no. 7699) was described as follows:

"The hemorrhagic areas on the anterior endometrial surface were comparable in arrangement, number, and size to those on the posterior endometrial surface and formed a fairly exact mirror image of them. This suggested a local action of some substance elaborated by the ovum, affecting not only the endometrium immediately around the ovum, but also that of the wall opposite the implantation site."

This unique observation indicates that some substance has the power of affecting the vascular beds immediately surrounding the blastocyst implantation site. Two substances produced by human blastocysts *in vitro* have been reported to date, PGE2 (Holmes et al., 1989) and PDGF (Svalander et al., 1991), both of which can affect the endometrial vasculature. In today's clinical IVF-ET treatment of infertility, normal blastocyst signalling can only be assured by an optimization of the culture system, thereby providing maximum viability and physiological function of the conceptus.

The human uterine environment

When characterizing the mechanisms of implantation, it is important to take into consideration the influence of the uterine milieu on the capacity of the conceptus to undergo blastulation, activation, hatching and trophoblast proliferation. In the clinical situation of IVF-ET treatment, the conceptus is transferred to the patient after reaching the early morula stage. Consequently, a failure of implantation may indeed be a growth inhibition induced by a non-receptive uterine environment rather than a failure in the blastocyst-uterine interactions, *per se*. Molecules able to exert such inhibitory actions can be found among the cytokine-family, some of which exert deleterious effects on preimplantation embryos *in vitro*. Hill and co-workers (1987) studied the influence of some cytokines on early mouse development and found that Granulocyte-macrophage colony stimulating factor (GM-CSF) and Interferon-γ (IFN-γ) exerted embryotoxic effects *in vitro*. This being the case, it seems of vital importance to study the uterine luminal environment in the early- and mid-luteal phases. In fact, studies of the uterine secretions during relevant phases of the menstrual cycle must be undertaken in order to shed more light onto this problem.

Efforts in the direction of defining the uterine milieu are, in fact, underway. We have begun an extensive investigation to measure the changes in pH, viscosity and colloid osmotic pressure of the uterine luminal secretions during the menstrual cycle in normally cycling females. The accumulated *in utero* measurements, so far, indicate that at no time during the cycle does the pH rise above 7.2, this value being at the time of ovulation. Just before and after menstruation, the pH is as low as 6.2 to 6.5 and, during the implantation window, it is approximately 6.9. Although these data are still preliminary, they indicate that our present day assumptions of the correct pH for *in vitro* blastocyst development are quite wrong. These data are also supported by our improved *in vitro* development of mouse blastocysts at lower pH values in new media formulations.

Endometrial maturation for implantation

The menstrual cycle begins with the breakdown, sloughing-off and extrusion of the uterine mucosa that has been formed during the foregoing cycle. Typical histological features prior to menstruation are the infiltration of leukocytes and the congestion of blood vessels (Noyes et al., 1950). When the process of menstruation is completed, active wound-healing and tissue formation takes place. During this proliferative phase a primary biological force is represented by the steroid hormone estrogen. Recent research results emphasize the interaction of ovarian steroid hormones and cytokines within the endometrium (Robertson et al., 1992). Taking into account the fact that numerous leukocytes are present in the human endometrium throughout the cycle, it seems likely that these resident bone-marrow derived cells (Bulmer et al., 1988) may be actively involved in the developmental and stimulatory processes occurring during the proliferative and secretive phases, respectively. Leukocytes are most likely involved after ovulation during the luteal phase, where progesterone induces a differentiation of the endometrium and produces the secretory changes typical of the luteal-secretive phase (Noyes et al., 1950). It is during the mid-luteal phase, approximately 7 days after the LH-surge ("implantation window"), when the optimum receptive conditions exist for the attachment of a blastocyst to the luminal epithelium of the endometrium.

Without progesterone, implantation is not possible. However, the question remains as to what effect progesterone has on the endometrium. One well-known and established progesterone-effect is the induction of secretory activity in the endometrial glands, an activity which peaks during the implantation window. The stromal edema reaches its maximum in the human endometrium around day LH+10/+11 (Johannisson et al., 1987). How, then, is this increase in vascular permeability induced? In order to answer such a question, the effect of progesterone on the human endometrium needs to be characterized in much greater detail. We have initiated such studies by taking advantage of the powerful progesterone-receptor antagonist RU486 (Rousell-Unclaf, Paris, France) and, with the help of this compound, the effect of progesterone withdrawal was studied in a group of normal, fertile women. Cryostat sections of endometrial biopsies, obtained 7-8 days after the LH-surge, were evaluated by a number of immunohistochemical variables (see below). The study was comprised of a control cycle and a treatment cycle with 200 mg RU486 administered orally at the time of ovulation, this being standardized as 2 days after the LH-surge. This dose was chosen in consideration of recent clinical trials in 350 cycles (Gemzell-Danielsson et al., 1993).

MATERIALS AND METHODS

Cycle monitoring

The first day of the patient's menstrual period was considered as cycle day 1. To detect the LH-surge, urine samples were analysed twice a day (morning and evening) using a self-test method (Ovu-Quick, Monoclonal antibodies Inc., CA) and confirmed by an automated immunoassay (Abbott IMX, Abbott, Chicago, IL). From cycle-day 9 onward throughout the cycle, the serum levels of estrogen and progesterone were assayed by radioimmunoassay (Diagnostic Products Corp., Los Angeles, CA).

Endometrial specimens

Endometrial biopsy specimens were obtained using a Randall currette. One piece of each biopsy was fixed in Bouin's solution and a second piece was fast-frozen in liquid nitrogen for storage. All frozen sections were cut at approximately 10 microns thick on a Reichert-Jung cryostat (Cryocut 1800, Reichert-Jung GmbH, Nussloch, Germany). After drying on microscope slides, the specimens were fixed in acetone for 10 minutes, air-dried, wrapped in parafilm and stored at -70 degrees until used.

Immunohistochemistry

The immunohistochemistry, essentially as described previously (Svalander et al., 1990), utilized a panel of monoclonal antibodies directed against a variety of antigens (Table 1), the binding being visualised with the avidin-biotin-immunoperoxidase detection system (Vectastain, Vector, Inc., Burlingame, CA). Control mid-luteal phase biopsies were compared with treatment-cycle biopsies obtained on the same day following the LH-surge. For comparison, biopsies were evaluated and dated according to both histological (Noyes et al., 1950) and morphometric criteria (Johannesson et al., 1987).

Treatment cycles

The patients assigned to the treatment group received one single dose of 200 mg RU486, the progesterone receptor-blocker, immediately post-ovulation, their time of ovulation being represented by the pre-ovulatory peak of LH plus 2 days.

Table 1. Monoclonal antibody panel used for assessing human implantation-phase endometrium

Antigen	Description	Antibody source
E2R	Nuclear estrogen receptor protein	Abbott, Inc.
P4R	Nuclear progesterone receptor protein	Abbott, Inc.
PDGFR	Platelet-derived growth factor AB receptor protein	R&D, Inc.
EGFR	Epidermal growth factor receptor protein	Oncogene Science, Inc.
IGF-1R	Insulin-like growth factor-1 receptor protein	Oncogene Science, Inc.
CD44	Leukocyte homing receptor	T Cell Diagnostics, Inc.
CD45	Pan-leukocyte marker	Dakopatts, Inc.
CD56	Natural killer cell marker	Becton Dickinson, Inc.
BerMac3	Activated macrophages	Dakopatts, Inc.
Vitronectin-R	Vitronectin receptor, β-chain	Takara Shuzo Co., Ltd.
CD31	Platelet-endothelial cell adhesion molecule	Dakopatts, Inc.
ELAM-1	Endothelial-leukocyte adhesion molecule	Genzyme, Inc.
ICAM-1	Intercellular cell adhesion molecule	Genzyme, Inc.
VCAM-1	Vascular cell adhesion molecule	Genzyme, Inc.
EN 7/44	Angiogenesis marker	Dianova GmbH

RESULTS

One single dose of 200 mg RU486 administered immediately post-ovulation affected the endometrium in all treated subjects. It retarded endometrial development and this retardation could be observed as immature glands, impaired secretory activity and diminished stromal edema. Although RU486 has a profound anti-progestational effect which is manifest in a retarded endometrial morphology, surprisingly few changes were seen at the immunohistochemical level for the variables studied, as listed in Table 2.

Progesterone and estrogen receptors

Progesterone and estrogen receptors are maximally expressed in the human uterus during the mid- to late proliferative phase of the menstrual cycle (Garcia et al., 1988; Lessey et al., 1988). The most profound effect of RU486 was seen on the expression of progesterone receptors, which exhibited impaired down-regulation in the glandular epithelial cells. Progesterone receptors remained strongly positive in the glandular epithelium and exhibited an irregular stromal distribution, with negative areas adjacent to strongly positive areas. Estrogen receptors, however, were only weakly detected in the endometrium and did not show any change in pattern after the RU486-treatment.

131

Growth factor receptors

The three growth factor receptors studied (PDGFR, EGFR and IGF-1R) showed different localization in the tissues. PDGFR exhibited a granular staining pattern present around vascular structures, and specifically, in the endothelial and smooth muscle cells of the vessels. This newly discovered labelling pattern was revealed only after using the monoclonal antibody at a concentration which permitted detection of the ligand-bound receptors. EGFR was present in the stroma and in the basal part of the glandular epithelium, whereas IGF-1R was detected in the glandular epithelial cells. Furthermore, no change was seen in these different staining patterns or their staining intensity after the RU486-treatment. These observations lead us to conclude that none of these growth factor receptors are under progesterone control in the human endometrium.

Leukocyte markers

CD44 is broadly distributed and believed to be involved in leukocyte chemotaxic mechanisms. This molecule was extensively distributed in the endometrial stroma, both in the control and the RU486-treated specimens, without any differences between the two groups.

CD45 is usually present on all cells of haematopoetic origin, except erythrocytes. A prominent staining of scattered cells was observed, sometimes in aggregates, without any specific localization within the endometrium. Moreover, there was no difference in staining between the control and RU486-treated endometrial sections. This pan-leukocyte staining was in the order of 20% of the total number of cells in the endometrial stroma, which coincides well with earlier studies.

CD56 is an isoform of the neuronal cell adhesion molecule (NCAM) and is expressed by all lymphocytes mediating non-MHC-restricted cytotoxicity. Monoclonal antibodies against this CD56-isoform are used as pan-natural killer (NK) cell markers. In the mid-luteal phase, positive cells were detected in a scattered pattern throughout the endometrium, without any difference between the two groups studied.

BerMac3 labels activated macrophages and this marker showed a distribution coinciding with the CD56 positive cells. Similarly, this marker did not differ between the groups.

It can be concluded from the observations above that the presence, number and distribution of haemopoietic cells in the human endometrium are not under the control of progesterone.

Cell adhesion molecules and endothelial cell markers

The $\alpha v \beta 3$ integrin (Vitronectin-R) has recently been attributed to being involved in the adhesion of the blastocyst to the endometrium during the initiation of implantation (Lessey et al., presented at this meeting). However, our studies did not show any staining for this molecule.

CD31 (PECAM) was present in all vascular structures in the endometrium and had a similar distribution and staining intensity in all specimens. Since we were unable to detect any change in the vascular structures, we used a number of endothelial cell markers which are related to the function of the vessels. These are usually detected during an inflammatory response, which, hypothetically, could be related to the development of the stromal edema during the implantation-phase. Despite this possibility, we were unable to detect ELAM-1, ICAM-1, VCAM-1 or EN 7/44 (a marker of angiogenesis) in the human endometrium, regardless of state. In conclusion, the endometrial vascular beds did not show any of the typical signs of inflammation at the molecular level. Obviously, the influence of progesterone during the luteal phase does not affect these inflammatory markers.

Table 2. Immunohistochemical localization of molecular markers

Antigen	Principal cellular localization	Alteration after RU486
E2R	Weak staining in stroma and glands	None
P4R	Stroma	Glandular epithelial staining
PDGFR	Granular staining around vascular structures	None
EGFR	Stroma	None
IGF-1R	Glandular epithelial staining	None
CD44	Stroma	None
CD45	Scattered cells throughout the endometrium	None
CD56	Scattered cells throughout the endometrium	None
BerMac3	Scattered cells throughout the endometrium	None
Vitronectin-R	Negative	None
CD31	Vascular structures	None
ELAM-1	Negative	None
ICAM-1	Negative	None
VCAM-1	Negative	None
EN7/44	Negative	None

DISCUSSION

The experimental findings in the work presented show that the untreated endometria exhibited progesterone receptors in the stroma and none in the glandular epithelium. The RU 486-blocked endometria, to the contrary, exhibited progesterone receptors in the glandular epithelium and none in the stroma. Otherwise, the quantity of staining for P4 receptors was similar in both situations.

It is an accepted fact that progesterone induces those changes in the estrogen-primed, human endometrium which are essential for implantation and pregnancy. Of course, neither the initial estrogen stimulus of the uterus just prior to ovulation nor the increasing progesterone levels after ovulation were different in the two groups. Therefore, the stimulation for progesterone-receptor development must also have been the same for both patient groups. It seems that the effect of RU486 was to mask the functionality of the receptors and, in so doing, hinder the normal functional and morphological development of the endometrium. However, numerous other immunohistochemical parameters appeared unaffected.

In using this progesterone-receptor antagonist, we failed to reveal any dramatic alterations in the presence and distribution of growth factor receptors, leucocyte markers and endothelial cell markers. The explanation for this could not be that we were unsuccessful in totally blocking the effect of progesterone, since the same dose of RU486 was used successfully in a recent contraceptive trial comprised of 350 unprotected cycles (Gemzell-Danielsson et al., 1993). Consequently, the most likely explanation is that these variables are not controlled by progesterone at all.

The present findings open the field for speculation regarding the mechanisms of implantation. It may well be that some receptors are regulated by mechanisms other than the ovarian steroid hormones, perhaps through the resident endometrial leukocytes or via direct influences from the pre-implantation conceptus.

Many leukocytes appear to reside in the endometrium without any alterations from the progesterone receptor antagonism. If they affect important functions involved in preparation of the endometrium for implantation, possibly through local regulation of the glandular and endothelial functions, a new panorama of research problems and possibilities opens up.

The pre-implantation conceptus, however, is known to produce and secrete active molecules, such as PGE2 (Holmes et al., 1989), CSF-1 (Robertson et al., 1992), hCG (Fishel et al., 1984; Lopata et al., 1993), IL-1a (Sheth et al., 1991) and IL-1b (Baranao et al., 1992), all of which may, in turn, have an effect on the progesterone-ripening of the endometrium. PGE2 is reknown as being an immunosuppressive prostaglandin that can also increase the life span of the corpus luteum, while hCG is definitely involved in stimulating the corpus luteum and increasing progesterone production. Of course, no conceptus was present in the normal patients of this study and, therefore, none had any influence on the endometrium. Moreover, it is quite probable that the endometrium of the pre-implantation pregnant uterus is quite different from that of the normally cycling female with respect to the immuno-histochemical parameters measured in this study.

ACKNOWLEDGMENTS

This study was supported by the Swedish Medical Research Council, Serono Nordic AB, and the Clinical and Psychosomatic Research Foundation. Mrs. Berit Ståbi provided excellent technical assistance.

REFERENCES

Baranao, R.I., Piazza, A., Rumi, L., Polak de Fried, E., 1992, Predictive value of Interleukin-1β in supernatants of human embryo cultures. 48th annual meeting of the American Fertility Society, O-007.

Bulmer, J.N., Lunny, D.P., and Hagin, S.V., 1988, Immunohistochemical characterization of stromal leukocytes in nonpregnant human endometrium. Am. J. Reprod. Immunol. 17:83-90.

Fishel, S.B., Edwards, R.G., and Evans, C.J., 1984, Human chorionic gonadotropin secreted by preimplantation embryos cultured in vitro. Science 223:816-818.

Garcia, E., Bouchard, P., De Brux, J., Berdah, J., Frydman, R., Schaison, G., Milgrom, E., and Perrot-Applanat, M., 1988, Use of immunocytochemistry of progesterone and estrogen receptors for endometrial dating, J Clin Endocrin Metab. 67:80-87.

Gemzell-Danielsson, K., Swahn, M.-L., Svalander, P. and Bygdeman, M., 1993, Early luteal phase treatment with mifepristone (RU 486) for fertility regulation. Human Reprod. 8:870-873.

Hertig, A., and Rock, J., 1941, Two human ova of the pre-villous stage, having an ovulation age of about eleven and twelve days respectively, in Contributions to Embryology, No. 184, Carnegie Institution of Washington, Baltimore.

Hill, J.A., Haimovici, F., and Andersson, D.J., 1987, Products of activated lymphocytes and macrophages inhibit mouse embryo development in vitro. J. Immunol. 139:2250-2254.

Holmes, P.V., Sjögren, A., and Hamberger, L., 1989, The immuno-modulatory compound prostaglandin-E2 is released by the preimplantation human conceptus. J Reprod Immunol. 17:79-86.

Johannisson, E., Landgren, B.M., Rohr, H.P., and Diczfalusy, E., 1987, Endometrial morphology and peripheral hormone levels in women with regular menstrual cycles, Fertil Steril. 48:401-408.

Lessey, B.A., Killam, A.P., Metzger, D.A., Haney, A.F., Greene, G.L. and McCarty, Jr., K.S., 1988, Immunohistochemical analysis of human uterine estrogen and progesterone receptors throughout the menstrual cycle, J Clin Endocrin Metab. 67:334-340.

Navot, D., Scott, R.T., Droesch, K., Veeck, L.L., Liu, H-C., and Rosenwaks, Z., 1991, The window of embryo transfer and the efficiency of human conception in vitro, Fertil. Steril. 55:114-118.

Noyes, R.W., Hertig, A.T., and Rock, J., 1950, Dating the endometrial biopsy, Fertil Steril. 1:3-25.

Robertson, S.A., Brännström, M., and Seamark, R.F., 1992, Cytokines in rodent reproduction and the cytokine-endocrine interaction, Curr. Opin. Immunol. 4:585-590.

Sheth, K.V., Roca, G.L., Al-Sedairy, S.T., Parhar, R.S., Hamilton, C.J. and Al-Abdul Jabbar, F., 1991, Prediction of successful embryo implantation by measuring interleukin-1-α and immunosuppressive factor(s) in preimplantation embryo culture fluid. Fertil. Steril. 55:952-957.

Svalander, P.C., Odin, P., Nilsson, B.O., and Öbrink, B., 1990, Expression of Cell-CAM 105 in the apical surface of rat uterine epithelium is controlled by ovarian steroid hormones, J. Reprod. Fert. 88:213-221.

Svalander, P.C., Holmes, P.V., Olovsson, M., Wikland, M., Gemzell-Danielsson, K., and Bygdeman, M., 1991, Platelet-derived growth factor is detected in human blastocyst culture medium but not in human follicle fluid - a preliminary report, Fertil. Steril. 56:367-369.

Lopata, A., and Oliva, K., 1993, Regulation of chorionic gonadotropin secretion by cultured human blastocysts, in Preimplantation Embryo Development, B.D. Bavister, ed., Springer-Verlag, New York.

RU486: AFTER TEN YEARS
NOVEL MOLECULES AND REPRODUCTIVE MEDICINE

Etienne-Emile BAULIEU

Unité de Recherches sur les Communications Hormonales
(INSERM U 33) and Faculté de Médecine Paris-Sud, 80 rue du
Général Leclerc, 94276 Bicêtre Cedex, France

The story of RU486* (Figure 1) may be seen as the combined result of the women's movement during the 20th century to control their reproductive life and the contemporary scientific bio medical revolution of the last few decades.

This conjunction was exemplified in 1950 when Margaret Sanger went to see Gregory Pincus to request a medical method to achieve "planned parenthood". As a result, "the" contraceptive pill (Pincus, 1965) was born and it remains at least as important symbolically as it is useful practically. Here merged science (hormone research) and the "cause des femmes". Scientifically, it was based on the physiological concept that sex steroid hormones exert a negative feed-back control on ovulation, which could be applied with precise steroid chemistry to provide orally active compounds (Djerassi, 1970).

In the sixties and seventies, it became clear that the available contraceptive methods could not completely cover the reproductive choices desired by the women of the world and their families, nor have the needed demographic effects occurred which could limit the population explosion. During these decades, ideas for alternative methods of contraception emerged as biology became focused more on cellular and molecular elements. The hormone responsive proteins of target cells of the reproductive tract, termed receptors, were discovered while

---

*Mifepristone, RU38486, 17β hydroxy 11β (4 dimethyl aminophenyl-1) 17α-(prop-1-ynyl)-estra-4,9-diene-3-one. Many publications are already available: the first paper (Herrmann et al., 1982) summarized the first laboratory and clinical data, the book edited with S. Segal (Baulieu and Segal, 1985) reported on the Bellagio meeting grouping almost all contributors known at that time, and since, many reviews have partially covered the field which has become very large: Henderson, 1987; Neef, 1987; Baulieu, 1989a,b,1991a,b; Laue et al., 1989; Avrech et al., 1991; Philibert et al., 1991; Ulmann et al, 1990; Cook and Grimes, 1992; Horwitz, 1992; Mao et al., 1992; Brodgen et al, 1993.

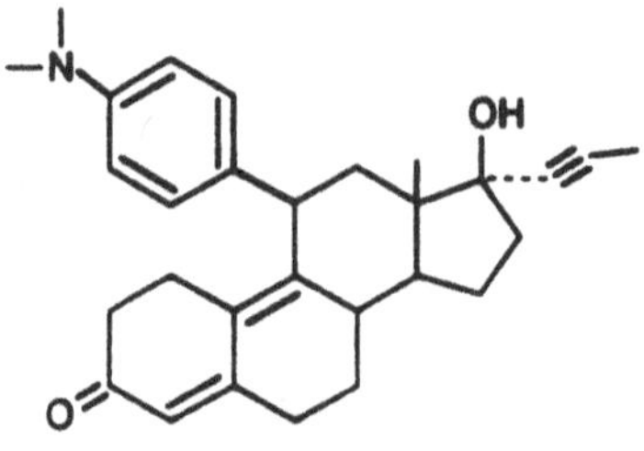

Figure 1

progesterone (P), designated by Corner (in 1932!; see Corner, 1963) as the hormone of gestation (pro gestare), was now easy to quantitate by radioimmunoassays (Lieberman, 1959). The uterine progesterone receptor (PR) (Milgrom et al., 1970) and the synthesis and action of prostaglandins (PG) (Bergström et al., 1972) were described, while the role of progesterone in the establishment and maintenance of pregnancy in women was demonstrated (Csapo and Pulkkinen, 1977). As it became clear that progesterone is involved at all steps of the reproductive processes, antagonists to progesterone were actively sought. As early as 1975, a "mid-cycle" contraception, a method based on progesterone receptor down-regulation obtained with an "antiprogesterone" ligand was proposed (Baulieu, 1975). Now, we have efficient antiprogestins. If abortion has been the most immediate application of such a compound, then delivery and contraception can also benefit from them, not to mention several hormone dependent diseases.

When developing a procedure for the termination of pregnancy in women, it is important to be aware of both moral and physiological ideals, as well as psychological concerns. For centuries abortion has been, not only a morally difficult event for women, but also a physically painful and dangerous procedure. A medical means should relieve this threat to women's health and, in turn, maintain their dignity. Furthermore, the distinction between abortion and contraception has become significantly less distinctive, because the beginning of pregnancy is now understood, in physiological terms, to be a progressive succession of steps. Hence, I have proposed the term *contragestion* (Baulieu, 1985, 1989a,b) to clearly designate a method which can provoke pregnancy interruption (contra gestation) and operates as soon as possible after fertilization might have occurred, before the word abortion is appropriate (is an IUD considered an abortifacient?) (see later discussion). This change in concept may be one of the most important outcomes of RU486 usage.

ANTIHORMONES: THE 20 YEARS BEFORE RU486

The aim of suppressing hormone activity is almost as old as the word hormone (ωρμειv: to excite) itself.

If a hormonal molecule is excitatory for the target cells, then suppression of its effects can be attained by i) abolition of its production, ii) blockade of its transport from the producing gland to target organs, or iii) in the case of small and lipophilic steroids such as P, which act intracellularly, prevention of its entry into potentially responsive cells. At present, only the first of these three possibilities, suppression of biosynthesis, seems feasible in humans. For example, the use of enzymatic inhibitors such as the drug Epostane (4,5-epoxy-17β-hydroxy-4,17α-dimethyl-3-oxo-5α-androstane-2-carbonitrile, an inhibitor of 3β-hydroxy steroid dehydrogenase) has been tested with some success in abortion (Birgerson and Odlind, 1987, Crooij and Janssens, 1988). However, an antihormone at the receptor level may act more rapidly and be more specific than an inhibitor of a key enzyme involved in the synthesis of many steroids. An approach to block the action of hormones, the use of specific antibodies for instance, interacting with P in the blood or in target organs (Wang et al., 1989) does not seem easily applicable to humans. In fact, the center of hormone action and thus the best molecular target for antihormonal action is the receptor (R) protein molecule, a mandatory element for cellular responses to hormone[†]. The image of receptors portrayed as a lock whose key is the hormone and whose keyhole (in fact a "binding site") can be competitively occupied and consequently put out of order by a false key (an antihormone) has been popular for decades. Because steroids are rigid molecules of well defined conformation, as also should be the high affinity binding site of the receptor, it seemed logical to expect a breakthrough in the hormone antagonism field would occur via the antisteroid field[‡]. Initially, receptors were detected by the binding of a traceable (labeled) hormone to tissue extracts. The first of these so-called radio-receptor experiments was performed with tritiated estradiol (the natural estrogen), and an antiestrogen such as MER 25 competed efficiently for the radioactive hormone uptake and retention in the uterus (Jensen and Jacobson, 1962). The structure of MER 25 (Figure 2) is not that of a steroid, but a triphenyl-ethylene derivative of stilbene with two phenyl cycles mimicking ring A and D of the steroids. X-ray crystallographic studies of the non-steroidal estrogen diethylstilbestrol (DES) and estradiol (E), have delineated their similarity (Hospital et al., 1972). The third cycle of triphenyl-ethylene derivatives is perpendicular to the rest of the steroid mimicking skeleton (Figures 2 and 3). This was of great importance; considering the high affinity that molecules such as E and DES show for the receptor, it was not surprising that triphenyl-ethylene derivatives such as MER 25 and tamoxifen (Figure 2) had lower affinity than the agonists. In fact, the presence or absence of the third cycle is not the critical factor for determining binding affinity, since 4-hydroxytamoxifen, with an additional hydroxyl to the ring A-equivalent of the tamoxifen molecule and mimicking the 3-hydroxyl group of estradiol, renders the compound with high affinity for the receptor and a resulting strong antiestrogenic effect ("pure", with no agonist activity in the chick (Sutherland et al., 1977). I was very impressed by these data, because they contradicted the current thinking, in that all known antagonists tested at that time had low affinity for their respective receptor. Antiestrogens (e.g. tamoxifen), antiandrogens (e.g. cyproterone acetate, flutamid), antiglucocorticosteroids (e.g. P), antialdosterone (e.g. spironolactone, P), and screening for antihormonal steroids tended to eliminate compounds demonstrating a high affinity for the receptor. Indeed, there was no adequate

[†]RU486 can be accommodated between partially unwound, double-stranded-DNA bases by computer modeling. Yet the altered conformation of DNA cannot be correlated to the pharmacological properties of antiprogestins (Hendry and Mahesh, 1992).

[‡]Thyroid hormones are other small, rigid and lipophilic hormones, and the corresponding antihormones "logically" could have been found earlier. However, we do not yet know of an antithyroid hormone compound acting at the receptor level.

theoretical reason to relate the quantitative notion of high affinity to the qualitative property of hormone antagonism, since the latter was predictably due to specific transconformation of receptor domain(s) involved in the transcription activation functions (TAFs) of the receptor, in particular in the ligand binding domain (LBD), while steroid binding affinity, reflecting interaction with the binding site also located in the LBD, is only important for kinetic quantitative aspects of the activity. I presented this scenario to Robert Bucourt, then head of the chemistry at Roussel-Uclaf in the early seventies.

Interestingly, Dr Bucourt and his colleagues also collaborated with us to purify the estrogen receptor by affinity chromatography. Initially, this involved the screening of potential receptor ligands and among the synthetic derivatives tested, by Hélène Richard-Foy, were estrogens with a long side chain grafted at the 7α position. We selected one of them for receptor purification (Bucourt et al., 1978); however, Roussel did not test its biological activity. It was found to be an antagonist of estrogens by the ICI researchers about ten years later! (Wakeling and Bowler, 1988). It is important to note that the 7α-substitution on the steroid skeleton is somewhat symmetrical to an 11β-substitution, consistent with the structures of the triphenyl-ethylene antiestrogens and of the 11β-phenyl derivative compounds of the RU486 series.

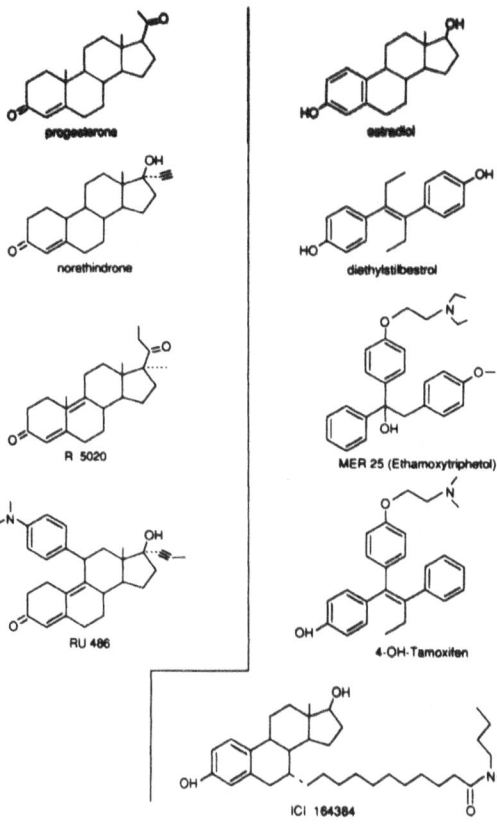

PROGESTINS, ESTROGENS AND ANTAGONISTS

Figure 2

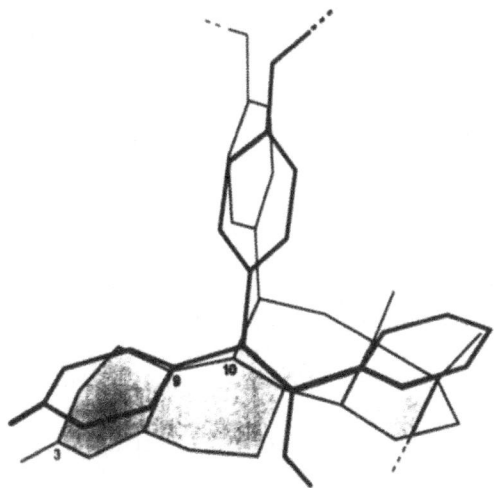

Figure 3. Calculations have been made from X Ray crystallographic data by Jean-Paul Mornon (Laboratoire de Crystallographie, CNRS URA 09 and Universités Paris VI and Paris VII).

Also in the early seventies, the Roussel chemists, while working to improve synthesis of new glucocorticosteroids, found a new way to produce 11β-derivatives of steroids. They discovered that 5α, 10α epoxides obtained by metachloroperbenzoic acid treatment of 5(10), 9(11)-estradienes are prone to nucleophilic opening with Grignard reagents (Nédélec and Gasc, 1970). In addition, either copper chloride catalyzed Grignard reagents or lithium organocuprates efficiently gave the corresponding regio- and stereo-specific 11β-substituted 4,9-estradienes (Teutsch and Bélanger, 1979; Bélanger et al., 1981). Interestingly, the size of the substituent appears to largely determine agonistic or antagonistic activities.

Thus chemical research for the synthesis of glucocorticosteroids and biological studies of estrogens/antiestrogens had converged when the RU486 series of compounds were synthesized by Georges Teutsch and colleagues (Teutsch et al., 1988). The remarkable analogy of orientation of the third cycle of tamoxifen and fifth cycle of RU486 (approximately coplanar with C9-C11 bound), both perpendicular to the basic stilbene or steroid skeletons is shown in Figure 3. Indeed, the 11β-phenyl-N-dimethyl substituted estradiol is a strong antiestrogen (unpublished result).

Briefly, the rest of the RU486 story, which has been presented in several publications (see note* p. 1), continued with the antiglucocorticosteroid activity of RU486 being observed, and thereafter the demonstration of its antiprogesterone property. The decision to test it for human abortion was made after the endocrinological and pharmacological studies performed by Daniel Philibert and colleagues. The compound was active and probably safe, we proposed, but the idea of using RU486 in human beings was almost "killed" by toxicologists who did not correctly interpret the signs of cortisol insufficiency when the product was given at very high doses in monkeys for several consecutive weeks: RU486 was rescued by my insistance that it was just a beautiful demonstration of the activity of the compound in primates (Baulieu, 1991c).

The career of the compound also became the subject of political debate, and this is not relevant to this résumé. However, the scientific story is not finished, and should be pursued in order to improve and to extend the first discoveries.

NOVEL MOLECULES

All the powerful antiprogestins and antiglucocorticosteroids so far described are 11β phenyl substituted steroids. The relatively long half-life of RU486 in human beings (~ 20 hours) seems to be due to its ability to bind to plasma orosomucoid (an $α_1$-glucoprotein) (Moguilewski and Philibert, 1985; Grimaldi et al., 1992); this binding is not found in non-human primates or other animals. RU40555 (see Figure 4 for structure of this and other compounds discussed in this section) does not bind to the orosomucoid, and has a shorter half-life which may be of interest for the kinetic assessment of the hypothalamo-pituitary-adrenal axis in clinical endocrinology (Bertagna et al., 1984; Gaillard et al., 1984). However, the binding of RU486 and lilopristone (ZK98734) to orosomucoid may enhance the antisteroid activity since it protects the drug against metabolic inactivation and provides a reservoir system for sustained delivery to target cells.

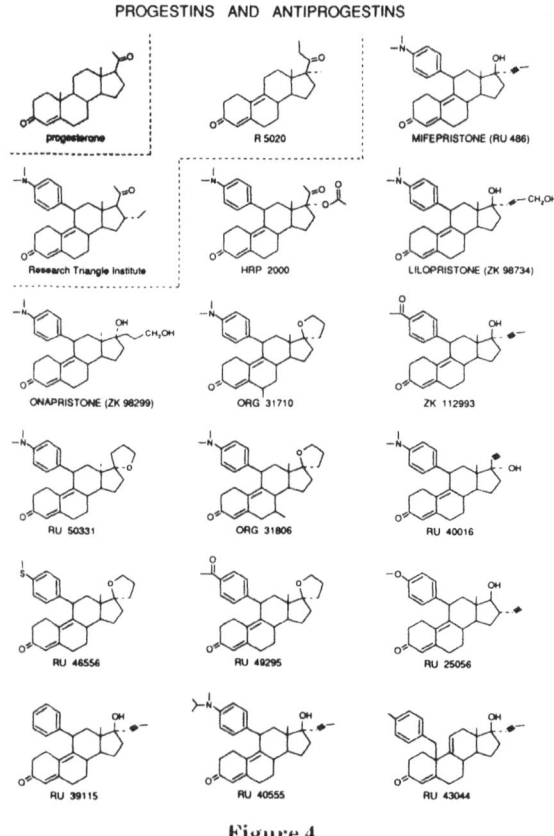

PROGESTINS AND ANTIPROGESTINS

Figure 4

Since the early studies with RU486, chemists have tried to dissociate the two main antihormonal activities of the compound and wished, for obvious medical reasons, to obtain "pure" antiprogestin(s) and "pure" antiglucocorticosteroid(s), which, in addition, would not display any other endocrine effects.

At the present time, there is no published account of a *pure antiprogestin* compound. However, it is important to note that for abortion the antiglucocorticosteroid effect is apparently neither necessary nor even useful, and that a single dose of ≤ 600 mg of RU486 does not create any medical problem related to corticosteroid insufficiency. RU486 derivatives[§], such as RU46556 and RU49295 are strong antiprogestins with limited antiglucocorticosteroid activity. ORG 31710 (more active) and ORG 31806 have less antiglucocorticosteroid activity than RU486 (Mizutani et al., 1992). A 17α-acetoxy derivative such as HRP2000 (Research Triangle Institute, Cook et al., 1992), with a 17β-progesterone side-chain and a 11β RU486-like substituent, is both an antiprogestin and an antiglucocorticosteroid. Curiously, 17β-acetyl, 16α-ethyl derivatives of 11β-phenyl-substituted steroids are progestin agonists (Cook et al., 1992). The Schering group has synthesized lilopristone, with a 17β side chain slightly different from that of RU486, with less antiglucocorticosteroid activity and higher binding to the androgen receptor. Another compound (ZK 112 993) also has reduced antiglucocorticosteroid activity in the rat, due to the acetyl group on the 11β-phenyl group. A significant change in the RU486 structure was obtained using onapristone (ZK 98 299); due to photochemical epimerization at C-13 there is inversion of the D ring and substitutions at the C17 position (Neef et al., 1984). Onapristone does not bind to orosomucoid (contrary to RU486), does not bind to the chick (c) PR (as RU486) (Nath et al., 1991), is an antiprogesterone (but less than RU486) and has weak antiglucocorticosteroid activity. Its mechanism of action may not involve binding to DNA (see later).

A *pure antiglucocorticosteroid* may be easier to use chronically in women. One possibility is RU40016, an RU486 like compound with inversion of substituents at the C17 position. While not very active, it has relatively more antiglucocorticosteroid and less antiprogestin effects than RU486. RU43044 is chemically very different, since the additional phenyl substituent is in the 10β position, and even if this ring is spacially partially superimposable to a phenyl group in 11β, there is no binding to the PR and the activity is purely antiglucocorticosteroid (even weaker than that of RU486). The compound, perhaps because of metabolic reasons, has no activity in vivo in animals, but being active in situ it provides some clues for the synthesis of a series of locally active therapeutic agents.

In conclusion, the 11β-phenyl substitution is essential in determining the antagonistic properties of most antisteroids, while an 11β-aliphatic chain may result in agonistic derivatives (Figure 5). However, *no steroidal structure carries per se an absolute intrinsic property of agonism or antagonism*, as demonstrated by steroid binding differences between the PR of different species, changes of activity when mutating the receptor LBD, and when the compound acts in different target cells themselves in different physiological states.

[§]RU46534 is a very active "contragestive" agent in dogs. The only structural difference with RU486 is its allylic 17α side chain.

RU486 and many corresponding Schering and ORG compounds do not bind to the cPR (Groyer et al., 1985), contrary to the binding found to the PR of humans (h) and most mammals. The change of a cysteine in the N-terminal region of the cPR LBD (Cys 575) to a corresponding glycine residue found in the hPR (cPR Cys 575 → Gly), permits the binding of RU486 and antisteroid activity. Interestingly, RU39115, that is RU486 minus N-dimethyl, is an antagonist of the hPR, but an agonist of the "humanized" cPR 575 → Gly, indicating that the interaction of steroid and receptor is more complex than just binding ability, and may depend on the overall structure of the LBD and consequently modify TAF2 function (see later). Systematic experiments indicate that according to the nature and positioning of the 11β-phenyl substituent, one may produce 11β-substituted steroids with progestin agonistic, antagonistic, or mixed agonistic and antagonistic activities (with, as expected, no relationship with binding affinity) (Benhamou et al., 1992; Garcia et al., 1992) (Figure 6).

Indeed it is just logical that the structure of the steroids and that of the LBD combine to ultimately direct the conformation and thus the function of TAF2, thus "deciding" if a compound will act as an agonist or an antagonist. This is potentially important for cancer treatment, since steroid receptor mutations observed in certain tumors may, therefore, radically change the properties and then the applicability of steroidal drugs.

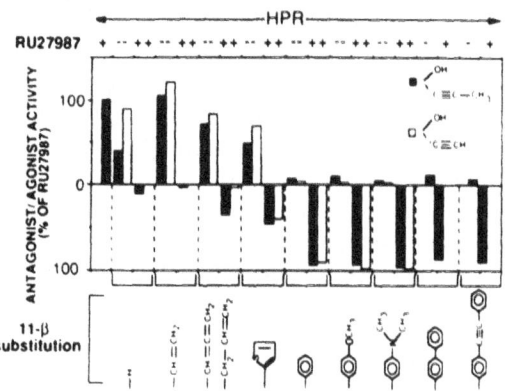

Figure 5. Agonistic and antagonistic potential of two series of 11β-substitutited steroids that differ from RU486 only in their 11β-substitutions or their 11β- and 17α-substitutions. The chemical symbols in the top right corner of the upper panels illustrate the 17α-substitutions. Tanscription activation was quantitated from normalized CAT assays in HeLa transiently transfected cells with the MMTV CAT reporter gene and the hPR expression vector hPR1 in the presence of the various compounds. The agonistic potential of ,the hPR in the presence of 1 µM of these steroids alone (-RU27,987 at the top) is expressed as a positive value relative to the activation seen with 10 nM RU27,987 (arbitrarily asssigned +100). Antagonistic potential was assayed by exposing transfected cells to 1 µM of a given steroid plus 10 nM RU27,987 (+RU27,987 at the top) and is expressed as a negative value, with -100 indicating complete inhibition of RU27,987 induced transcription. The individual 11β-substitutions are depicted. RU27,987 is a 17α,21-dimethyl 3,20 dioxo,21-hydroxy-19-nor pregna-4,9-diene progestin agonist. (Garcia et al., 1992).

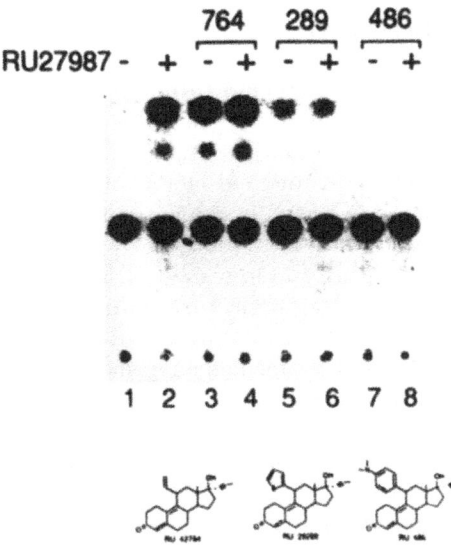

Figure 6. Ligands for the hPR may generate three distinct types of TAF-2-dependent transcriptional responses: they act as *agonists* with no antagonistic potential, or as *antagonists* with no agonistic potential, or they may generate a *mixed* response, since they both activate and antagonize transcription activation. HeLa cells were transiently transfected with a reporter gene and a PRE, and exposed to the steroids in the absence or presence of RU27,987 (Figure 5). Steroids were used at 1 μM (RU27,987; lanes 3, 5, and 7); in cases where the antagonistic potential was analyzed, activation was achieved with 10 nM RU27,987; lanes 4, 6, and 8). Note that RU28,289 acts as both agonist and antagonist (Garcia et al., 1992).

Table 1. RU486: use in reproductive medicine

CONTRACEPTION/CONTRAGESTION/ABORTION
        5-9 weeks of amenorrhea
        occasional luteal contragestion
        once a month menses induction
        emergency contraception
        once a month antiimplantation
        "endometrial" contraception
        suppression of ovulation

MEDICAL INTERUPTION OF PREGNANCY
        2nd trimester
        3rd trimester

LABOR INDUCTION

Voluntary Early Pregnancy Interruption

On the basis of animal experiments, we expected RU486 to meet the recognized need for an efficient medical means of early abortion, safer than an instrumental technique and relatively convenient and cheap (no anesthesia, no operating room). To develop a medical method of abortion was a must in terms of women's health, and potentially a step toward more privacy for those having taken the difficult decision of pregnancy termination.

Early abortion offered the first opportunity to test RU486 activity in humans, and this was initially performed on a small group of women volunteers, under the direction of Walter Herrmann at the Hôpital Cantonal in Geneva in 1982. Since the postulated mechanism of action (Figure 7) suggested that a single dose should be sufficient, and since the compound was rapidly eliminated from the body, only rather short toxicological studies were necessary.

The results were so striking that RU486 immediately got the nick-name of "abortion pill". This was in spite of the many other medical potentialities already predicted when the compound was announced (they progressively become reality). The first large scale studies indicated approximately 80% success in ≤ 42 days of amenorrhea pregnancies (Couzinet et al., 1986), and the compound was presented for registration to the French Ministry of Health. However, following the Bygdeman trial (Bygdeman and Swahn, 1985) (Table 2), the improved results obtained with the subsequent (approximately 48 hours later) administration of a small dose of prostaglandin 48 h after RU486 (Baird et al.,

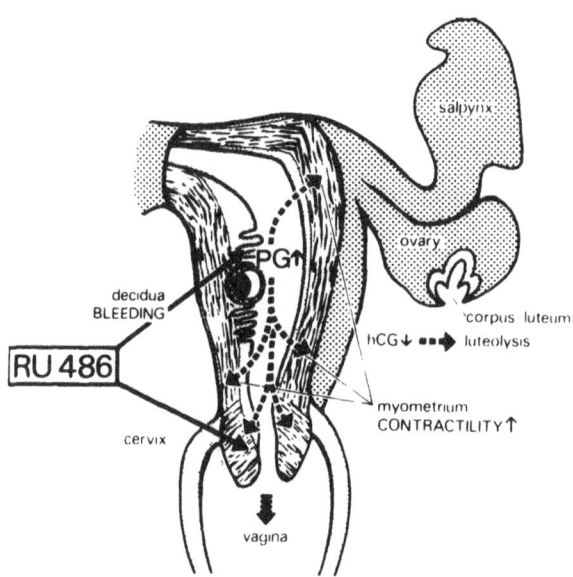

Figure 7. Physio pharmacological mechanism of action of RU486 on the implanted blastocyst. Temporally, the antiprogesterone effect of RU486 comes first, and then an increase of PG concentration and action, and a decrease of hCG sustained corpus luteum function.

Table 2. Uterine activity during early pregnancy in control and RU486 treated patients (Montevideo Units; mean ± SEM)

	Control	36 h after RU486 (50 mg)
mean uterine activity	6 ± 4	222 ± 93
sulprostone 0.05 mg*	49 ± 24	711 ± 136

*administered 0.5 hour after start of recording
(Swahn ML and Bygdeman, M. 1988)

1988; Dubois et al., 1988a) led to the approval ("AMM": autorisation de mise sur le marché) of RU486 + prostaglandin for up to 49 days of amenorrhea pregnancies, in September 1988, in the context of the French law of abortion. Sulprostone (a PGE_2 analog, 250 µg intramuscularly injected) was the prostaglandin mostly used. In Great Britain, the trials with gemeprost (a PGE_1 analog, 0.5 mg administered vaginally) were satisfactory when given up to 63 days of amenorrhea, and registration was acquired in July 1991, as in Sweden in 1992 with the same protocol. The largest study (Ulmann et al., 1992) indicated 95% complete efficacy, with 1% of ongoing pregnancies, emphasizing the obligation to instrumentally evacuate the uterus in case of failure. There were no particular bleeding problems with only approximately 1/1000 patients receiving a transfusion. However, three myocardial infarctions, including one fatal case (a medical mistake occured when sulprostone was injected into a woman at great risk), were recorded after more than 60,000 cases. Sulprostone for intramuscular administration, responsible for the accidents, has since been withdrawn from the market in France.

Coincidently, the death of a patient was reported at the same time as the first trial of RU486 plus orally active misoprostol (a PGE_1 derivative) instead of Sulprostone, was reported (Aubeny and Baulieu, 1991). We had hoped to use a safer prostaglandin (misoprostol has a record of millions of users for prevention and treatment of gastrointestinal ulcers in individuals often more at cardiovascular risk than normal pregnant women). It was also obvious that an orally effective, already available, cheap, and easy to store prostaglandin, had the potential to be an important improvement, leading to a more convenient and private method. Most results (Peyron et al., 1993) have been obtained using 600 mg of RU486 and misoprostol 400 µg (two tablets) 48 h later. Four hours after misoprostol, approx 70% of women expelled; if expulsion did not occur, a third misoprostol tablet (200 µg) was given, and this method provided > 98% efficacy, according to the current trials in ≤ 49 days amenorrhea pregnancies (Table 3) and, besides safety, was well tolerated (uterine cramps were minimized). Details and discussion of the method are found elsewhere (Peyron et al., 1993).

Currently in France, a woman suspecting an unwanted pregnancy sees a physician for a first visit, and after a delay for reflection she may return (second visit) to take the RU486 pills. Two days later, the third visit to the Center is to receive prostaglandin (she stays for four hours), and a control visit (fourth)

Table 3. Interruption of early pregnancy with RU486 and one or two doses of misoprostol

Treatment	Total	Success* number (% of total)	Failures* number (% of total)	
• Mifepristone 600 mg	385			
• Abortion before the time of administration of misoprostol (400µg)		21 (5.4%)		
• Administration of first dose of misoprostol followed by abortion within 4 h	364	266 (69.1%)		
• Later outcome — Refused more misoprostol	27	26 (6.8%)	Ongoing pregnancy	1 (≪ 1%)
Second dose of misoprostol (200 µg)	71	67 (17.4%)	Partial retention	2 (< 1%)
			Synechiae with ongoing pregnancy	1 (≪ 1%)
			Ectopic pregnancy	1 (≪ 1%)
		380 (98.7%)		5 (1.3%)

*Success was defined as interruption of pregnancy and complete expulsion of the ovum. Failures defined as indicated. Peyron et al., 1993.

should take place 10-15 days later. This method is currently not applicable to heavy smokers and women older than 35. All these precautions need to be reexamined, and most appear to be dispensable. In the future, it is hoped that women consult their physician as early as possible in case of missed menses, then receive, if this is their choice, RU486 from a medically competent person who will have examined her. She then will take home the pills of misoprostol for self administration 48 h later and consult for a check up approximately 2 weeks later.

While RU486 and misoprostol are safe drugs, pregnancy itself is a risk for women no matter whether they wish to interrupt or to continue their pregnancy (for example: ectopic pregnancy is not aborted by RU486 and may be fatal if not mechanically treated). To maintain contact with a physician is mandatory and there should be an appropriate permanent connection with a competent medical center in case of complications. While it may be sufficient, in the vast majority of cases, that physicians (preferably gynecologists) see patients privately, it has been reported that many women prefer to be treated within a group at a medical center (Thong et al., 1992). Research should be conducted to define the best ways to administer the medications under specific conditions. It is certain that requirements for skilled personnel and sterilized surgical facilities will be decreased (El-Refaey et al., 1992). The mechanism of action of prostaglandin at low doses indicates that their effect takes place only when progesterone activity has been much decreased by the antisteroid, after more than 24 hours. Whether a combination of RU486 and prostaglandin, to be administered simultaneously, will become available is not predictable, no technology for delayed prostaglandin release being in sight.

Application of the method in developing countries is necessarily more difficult and local conditions must be considered, including the availability of medical facilities and personnel, the cultural traditions (bleeding for several days may be a problem), social context, etc... However, women have the right to obtain medical assistance in case of suspected pregnancy, to have the choice to decide to abort and if so, to also have the choice of either a surgical or a medical method. We ought, whether in developing or industrialized countries, to offer a complete medical choice to women. Even the RU486 plus misoprostol method may be imperfectly applied for a period of time in certain countries, but it can only be a definite improvement of the present situation. It is also successfully applicable to missed abortions and anembryonic pregnancies (El-Refaey et al., 1992). Whether RU486 plus misoprostol may be used to compensate for the lack of access to family planning and to health facilities is another question. Generally speaking, the best solution is to make available widely accepted and very efficient methods of contraception.

Pregnancy Interruption After 9 Weeks of Amenorrhea

The effects of progesterone, essentially on the decidua (implantation), the myometrium (calming effect) and LH secretion (depressed with lack of ovulation) are observed throughout the course of pregnancy. Thus it is not surprising that an antiprogesterone is potentially useful for pregnancy interruption and labor induction.

In France *voluntary pregnancy interruption* is legally permitted *until 12 weeks* of amenorrhea. When women have passed beyond the current legal limit for RU486 plus misoprostol treatment (7 weeks), vacuum aspiration is performed and this can be greatly facilitated by RU486 taken 24-48 h before. This is *preoperative cervical preparation* (ripening) (Henshaw and Templeton, 1991; Urquhart and Templeton, 1990). It may not be due to a change of prostaglandin metabolism in the cervix (Rådestad and Bygdeman, 1992), but to a decrease of α_2-adrenoreceptors (Kovacs and Falkay, 1993). A dose of 200-600 mg of RU486 decreases the force required to dilate the cervix, and has significantly fewer side effects than gemeprost (vaginal pessary). It also compares favorably with mechanical dilators such as lamicel or dilapan (Cohn and Stewart, 1991; Henshaw and Templeton, 1991; Gupta and Johnson, 1992; Thong and Baird, 1992).

In *therapeutic second and third trimester abortions*, RU486 is most often used before the administration of prostaglandin, whose dose can be decreased, while pain and other side effects are reduced and expulsion accelerated (Rodger and Baird, 1990). RU486 also is decreasing the waiting time and thus pain and psychological suffering in cases of *dead fetus* in utero (Cabrol et al., 1985).

Initiation of Labor

A decrease of progesterone activity occurs during parturition, but its precise role in successful delivery is unclear, particularly in primates (including humans) where it does not seem to be the primary event. In rats RU486 can synchronize delivery (Bose et al., 1985), and in cattle (Li et al., 1991) it is also very efficient in facilitating parturition. In rhesus macaques near term, RU486 provokes changes of prostaglandins and of the cervical status, but these modifications do not follow the same orderly sequence as these found during spontaneous delivery (Haluska et al., 1987; Wolf et al., 1993). It is not known if RU486 increases gap junctions between myometrial cells in women as occurs in treated rats (Garfield et al., 1987) while β_2-adrenoreceptor levels are unchanged in the myometrium (El Alj et al., 1989).

RU486 has been tested in women at term who require labor induction for various medical indications (post date, preeclampsia, etc...). When compared to placebo controls, the number of spontaneous deliveries is significantly increased, and the amount of required oxytocin is efficaceous at much lower levels after mifepristone, and the time to induce labor is shortened (Frydman et al., 1992). No undesirable incident, in mothers and newborns, was observed with the dose used (two times 200 mg), similar to observations made in monkey studies (Wolf et al., 1989).

Summarizing this section, RU486 appears to be a *safe inducing agent for labor* when the continuation of pregnancy is a risk for the fetus and/or the mother. Systematic studies should now follow the development of babies born after RU486 treatment, since it is known that in primates RU486 passes from mother to fetus during pregnancy (Wolf et al., 1988) and effects of neonatal and even embryonic (Wolf et al., 1990) exposure need to be carefully assessed (Weinstein et al., 1991). Until absolute safety is proved in cases where there is a medical indication, the use of the antiprogestin to induce labor for *conveniency* should be *forbidden*.

Contragestive and Contraceptive Methods Using RU486

The previous considerations apply to pregnancy, suspected by menses suspension and demonstrated by a positive pregnancy test, a situation clear to all women. Before this well defined state, even if the biological steps are known, there is still confusion and ignorance as to when does pregnancy begin. As a result, the vocabulary used to define the possible antihormonal interventions along the processes of the establishment of pregnancy, needs to be clarified.

If menses do not occur and the pregnancy test is positive, interruption at this stage is clearly defined as *abortion*. However, vacuum aspiration practiced very early, within approximately two weeks of menses delay, is called "menses regulation" (MR) (e.g. officially in Bangladesh and Turkey) or menstrual induction, and can be considered as contragestive. Differently, any manouver precluding fertilization is called *contraception*, for contra-conception. The word conception is generally understood as fertilization; this is etymologically wrong, since *concipire* (latin) means to retain (originally retain sperm and mother blood in the uterus to make the child).

Contraception is, therefore, commonly understood as a method to preclude fertilization, for instance by suppression of ovulation or preventing sperm to attain the ovum. However, physicians also designate, as contraceptive, the methods which are applied before implantation is completed. They argue that a fertilized ovum not implanted after in vitro fertilization does not define a pregnancy. In fact, the available pregnancy tests are based on the passage of human chorionic gonadotropin (hCG), produced by the embryonic chorion, into the woman's blood, which occurs only after implantation has been initiated. IUDs can be defined as "contraceptive" tools because they work, in part, as anti-implantation agents. The word post-coital contraception is also largely accepted and applies to a possibly fertilized ovum. Note also that the process of implantation is not instantaneous and takes several days during the last week of the fertile menstrual cycle, just before the time at which menses would occur. Coincidently, the development of the embryo is characterized by the streak (a marker indicating that an individual embryo has been formed, and there is no further risk of twins), which should occur at approximately the same time, about 15 days after fertilization. Before, not only may genetic abnormalities or defect of implantation stop the process leading to pregnancy, but the very definition of a single potential person cannot be rigorously applied. In short, during the period between fertilization and the time at which menses should occur, interrupting methods are *contragestive*, differing from both abortion and contraception as defined above, and not hiding the fact that they oppose pregnancy.

In summary, contragestion includes all treatments that operate over a period of approximately four weeks post-fertilization. Such treatments include RU486, other "morning-after" pills, early vacuum aspiration and IUDs (as well as a potentially anti-hCG "vaccine").

According to the previous discussion, *six methods* using RU486 can be classified under the contraceptive/contragestive terms (Figure 8).

A) Emergency post-coital contraception (contraceptive or contragestive according to when it is applied)

Recent studies have shown that RU486 (single 600 mg dose) given to women after unprotected intercourse within the preceding 72 hours is highly efficient at preventing pregnancy (Glasier et al., 1992; Grimes, 1992; Webb et al., 1992), most likely acting as an anti-implantation agent. It is at least as effective as the high-dose combined estrogen and progestin preparations, and better tolerated by the women. Research needs to be pursued in order to determine the appropriate dose, and whether the administration can be repeated, how many times, and for how long, a possibility which appears rather remote because of probable changes in menstrual cyclicity.

B) Late luteal phase administration (occasional use) (contragestive)

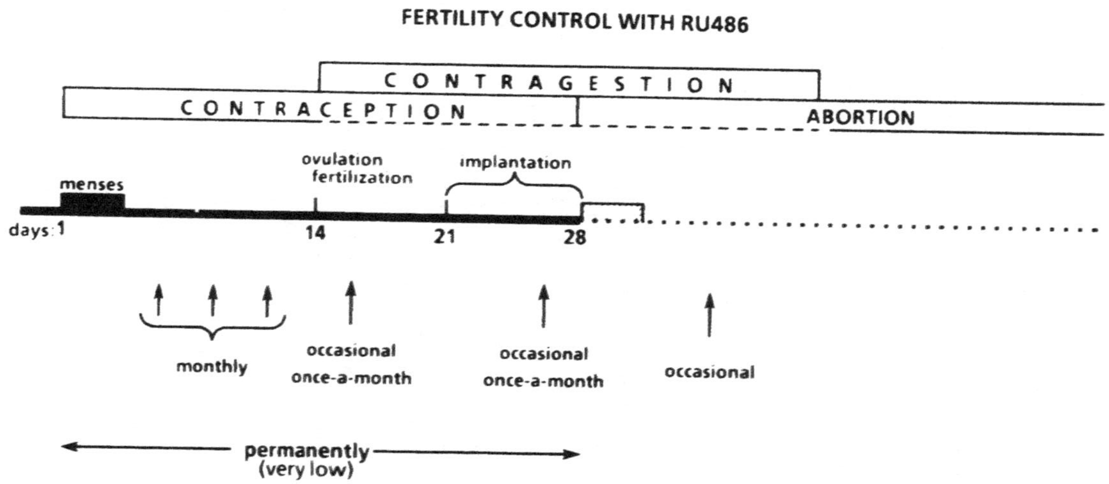
FERTILITY CONTROL WITH RU486
CONTRAGESTION
CONTRACEPTION
ABORTION
ovulation
fertilization
implantation
menses
days: 1
14
21
28
monthly
occasional
once-a-month
occasional
once-a-month
occasional
permanently
(very low)

Figure 8

The administration of 400 or 600 mg of RU486, once or twice, two days before the expected menses time, gives ~ 20% failure in terms of initiating pregnancy. Since the probability of being pregnant is ~ 20% in normal couples, this leaves only 4% of women being pregnant (to be secondarily interrupted by other means) (Dubois et al., 1988b; Lähteenmäki et al., 1988; Grimes et al., 1992). The method is well tolerated and may be improved with the associated use of Misoprostol.

C) Monthly premenstrual (late luteal phase administration, repeated use) (contragestive)

To give RU486 approximately two days before the expected day of menses, over several months, has not been successful (Van Santen and Haspels, 1987), the main reason being the irregularity of cycles, often due to the retardation of ovulation which is induced by the compound. In order to try to overcome this difficulty, a trial using a lower dose of RU486 plus misoprostol, two days before and on the expected day of menses, has been undertaken. The prostaglandin may permit more rapid evacuation of the blastocyst and therefore a faster decline of hCG which disturbs folliculogenesis during the next cycle (Aubény and Baulieu, work in progress).

D) Anti-implantation once per month (early luteal phase administration)

P acts on the endometrium to prepare for implantation. This has been studied in detail in the ovariectomized rhesus monkey by analyzing the decidual transformation and the epithelial plaque formed in response to deciduogenic stimulus (Ghosh and Sengupta, 1992). An immunocytochemical study, assessing an endometrial secretory glycan (sialo-oligosaccharide), has shown that its production is P-dependent (Graham et al., 1991). Moreover, experiments in rabbits have shown that endometrial receptivity and embryo implantation can be modified by antiprogestins (Hegele-Hartung et al., 1992) and RU486 induces epithelium apoptosis (Rotello, 1992). Treatment with 200 mg of RU486 was performed on women at the 2 days post LH stage who had had unprotected intercourse at least once during the period three days before to one day after ovulation. Over 157 cycles, only one pregnancy occured (Gemzell-Danielson et al., 1993). The main drawback of the method is the timing of the treatment, and to use the method monthly, ovulation detection needs to become routine in the future. Ten mg doses of RU486 administered on days 5 and 8 after the LH surge do not provoke hormonal change but interrupt endometrial maturation with lowered PR-levels (Greene et al., 1992).

E) "Endometrial contraception" (daily delivery of very low dose) (contraceptive and contragestive)

This provisional denomination is given to the continuous exposure to a very low dose of RU486, which may modify the genital tract in such a way that implantation and possibly fertilization does not occur, while ovulation and the pattern of estrogen and progesterone secretion are unchanged, and adrenal function is not modified. Daily use in cycling guinea-pig prevents implantation (Batista et al., 1991).

After administration of RU486 a low dose (12.5 mg to rhesus monkeys once per week), there was complete temporary infertility, while the modifications of the cycle were very limited during three cycles of the experiment (Gary Hodgen, unpublished). In five women, 1 mg of RU486 given daily for 30 days established a plasma concentration of ≥ 50 ng/ml; in one case ovulation was suppressed (Croxatto et al., 1993). In a similar study, RU486 (1 mg) given daily to nine women, delayed ovulation and endometrial maturation, with a reduced peak of placental protein 14 (a glycoprotein marker for endometrium function) (Batista et al., 1992).

Currently studies are being organized using even lower daily doses of RU486 given to women. An international comparative study will define the highest dose of RU486 (actually very small) which can be administered permanently to women without perturbation of the cycle. Furthermore, this dose will be tested for contraceptive efficacy. Initially, these studies will be conducted using RU486-containing pills administered daily. Secondarily, in case of success, we will move from pills to injectable microspheres that will allow the slow release of RU486 for several months.

F) Ovulation suppression (daily delivery of low dose) (contraceptive)

A number of observations demonstrate that P contributes to ovulation (Collins and Hodgen, 1986; Liu et al., 1987; Shoupe et al., 1987; Luukkainen et al., 1988; Danforth et al., 1989). The administration of RU486 during the follicular phase delays or suppresses ovulation. This may designate a new method of estrogen-free contraception, the main problem being to find an effective, well tolerated dose. Using 5 mg of RU486 per day, there may be ovulation blockade and no change in adrenal function (Ledger et al., 1992; Croxatto et al., 1993). The sequential administration of RU486 and progestin in order to maintain menstrual bleeding has been proposed (Croxatto and Salvatierra, 1991). We submit that this method with ovulation suppression will be more difficult to create relative to continuous administration of a very low dose.

In conclusion, there is already hope, not to say certainty, for an occasional contraceptive/contragestive method, and "endometrial contraception" is a very appealing possibility. However, any contraceptive method should be studied carefully for a rather long period of time in order to delineate the possiblity of side-effects for the women and alteration of the embryo in case of failure (Wu, 1992).

Male Contraception

Progesterone increases calcium uptake by human sperm and favors the acromosal reaction (Baldi et al., 1991; Blackmore et al., 1991; Parinaud et al., 1992; Uhler et al., 1992). There is likely to be a membrane receptor mediating its action, as in the progesterone-induced reinitiation of meiosis in *Xenopus laevis* oocytes (Blondeau and Baulieu, 1985). However, in sperm, as well as in oocytes, RU486 does not act as an antiprogestin at the membrane receptor level. A preliminary report (CP Puri. patent preview WO 9210194A1, 1992) of a contraceptive effect of RU486 in monkeys with a decrease of sperm counts has not, to my current knowledge, been confirmed.

ACKNOWLEDGEMENTS

I would like to acknowledge the editorial work of Rod Fiddes, PhD and the contribution to the manuscript by Françoise Boussac, Jean-Claude Lambert, Philippe Leclerc, Corinne Legris, Luc Outin and Claude Secco.

This work could not have been presented without the long-time collaboration of my INSERM colleagues and of the researchers at Roussel-Uclaf (Romainville, France).

REFERENCES

Aubény, E., and Baulieu, E.E.,1991, Activité contragestive de l'association au RU486 d'une prostaglandine active par voie orale. *C. R. Acad. Sci. Paris* 312:539-545.

Avrech, O.M., Bukovsky, I., Golan, A., Caspi, E., and Weinraub, Z., 1991, Mifepristone (RU486) alone or in combination with a prostaglandin analogue for termination of early pregnancy: a review. *Fertil. Steril.* 56:385-393.

Baird, D.T., Rodger, M., Cameron, I.T., and Roberts, I., 1988, Prostaglandins and antiestrogens for the interruption of early pregnancy. *J. Reprod. Fertil.* 36:173-179.

Baldi, E., Casano, R., Falsetti, C., Krausz, C., Maggi, M., and Forti, G., 1991, Intracellular calcium accumulation and responsiveness to progesterone in capacitating human spermatozoa. *J. Androl.* 12:323-330.

Batista, M.C., Bristow, T.L., Mathews, J., Stokes, W.S., Loriaux, D.L., and Nieman, L.D., 1991, Daily administration of the progesterone antagonist RU486 prevents implantation in the cycling guinea pig. *Am. J. Obstet. Gynecol.* 165:82-86.

Batista, M.C., Cartledge, T.P., Zellmer, A.W., Nieman, L.K., Merriam, G.R., and Loriaux, D.L., 1992, Evidence for a critical role of progesterone in the regulation of the midcycle gonadotropin surge and ovulation. *J. Clin. Endocrinol. Metab.* 74:565-570.

Baulieu, E.E., 1975, Antiprogesterone effect and midcycle (periovulatory) contraception. *Eur. J. Obstet. Gynecol. Reprod. Biol.* 4:161-166.

Baulieu, E.E., 1985, RU486: an antiprogestin steroid with contragestive activitiy in women, *in:* "The Antiprogestin Steroid RU486 and Human Fertility Control", E.E. Baulieu, and S.J. Segal, eds., Plenum Press, New York.

Baulieu, E.E., and Segal, S.J., 1985, "The Antiprogestin Steroid RU486 and Human Fertility Control", Plenum Press, New York and London.

Baulieu, E.E., 1989a, Contragestion and other clinical applications of RU486, an antiprogesterone at the receptor. *Science* 245:1351-1357.

Baulieu, E.E., 1989b, RU486 as an antiprogesterone steroid, from receptor to contragestion and beyond. *J. Am. Med. Assoc.* 262:1808-1811.

Baulieu, E.E., 1991a, RU486 and the early nineties. *Adv. Contracept.* 7:345-351.

Baulieu, E.E., 1991b, The Antisteroid RU486: Its cellular and molecular mode of action. *Trends Endocrinol. Metab.* 2:233-239.

Baulieu, E.E., 1991c, "The Abortion Pill", Simon & Schuster, New York.

Bélanger, A., Philibert, D., and Teutsch, G., 1981, Regio and stereospecific synthesis of 11β-substitution on progesterone receptor affinity. *Steroids* 37:361-383.

Benhamou, B., Garcia, T., Lerouge, T., Vergezac, A., Gofflo, D., Bigogne, C., Chambon, P. and Gronemeyer, H., 1992, A single amino acid that determines the sensitivity of progesterone receptors to RU486. *Science* 255:206-209.

Bergström, S., Diczfalusy, E., Borell, U., Karim, S., Samuelson, B., Uvnas, B., Wiqvist, N., and Bygdeman, M., 1972, Prostaglandins in fertility control. *Science* 175:1280-1287.

Bertagna, X., Bertagna, C., Luton, J.P., Husson, J.M., and Girard, F., 1984, The new steroid analog RU486 inhibits glucocorticoid action in man. *J. Clin. Endocrinol. Metab.* **59**:25-38.

Birgerson, L., and Odlind, V., 1987, Early pregnancy termination with antiprogestins: a comparative clinical study of RU486 given in two dose regimens and Epostane. *Fertil. Steril.* 48:565-570.

Blackmore, P.F., Neulen, J., Lattanzio, F., and Beebe, S.J. 1991, Cell surface-binding sites for progesterone mediate calcium uptake in human sperm. *J. Biol. Chem.* **266**:18655-18659.

Blondeau, J.P., and Baulieu, E.E., 1985, Progesterone inhibited phosphorylation of an unique Mr 48,000 protein in the plasma membrane of Xenopus laevis oocytes. *J. Biol. Chem.* 260:3617 3625.

Bosc, M.J., Germain, G., Nicolle, A. Philibert, D., and Baulieu, E.E., 1985, The use of antiprogesterone compound RU486 to control timing of parturition in rats. *in:* "The Antiprogestin Steroid RU486 and Human Fertility Control", E.E. Baulieu, and S.J. Segal, eds., Plenum Press, New York.

Brogden, R.N., Goa, K.L., and Faulds, D., 1993, Mifepristone: a review of its pharmacodynamic and pharmacokinetic properties, and therapeutic potential. *Drugs* 45:384-409.

Bucourt, R., Vignau, M., Torelli, V., Richard Foy, H., Geynet, C., Secco-Millet, C., Redeuilh, G., and Baulieu, E.E., 1978, New biospecific adsorbents for the purification of estradiol receptor. *J. Biol. Chem.* 25:8221 8228

Bygdeman, M., and Swahn, M.L., 1985, Progesterone receptor blockage: effect on uterine contractility and early pregnancy *Contraception* 32:45 51.

Cabrol, D., Bouvier d'Yvoire, M., Mermet, E., Cedard, L., Sureau, C., and Baulieu, E.E., 1985, Induction of labour with mifepristone after intrauterine fetal death. *Lancet* 8462:1019.

Cohn, M., and Stewart, P., 1991, Pretreatment of the primigravid uterine cervix with mifepristone 30 h prior to termination of pregnancy: a double blind study. *Br. J. Obstet. Gynaecol.* 98:778-782.

Collins, R.L., and Hodgen, G.D., 1986, Blockade of the spontaneous midcycle gonadotropin surge in monkeys by RU486. A progesterone antagonist? *J. Clin. Endocrin. Metab.* 63:1270-1276.

Cook, C.E., Wani, M.C., Lee, Y-W., Fail, P.A., and Petrow, V., 1992, Reversal of activity profile in analogs of the antiprogestin RU486: effect of a 16α-substituent of progestational (agonist) activity. *Life Sci.* 52.155 162

Cook, R.J., and Grimes, D.A., 1992, Antiprogestin Drugs. Ethical, Legal and Medical Issues. *Law, Medicine & Health Care* 20 (3).

Corner, G.W., 1963, "The Hormones in Human Reproduction", Athenum, New York.

Couzinet, B., Le Strat, N., Ulmann, A., Baulieu, E.E., and Schaison, G., 1986, Termination of early pregnancy by the progesterone antagonist RU486 (mifepristone). *New Engl. J. Med.* 315:1565-1570.

Crooij, M.J., and Janssens, J., 1988, Termination of early pregnancy by the 3β-hydroxysteroid dehydrogenase inhibitor epostane. *New Engl. J. Med.* 319, 813-817.

Croxatto, H B, and Salvatierra, 1991, Cyclic use of antigestagens for fertility control. *in:*

"Female Contraception and Male Fertility Regulation", B. Runnebaum, T. Rabe, and L. Kiesel, eds., The Parthenon Publishing Group, Casterton.

Croxatto, H.B., Salvatierra, A.M., Croxatto, H.D., and Fuentealba, B., 1993, Effects of continuous treatment with low dose mifepristone throughout one menstrual cycle. *Hum. Reprod.* 8:201-207.

Csapo, A.I., and Pulkkinen, M.O., 1977, Control of human parturition. *in:* "Progress in Perinatology", H.A. Kaminetzky, and L. Iffy, eds., G.F. Stickley Company, Philadelphia.

Danforth, D.R., Dubois, C., Ulmann, A., Baulieu, E.E., and Hodgen, G.D., 1989, Contraceptive potential of RU486 by ovulation inhibition. III. Preliminary observations on once weekly oral administration. *Contraception* 40:195-200.

Djerassi, C., 1970, Birth control after 1984. *Science* 169:941-951.

Dubois, C., Ulmann, A., Aubeny, E., Elia, D., Jourdan, M.C., Van Den Bosch, M.C., Leton, M., and Baulieu, E.E., 1988a, Contragestion par le RU486: intérêt de l'association à un dérivé prostaglandine. *C. R. Acad. Sci. Paris* 306:57-61.

Dubois, C., Ulmann, A., and Baulieu, E.E., 1988b, Contragestion with late luteal administration of RU486 (Mifepristone). *Fertil. Steril.* 50:593-596.

El Alj, A., Breuiller, M., Jolivet, A., Ferre, F., and Germain, G., 1989, β_2-adrenoceptor response in the rat uterus at the end of gestation and after induction of labor with RU486. *Can. J. Physiol. Pharmacol.* 67.1051-1057.

El-Refaey, H., Hinshaw, K., Henshaw, R., Smith, and Templeton, A., 1992, Medical management of missed abortion and anembryonic pregnancy. *Br. Med. J.* 305:1399.

Frydman, R., Baton, C., Lelaidier, C., Fernandez, H., Vial, M., and Bourget, P., 1991, Mifepristone for induction of labour. *Lancet* 337:488-489.

Frydman, R., Lelaidier, C., Baton Saint Mleux, C., Fernandez, H., Vial, M., and Bourget, P., 1992, Labor induction in women at term with mifepristone (RU486): a double-blind randomized, placebo controlled study. *Obstet. Gynecol.* 80:972-975.

Gaillard, R.C., Riondel, A., Muller, M.F., Herrmann, W., and Baulieu, E.E., 1984, RU486: a steroid with antiglucocorticosteroid activity that only disinhibits the human pituitary-adrenal system at a specific time of day. *Proc. Natl. Acad. Sci. USA* 81:3879-3882.

Garcia, T., Benhamou, B., Gofflo, D., Vergezac, A., Philibert, D., Chambon, P., and Gronemeyer, H., 1992, Switching agonistic, antagonistic, and mixed transcriptional responses to 11β-substituted progestins by mutation of the progesterone receptor. *Mol. Endocrinol.* 6:2071-2078.

Garfield, R.E., Gasc, J.M., and Baulieu, E.E., 1987, Effects of the antiprogesterone RU486 on preterm birth in the rat. *Am. J. Obstet. Gynecol.* 157:1281-1285.

Gemzell-Danielsson, K., Swahn, M.L., Svalander, P., and Bygdeman, M., 1993, Early luteal phase treatment with RU486 fertility regulation. (submitted).

Ghosh, D., De, P., and Sengupta, J., 1992, Effect of RU486 on the endometrial response to deciduogenic stimulus in ovariectomized rhesus monkeys treated with oestrogen and progesterone. *Hum. Reprod.* 7:1048-1060.

Glasier, A., Thong, K.J., Dewar, M., Mackie, M., and Baird, D.T., 1992, Mifepristone (RU486) compared with high dose estrogen and progestogen for emergency postcoital contraception. *New Engl. J. Med.* 327:1041-1044.

Graham, R.A., Aplin, J.D., Li, T.C., Cooke, I.D. and Seif, M.W., 1991, The effects of the antiprogesterone RU486 (Mifepristone) on an endometrial secretory glycan: an immunocytochemical study. *Fertil. Steril.* 55:1132-1136.

Greene, K.E., Kettel, L.M., and Yen, S.S.C., 1992, Interruption of endometrial maturation without hormonal changes by an antiprogesterone during the first half of luteal phase of the menstrual cycle: a contraceptive potential. *Fert. Steril.* 58:338-343.

Grimaldi, B., Barre, J., and Tillement, J.P., 1992, Les rôles respectifs des liaisons aux protéines sanguines et tissulaires de la mifepristone (RU486) dans sa distribution quantitative dans l'organisme. *C. R. Acad. Sci. Paris* t 315:93-99.

Grimes, D.A., 1992, Mifepristone (RU486) an abortifacient to prevent abortion? *New Engl. J. Med.* 327:1088-1089.

Grimes, D.A., Mishell, D.R., and David, H.P., 1992, A randomized clinical trial of Mifepristone (RU486) for induction of delayed menses: efficacy and acceptability. *Contraception* 46:1-10.

Groyer, A., Le Bouc, Y., Joab, I., Radanyi, C., Renoir, J.M., Robel, P., and Baulieu, E.E., 1985, Chick oviduct glucocorticosteroid receptor. Specific binding of RU486 and immunological studies with antibodies to chick oviduct progesterone receptor. *Eur. J. Biochem.* 149:445-451.

Gupta, J.K., and Johnson, N., 1992, Should we use prostaglandins, tents or progesterone antagonists for cervical ripening before first trimester abortion? *Contraception* 46:489-497.

Haluska, G.J., Stanczyk, F.Z., Cook, M.J., and Novy, M.J., 1987, Temporal changes in uterine activity and prostaglandin response to RU486 in rhesus macaques in late gestation. *Am. J. Obstet. Gynecol.* 157:1487-1495.

Hegele-Hartung, C., Mootz, U., and Beier, H.M., 1992, Luteal control of endometrial receptivity and its modification by progesterone antagonists. *Endocrinology* 131:2446-2460.

Henderson, D., 1987, Antiprogestational and antiglucocorticoid activities of some novel 11β-aryl substituted steroids *in* "Pharmacology and Clinical Uses of Inhibitors of Hormone Secretion and Action", F.J.A. Furr, and A.E. Wakeling, eds., Baillière Tindall, London.

Hendry, L.B., and Mahesh, V.B., 1992, Stereochemical complementarity of progesterone, RU486 and cavities between base pairs in partially unwound double stranded DNA assessed by computer modeling and energy calculations. *J. Steroid Biochem. Mol. Biol.* 41:647-651.

Henshaw, R.C., and Templeton A.A., 1991, Pre-operative cervical preparation before first trimester vacuum aspiration: a randomized controlled comparision between gemeprost and mifepristone (RU486) *Br. J. Obstet. Gynaecol.* 98:1025-1030.

Herrmann, W., Wyss, R., Riondel, A., Philibert, D., Teutsch, G., Sakiz, E., and Baulieu, E.E., 1982, Effet d'un stéroïde anti-progesterone chez la femme : interruption du cycle menstruel et de la grossesse au début. *C. R. Acad. Sci. Paris* 294:933-938.

Horwitz, K.B., 1992, The molecular biology of RU486. Is there a role for antiprogestins in the treatment of breast cancer? *Endocr. Rev.* 13:146-163.

Hospital, M., Busetta, B., Bucourt, R., Weintraub, H., and Baulieu, E.E., 1972, X ray crystallography of estrogens and their binding to receptor sites. *Mol. Pharmacol.* 8:438-445.

Jensen, E.V., and Jacobson, H.I., 1962, Basic guides to the mechanism of estrogen action. *in:* "Recent Progress in Hormone Research", G. Pincus, ed., Academic Press, New York and London.

Kovács, L., and Falkay, G., 1993, Changes in adrenergic receptors in the pregnant human uterine cervix following mifepristone or placebo treatment in the first trimester. *Hum. Reprod.* 8:119-121.

Lähteenmäki, P., Rapeli, T., Kääriäinen, M., Alfthan, H., and Ylikorkala, O., 1988, Late postcoital treatment against pregnancy with antiprogesterone RU486. *Fertil. Steril.* 50:36-38.

Laue, L., Kawai, Udelsman, R., Nieman, L.K., Brandon, D.D., Gallucci, W.T., Gold, P.W., Loriaux, D.L., and Chrousos, G.P., 1989, Glucocorticoid antagonists: pharmacological attributes of a prototype antiglucocorticoid (RU486). *in.* "Anti-inflammatory Steroid Action: Basic and Clinical Aspects," Academic Press, New York.

Ledger, W.L., Sweeting, V.M., Hillier, H., and Baird, D.T., 1992, Inhibition of ovulation by low-dose mifepristone (RU486). *Hum. Reprod.* 7:945-950.

Li, Y., Perezgrovas, R., Gazal, O.S., Schwabe, C., and Anderson, L.L., 1991, Antiprogesterone, RU486, facilitates parturition in cattle. *Endocrinology* 129:765-770.

Lieberman, S., Erlanger, B.F., Beiser, S.M., and Agate, F.J. Jr., 1959, Steroid-protein conjugates: their chemical, immunochemical, and endocrinological properties. *in:* "Recent Progress in Hormone Research", G. Pincus, ed., Academic Press, New York and London.

Liu, J.H., Garzo, G., Morris, S., Stuenkel, C., Ulmann, A., and Yen, S.S.C., 1987, Disruption of follicular maturation and delay of ovulation after administration of the antiprogesterone RU486. *J. Clin. Endocrinol. Metab.* 65:1135-1140.

Luukkainen, T., Heikinheimo, O., Haukkamaa, M., and Lahteenmaki, P., 1988, Inhibition of folliculogenesis and ovulation by the antiprogesterone RU486. *Fertil. Steril.* 49:961-963.

Mao, J., Regelson, W., and Kalimi, M., 1992, Molecular mechanism of RU486 action: a review. *Mol. Cell. Biochem.* 109:1-8.

Milgrom, E., Atger, M., and Baulieu, E.E., 1970, Progesterone in uterus plasma. IV Progesterone receptor(s) in guinea pig uterus cytosol. *Steroids* 16:741-754.

Mizutani, T., Bhakta, A., Kloosterboer, H.J., and Moudgil, V.K., 1992, Novel antiprogestins ORG 31806 and 31710: interaction with mammalian progesterone receptor and DNA binding of antisteroid receptor complexes. *J. Steroid Biochem. Mol. Biol.* 42:695-704.

Moguilewsky, M., and Philibert, D., 1985, Biochemical profile of RU486. *in:* "The Antiprogestin Steroid RU486 and Human Fertility Control", E.E. Baulieu, and S.J. Segal, eds., Plenum Press, New York.

Nath, R., Bhakta, A., and Moudgil, V.K., 1991, ZK98299 - a new antiprogesterone: biochemical characterization of steroid binding parameters in the calf uterine cytosol. *Arch. Biochem. Biophys.* 292:303-310.

Nedelec, L., and Gase, J.C., 1970, Alcoylation angulaire par condensation des organo magnésiens sur les époxy-5,10 estrènes-9(11). *Bull. Societe Chim. F.* 2556-2564.

Neef, G., Beier, S., Elger, W., Henderson, and Wiechert, R., 1984, New steroids with antiprogestational and antiglucocorticoid activities. *Steroids* 44:349-372.

Neef, G., 1987, Antiprogestins. A new approach to contraception. *in* "Trends in Medicinal Chemistry", E. Mutschler, and E. Winterfeldt, eds., VCH, Veinheim.

Parinaud, J., Labal, B., and Vieitez, G., 1992, High progesterone concentrations induce acrosome reaction with a low cytotoxic effect. *Fertil. Steril.* 58.599-602.

Peyron, R., Aubény, E., Targosz, V., Silvestre, L., Renault, M., Elkik, F., Leclerc, P., Ulmann, A., and Baulieu, E.E., 1993, Early pregnancy interruption with mifepristone (RU486) and the orally active prostaglandin misoprostol. *New Engl. J. Med.* 328:1509-1513.

Philibert, D., Costerousse, G., Gaillard-Moguilewsky, M., Nedelec, L., Nique, F., Tournemine, C., and Teutsch, G., 1991, From RU38486 towards dissociated antiglucocorticoid and antiprogesterone. *in:* "Antihormones in Health and Disease", M.K. Agarwal, ed., Karger, Basel.

Pincus, G., 1965, "The Control of Fertility", Academic Press, New York and London.

Rådestad, A., and Bygdeman, M., 1992, Cervical softening with mifepristone (RU486) after pretreatment with naproxen. A double-blind randomized study. *Contraception* 45:221-227.

Rodger, M.W., and Baird, D.T., 1990, Pretreatment with mifepristone (RU486) reduces interval between prostaglandin administration and expulsion in second trimester abortion. *Br. J. Obstet. Gynaecol.* 97.41-45.

Rotello, R.J., Lieberman, R.C., Lepoff, R.B., and Gerschenson, L.E., 1992, Characterization of uterine epithelium apoptotic cell death kinetics and regulation by progesterone and RU486. *Am. J. Pathol.* 140:449-456.

Shoupe, D., Mishell, D.R., Page, M.A., Madkour, H., Spitz, I.M., and Lobo, R.A., 1987, Effects of an antiprogesterone RU486 in normal women. II. Administration in the late follicular phase. *Am. J. Obstet. Gynecol.* 157.1421-1426.

Sutherland, R., Mester, J., and Baulieu, E.E., 1977, Tamoxifen is a potent "pure" anti-oestrogen in chick oviduct. *Nature* 267:434-435.

Swahn, M.L. and Bygdeman, M., 1988, The effect of the antiprogestin RU486 on uterine contractility and sensitivity to prostaglandin and oxytocin. *Br. J. Obstet. Gynaecol.* 95:126-134.

Teutsch, G., and Bélanger, A., 1979, Regio and stereospecific synthesis for 11β-substituted 19-norsteroids. *Tetrahedron Lett.* 22:2051-2054.

Teutsch, G., Ojasoo, T., and Raynaud, J.P., 1988, 11β substituted steroids, an original pathway to antihormones. *J. Steroid Biochem.* 31:549-565.

Thong, K.J., and Baird, D.T., 1992, A study of Gemeprost alone, Dilapan or Mifepristone in combination with Gemeprost for the termination of second trimester pregnancy. *Contraception* 46:11-17.

Thong, K.J., Dewar, M.H., and Baird, D.T., 1992, What do women want during medical abortion? *Contraception* 46:435-442.

Uhler, M.L., Leung, A., Chan, S.Y.W., and Wang, C., 1992, Direct effects of progesterone and antiprogesterone on human sperm hyperactivated motility and acrosome reaction. *Fertil. Steril.* 58:1191-1198.

Ulmann, A., Teutsch, G., and Philibert, D., 1990, RU486. *Scient. Am.* 262:18-24.

Ulmann, A., Silvestre, L., Chemama, L., Rezvani, Y., Renault, M., Aguillaume, C.J., and Baulieu, E.E., 1992, Medical termination of early pregnancy with mifepristone (RU486) followed by a prostaglandin analogue: study in 16,369 women. *Acta Obstet. Gynecol. Scand.* 71:278-283.

Urquhart, D.R., and Templeton, A.A., 1990, Mifepristone (RU486) for cervical priming prior to surgically induced abortion in the late first trimester. *Contraception* 42:191-199.

Van Santen, M.R., and Haspels, A.A., 1987, Failure of mifepristone (RU486) as a monthly contragestive, "Lunarette". *Contraception* 35:433-438.

Wakeling, A.E., and Bowler, J., 1988, Novel antioestrogens without partial agonist activity. *J. Steroid Biochem.* 31:645-659.

Wang, W.M., Heap, R.B., and Taussig, M.J., 1989, Blocking of pregnancy in mice by immunization with anti-idiotype directed against monoclonal anti-progesterone antibody. *Proc. Natl. Acad. Sci. USA* 86:7098-7102.

Webb, A.M.C., Russell, J., and Elstein, M., 1992, Comparison of Yuzpe regimen, danazol, and mifepristone (RU486) in oral postcoital contraception. *Br. Med. J.* 305:927-931.

Weinstein, M.A., Pleim, E.R., and Barfield, R.J., 1991, Effects of neonatal exposure to the antiprogestin mifepristone, RU486, on the sexual development of the rat. *Pharmacol. Biochem. Behav.* 41:69-74.

Wolf, J.P., Chillik, C.F., Itskovitz, J., Weyman, D., Anderson, T.L., Ulman, A., Baulieu, E.E., and Hodgen, G.D., 1988, Transplacental passage of a progesterone antagonist in monkeys. *Am. J. Obstet. Gynecol.* 159:238-242.

Wolf, J.P., Sinosich, M., Anderson T.L., Ulmann, A., Baulieu, E.E., and Hodgen, G.D., 1989, Progesterone antagonist (RU 486) for cervical dilatation, labor induction, and delivery in monkeys : effectiveness in combination with oxytocin. *Am. J. Obstet. Gynecol.* 160:45-47.

Wolf, J.P., Chillik, C.F., Dubois, C., Ulmann, A., Baulieu, E.E., and Hodgen, G.D., 1990, Tolerance of perinidatory primate embryos to RU486 exposure *in vitro* and *in vivo*. *Contraception* 41:85-92

Wolf, J.P., Simon, J., Itskovitz, J., Sinosich, M.J., Ulman, A., Baulieu, E.E., and Hodgen, G.D., 1993, Progesterone antagonist RU486 accommodates but does not induce labor and delivery in primates. *Hum Reprod.* 8:759-763.

Wu, J.T., 1992, Failure of corticosterone (B) and progesterone (P) to reverse the inhibitory effect of RU486 (RU) on the development of two cell mouse embryos in vitro. *Biol. Reprod.* 46:77, Abstract 106.

PREGNANCY RECOGNITION

Takahide Mori and Hideharu Kanzaki

Department of Gynecology and Obstetrics, Faculty of Medicine
Kyoto University
54 Shogoin Kawahara-cho, Sakyo-ku, Kyoto 606-01, Japan

CONCEPT OF PREGNANCY RECOGNITION

The descriptive term "Maternal Recognition of Pregnancy" was first coined by Short in 1969 at a symposium on fetal autonomy while making a radical question on how an animal can know that it is pregnant[1]. Ever since the discovery of an indispensable role of the corpus luteum for maintenance of pregnancy in rabbits by Fraenkel around the turn of the century, numerous scientific endeavours have continued to elucidate what and how signals from the conceptus operate to prolong functional life-span of the corpus luteum during pregnancy. To date, an abundant knowledge has accumulated to indicate that a wide variety of endocrine principles with species difference are involved in this mechanism. Apart from these endocrine recognition, owing to the great stride of advancement in reproductive immunology, it is now apparent that the mother recognizes the allogeneic embryos by responding to the paternal antigen(s) in terms of humoral and cellular immunity such as production of alloantibodies, induction of suppressor T cells and recruitment of natural killer (NK) cells at the site of implantation. Generally speaking, these immune responses are thought to alter the maternal immune system to accept the allogeneic conceptus. Therefore, the concept of pregnancy recognition should now implicate in two categories of endocrine and immune recognitions.

A background evidence for the concept was provided by a series of biological experiments in sheep by Moor and Rowson in 1960's to show that the presence of the conceptus was required for establishment of pregnancy and subsequent continuation[2]. In this sense, the signal should come from preimplantation embryos for maternal recognition even before implantation, but the timing of signal emission is different among mammalian species in relation to implantation. Therefore, on considering the endocrine recognition the relationship between the corpus luteum and the placenta in progesterone (P) production is needed. Two distinct events appear to occur sequentially for continual production of P ; one is luteal activation of cyclic to pregnancy corpus luteum and the other luteoplacental shift. The modes of involvement of embryonic signals in P production by these two sources seem to be greatly different among mammalian species depending on the mechanism of luteal activation and on the timing of luteoplacental sift. We must look back the mechanism and temporal relationship between these two events to identity the physiological role of embryonic signals in pregnancy recognition from the view point of comparative endocrinology.

Endocrinology of Embryo-Endometrium Interactions
Edited by S.R. Glasser *et al.*, Plenum Press, New York, 1994

Since uterine PGF2α is known to be the major, if not only, luteolytic factor in sheep, some anti-luteolytic principle must be delivered from preimplantation embryos in utero on day 13 after estrus to overcome the lytic action of PGF2α , because the cyclic corpus luteum starts regressing on day 14 but implantation usually begins to take place on day 18. As luteoplacental shift of P production is shown to be on day 50 in this species, the signal for maternal recognition should come from the conceptus in the pre-and postimplantation period. The substantial basis of this signal has been identified as ovine Trophoblast Protein-1 (oTP-1) or ovine Trophoblastin (oTP), which exerts its luteotrophic action by suppressing uterine production of PGF2α. The homologous type of signal transmission even before implantation is believed to occur in cows by bovine Trophoblast Protein (bTP) and in pigs.

The situation of luteal activation and luteoplacental shift in primates is quite different as compared to that of sheep. An initial and marginal rise of chorionic gonadotropin (CG) secreted from the conceptus immediately after implantation rescues the function of the cyclic corpus luteum at the stage of impending regression as exemplified in monkeys. Since PGF2α of corpus luteal but not uterine origin is believed to be the luteolytic factor in primates, the principal motivation of luteal activation is the predominance of luteotrophic action of CG over luteolytic action of luteal PGF2α ; thereby necessitating direct action of CG on cyclic corpus luteum for its activation. However, functional life-span of the activated corpus luteum seems limited approximately to 14 days in humans as assessed by circulating levels of 17HO-P which the placenta is not capable of producing[3]. As luteoplacental shift occurs around 10th weeks of gestation, namely 7 weeks after implantation, in humans, regulatory mechanisms of P production during the transient period of 5 weeks between functional decline of activated corpus luteum and commencement of placental function remains as the major question. It is likely that some other embryonic signals are required to maintain P production during this period.

A further regulation unique to rats and perhaps also to mice can be noted with luteal activation. Following mating stimuli or pseudopregnant treatments, pituitary prolactin is destined to be released with regular rhythms to enhance P accumulation as the result of blocking the conversion of P to biologically inactive 20αHO-P. Since no discernible difference in circulating level of P between pregnant and pseudopregnant rats before luteoplacental shift on day 15, any embryo-derived signal may not necessary to maintain P production. Though hysterectomy can prolong the functional life-span of pseudopregnant corpus luteum in this species up until parturition, the major stimulatory factor of P production by the placenta in the last third of pregnancy period is likely to be placental lactogen, because P level is much higher in pregnant rats than in hysterectomized pseudopregnant rats after the time of luteoplacental shift. If a signal from the conceptus has any effect on P production during pregnancy, it is an anti-luteolytic factor suppressing production of uterine PGF2α after the luteoplacental shift. Unfortunately, biochemical nature of this substance has not yet been defined.

POTENTIAL AGENTS FOR PREGNANCY RECOGNITION

Since thrombocytopenia due to platelet consumption is an initial maternal response to conception in mice, platelet-activating factor (PAF) has been proposed as a mediator for pregnancy recognition[4]. Though production of PAF is reported in human, mice and cattle preimplantation embryos with presumed physiological significance for promoting implantation as a embryonic autacoid, direct luteotrophic action of this substance is less well documented. Rather, it may serve as luteotrophic factor indirectly via platelet-activated compounds including platelet-derived growth factor (PDGF) and serotonin which has recently been demonstrated to enhance P production by bovine luteal cells[5]. A further maternal response to the embryonic PAF is the production of early pregnancy factor (EPF), of which an immunosuppressive role has been proclaimed with still ambiguity. Since PAF itself reduces proliferation of lymphocytes and interleukin-2 production while enhancing suppressor and NK activity, this substance may have a potential role in immune recognition of pregnancy. As antigenically specific reactions can no longer be defined in PAF-related maternal immune response, the only conceivable mode for PAF to involve in this way is to convey an inflammatory stimulus of the embryo to maternal immune system. Thus, PAF could be involved both in endocrine and non-specific immune recognition of pregnancy.

Though oTP-1 was first recognized as an anti-luteolytic principle secreted by the trophectoderm of sheep conceptuses[6], a remarkably high homology in amino acid sequence and molecular structure of cDNA with interferon (IFN)-α class II was subsequently noted[7] and now is duely called embryonic IFN, because it possesses anti-viral activity. A similar bovine Trophoblast Protein IFN and its gene have also identified. The mechanism by which oTP-1 reveals its anti-luteolytic action is possibly thought to inhibit uterine production of PGF2α, the production of which is usually stimulated by oxytocin though this cytokine is proved to reduce P production on porcine granulosa cells *in vitro*[8]. However, molecular mechanism of the regulation of oTP-1 production is far from clear and a detailed information including proto-oncogenes, cytokines is expected to become available in this session. Another aspect of trophoblast IFNαII concerning with pregnancy recognition is its immunosuppressive potency. Both decidualized stromal cells and activated macrophages in response to implantation stimuli are capable of secreting PGE2 which, in turn, is reported to suppress NK activity. This is an explanation why decidual NK cells consisting of a great majority of decidual mononuclear cells stay dormant in the decidua[9]. Further informations on involvement of oTP-1 in feto-maternal immune system is anticipted to be presented. As pleiotropic action is generally observed with cytokines, this trophoblastic IFN may participate in pregnancy recognition through endocrine and immune systems.

REFERENCES

1. Short, R.V., 1969, Implantation and maternal recognition of pregnancy, *in* Fetal Autonomy (Chiba Foundation Symposium), pp 2-26, Churchill, London..
2. Moor, R.M., and Rowson, L.E.A., 1964, Influence of the embryo and uterus on luteal function of the sheep, *Nature* 201:522.
3. Yoshimi, T,. Strott, C.A., Marshall, J.R. and Lipstt, M.B.,1969, Corpus luteum function in early pregnancy, J Clin Endocrinol Metab 29:225
4. O'Neill, C., Collier, M., Ryan, J.P. and Spinks, N.R., 1989, Embryo-derived platelet-activating factor, *J Reprod Fert Suppl* 37:19
5. Battista, P.J., Alila, H.W., Rexroad, Jr C.E. and Hansel, W., 1989, The effects of platelet-activating factor and platelet-derived compounds on bovine luteal cell progesterone production, *Biol Reprod* 40:769
6. Martal, J., Lacroix, M-C., Loundes, C., Saunier, M. and Wintenberger-Torres, S.R., 1979, Trophoblastin, an antiluteolytic protein present in early pregnancy in sheep, *J Reprod Fert* 56:63
7. Imakawa, K., Anthony, R.V., Kazemi, M., Marotti, K.R. Polites, H.G. and Roberts, R.M., 1987, Interferon-like sequence og ovine trophoblast protein secreted by embryonic trophectoderm, *Nature Lond* 330:377
8. Yasuda, K., Fukuoka, M., Fujiwara, H. and Mori, T., 1992, Effects of interferon on steroidogenic functions and proliferation of immature porcine granulosa cells in culture, *Biol Reprod* 47:931
9. Mori, T., Takakura, K., Narimoto, K., Kariya, M., Imai, K., Fujiwara, H., Okamoto, N., Kariya, M., Shiotani, M., Umaoka, Y., Kanzaki, H., Noda, Y., and Uchida, A., 1991, Endocrine and immune implications of human endometrial decidualization in implantation, *in* Frontiers in Human Reproduction (eds) Seppälä, M. and Hamberger, L., pp 321-330, Ann NY Acad Sci. Vol 626, New York

HEMOPOIETIC CYTOKINE REGULATION OF TROPHOBLAST INTERFERON, OVINE TROPHOBLAST PROTEIN-1

K. Imakawa[1], S.D. Helmer[2], L.A. Harbison[1], C.S.R. Meka[1] and R.K. Christenson[3]

[1]The Women's Research Institute, Department of Obstetrics and Gynecology, The University of Kansas School of Medicine-Wichita, Wichita, KS 67214; Department of Biological Science, [2]The Wichita State University, Wichita KS 67260; and [3]USDA, Agricultural Research Service, U.S. Meat Animal Research Center, Clay Center, NE 68933

Introduction

In early pregnancy, local signals derived from the uterine epithelium regulate extensive differentiation and reorganization of stroma cells. However, molecular mechanisms associated with rapid cellular reorganization of the uterus, particularly biochemical events associated with uterine receptivity, are poorly understood. Only a number of potential mediators that are involved in embryo-maternal, and in inter-uterine cellular communication have been identified, but their physiological significance *in vivo* remains to be determined. Recently, it has become apparent that lympho-hemopoietic growth factors derived from maternal endometrial cells are required for proper placental cell growth and differentiation.

There is accumulating evidence that failure in embryo-maternal signalling during pregnancy recognition is a significant source of embryonic loss in mammalian species. Ovine trophoblast interferon (IFN; ovine trophoblast protein-1, oTP-1), a low molecular weight secretory polypeptide produced by the preattached blastocysts, appears to be involved in maternal-fetal communication in sheep. It has been suggested that induction of trophoblast IFN expression may be a developmentally preprogrammed event involving autocrine factors and may not require uterine factors; however, the sufficient production of trophoblast IFN may be stimulated by uterine factors. Molecular events that maintain the massive production of trophoblast IFN required for pregnancy establishment *in vivo* remain obscure. It is our hypothesis that the sufficient expression and termination of trophoblast IFN genes are controlled by uterine factors. Presented are our recent findings regarding possible roles of hemopoietic growth factors as means of maternal-fetal communication that are associated with differential expression of trophoblast IFN genes.

Ovine trophoblast protein-1 is an interferon, oIFNτ

During the process of pregnancy establishment, biochemical communication between the mother and the conceptus is required to prevent the normal cyclic regression of the corpus luteum and to ensure continued production of progesterone. This phenomenon, maternal recognition of pregnancy (Short, 1969), has led to discovery of an antiluteolytic hormone, oTP-1 (Godkin et al., 1982), also called trophoblastin (Martal et al., 1979), which is transiently secreted from the trophectoderm of sheep concepti between days 12-21 of pregnancy (day 0 = day of estrus). Isolation and analysis of cDNA sequences corresponding to this polypeptide revealed that amino acid sequence of oTP-1 is similar to that of IFN alpha family (Imakawa et al., 1987; Stewart et al., 1987), more specifically αII (Capon et al., 1985) or omega (Hauptmann and Swetly, 1985) subfamily. This observation was independently confirmed by N-terminal amino acid sequences analyzing trophoblastin (Charpigny et al., 1988). Because oTP-1 is antigenically distinct from other IFNαs and omega and is produced by non-lymphatic, trophoblast cells, this IFN is named as ovine IFNτ (tau; Roberts et al., 1992). oIFNτ will be used to describe the antiluteolytic polypeptide, oTP-1 and trophoblastin, throughout this communication. Although several lines of evidence suggest that the existence of this particular IFN, IFNτ, is limited to only ruminants (Leaman and Roberts, 1992), the presence of this type of polypeptide in other species remains to be investigated.

Ovine IFNτ belongs to a gene family

In the human and bovine, IFN-omega is encoded by multiple, functional genes (Capon et al., 1985). Several observations suggest that this may also be true for oIFNτ: 1) at least five isoforms of oIFNτ exist (Godkin et al., 1982); 2) cell-free translation of poly $(A)^+$ RNA extracted from ovine concepti produces polypeptides with multiple isoelectric variants (Anthony et al., 1988); 3) multiple oIFNτ cDNAs have been isolated (Imakawa et al., 1987; Klemann et al., 1990); 4) Southern blot analysis of high molecular weight (HMW) ovine DNA reveals numerous hybridizing bands that most likely represent multiple genes (Charlier et al., 1991).

Ovine HMW DNA from adult lymphocytes has been isolated (Nephew et al., 1993). The HMW DNA was subjected to Southern blot analysis using various restriction endonucleases and a specific oIFNτ probe (Fig. 1). There are multiple bands regardless of restriction endonucleases which clearly suggest multiple genes exist for oIFNτ.

Isolation and characterization of multiple oIFNτ genes

Based on the observation described in the previous section, multiple oIFNτ genes are potentially transcribed in order to maintain sufficient production of oIFNτ during the period of maternal recognition of pregnancy. Objectives of this study were to isolate multiple oIFNτ genes using oIFNτ cDNA probe and to characterize the nucleotide sequences of IFNτ particularly in the 5'-promoter regions (Nephew et al., 1993). A 10 ug sample of HMW DNA extracted from day 25 concepti was digested with EcoRI restriction endonuclease and was subjected to electrophoretic analysis. Nine DNA fragments ranging from 1 to

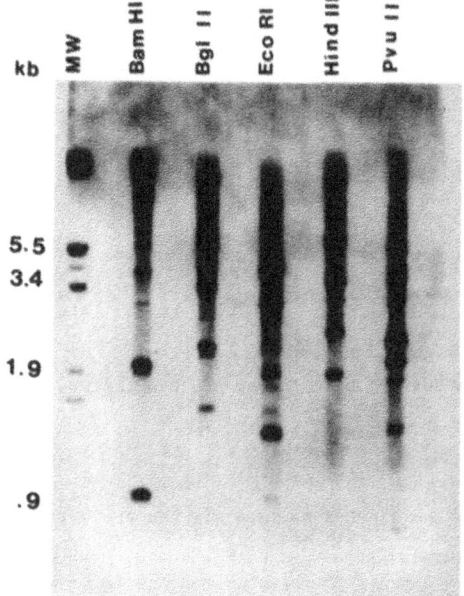

kb

MW Bam HI Bgl II Eco RI Hind III Pvu II

5.5
3.4

1.9

.9

Figure 1. Southern blot analysis of HMW DNA extracted from sheep lymphocytes with the IFNτ probe. Five ug DNA were digested with restriction endonucleases, electrophoresed, and transferred to a nylon membrane for hybridization. Blot was hybridized with a specific oIFNτ probe derived from the 3'-nontranslated region of the oIFNτ cDNA (Imakawa et al., 1987). Molecular sizes are indicated in kilobase pairs.

10 kb (approximately 1 kb/range) extracted from the agarose gel were each analyzed with polymerase chain reaction (PCR) using oIFNτ specific oligonucleotide primers (Nephew et al., 1993). The DNA fragment of 3.0 to 4.2 kb size range which strongly responded to a PCR analysis was cloned into λgt 11 vector. This subgenomic library was subsequently screened to identify multiple genes. Four oIFNτ genes were isolated and their nucleotides between -1 to +1 kb (+1 is a cap site) were completely sequenced. Although their coding regions were similarly oriented as was seen in the inferred amino acid sequences (Fig. 2), the 5'-flanking regions of these clones revealed a wide range of disparity (Fig. 3). It was of interest that clone o10 contained an AP-1, transcription enhancer element, in its promoter region. It was hypothesized that variations among the 5' non-coding regions may contribute to differential expression of oIFNτ genes within various uterine environments.

Are these oIFNτ mRNAs similarly expressed?

The objective of the following experiments was to investigate whether a particular oIFNτ gene(s) isolated in the previous section was transcribed during conceptus development (Nephew et al., 1993). Unsuccessful attempts were made to distinguish the levels of different oIFNτ mRNAs using northern blot analysis with various probes including oligonucleotides. Next, quantitative reverse transcription (RT)-PCR analyses with specific oligonucleotide primers were used to distinguish various oIFNτ mRNAs. Because the oIFNτ clones isolated were so highly related, only two of the four could be distinguished using specific primers. The third set of primers was designed to detect all of the available oIFNτ mRNAs and therefore served as a positive control. While quantitative RT-PCR reactions for oIFNτ mRNAs were performed, titrated amounts of cRNA corresponding to various oIFNτ mRNAs were always used as an

```
oTP   Met Ala Phe Val Leu Ser Leu Leu Met Ala Leu Val Leu Val Ser Tyr Gly Pro Gly Gly
o2     .   .   .   .   .   .   .   .   .   .   .   .   .   .   .   .   .   .   .   .
o7     .   .   .   .   .   .   .   .   .   .   .   .   .   .   .   .   .   .   .   .
o8     .   .   .   .   .   .   .   .   .   .   .   .   .   .   .   .   .   .   .   .
o10    .   .   .   .   .   .   .   .   .   .   .   .   .   .   .   .   .   .   .   .
      S1                                                                          S20

oTP   Ser Leu Gly Cys Tyr Leu Ser Glu Arg Leu Met Leu Asp Ala Arg Glu Asn Leu Lys Leu
o2     .   .   .   .   .   .   .  Gln  .   .   .   .   .   .   .   .   .   .  Arg  .
o7     .   .   .   .   .   .   .  Arg  .   .   .   .   .   .   .   .   .   .  Arg  .
o8     .   .   .   .   .   .   .  Gln  .   .   .   .   .   .   .   .   .   .  Arg  .
o10    .   .   .   .   .   .   .  Gln  .   .   .   .   .   .   .   .   .   .   .   .
      S21         1                                                              17

oTP   Leu Asp Arg Met Asn Arg Leu Ser Pro His Ser Cyc Leu Gln Asp Arg Lys Asp Phe Gly
o2     .   .   .   .   .   .   .   .   .   .   .   .   .   .   .   .   .   .   .   .
o7     .   .   .   .   .   .   .   .   .   .   .   .   .   .   .   .   .   .   .   .
o8     .   .   .   .   .   .   .   .   .   .   .   .   .   .   .   .   .   .   .   .
o10    .  Glu Pro  .   .   .   .   .   .   .   .   .   .   .   .   .   .   .   .   .
      18                                                                          37

oTP   Leu Pro Gln Glu Met Val Glu Gly Asp Gln Leu Gln Lys Asp Gln Ala Phe Pro Val Leu
o2     .   .   .   .   .   .   .   .   .   .   .   .  Glu Ala  .   .  Cys  .   .   .
o7     .   .   .   .   .   .   .   .   .   .   .   .   .   .   .   .   .   .   .   .
o8     .   .   .   .   .   .   .   .   .   .   .   .  Glu  .   .   .   .   .   .   .
o10    .   .   .   .   .   .   .   .   .   .   .   .   .   .   .   .   .   .   .   .
      38                                                                          57

oTP   Tyr Glu Met Leu Gln Gln Ser Phe Asn Leu Phe Tyr Thr Glu His Ser Ser Ala Ala Trp
o2     .   .   .   .   .   .   .   .   .   .  His  .   .  Arg  .   .   .   .   .
o7     .   .   .   .   .   .   .   .   .   .   .   .   .   .   .   .   .   .   .   .
o8     .   .   .   .   .   .   .   .   .   .   .   .   .   .   .   .   .   .   .   .
o10    .   .   .   .   .  Thr  .   .   .   .  His  .   .   .   .   .   .   .   .
      58                                                                          77

oTP   Asp Thr Thr Leu Leu Glu Gln Leu Cys Thr Gly Leu Gln Gln Gln Leu Asp His Leu Asp
o2    Asn  .   .   .   .   .   .   .   .   .   .   .   .   .   .   .  Glu Asp  .   .
o7     .   .   .   .   .   .   .   .   .   .   .   .   .   .   .   .   .   .   .   .
o8     .   .   .   .  Asp  .   .   .   .   .   .   .   .   .   .   .  Glu Asp  .   .
o10    .   .   .   .   .   .   .   .   .   .   .   .   .   .   .   .  Glu Asp  .   .
      78                                                                          97

oTP   Thr Cys Arg Gly Gln Val Met Gly Glu Glu Asp Ser Glu Leu Gly Asn Met Asp Pro Ile
o2     .   .   .   .  Pro  .   .   .  Lys  .   .   .   .   .  Lys  .   .   .   .
o7     .   .   .   .   .   .   .   .  Lys  .   .   .   .   .  Lys  .   .   .   .
o8     .   .   .   .   .   .   .   .   .   .   .   .   .   .   .   .   .   .   .
o10    .   .   .   .   .   .   .   .   .   .   .   .   .   .   .   .   .   .   .
      98                                                                          117

oTP   Val Thr Val Lys Lys Tyr Phe Gln Gly Ile Tyr Asp Tyr Leu Gln Glu Lys Gly Tyr Ser
o2     .   .   .   .   .   .   .   .   .   .  His  .   .   .   .   .   .   .   .   .
o7     .   .   .   .   .   .   .   .   .   .  His  .   .   .   .   .   .   .   .   .
o8     .   .   .   .   .   .   .   .   .   .   .   .   .   .   .   .   .   .   .   .
o10    .   .   .   .   .   .   .   .   .   .   .   .   .   .   .   .   .   .   .   .
      118                                                                         137

oTP   Asp Cys Ala Trp Glu Ile Val Arg Val Glu Met Met Arg Ala Leu Thr Val Ser Thr Thr
o2     .   .   .   .   .  Thr  .   .   .   .   .   .   .   .   .   .  Ser  .   .
o7     .   .   .   .   .   .   .   .   .   .   .   .   .   .   .   .  Ser  .   .
o8     .   .   .   .   .   .   .   .   .   .   .   .   .   .   .   .  Ser  .   .
o10    .   .   .   .   .   .   .   .   .   .   .   .   .   .   .   .   .   .   .
      138                                                                         157

oTP   Leu Gln Lys Arg Leu Thr Lys Met Gly Gly Asp Leu Asn Ser Pro
o2     .   .   .   .   .   .   .  Thr  .   .   .   .   .   .   .
o7     .   .   .   .   .   .   .   .   .   .   .   .   .   .
o8     .   .   .   .   .   .   .  Thr  .   .   .   .   .   .
o10    .   .   .   .   .   .   .   .   .   .   .   .   .   .
      158                                          172
```

Figure 2. An alignment of the amino acid sequence deduced from oIFNτ cDNA and IFNτ genes (o2, o7, o8 and o10). Data are from Imakawa et al. (1987) and Nephew et al. (1993). A potential glycosylation site (Asn-x-Thr) on IFNτ o2 is located at positions 78-80.

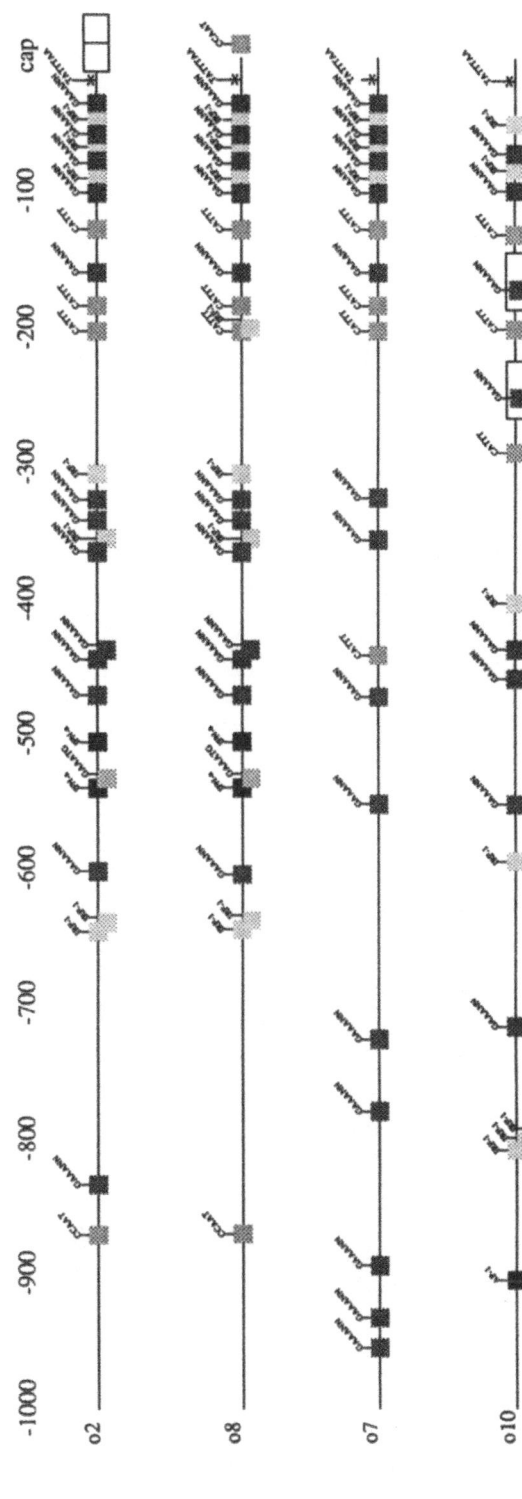

Figure 3. An illustration showing disparity in the 5'-flanking regions among oIFNτ genes. Data are from Nephew et al. (1993). Boxed regions on IFNτ o2 and o10 are the repeated sequence. GAAANN motifs (N signifies any base) are viral responsive elements and AAATGA/AAAGGA/AACTGA represent binding sites for interferon regulatory factor-1 (IRF-1). Unique to IFNτ o10 is an AP-1 like sequence located at -920.

171

internal standard in the same RT-PCR reaction where unknown samples were examined. Furthermore, PCR products were cloned and their nucleotides were sequenced in order to confirm primer specificity (Nephew et al., 1993). As shown in Fig. 4, the mRNA corresponding to one of the oIFNτ genes contains an AP-1 site in its 5'-flanking region and was expressed predominantly in day 13-45 concepti, suggesting that differential expression of oIFNτ genes exists at least at the transcriptional level, and that these differences are due to morphological and endocrinological changes in the uterine environment.

Growth factors in pregnancy

In the bovine, *in vitro* fertilized concepti cultured for 11 days start to produce bovine trophoblast protein-1 (bTP-1, bIFNτ) without maternal uterine influences (Hernandez-Ledezma et al., 1992). However, the concepti co-cultured with endometrium explants produce ~2000 times more IFNτ, suggesting that IFNτ production *in vivo* is supported by maternal uterine factors. In sheep, it is well documented that morphological changes of concepti from spherical to tubular to filamentous forms coincide with drastic increases in oIFNτ production (Ashworth and Bazer, 1989). It is conceivable that while oIFNτ production can be initiated developmentally without maternal factors, a progesterone dominated uterine environment is required either for massive amounts of oIFNτ production or for sufficient production of morphological changes of the conceptus, leading to the production of oIFNτ levels required for the maternal recognition of pregnancy.

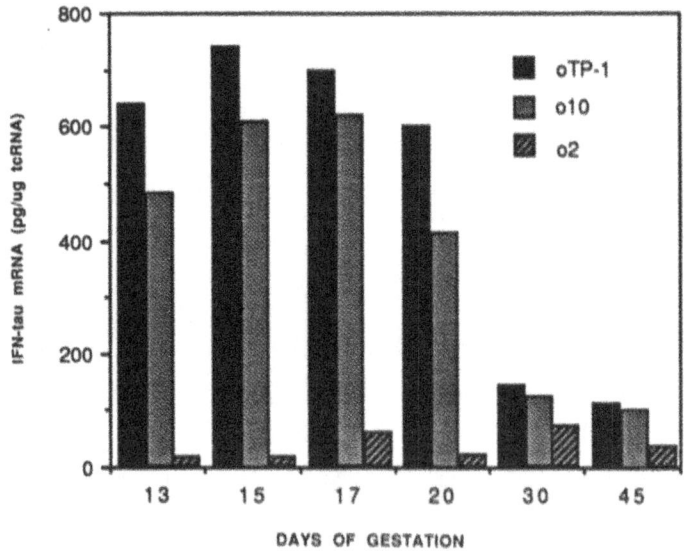

Figure 4. Amounts of oIFNτ mRNAs between 13 and 45 days of pregnancy. RT-PCR was used extensively as means to detect total oIFNτ (oTP-1), o10 and o2 mRNAs. Five ug of tcRNA were reverse-transcribed and subjected to 25 cycles of PCR analyses (Nephew et al., 1993). One of the oIFNτ genes (o10) containing an AP-1 site in the promoter region (Fig. 3) had the highest amounts of transcripts during the period of maternal recognition of pregnancy.

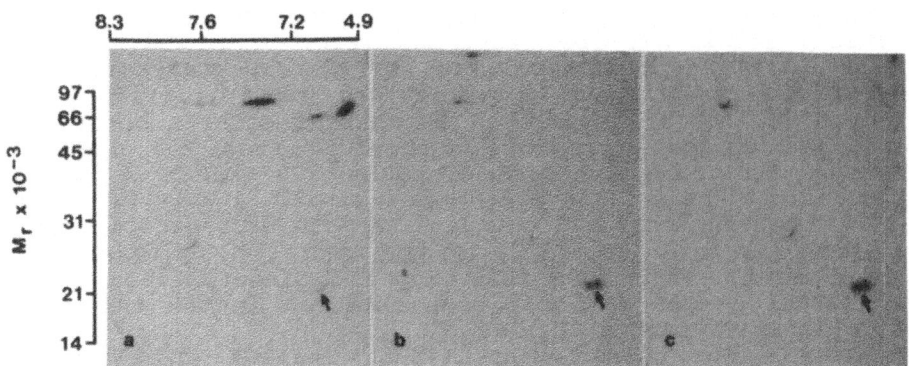

Figure 5. Fluorography illustrating radiolabeled proteins detected by 2D SDS-PAGE in day 17 ovine conceptus (24 hr) culture: (a) control, no cytokines; (b) culture treated with 300 U/ml hPDGF; (c) 300 U/ml hGM-CSF. Enhanced production of oIFNτ can be seen in the two protein isoforms at 21,000D, as indicated by the arrow.

Numerous growth factors (both mRNA and polypeptides) have been detected in uterine, fetal and placental tissues and the role of endometrial growth factors in maternal-fetal communications has been proposed (Armstrong and Chaouat, 1989; Brown et al., 1989; Paria and Dey, 1990; Simmen and Simmen, 1991). There is also ample evidence that the conceptus may benefit from the local abundance of inflammatory cytokines. Both preimplantation conceptus and trophoblast cells have been shown to express receptors for cytokines derived from uterine epithelial cells. This supports the idea that the epithelium may be a primary source of paracrine regulators in early embryonic and placental development (Pollard, 1990). Furthermore, conceptus production of cytokines and growth factors during the period of early conceptus development is considered to affect the progressive, endocrinological and immunological changes of uterine endometrium necessary for pregnancy establishment. Various cytokines and growth factors including EGF (Brown et al., 1989) IGF I (Ko et al., 1991), IGF II (Ko et al., 1991), G-CSF (Nicola et al., 1979), GM-CSF (Wegmann et al., 1989; Robertson et al., 1992), M-CSF (Pollard et al., 1987), IL-6 (Robertson et al., 1992) and PDGF (Goustin et al., 1985; Rappolee et al., 1988) are known to be present at the maternal-fetal interface.

To determine which growth factors/cytokines affect oIFNτ production in *in vitro* culture system, the following study was conducted. Conceptus tissues from ewes on day 17 of gestation (n = 5) were divided into five roughly equal masses (200 mg, wet weight) and placed in 7 ml Eagle's Minimum Essential Medium (MEM) containing 50 uCi ^{3}H-leucine. To each culture dish was added: (1) no treatment, (2) insulin [0.2 U/ml (12.7 nM)], (3) PDGF [300 U/ml (150 pM)], (4) M-CSF [300 U/ml (60 pM)], or (5) GM-CSF [300 U/ml (130 pM)]. Because this was the initial experiment, the effect of single factors (not combination of factors) on oIFNτ production was examined in five replicates (Fig. 5). Increase in oIFNτ was observed in 0 (no treatment, 0%), 2 (insulin, 40%), 1 (PDGF, 20%), 2 (M-CSF, 40%) and 4 (GM-CSF, 80%) out of five replicates when the culture media were analyzed by two dimensional sodium dodecyl sulfate polyacrylamide gel electrophoresis (2D SDS-PAGE) and fluorography (Imakawa et al., 1993).

The results obtained from the initial experiment were somewhat surprising because several cytokines including M-CSF provided relatively low responses (Fig. 5). The effect of GM-CSF on oIFNτ production (Fig. 6) was expected because (1) GM-CSF has been shown to stimulate the proliferation of trophoblast cells *in vitro*, with maximal effects observed in ectoplacental cone trophoblast derived from 7.5 day mouse embryos (Armstrong and Chaouat, 1989), (2) GM-CSF is released from decidual cell cultures (Wegmann et al., 1989), (3) GM-CSF mRNA of conventional and unique sizes are expressed in murine decidua (Crainie et al., 1990), and (4) using *in situ* hybridization techniques (Lawrence and Singer, 1985; Hunt et al., 1992) GM-CSF mRNA has been detected in the endometrium (Fig. 7; Imakawa et al., 1993).

Apart from those observations, Imakawa et al. (1993) hypothesized that GM-CSF may enhance oIFNτ production because (1) one of five oIFNτ genes isolated contains an AP-1 site in the 5'-flanking region (Fig. 3; Nephew et al., 1993), (2) this oIFNτ gene is highly expressed (up to 75% of total oIFNτ mRNAs) between days 13 and 20 of pregnancy (Nephew et al., 1993), (3) c-fos mRNA has been detected within the trophoblast cells of ovine concepti and its expression coincides with that of oIFNτ mRNA (Xavier et al., 1991), and (4) GM-CSF increases expression of nuclear protooncogenes, c-fos and c-jun (Adunyah et al., 1991) and these oncogenic proteins, Fos and Jun, form a heterodimer which interacts with a cis-acting element, an AP-1 site, resulting in the enhancement of the gene transcription (Bohmann et al., 1987).

Is the effect of GM-CSF limited to oIFNτ?

GM-CSF appears to be a component of CSF bioactivity in supernatants from cultures of human placenta and murine decidual cells (Wegmann et al., 1989), and conventional and unique sizes of GM-CSF mRNAs have been detected in murine placental tissues (Crainie et al., 1990). It has been proposed that T lymphocytes responding to fetal alloantigens are the origin of decidual GM-CSF (Wegmann et al., 1989). Recently, it

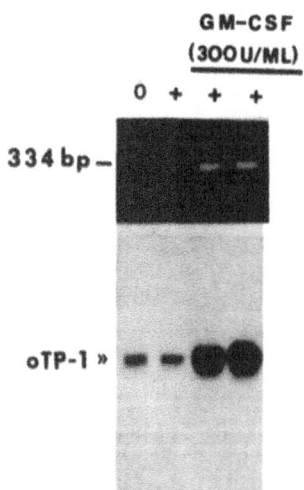

GM-CSF
(300 U/ML)

0 + + +

334 bp —

oTP-1 »

Figure 6. RT-PCR and Southern blot analysis of oIFNτ (oTP-1) mRNA. Concepti (day 17, 200 mg wet weight/culture dish) were cultured with 300 U/ml GM-CSF (lane 2-4) or without GM-CSF (lane 1) for 18 hrs. The media were analyzed by 2D SDS-PAGE and the tissues were subjected to tcRNA extraction and RT-PCR analysis using the oIFNτ primers. The top panel represents an ethidium bromide-stained gel of the PCR amplified products. The bottom panel is an autoradiograph of these products hybridized with ^{32}P-labelled oIFNτ cDNA.

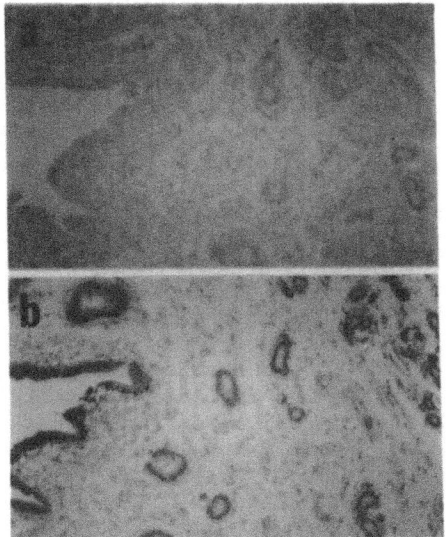

Figure 7. Localization of GM-CSF mRNA in uterine endometrium (day 17) utilizing (a) sense and (b) antisense GM-CSF cRNA probes in *in situ* hybridization experiments. Concepti (day 17) analyzed for the presence of GM-CSF mRNA did not respond to the probe (data not shown).

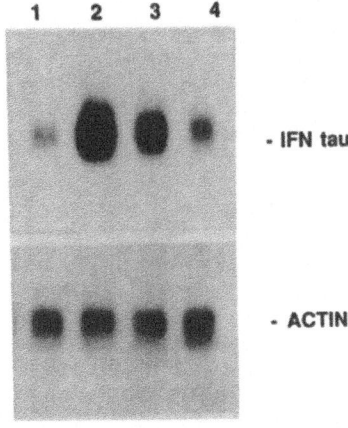

Figure 8. Northern blot analysis of conceptus tcRNA for the enhancement of oIFNτ mRNA by IL-3 treatment. Five ug tcRNA extracted from concepti were electrophoresed, transferred to a nylon membrane and hybridized with a full length oIFNτ cDNA probe (top panel): (1) control no hIL-3; (2) culture treated with 75 U/ml hIL-3; (3) 150 U/ml hIL-3; (4) 300 U/ml hIL-3. Northern blot analysis with γ-actin probe is shown at the bottom.

was reported that endometrial derived T cells are not a major source of GM-CSF and that uterine glandular and/or luminal epithelial cells harvested from the murine uterus in early pregnancy, and in mid gestation are a potent source of GM-CSF (Robertson and Seamark, 1990). GM-CSF derived from endometrial cells is not distinguishable from GM-CSF of other cellular sources and the supernatant from endometrial cell cultures stimulates granulocyte and macrophage colony formation in a bone marrow colony assay (Robertson and Seamark, 1992), suggesting that both T lymphocytes and endometrial cells derived GM-CSF polypeptides are the same, at least in their bioactivities.

GM-CSF stimulates proliferation of placental trophoblast cells. Tartakovsky et al. (1991) reported that GM-CSF administered to pregnant CBA/J female mice (mating to DBA/2 males is known to create high incidence of fetal loss through resorption during the first four days of pregnancy) was able to restore normal development of fertilized eggs, and administration between days 6.5 and 10.5 increased fetal and placental weights. It is therefore possible that in addition to stimulatory effect on oIFNτ, GM-CSF may play a role in promoting trophoblast and placental cell growth. The presence and physiological significance of uterine GM-CSF during mid-late gestation in domestic animals awaits further investigation.

Is GM-CSF the only cytokine enhancing oIFNτ production?

The presence of other hemopoietic growth factors such as M-CSF, IL-3 and IL-6 has been detected in the feto-maternal compartment. Because the addition of M-CSF in *in vitro* conceptus culture system minimally affected the production of oIFNτ (Imakawa et al., 1993), M-CSF was excluded from our subsequent studies.

Both IL-3 and GM-CSF are potent hemopoietic growth factors that stimulate the proliferation and differentiation of various lineages of hemopoietic cells including multipotent hemopoietic stem cells and lineage-committed cells (Schrader, 1986; Metcalf, 1986). These overlapping biological activities of IL-3 and GM-CSF are consistent with the observations that both IL-3 and GM-CSF activate similar signal transduction pathways: tyrosine phosphorylation of a similar set of proteins (Kanakura et al., 1990; Isfort and Ilhe, 1990) and activation of p21 ras (Satoh et al., 1991). In addition, it has been reported that binding of IL-3 to hemopoietic cells is partially inhibited by GM-CSF and that the high affinity binding of GM-CSF is inhibited by IL-3 (Gesner et al., 1988). Because GM-CSF enhanced IFNτ production *in vitro*, it was questioned whether IL-3 would also stimulate oIFNτ expression. IL-3 (doses of IL-3 in this study were the same as GM-CSF) enhanced oIFNτ production at both mRNA (Fig. 8) and polypeptide levels (not shown). The degree of oIFNτ enhancement achieved by IL-3 was very similar to that of GM-CSF. However, at present, it is not possible to clarify whether IL-3 has exhibited effects on oIFNτ production through its own receptor or GM-CSF receptor (Gearing et al., 1989; Kitamura et al., 1991). Furthermore, whether sheep uterine epithelial cells produce IL-3 needs further investigation.

IL-6 is a multifunctional cytokine with diverse biological properties including regulation of hemopoiesis, proliferation of B and T lymphocytes and macrophages, and participation in acute-phase reactions (Van Snick, 1990). It has been noted that human placental trophoblast cells synthesize IL-6 and IL-6 receptors, and IL-6 mRNA expression is regulated by estrogen in human endometrial stromal cells (Nishino et al., 1990). Recently, Robertson et al. (1992) reported *in vitro* production of IL-6 by uterine glandular and luminal epithelial cells and Mathialagan et al. (1992) found the expression of IL-6 mRNA in porcine, ovine and bovine preimplantation concepti. Therefore, experiments examining the effect of IL-6 on oIFNτ production *in vitro* were conducted. IL-6, regardless of doses tested, enhanced both oIFNτ mRNA and oIFNτ production (unpublished observation, Imakawa, K., and Christenson, R.K.).

Is endometrial production of GM-CSF regulated by steroids?

Robertson and Seamark (1990) have demonstrated that GM-CSF synthesis is almost completely abrogated in ovariectomized mice. These investigators (Robertson and Seamark, 1992) subsequently show that ovariectomized mice treated with estrogen, but not progesterone, secrete more GM-CSF in utero than untreated control animals. In addition, GM-CSF synthesis is not inhibited by treatment with the anti-progestin RU 486, while mice treated with ZK 119010 (estrogen antagonist) yield endometrial cells that synthesize significantly less GM-CSF. These observations strongly suggest that the production of uterine GM-CSF is regulated by estrogen in mice.

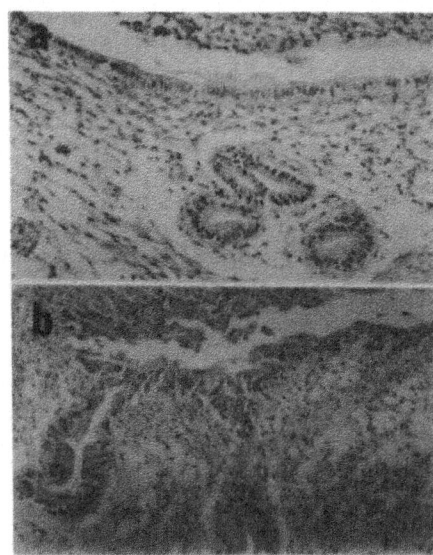

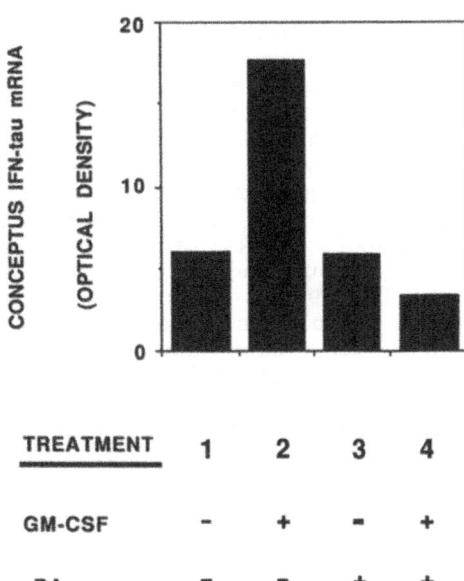

Figure 9. Localization of GM-CSF mRNA in uterine endometrium using antisense GM-CSF cRNA probe: (a) the endometrium from luteal phase (day 8), and (b) pregnant (day 17). Tissues were fixed within the same block.

Figure 10. Denstometric analysis of oIFNτ mRNA detected by Northern blot analysis. Concepti (200 mg wet weight) were cultured with (1) no hGM-CSF, no RA, (2) 150 U/ml hGM-CSF, (3) 1 uM RA, or (4) a combination of 150 U/ml hGM-CSF and 1 uM RA.

Again, using *in situ* hybridization techniques (Imakawa et al., 1993), the presence of GM-CSF mRNA in endometrial tissue obtained from day 8, 10 and 12 cyclic ewes was examined. For the purpose of qualitative comparison, endometrial tissues from both cyclic and pregnant ewes were fixed within the same block. It was found that the intensity of hybridization signal was lower in cyclic animals than those in pregnant animals (Fig. 9), further suggesting that the presence of conceptus and/or conceptus products enhances GM-CSF mRNA transcripts by the endometrium.

Termination of oIFNτ production

Within one week of the peri-implantation period (day 13-19 concepti), massive production of oIFNτ is abruptly terminated (Godkin et al., 1982). The molecular mechanism(s) responsible for the termination of transcription and/or translation of oIFNτ genes has not been adequately studied. Guillomot et al. (1991) reported that oIFNτ expression was reduced at the attachment sites of the trophoblast to the uterine epithelium and that by day 22, when most of the trophoblast was attached, expression of oIFNτ was virtually undetectable. Thus, contact between the trophoblast and the uterine epithelium appeared to terminate oIFNτ gene expression (Guillomot et al., 1990; Xavier et al., 1991).

It is known that vitamin A (retinol) plays an essential role in vision, in growth, and in the maintenance of differentiated epithelium, and that it possesses properties of an embryonic morphogen, capable of distinctive regulation of pattern formation in developing embryo (Wolf, 1984). It is well established that retinoic acid (RA) affects gene transcription, steroidogenesis, hematopoiesis and immune cell function and interferon production. RA has also been shown to suppress carcinogenesis and metastasis, and RA or its analogs have been used clinically to treat head and neck tumors, promyelocytic leukemia and melanoma. Studies on the inhibition of collagenase gene transcription by RA revealed that this inhibition is due to an interaction between retinoic acid receptors (RARs) and AP-1 proteins (a heterodimer of Fos and Jun), resulting in mutual loss of DNA binding activity (Schule et al., 1991). Recently, Busam et al. (1992) have demonstrated that RA blocks c-fos transcription and that this suppression involves a serum response element of the c-fos promoter.

As stated previously, one of four oIFNτ genes contains an AP-1 site in the 5'-flanking region and is expressed predominantly at the transcriptional level between day 13-20 of pregnancy. If in fact the AP-1 site of the oIFNτ gene represents one of major transcriptional regulators, it is possible that ovine concepti cultured with RA might have reduced c-fos transcription and thus the AP-1 site no longer serves as a transcriptional enhancer, thereby resulting in the reduction of oIFNτ mRNA. It was therefore hypothesized that an abrupt drop in oIFNτ production around day 20-22 of pregnancy was at least in part controlled by RA. Conceptus (day 17) tissues were divided into four roughly equal masses (200 mg, wet weight) and placed in 7 ml MEM containing 50 u Ci ^{3}H-leucine. To each culture dish was added: (1) no treatment, (2) GM-CSF [150 U/ml (65 pM)], (3) RA (1 uM) or (4) a combination of GM-CSF (150 U/ml) and RA (1 uM). As shown in Fig. 10, RA reduced oIFNτ mRNA and GM-CSF treatment did not overcome the effect of RA, suggesting that the suppression of oIFNτ mRNA involved the AP-1 site located in the oIFNτ promoter.

The mechanism by which vitamin A is transported from the mother to the fetus is not fully understood, but it appears to involve maternal and/or fetal retinol binding proteins (RBPs) (Harney et al., 1990). RBPs are major components of porcine and ovine conceptus secretory proteins and are also secreted by endometrial epithelium of pigs in response to progesterone. It has been discovered that RBP is synthesized by ovine placenta as early as day 23 (Liu et al., 1992). In fact, bovine IFNτ production diminishes around the period when conceptus production of RBP becomes prominent (Liu and Godkin, 1992), further suggesting a possible role of RA on IFNτ expression.

Concluding Remarks

IFNs are known to possess functional properties such as effects on cell proliferation and differentiation, and the induction of specific gene transcription in target cells. Coexpression of hemopoietic cytokine and trophoblast IFN during the period of pregnancy establishment strongly suggests that both GM-CSF and oIFNτ participate in local regulation of the uterine immune system.

The phenomenon of maternal recognition of pregnancy is generally implied to explain the process of prolongation of corpus luteum function, resulting in a continued secretion of progesterone. It is also considered as a one time, two-way communication from the fetus to the mother. However, the facts that even non-pregnant (cyclic) animals possess GM-CSF mRNA and that the endometrium from pregnant animals contains the increased amounts of GM-CSF mRNA support the idea of an autocrine-paracrine network of cytokines during early pregnancy. Expression of various endometrial growth factors is carefully regulated and the degree of their expression is controlled by substances secreted by the conceptus and endometrium. Therefore, the degree and timing of their expression in utero are orchestrated by frequent and progressive communication between the uterus and developing conceptus.

Acknowledgements

We thank Dr. K.P. Nephew, D. Sypherd, A.E. Whaley and R.S. Carroll for their data analyses. We also appreciate A. Moore for her careful preparation of the manuscript. These studies were supported in part by funds from The Wesley Foundation, Wesley Medical Research Institutes, and the Maizie Wilkonson Memorial Endowed Funds for Cancer Research. Proprietary or brand names are necessary to report factually on available data; however, the USDA neither guarantees nor warrants the standard of the product, and the use of the name by USDA implies no approval of the product to the exclusion of others that may also be suitable.

References

Adunyah, S.E., Unlap, T.M., Wagner F., and Kraft, A.S., 1991, Regulation of c-jun expression and AP-1 enhancer activity by granulocyte-macrophage colony stimulating factor, J Biol Chem. 266:5670-5675.

Anthony, R.V., Helmer, S.D., Sharif, S.F., Roberts, R.M., Hansen, P.J., Thatcher, W.W., and Bazer, F.W., 1988, Synthesis and processing of ovine trophoblast protein-1 and bovine trophoblast protein-1, conceptus secretory proteins involved in the maternal recognition of pregnancy, Endocrinology 123:1274-1280.

Armstrong, D.T., and Chaouat, G., 1989, Effect of cytokines and immune complexes on murine placental growth in vitro, Biol Reprod. 40:466-474.

Ashworth, C.J., and Bazer, F.W., 1989, Changes in ovine conceptus and endometrial function following asynchronous embryo transfer or administration of progesterone, Biol Reprod. 40:425-434.

Bohmann, D., Bos, T.J., Admon, A., Nishimura, T., Vogt, P.K., and Tjian, R., 1987, Human proto-oncogene c-jun encodes a DNA binding protein with structural and functional properties of transcription factor AP-1, Science 238:1386-1392.

Brown, M.J., Zogg, J.L., Schultz, G.S., and Hilton, F.K., 1989, Increased binding of epidermal growth factor at preimplantation sites in mouse uteri, Endocrinology 124:2882-2888.

Busam, K.J., Roberts, A.B., and Sporn, M.B., 1992, Inhibition of mitogen-induced c-fos expression in melanoma cells by retinoic acid involves the serum response element, J Biol Chem. 267:19971-19977.

Capon, D.J., Shepard, H.M., and Goeddel, D.V., 1985, Two distinct families of human and bovine interferon-α genes are coordinately expressed and encode functional polypeptides, Mol Cell Biol. 5:768-779.

Charlier, M., Hue, D., Boisnard, M., Martal, J., and Gaye, P., 1991, Cloning and structural analysis of two distinct families of ovine interferon-α genes encoding functional class II and trophoblast (oTP) α-interferons, Mol Cell Endocrinol. 71:161-171.

Charpigny, G., Reinaud, P., Huet, J.C., Guillomot, M., Charlier, M., Pernollet, J.C., and Martal, J., 1988, High homology between a trophoblast protein (trophoblastin) isolated from ovine embryo and α-interferons, FEBS Letters 228:12-16.

Crainie, M., Guilbert, L., and Wegmann, T.G., 1990, Expression of novel cytokine transcripts in the murine placenta, Biol Reprod. 43:999-1005.

Gearing, D.P., King, J.A., Gough, N.M. and Nicola, N.A., 1989, Expression cloning of a receptor for human granulocyte-macrophage colony stimulating factor, EMBO J. 8:3667-3676.

Gesner, T.G., Mufson, R.A., Norton, C.R., Turner, K.J., Yang, Y.C., and Clark, S.C., 1988, Specific binding, internalization, and degradation of human recombinant interleukin-3 by cells of the acute myelogenous leukemia line, KG-1, J Cell Physiol. 136:493-499.

Godkin, J.D., Bazer, F.W., Moffat, J., Sessions, F., and Roberts, R.M., 1982, Purification and properties of a major, low molecular weight protein released by the trophoblast of sheep blastocysts on day 13-21, J Reprod Fertil. 65:141-150.

Goustin, A.S., Betsholtz, C., Pfeifer-Ohlsson, S., Persson, H., Rydnert, J., Bywater, M., Holmgren, G., Heldin, C-H, Westermark, B., and Ohlsson, R., 1985, Coexpression of the sis and myc proto-oncogenes in developing human placenta suggests autocrine control of trophoblast growth, Cell 41:301-312.

Harney, J.P., Mirando, M.A., Smith, L.C., and Bazer, F.W., 1990, Retinol-binding protein: a major secretory product of the pig conceptus, Biol Reprod. 42:523-532.

Hauptmann, R., and Swetly, P., 1985, A novel class of human type 1 interferons, Nucl Acids Res. 13:4739-4749.

Hernandez-Ledezma, J.J., Sikes, J.D., Murphy, C.N., Watson, A.J., Schultz, G.A., and Roberts, R.M., 1992, Expression of bovine trophoblast interferon in conceptuses derived by in vitro technique, Biol Reprod. 47:374-380.

Hunt, J.S., Chen, H-L, Hu, X-L, Tabibzadeh, S., 1992, Tumor necrosis factor-α messenger ribonucleic acid and protein in human endometrium, Biol Reprod. 47:141-147.

Imakawa, K., Anthony, R.V., Kazemi, M., Marotti, K.R., Polites, H.G., and Roberts, R.M., 1987, Interferon like sequence of ovine trophoblast protein secreted by embryonic trophectoderm, Nature (Lond.) 330:377-379.

Imakawa, K., Helmer, S.D., Nephew, K.P., Meka, C.S.R., and Christenson, R.K., 1993, A novel role for GM-CSF: Enhancement of pregnancy specific interferon production, ovine trophoblast protein-1, Endocrinology 132:1869-1871.

Isfort, R.J., and Ilhe, J.N., 1990, Multiple hemopoietic growth factors signal through tyrosine phosphorylation, Growth Factors 2:213-220.

Kanakura, Y., Druker, B., Cannistra, S.A., Furukawa, Y., Torimoto, Y., and Griffin, J.D., 1990, Signal transduction of the human granulocyte-macrophage colony-stimulating factor and interleukin-3 receptors involve tyrosine phosphorylation of a common set of cytoplasmic protein, Blood 76:706-715.

Kitamura, T., Sato, N., Arai, K., and Miyajima, A., 1991, Expression cloning of the human IL-3 receptor cDNA reveals a shared ß subunit for the human IL-3 and GM-CSF receptors, Cell 66:1165-1174.

Klemann, S.W., Imakawa, K., and Roberts, R.M., 1990, Sequence variability among ovine trophoblast interferon cDNA, Nucl Acids Res. 18:6724.

Ko, Y., Lee, C.Y., Ott, T.L., Davis, M.A., Simmen, R.C.M., Bazer, F.W., and Simmen, F.A., 1991, Insulin-like growth factors in sheep uterine fluids: Concentrations and relationship to ovine trophoblast protein-1 production during early pregnancy, Biol Reprod. 45:135-142.

Lawrence, J.B., and Singer, R.H., 1985, Quantitative analysis of in situ hybridization methods for the detection of actin gene expression, Nucl Acids Res. 13:1777-1799.

Leaman, D.W., and Roberts, R.M., 1992, Genes for the trophoblast interferons in sheep, goat and musk ox and distribution of related genes among mammals, J Interferon Res. 12:1-11.

Liu, K.H., Gao, K., Baumbach, G.A., and Godkin, J.D., 1992, Purification and immunolocalization of ovine placental retinal-binding protein, Biol Reprod. 46:23-29.

Liu, K.H., and Godkin, J.D., 1992, Characterization and immunolocalization of bovine uterine retinol-binding protein, Biol Reprod. 47:1099-1104.

Martal, J., Lacroix, C., Loudes, C., Saunier, M., and Wintenberger-Torres, S., 1979, Trophoblastin, an antiluteolytic protein present in early pregnancy in sheep, J Reprod Fertil. 56:63-73.

Mathialagan, N., Bixby, J.A., and Roberts, R.M., 1992, Expression of interleukin-6 in porcine, ovine and bovine preimplantation conceptuses, Mol Reprod Dev. 32:324-330.

Metcalf, D., 1986, The molecular biology and functions of the granulocyte-macrophage colony-stimulating factors, Blood 67:257-267.

Nephew, K.P., Whaley, A.E., Christenson, R.K., and Imakawa, K., 1993, Differential expression of distinct mRNAs for ovine trophoblast protein-1 and related sheep type I interferons, Biol Reprod. 48:768-778.

Nicola, N.A., Burgess, A.W., and Metcalf, D., 1979, Similar properties of granulocyte-macrophage colony-stimulating factors produced by different mouse organs in vitro and in vivo, J Biol Chem. 254:5290-5299.

Nishino, E., Matsuzaki, N., Masuhiro, K., Kameda, T., Taniguchi, T., Saji, F., Tanizawa, O., 1990, Trophoblast-derived interluekin-6 (IL-6) regulates human chorionic gonadotropin release through IL-6 receptor on human trophoblast, J Clin Endocrinol Metab. 71:436-441.

Paria, B.C., and Dey, S.K., 1990, Preimplantation embryo development *in vitro*: Cooperative interactions among embryos and role of growth factors, Proc Natl Acad Sci USA 87:4756-4760.

Pollard, J.W., Bartocci, A., Arceci, R., Orlosfsky, A., Ladner, M.B., and Stanley, E.R., 1987, Apparent role of the macrophage growth factor CSF-1, in placental development, Nature (Lond.) 330:484-486.

Pollard, J.W., 1990, Regulation of polypeptide growth factor synthesis and growth factor-related gene expression in the rat and mouse uterus before and after implantation, J Reprod Fertil. 88:721-731.

Rappolee, D.A., Brenner, C.A., Schultz, R., Mark, D., Werb, Z., 1988, Developmental expression of PDGF, TGF-α, and TGF-β genes in preimplantation mouse embryo, Science 241:1823-1825.

Roberts, R.M., Cross, J.C., and Leaman, D.W., 1992, Interferons as hormones of pregnancy, Endo Rev. 13:432-452.

Robertson, S.A., and Seamark, R.F., 1990, Granulocyte macrophage colony stimulating factor (GM-CSF) in the murine reproductive tract: Stimulation by seminal factors, Reprod Fertil Dev. 2:359-368.

Robertson, S.A., Mayrhofer, G., and Seamark, R.F., 1992, Uterine epithelial cells synthesize granulocyte-macrophage colony-stimulating factor (GM-CSF) and interleukin-6 (IL-6) in pregnant and non-pregnant mice, Biol Reprod. 46:1069-1079.

Robertson, S.A., and Seamark, R.F., 1992, Granulocyte-macrophage colony stimulating factor (GM-CSF): One of a family of epithelial cell-derived cytokines in the preimplantation uterus, Reprod Fertil Dev. 4:435-448.

Satoh, T., Nakafuku, M., Miyajima, A., and Kaziro, Y., 1991, Involvement of ras p21 protein in signal-transduction pathways from interleukin 2, interleukin 3, and granulocyte/macrophage colony-stimulating factor, but not from interleukin 4, Proc Natl Acad Sci USA 88:3314-3318.

Schrader, J.W., 1986, The panspecific hemopoietin of activated T lymphocytes (interleukin 3), Annu Rev Immunol. 4:205-230.

Schule, R., Rangarajan, P., Yang, N., Kliewer, S., Ransone, L.J., Bolado, J., Verma, I.M., and Evans, R.M., 1991, Retinoic acid is a negative regulator of AP-1-responsive genes, Proc Natl Acad Sci USA 88:6092-6096.

Short, R.V., 1969, Implantation and the maternal recognition of pregnancy, in: Foetal Anatomy, Ciba Foundation Symposium, G.E.W. Wolstenhome, and M. O'Connor, eds., J&A Churchill, London.

Simmen, F.A., and Simmen, R.C.M., 1991, Peptide growth factors and proto-oncogenes in mammalian conceptus development, Biol Reprod. 44:1-5.

Stewart, H.J., McCann, S.H.E., Barker, P.J., Lee, K.E., Lamming, G.E., and Flint, A.P.F., 1987, Interferon sequence homology and receptor binding activity of ovine trophoblast antiluteolytic protein, J Endocrinol. 115:R13-R15.

Tartakovsky, B., Goldstein, O., Ben-Yair, E., 1991, *In vivo* modulation of pre-implantation embryonic development by cytokines, in: Molecular and Cellular Biology of Feto-Maternal Relationship, G. Chaouat, J. Mowbray, eds., Colloque INSERM/John Libbey Eurotext, Paris.

Van Snick, J., 1990, Interleukin-6: an overview, Annu Rev Immunol. 8:253-267.

Wegmann, T.G., Athanassakis, I., Guilbert, L., Branch, D., Dy, M., Menu, E., and Chaouat, G., 1989, The role of M-CSF and GM-CSF in placental growth, fetal growth, and fetal survival, Transplant Proc. 21:566-568.

Wolf, G., 1984, Multiple functions of vitamin A, Physiol Rev. 64:873-937.

Xavier, F., Guillomot, M., Charlier, M., Martal, J., and Gaye, P., 1991, Co-expression of the protooncogene FOS(c-fos) and an embryonic interferon (ovine trophoblastin) by sheep conceptuses during implantation, Biol Cell. 73:27-33.

INVOLVEMENT OF LOCAL MEDIATORS IN BLASTOCYST IMPLANTATION

Thomas G. Kennedy

Departments of Obstetrics & Gynaecology and of Physiology
The University of Western Ontario
London, ON Canada N6A 5A5

INTRODUCTION

In most mammalian species, the earliest macroscopically identifiable event in the process of blastocyst implantation is an increase in endometrial vascular permeability which is localized to implantation sites (Psychoyos, 1973). In the rat this increase in permeability can be first detected some 20-24 hours before trophoblastic invasion of the endometrium (Psychoyos, 1973). In those species in which stimulation of the uterus results in decidualization, an increase in endometrial vascular permeability precedes the decidualization; depending on the nature of the stimulus applied, the permeability response and subsequent decidualization may be localized or generalized.

The localized nature of the endometrial response at implantation infers that there is a localized interaction between the implanting blastocyst and the endometrium. What is the nature of this interaction? It is possible that the interactions are entirely physical in nature (Kennedy and Armstrong, 1981); the blastocyst, by virtue of its close contact with the endometrial luminal epithelium (Psychoyos, 1973; Enders and Schlafke, 1979) possibly augmented by rhythmic contractions and expansions (Kennedy, 1983), stimulates a change in the metabolism of the luminal epithelial cells. This results in the production of a signal which diffuses to the endometrial stroma where it initiates responses which ultimately result in increased endometrial vascular permeability followed, in many species, by proliferation and differentiation of endometrial stromal cells to give rise, ultimately, to the maternal component of the placenta. Alternatively, the signal from the blastocyst may be chemical in nature. It may act on the luminal epithelium to produce another signal, or it may diffuse across the epithelium to act within the stroma. Because of the evidence for an essential transductive role of the luminal epithelium in implantation and decidualization (Ferrando and Nalbandov, 1968; Lejeune et al., 1981), it seems likely that the epithelial cells are a source of essential components of the cascade.

This chapter will review the current state of knowledge of putative mediators involved in blastocyst implantation, with particular emphasis on recent developments.

LOCAL MEDIATORS - CRITERIA

Recently there has been an escalated interest in potential local mediators of implantation and decidualization. It is therefore appropriate to establish criteria which should be met if a compound is to be implicated as a mediator. The following criteria have been suggested (Yee et al., 1993):

1. The compound should be produced by either the conceptus or the uterus at the appropriate time.
2. Receptors for the compound should be present within the uterus.
3. Treatment with inhibitors of synthesis or antagonists of the compound should inhibit implantation and/or decidualization.
4. Treatment with exogenous agonists should override the inhibition.

It is quite likely that two or more compounds interact to bring about the endometrial responses.

Strategies for implicating compounds as mediators follow from the above criteria. In addition to investigations in pregnancy, the use of artificially induced decidualization in pseudopregnant or ovariectomized, steroid-treated animals has been profitable, particularly for investigating criterion 4. Because it is difficult, if not impossible, to deliver exogenous compounds into the uterus in a manner which mimics that which occurs in pregnancy, generalized, rather than localized, endometrial responses are obtained in these experiments.

PUTATIVE LOCAL MEDIATORS

An ever-lengthening list of potential mediators has arisen; it includes histamine, blastocyst-produced estrogen, prostaglandins, leukotrienes, platelet-activating factor, angiotensin II, and neutrophil-derived factors.

Histamine and Blastocyst-Produced Estrogen

Evidence for the role of histamine and of blastocyst-produced estrogen has been reviewed previously (Kennedy, 1983; Kennedy et al., 1989) and because there have not been any new developments, the interested reader is referred to these earlier reviews. In terms of the criteria given above, there is some evidence for histamine production by blastocysts, and for histamine receptors within the endometrium. However, the effects of inhibitors of synthesis, and particularly those of antagonists, are very controversial.

In some species there is good evidence for the production of estrogens by blastocysts. However, this estrogen has been linked to the maternal recognition of pregnancy (Flint et al., 1979; Bazer et al., 1989) rather than to implantation. In species where blastocysts remain small prior to implantation, there is little direct evidence of production of estrogens by blastocysts. Estrogen receptors are present in the endometrium (Korach et al., 1988; Brenner et al., 1990). Data from studies in which inhibitors of estrogen synthesis or estrogen antagonists have been utilized have been difficult to interpret because of the well-established requirement in some species for estrogen in addition to progesterone in order to obtain an endometrium which will allow implantation. In the hamster which does not require a maternal source of estrogen for implantation (Prasad et al., 1960; Harper et al., 1969), inhibition of estrogen biosynthesis did not affect the initiation of implantation in ovariectomized-adrenalectomized, progestin-treated animals (Evans and Kennedy, 1980).

More recently, attention has focused on the possibility that catechol-estrogens may be involved in the initiation of implantation (Hoversland et al., 1982; Kantor et al., 1985;

Dey et al., 1986). Implantation can be induced in ovariectomized pregnant mice by the systemic administration of either 4-hydroxy- or 2-hydroxy-estradiol, although the latter is substantially less potent (Hoverland et al., 1982). Studies utilizing fluorinated estrogens, potent estrogens with limited capacity to form catechol-estrogens, have suggested that the formation of catechol-estrogens is required for the initiation of implantation and decidualization. However, the results obtained with the fluorinated estrogens are contrary to what would be expected based on the potencies of the catechol-estrogens; Dey et al. (1986) initiated implantation in ovariectomized pregnant rats with 4-fluoro-estradiol but not with 2-fluoro-estradiol. These data have been interpreted as indicating that the vasoactive component of estrogen action in decidualization requires the conversion of estrogen to a catechol-estrogen (Dey and Johnson, 1987). Clearly more work needs to be done, in particular to assess the local effects of catechol-estrogens in the endometrium.

Prostaglandins

There is now considerable evidence from a variety of species that prostaglandins are involved, at least partially, in the increase in endometrial vascular permeability and subsequent decidualization (reviewed by Kennedy, 1990; Kennedy et al., 1989). During pregnancy, the concentrations of prostaglandins are elevated at implantation sites. Binding sites, presumably representing receptors, are present in the endometrium. Inhibitors of prostaglandin synthesis and prostaglandin antagonists inhibit or delay the initiation of implantation, and this inhibition can be overridden, at least partially, by exogenous prostaglandins.

During artificially induced decidualization, uterine concentrations of prostaglandins are elevated, suggesting increased uterine prostaglandin production, with a time-course consistent with the prostaglandin concentrations being elevated as the cause, rather than the consequence, of the endometrial responses. Inhibitors of prostaglandin synthesis substantially attenuate the increases in endometrial vascular permeability, alkaline phosphatase activity (Yee and Kennedy, 1988) and subsequent decidualization that otherwise occur in response to an artificial deciduogenic stimulus. These responses can be restored by the intrauterine administration of prostaglandins. At present, it is thought that prostaglandin E_2 is the prostaglandin most likely involved. Although both prostaglandin E_2 and prostaglandin $F_{2\alpha}$ are able to restore the endometrial responses after inhibition of prostaglandin production, prostaglandin E_2 is in general more effective. In addition, the endometrial binding sites are specific for prostaglandins of the E-series (Kennedy et al., 1983); no binding sites for prostaglandin $F_{2\alpha}$ have been detected (Martel et al., 1985). When analogues of prostaglandin E_2 or prostaglandin $F_{2\alpha}$ were infused into the uterine lumen of rats in which endogenous prostaglandin production had been inhibited, decidualization was obtained in response to the prostaglandin E_2 analogue only (Kennedy and Doktorcik, 1988). Finally, rat endometrial stromal cells in vitro respond to prostaglandin E_2 but not prostaglandin $F_{2\alpha}$ with accelerated decidualization, as assessed by alkaline phosphatase activity (Yee and Kennedy, 1991).

Within the uterus the sites of synthesis of the prostaglandins involved in implantation and decidualization are at present unknown. Blastocysts are capable of prostaglandin synthesis (reviewed by Kennedy et al., 1989; Kennedy, 1990), but as yet there is no good evidence that it is blastocyst-produced prostaglandins which mediate the endometrial responses. In contrast, embryo transfer experiments in rabbits suggested that an endometrial source was more likely (Snabes and Harper, 1984). In many species both the epithelial and stromal cells are capable of prostaglandin synthesis. Indirect evidence that the luminal epithelial cells may be an important source comes from the observations of Parr et al. (1988) of immunohistochemical localization of prostaglandin synthase in these cells at the time of implantation, and of decreased arachidonic acid content of epithelial cells after the application of an artificial deciduogenic stimulus (Moulton and Russell, 1989). In addition, polarized cultures of luminal epithelial cells secrete prostaglandins preferentially in the basal direction (Cherny and Findlay, 1990; Jacobs et al., 1990), a characteristic which, if operating in vivo, would facilitate the accumulation of prostaglandins in the stroma, their presumed site of action.

The sites of action of the prostaglandins within the uterus are also uncertain. In the rat, prostaglandin E binding sites are detectable in the endometrial stroma, but not luminal epithelium (Kennedy et al., 1983) while in the human binding has been localized by autoradiography to stromal cells, glandular epithelium, arterioles and erythrocytes (Chegini et al., 1986). The binding sites on the stromal cells are apparently functional because isolated endometrial cells cultured in vitro respond to prostaglandin E_2 with an accelerated rate of decidualization, as indicated by the time-course of changes in alkaline phosphatase activity, and this response is mediated at least in part by increased production of cAMP (Yee and Kennedy, 1991).

Leukotrienes

There is interest in the potential involvement of leukotrienes in implantation and decidualization because (i) of their vasoactive properties (Hammarström, 1983), (ii) the 5-lipoxygenase pathway which leads to the formation of the leukotrienes is present in the uterus (Pakrasi et al., 1985; Pakrasi and Dey, 1985; Malathay et al., 1986; Tawfik et al., 1987; Rees et al., 1987), and (iii) specific leukotriene C_4 binding sites are present in endometrium (Chegini and Rao, 1988a,b). Evidence for the involvement of leukotrienes in implantation has come from studies employing nordihydroguaretic acid, an inhibitor of leukotriene synthesis; this compound inhibits the ability of exogenous estrogen to induce implantation in pregnant mice ovariectomized prior to the nidatory estrogen surge (Gupta et al., 1989). Simultaneous treatment with leukotriene C_4 overrode the inhibition. When infused into the uterus, nordihydroguaretic acid inhibited decidualization in pseudopregnant rats (Tawfik and Dey, 1988); this was overridden in part by the simultaneous infusion of leukotriene C_4, and completely by a combination of leukotriene C_4 and prostaglandin E_2. In our hands the inhibitory effect of nordihydroguaretic acid on both the endometrial vascular permeability response (assessed with ^{125}I-labelled bovine serum albumin 10 hours after uterine stimulation, the intrauterine infusion at 1 μl/hour of phosphate-buffered saline; Kennedy and Lukash, 1982) and decidualization (assessed by uterine weights 5 days after uterine stimulation) in rats were completely restored by the simultaneous infusion of prostaglandin E_2 at the rate of 1 μg/hour (T.G. Kennedy and H.E. Ross, unpublished observations). Because nordihydroguaretic acid inhibits both 5-lipoxygenase and prostaglandin synthase, we have subsequently utilized more specific inhibitors: piriprost for 5-lipoxygenase (Bach et al., 1985), and indomethacin for prostaglandin synthase (Vane, 1971). When infused into the uterine lumen of rats at the rate of 2.3 μg/hour (5 nmol/hour), piriprost inhibited both the endometrial vascular permeability response and decidualization; at one-tenth the rate (0.23 μg/hour) it was without effect (J. Little and T.G. Kennedy, unpublished observations). Unfortunately, because of cross-reactivity of piriprost in our prostaglandin E_2 radioimmunoassay, we have been unable to determine if piriprost, when infused at the rate which is effective, affects uterine prostaglandin concentrations. When rats were treated with both piriprost and indomethacin and then given exogenous leukotriene C_4 with or without prostaglandin E_2, leukotriene C_4 had no significant effect, either alone or in combination with prostaglandin E_2, on either endometrial vascular permeability (assessed with ^{125}I-labelled bovine serum albumin) or decidualization (assessed by uterine weights). By contrast, prostaglandin E_2 infusion restored the endometrial vascular permeability response to a level not statistically different from vehicle-treated animals, and decidualization to a level almost equivalent (Figure 1; Little and Kennedy, unpublished observations). Although these data do not support the notion that leukotrienes are involved in decidualization, they should be interpreted cautiously. It is possible, for example, that interactions between prostaglandin E_2 and the leukotrienes were not observed because a supramaximal rate of infusion of the prostaglandin was used.

For rat endometrial stromal cells cultured in vitro, leukotrienes and piriprost have only marginal effects on the rate of decidualization, as assessed by alkaline phosphatase activity (Cejic and Kennedy, 1991a,b).

Additional work needs to be done on the possible involvement of leukotrienes in implantation and decidualization, and this will be aided by the development of specific antagonists for the leukotrienes.

Platelet-Activating Factor

Evidence for the involvement of platelet-activating factor in implantation and decidualization is controversial. There are contradictory reports on the ability of the preimplantation conceptus to produce this factor, with some studies producing evidence for production (O'Neill, 1985; O'Neill et al., 1987; Collier et al., 1988, 1990; Ryan et al., 1989; Battye et al., 1991), and others finding no evidence for production (Angle et al., 1988; Amiel et al., 1989; Smal et al., 1990; Kasamo et al., 1992). By contrast there is more general agreement that the uterus is a potential source of platelet-activating factor (Yasuda et al., 1986; Angle et al,. 1988; Alecozay et al., 1991; Kasamo et al., 1992).

Platelet-activating factor receptors are present in the endometrium (Kudolo and Harper, 1989; Kasamo et al., 1992) although the ability of the biologically inactive platelet-activating factor precursor/metabolite, lyso-platelet-activating factor, to compete for binding raises questions about their physiological significance. Based on two lines of evidence it seems likely that the receptors are located on the luminal epithelial cells. By autoradiography [^{3}H]-platelet-activating factor binding was localized to these cells (Kudolo et al., 1991), and Squires et al. (1991) found that epithelial, but not stromal cells from rat uteri responded to platelet-activating factor with rapid, transient increases in cytosolic free calcium concentrations.

The effects of antagonists of platelet-activating factor on implantation and artificially induced decidualization are also controversial. Spinks and O'Neill (1988) reported that platelet-activating factor antagonists impaired implantation in mice. Subsequently, based on the results of embryo transfer experiments and lack of effect of antagonists on artificially induced decidualization, Spinks et al. (1990) concluded that the antagonists act on the embryo rather than on the uterus. By contrast, Ando et al. (1990) reported that a platelet-activating factor antagonist inhibited implantation when administered to recipients in embryo transfer experiments, without any apparent effect on embryonic development. Milligan and Finn (1990) were unable to inhibit implantation in mice with a number of platelet-activating factor antagonists. In addition, platelet-activating factor did not induce decidualization when introduced into the uterus. Again by contrast, Acker et al. (1989) reported that the intrauterine injection of platelet-activating factor induced a decidual response, inhibitable by a platelet-activating factor antagonist, in pseudopregnant rats.

We have tested the hypothesis that artificial deciduogenic stimuli result in elevated platelet-activating factor synthesis with the factor then mediating the endometrial responses by infusing platelet-activating factor antagonists into the uterine lumen of rats. Under the conditions of the experiments the vehicle for the infusion of the antagonists is itself a

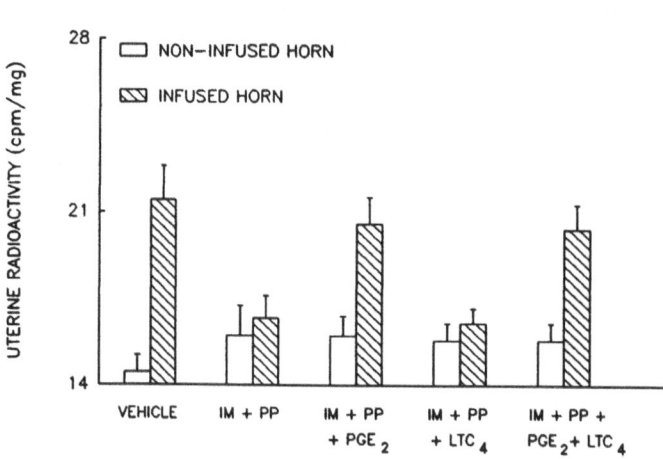

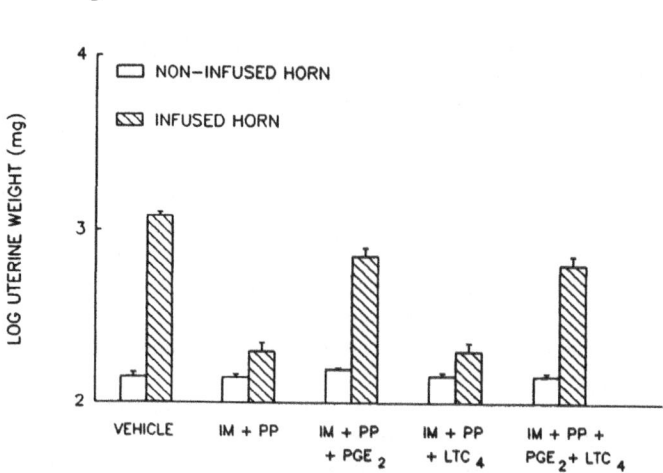

Figure 1. Determination of the ability of intrauterine-infused prostaglandin E_2 (PGE$_2$; 1 μg/h) and leukotriene C_4 (LTC$_4$; 20 ng/h) to override the inhibitory effects of combined treatment with indomethacin (IM; 2 mg sc prior to pump insertion and 0.5 nmol/h in infusate) and piriprost (PP; 5 nmol/h) on endometrial vascular permeability (A) and decidualization (B). Ovariectomized rats were treated with estradiol and progesterone to sensitize the uteri for decidualization, and then received unilateral intrauterine infusions. Endometrial vascular permeability was assessed 10 h later by uterine concentrations of radioactivity 15 min after an iv injection of ^{125}I-labelled bovine serum albumin. Decidualization was assessed 5 days later by uterine weights. Control animals received an intrauterine infusion of the vehicle, 2% ethanol in 0.05 M sodium phosphate, 0.154 M sodium chloride, pH 7.4, 1 μl/h. For detailed methodology, see Kennedy and Lukash (1982).

188

deciduogenic stimulus. The results of experiments utilizing WEB-2086 are shown in Figure 2. At no rate of infusion of the antagonist was the endometrial vascular permeability response, assessed by the concentration of radioactivity in the uterine horns 15 min after the intravenous injection of ^{125}I-labelled bovine serum albumin, or decidualization, assessed by uterine weight, affected.

Angiotensin

There is evidence for the local production of angiotensin II, and for the presence of the enzymes involved in its synthesis, renin and angiotensin-converting enzyme, in the uteri of a number of species (Cushman and Cheung, 1971; Deboben et al., 1983; Naruse et al., 1985). Specific and saturable angiotensin II binding, consistent with the presence of receptors, has also been demonstrated in the uterus (Lin and Goodfriend, 1970). The receptors are probably present on endometrial stromal cells; angiotensin II causes a transient increase in cytosolic intracellular calcium concentrations in stromal, but not epithelial cells (Squires et al., 1993).

An indication that angiotensin II may be involved in decidualization comes from the results of Squires and Kennedy (1992) who found that the infusion of enalaprilat, an angiotensin-converting enzyme inhibitor, into the uterine lumen of rats inhibited the vascular permeability response and subsequent decidualization. However, the inhibition was not reversed by angiotensin II infusion, possibly because the peptide was unable to gain access to the endometrial stroma when infused into the uterine lumen. This interpretation is supported by observations that in vitro decidualization of endometrial stromal cells is also inhibited by enalaprilat, and this inhibition can be reversed by angiotensin II (Squires and Kennedy, 1992). The effect of angiotensin is mediated in part, but not completely, by prostaglandins (Squires and Kennedy, 1992; Squires et al., 1993).

Neutrophils

In response to blastocyst implantation and to the application of artificial deciduogenic stimuli, there is an infiltration of neutrophils into the endometrium (Finn and Pope, 1991; Orlando-Mathur and Kennedy, 1993). In some vascular beds, increased permeability is a neutrophil-dependent process (Smith et al., 1987; Wallace et al., 1990), and consequently it seemed possible that in the uterus neutrophils, by releasing some vasoactive compound, might have an essential role in mediating the endometrial vascular responses at implantation and following the application of an artificial deciduogenic stimulus. This possibility has been tested by making rats neutropenic by treatment with either methotrexate or antineutrophil serum. In response to these treatments, endometrial neutrophil infiltration in response to an artificial deciduogenic stimulus was virtually abolished, but the increase in endometrial vascular permeability and subsequent decidualization was unaffected. In addition, in antineutrophil serum-treated rats, blastocyst implantation and the progression of pregnancy were not affected (Orlando-Mathur and Kennedy, 1993). These results, and those of Rogers et al. (1992) suggest that neutrophils do not have an essential role in implantation/decidualization.

SUMMARY AND CONCLUSIONS

There is experimental evidence for local interactions occurring between the blastocyst and endometrium at the time of implantation. The initial interaction probably initiates a cascade of events and these can be mimicked, at least in part in some species,

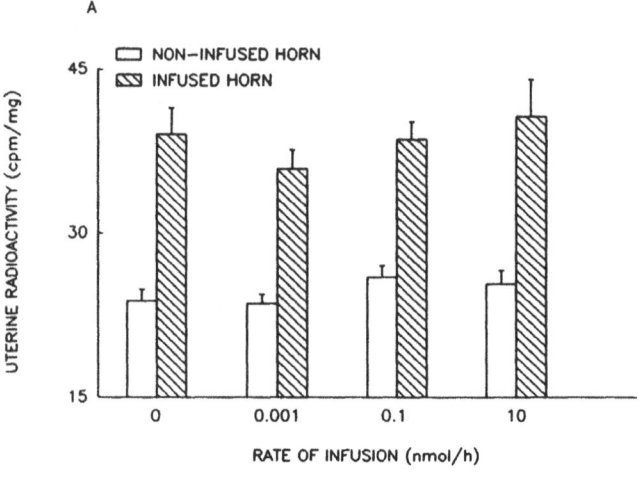

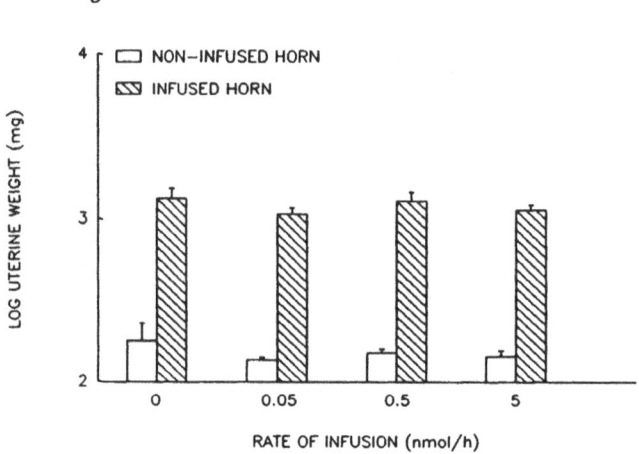

Figure 2. Effects of the intrauterine infusion of WEB-2086, a platelet-activating factor antagonist, at various rates on endometrial vascular permeability (A) and decidualization (B). Ovariectomized rats were treated with estradiol and progesterone to sensitize the uteri for decidualization, and then received unilateral intrauterine infusions of the vehicle (0.05 M sodium phosphate, 0.154 M sodium chloride, pH 7.4, 1 μl/h) or various amounts of WEB-2086. Endometrial vascular permeability was assessed 10 h later by uterine concentrations of radioactivity 15 min after an iv injection of ^{125}I-labelled bovine serum albumin. Decidualization was assessed 5 days later by uterine weights. For detailed methodology, see Kennedy and Lukash (1982).

by the application of an artificial deciduogenic stimulus to the appropriately sensitized uterus. A number of compounds have been implicated as mediators of the uterine responses. At present it is not possible to exclude any of the compounds with certainty, and it is likely that the responses may be a consequence of interactions between compounds.

Acknowledgments

The research performed by the author was supported by Medical Research Council of Canada grant no. MA-10414. Sincere thanks are extended to Barbara Lowery and Elizabeth Ross for assistance in preparing the manuscript.

REFERENCES

Acker, G., Braquet, P., and Mencia-Huerta, J.M., 1989, Role of platelet-activating factor (PAF) in the initiation of the decidual reaction in the rat, J. Reprod. Fertil. 85:623-629.

Alecozay, A.A., Harper, M.J.K., Schenken, R.S., and Hanahan, D.J., 1991, Paracrine interactions between platelet-activating factor and prostaglandins in hormonally-treated human luteal phase endometrium in vitro, J. Reprod. Fertil. 91:301-312.

Amiel, M.L., Duquenne, C., Benvenista, J., and Testart, J., 1989, Platelet aggregating activity in human embryo culture medium free of PAF-acether, Hum. Reprod. 4:327-330.

Ando, M., Suginami, H., and Matsuura, S., 1990, Pregnancy suppression by a structurally related antagonist for platelet activating factor, CV-6290, in mice, Asia Oceania J. Obstet. Gynaecol. 16:283-290.

Angle, M.J., Jones, M.A., McManus, L.M., Pinckard, R.N., and Harper, M.J.K., 1988, Platelet-activating factor in the rabbit uterus during early pregnancy, J. Reprod. Fertil. 83:711-722.

Bach, M.K., Bowman, B.J., Brashler, J.R., Fitzpatrick, F.A., Griffin, R.L., Johnson, H.G., Major, N.J., McGuire, J.C., McNee, M.L., Richards, I.M., Smith, H.W., Smith, R.J., Speziale, S.C., and Sun, F.F. 1985, Piriprost: a selective inhibitor of leukotriene synthesis, Adv. Prostagl. Thrombox. Leukotr. Res. 15:225-227.

Battye, K.M., Ammit, A.J., O'Neill, C., and Evans, G., 1991, Production of platelet-activating factor by the pre-implantation sheep embryo, J. Reprod. Fertil. 93:507-514.

Bazer, F.W., Vallet, J.L., Harney, J.P., Gross, T.S., and Thatcher, W.W., 1989, Comparative aspects of maternal recognition of pregnancy between sheep and pigs, J. Reprod. Fertil., Suppl. 37:85-89.

Brenner, R.M., West, N.B., and McClellan, M.C. 1990, Estrogen and progestin receptors in the reproductive tract of male and female primates, Biol. Reprod. 42:11-19.

Cejic, S.S., and Kennedy, T.G., 1991a, Effects of piriprost (U-60, 257B) and leukotrienes on alkaline phosphatase activity of rat endometrial stromal cells in vitro, Prostaglandins 42:163-177.

Cejic, S.S., and Kennedy, T.G., 1991b, Examination of the effects of piriprost (U-60, 257B) on alkaline phosphatase activity in rat endometrial stromal cells in vitro, Prostaglandins 42:179-189.

Chegini, N., and Rao, C.V., 1988a, Quantitative light microscopic autoradiographic study on [^{3}H] leukotriene C$_4$ binding to nonpregnant bovine uterine tissue, Endocrinology 122:1732-1736.

Chegini, N., and Rao, C.V., 1988b, The presence of leukotriene C$_4$- and prostacyclin-binding sites in nonpregnant human uterine tissue, J. Clin. Endocrinol. Metab. 66:76-87.

Chegini, N., Rao, C.V., Wakim, N., and Sanfilippo, J., 1986, Prostaglandin binding to different cell types of human uterus: quantitative light microscope autoradiographic study, Prostaglandins Leukotrienes Med. 22:129-138.

Cherny, R.A., and Findlay, J.K., 1990, Separation and culture of ovine endometrial epithelial and stromal cells: evidence of morphological and functional polarity, Biol. Reprod. 43:241-250.

Collier, M., O'Neill, C., Ammit, A.J., and Saunders, D.M., 1988, Biochemical and pharmacological characterization of human embryo-derived platelet-activating factor, Hum. Reprod. 3:993-998.

Collier, M., O'Neill, C., Ammit, A.J., and Saunders, D.M., 1990, Measurement of human embryo-derived platelet-activating factor (PAF) using a quantitative bioassay of platelet aggregation, Hum. Reprod. 5:323-328.

Cushman, D.W., and Cheung, H.S., 1971, Concentrations of angiotensin-converting enzyme in tissues of the rat, Biochim. Biophys. Acta 250:261-265.

Deboben, A., Inagami, T., and Ganten, D., 1983, Tissue renin, in "Hypertension, Pathophysiology and Treatment", J. Genest, O. Kuchel, P. Hamet, and M. Cantin, eds., McGraw Hill, New York, pp. 194-209.

Dey, S.K., and Johnson, D.C., 1987, Embryonic signals in pregnancy, Ann. NY Acad. Sci. 476:49-62.

Dey, S.K., Johnson, D.C., Pakrasi, P.L., and Liehr, J.G., 1986, Estrogens with reduced catechol-forming capacity fail to induce implantation in the rat, Proc. Soc. Exp. Biol. Med. 181:215-218.

Enders, A.C., and Schlafke, S., 1979, Comparative aspects of blastocyst-endometrial interactions at implantation, CIBA Found. Symp. 64:3-22.

Evans, C.A., and Kennedy, T.G., 1980, Blastocyst implantation in ovariectomized, adrenalectomized hamsters treated with inhibitors of steroidogenesis during the pre-implantation period, Steroids 36:41-52.

Ferrando, G., and Nalbandov, A.V., 1968, Relative importance of histamine and estrogen on implantation in rats, Endocrinology 83:933-937.

Finn, C.A., and Pope, M.D., 1991, Infiltration of neutrophil polymorphonuclear leucocytes into the endometrial stroma at the time of implantation of ova and the initiation of the oil decidual cell reaction in mice, J. Reprod. Fertil 91:365-369.

Flint, A.P.F., Burton, R.D., Gadsby, J.E., Saunders, P.T.K., and Heap, R.B., 1979, Blastocyst estrogen synthesis and the maternal recognition of pregnancy, CIBA Found. Symp. 64:209-228.

Gupta, A., Huet, Y.M., and Dey, S.K., 1989, Evidence for prostaglandins and leukotrienes as mediators of phase I of estrogen action in implantation in the mouse, Endocrinology 124:546-548.

Hammerström, S., 1983, Leukotrienes, Annu. Rev. Biochem. 52:355-377.

Harper, M.J.K., Dowd, D., and Elliott, A.S.W., 1969, Implantation and embryonic development in the ovariectomized-adrenalectomized hamster, Biol. Reprod. 1:253-257.

Hoversland, R.C., Dey, S.K., and Johnson, D.C., 1982, Catechol estradiol induced implantation in the mouse, Life Sci. 30:1801-1804.

Jacobs, A.L., Decker, G.L., Glasser, S.R., Julian J., and Carson, D.D., 1990, Vectorial secretion of prostaglandins by polarized rodent uterine epithelial cells, Endocrinology 126:2125-2136.

Kantor, B.S., Dey, S.K., and Johnson, D.C., 1985, Catechol estrogen induced initiation of implantation in the delayed implanting rat, Acta Endocrinol. 109:418-422.

Kasamo, M., Brandt, M., Ishikawa, M., Shimizu, T., and Harper, M.J.K., 1992, In vitro prostaglandin release from and platelet-activating factor accumulation in isolated endometrial cells from pregnant and pseudopregnant rabbits, Biol. Reprod. 46:829-845.

Kennedy, T.G., 1983, Embryonic signals and the initiation of blastocyst implantation, Aust. J. Biol. Sci. 36:531-543.

Kennedy, T.G., 1990, Eicosanoids and blastocyst implantation, in "Eicosanoids in Reproduction", M.D. Mitchell, ed., CRC Press, Boca Raton, pp. 123-138.

Kennedy, T.G., and Armstrong, D.T., 1981, The role of prostaglandins in endometrial vascular changes at implantation, in "Cellular and Molecular Aspects of Implantation," S.R. Glasser and D.W. Bullock, eds., Plenum, New York, pp. 349-363.

Kennedy, T.G., and Doktorcik, P.E., 1988, Effects of analogues of prostaglandin E_2 and $F_{2\alpha}$ on the decidual cell reaction in the rat, Prostaglandins 35:207-219.

Kennedy, T.G., and Lukash, L.A., 1982, Induction of decidualization in rats by the intrauterine infusion of prostaglandins, Biol. Reprod. 27:253-260.

Kennedy, T.G., Martel, D., and Psychoyos, A., 1983, Endometrial prostaglandin E_2 binding during the estrous cycle and its hormonal control in ovariectomized rats, Biol. Reprod. 29:565-571.

Kennedy, T.G., Squires, P.M. and Yee, G.M., 1989, Mediators involved in decidualization, in "Blastocyst Implantation," K. Yoshinaga, ed., Serono Symposia, Boston, pp. 135-143.

Korach, K.S., Horigoma, T., Tomooka, Y., Yamashita, S., Newbold, R.R., and McLachlan, J.A., 1988, Immunodetection of estrogen receptor in epithelial and stromal tissues of neonatal mouse uterus, Proc. Natl. Acad. Sci. USA 85:3334-3337.

Kudolo, G.B., and Harper, M.J.K., 1989, Characterization of platelet-activating factor binding sites on uterine membranes from pregnant rabbits, Biol. Reprod. 41:587-603.

Kudolo, G.B., Kasamo, M., and Harper, M.J.K., 1991, Autoradiographic localization of platelet-activating factor (PAF) binding sites in the rabbit endometrium during the peri-implantation period, Cell Tiss. Res. 265:231-241.

Lejeune, B., VanHoeck, J., and Leroy, F., 1981, Transmitter role of the luminal uterine epithelium in the induction of decidualization in rats, J. Reprod. Fertil. 61:235-240.

Lin, S.-Y., and Goodfriend, T.L., 1970, Angiotensin receptors, Am. J. Physiol. 218:1319-1328.

Malathy, P.V., Cheng, H.C., and Dey, S.K., 1986, Production of leukotrienes and prostaglandins in the rat uterus during periimplantation period, Prostaglandins 32:605-614.

Martel, D., Kennedy, T.G., Monier, M.N., and Psychoyos, A., 1985, Failure to detect specific binding sites for prostaglandin F-2α in membrane preparations from rat endometrium, J. Reprod. Fertil. 75:265-274.

Milligan, S.R., and Finn, C.A., 1990, Failure of platelet-activating factor (PAF-acether) to induce decidualization in mice and failure of antagonists of PAF to inhibit implantation, J. Reprod. Fertil. 88:105-112.

Moulton, B.C., and Russell, P.T., 1989, Arachidonic acid in uterine phospholipid during early pregnancy and following hormone treatment, Biol. Reprod. 41:821-826.

Naruse, M., Naruse, K., Kurimoto, F., Sakurai, H., Yoshida, S., Toma, H., Ishii, T., Obana, K., Demura, H., Inagami, T., and Shizume, K., 1985, Evidence for the existence of des-asp^1-angiotensin II in human uterine and adrenal tissues, J. Clin. Endocrinol. Metabl. 61:480-483.

O'Neill, C., 1985, Partial characterization of the embryo-derived platelet-activating factor in mice, J. Reprod. Fertil. 75:375-380.

O'Neill, C., Gidley-Baird, A.A., Pike, I.L., and Saunders, D.M., 1987, Use of a bioassay for embryo-derived platelet-activating factor as a means of assessing quality and pregnancy potential of human embryos, Fertil. Steril. 47:969-975.

Orlando-Mathur, C.E., and Kennedy, T.G., 1993, An investigation into the role of neutrophils in decidualization and early pregnancy in the rat, Biol. Reprod. 48:1258-1265.

Pakrasi, P.L., and Dey, S.K., 1985, Evidence for an inverse relationship between cyclooxygenase and lipoxygenase pathways in the pregnant rabbit endometrium, Prostaglandins Leukotrienes Med. 18:347-352.

Pakrasi, P.L., Becka, R., and Dey, S.K., 1985, Cyclooxygenase and lipoxygenase pathways in the preimplantation rabbit uterus and blastocyst, Prostaglandins 29:481-495.

Parr, M.B., Parr, E.L., Munaretto, K., Clark, M.R., and Dey, S.K., 1988, Immunohistochemical localization of prostaglandin synthase in the rat uterus and embryo during the peri-implantation period, Biol. Reprod. 38:333-343.

Prasad, M.R.N., Orsini, M.W., and Meyer, R.K., 1960, Nidation in progesterone-treated, estrogen-deficient hamsters, Masocricetus auratus (Waterhouse), Proc. Soc. Exp. Biol. Med. 104:48-51.

Psychoyos, A., 1973, Endocrine control of egg implantation, in "Handbook of Physiology," R.O. Greep, E.B. Astwood and S.R. Geiger, eds., Sect. 7, Vol. II, Part 2, American Physiological Society, Bethesda, pp. 187-215.

Rees, M.C.P., DiMarzo, V., Tippins, J.R., Morris, H.R., and Turnbull, A.C., 1987, Leukotriene release by endometrium and myometrium throughout the menstrual cycle and menorrhagia, J. Endocrinol. 113:291-295.

Rogers, P.A.W., Macpherson, A.M, and Beaton, L., 1992, Reduction in endometrial neutrophils in proximity to implanting rat blastocysts, J. Reprod. Fertil. 96:283-288.

Ryan, J.P., Spinks, N.R., O'Neill, C., Ammit, A.J., and Wales, R.G., 1989, Platelet activating factor (PAF) production by mouse embryos in vitro and its effect on embryonic metabolism, J. Cell. Biochem. 40:387-395.

Smal, M.A., Dziadek, M., Cooney, S.J., Attard, M., and Baldo, B.A., 1990, Examination for platelet-activating factor production by preimplantation mouse embryos using a specific radioimmunoassay, J. Reprod. Fertil. 90:419-425.

Smith, S.M., Holm-Rutili, L., Perry, M.A., Grisham, M.B., Arfors, K.-E., Granger, D.N., and Kvietys, P.R., 1987, Role of neutrophils in hemorrhagic shock-induced gastric mucosal injury in the rat, Gastroenterology 93:466-471.

Snabes, M.C., and Harper, M.J.K., 1984, Site of action of indomethacin on implantation in the rabbit, J. Reprod. Fertil. 71:559-565.

Spinks, N.R., and O'Neill, C., 1988, Antagonists of embryo-derived platelet-activating factor prevent implantation of mouse embryos, J. Reprod. Fertil. 84:89-98.

Spinks, N.R., Ryan, J.P., and O'Neill, C., 1990, Antagonists of embryo-derived platelet-activating factor act by inhibiting the ability of the mouse embryo to implant, J. Reprod. Fertil. 88:241-248.

Squires, P.M., and Kennedy, T.G., 1992, Evidence for a role for a uterine renin-angiotensin system in decidualization in rats, J. Reprod. Fertil. 95:791-802.

Squires, P.M., Dixon, S.J., and Kennedy, T.G., 1991, Platelet activating factor increases cytosolic free Ca^{2+} in rat endometrial cells from uteri sensitized for implantation, Proc. Can. Fed. Biol. Soc. 34:722 (Abstr.).

Squires, P.M., Dixon, S.J., and Kennedy, T.G., 1993, Maximal decidualization in rats requires both Ca^{2+} mobilization and increased prostaglandin production: evidence for involvement of angiotensin II, J. Reprod. Fertil., in press.

Tawfik, O.W., and Dey, S.K., 1988, Further evidence for role of leukotrienes as mediators of decidualization in the rat, Prostaglandins 35:279-286.

Tawfik, O.W., Huet, Y.M., Malathay, P.V., Johnson, D.C., and Dey, S.K., 1987, Release of prostaglandins and leukotrienes from the rat uterus is an early estrogenic response, Prostaglandins 34:805-815.

Vane, J.R., 1971, Inhibition of prostaglandin synthesis as a mechanism of action of aspirin-like drugs, Nature, New Biol. 231:232-235.

Wallace, J.L., Keenan, C.M., and Granger, D.N., 1990, Gastric ulceration induced by nonsteroidal anti-inflammatory drugs is a neutrophil-dependent process, Am J. Physiol. 259:G462-G467.

Yasuda, K., Satouchi, K., and Saito, K., 1986, Platelet-activating factor in normal rat uterus, Biochem. Biophys. Res. Commun. 138:1231-1236.

Yee, G.M., and Kennedy, T.G., 1988, Stimulatory effects of prostaglandins upon endometrial alkaline phosphatase activity during the decidual cell reaction in the rat, Biol. Reprod. 38:1129-1136.

Yee, G.M., and Kennedy, T.G., 1991, Role of adenosine 3',5'-monophosphate in mediating the effect of prostaglandin E_2 on decidualization in vitro, Biol. Reprod. 45:163-171.

Yee, G.M., Squires, P.M., Cejic, S.S., and Kennedy, T.G., 1993, Lipid mediators of implantation and decidualization, J. Lipid Med. 6:525-534.

IMMUNOENDOCRINE FUNCTIONS OF TROPHOBLAST INTERFERONS (IFN-τ OR TP-1 OR TROPHOBLASTINS) IN THE MATERNAL RECOGNITION OF PREGNANCY

Jacques Martal[1], Nasser-Eddine Assal[1], Aines Assal[2], Kamel Zouari[1], Louis Huynh[1], Nicole Chêne[1], Pierrette Reinaud[1], Gilles Charpigny[1], Madia Charlier[1], Gérard Chaouat[2]

[1] I.N.R.A., Unité d'Endocrinologie de l'Embryon, Station de Physiologie animale 78352 Jouy-en-Josas cédex, France
[2] Biologie cellulaire et moléculaire de la Relation materno-foetale, Gynécologie-Obstétrique, Hôpital Antoine Béclère, 92140 Clamart, France

INTRODUCTION

It has been recently shown in ruminants that interferons are constitutively produced by the trophoblast during the critical peri-implantation period. Ruminant trophoblast IFN gene sequences provide evidence that these IFNs recently called IFN-τ constitute a distinct family of other interferons of type I (α, β, ω). Embryonic IFNs are not confined in ruminants but have been found in other species though their structural characteristics can be different. In culture of ovine endometrial cells, ovine Trophoblast Protein (oTP) is able to inhibit the biosynthesis of prostaglandin $F_{2\alpha}$, even when stimulated by oxytocin. *In vivo*, recombinant oTP (r.oTP) inhibits the cyclic luteolysis of recipient ewes for 1 month or more as do trophoblast homogenates or trophoblastic vesicles and better than bovine recombinant IFNα. Furthermore, r.oTP is able to block the cyclic $PGF_{2\alpha}$ pulsatility thus showing the key role played by the ruminant trophoblastins in the mechanisms of maternal recognition of pregnancy. Besides displaying a paracrine antiluteolytic function, r.oTP exhibits immunoregulatory activities similar to the natural oTP isoforms. It blocks phytohemagglutinin A (PHA) driven lymphocyte proliferation on murine, human and ovine lymphocytes. oTP is acting especially on human, murine and ovine CD_4 T cells (helper lymphocytes). It does not block in mice interleukin-2 dependent cell proliferation when assayed on CTL-L2 cell line (CD_8^+) but inhibits in sheep PHA driven CD_8 T lymphocyte proliferation. oTP and r.oTP were found to be immunosuppressive on a murine and a human Mixed Lymphocyte Reaction which is usually considered as a maternal allo-recognition of paternal antigens of the conceptus. *In vivo*, r.oTP also exhibits immunosuppressive activity in mice on a Graft Versus Host (Popliteal Lymph Nodes) assay. Finally, r.oTP strongly inhibits immunological embryonic abortion in the usual model of CBA/J x DBA/2 mice if given early in pregnancy, at days 5 and 8 but not later (days 8 and 10). In conclusion, we propose a working hypothesis of the immunoendocrine function of ovine trophoblastin in the mechanisms of embryo immunotolerance.

A. UBIQUITOUS PROPERTY OF TROPHOBLAST IFNs

I. Trophoblastins or ruminant IFN-τ

1. Structural characteristics of trophoblastins (TP or IFN-τ). Interferons (IFN) are molecules induced by a viral infection (Isaacs and Lindenmann, 1957, see review by Lefèvre, 1989). They are secreted by leukocytes (IFN-α and IFN-ω) or by fibroblasts (IFN-ß) or by activated T lymphocytes (IFN-γ). IFN-α, -ß and -ω constitute the type I and IFN-γ the type II. Antiviral, antitumoral and immunoregulatory properties are well known (De Mayer and De Mayer-Guignard, 1988). In searching the biochemical structure of an ovine embryonic antiluteolytic protein, named trophoblastin (oTP) (Martal et al., 1979) and in characterizing a major protein secreted by the trophoblast (oTP-1) (Godkin et al., 1982, 1984a,b), we found that the trophoblast was constitutively producing interferons (Imakawa et al., 1987; Stewart et al., 1987; Charpigny et al., 1988a,b). Ovine trophoblastins (oTP) are holoproteins with apparent molecular weight (MW) around 20 kDa. On the opposite IFNs are usually glycoproteins. Bovine trophoblastins (bTP) have apparent MW of 20 and 22 kDa and exhibit a N-glycosylation site (Godkin et al., 1982; Martal et al., 1984; Helmer et al., 1987; Baumbach et al., 1990).

Both N-terminal sequence (Stewart et al., 1987; Charpigny et al., 1988a) and amino acid sequence deduced from oTP cDNA (Imakawa et al., 1987; Stewart et al., 1989; Charlier et al., 1989; Whaley et al., 1991) share high identity with IFN-α (55%) and even higher with IFN-ω (70%). Ovine and bovine IFNτ exhibit a higher homology between themselves (around 82%) than with their homologous IFN-ω (72% with oIFN-ω and 79% with bIFN-ω) (Imakawa et al., 1989; Charlier et al., 1989). Encoding sequences deduced from oTP and bTP genes confirm these similarities. Like other type I IFN genes, IFN-τ genes are intronless and constitute a multigenic family (Stewart et al., 1990; Charlier et al., 1991). In their 5' and 3' flanking sequences, oTP and bTP genes share much higher homology than with their homologous IFN-ω (respectively 89% and 81% between oTP and bTP, 58% and 62% between oTP and oIFN-ω, and 55% and 68% between bTP and bIFN-ω). These structural features added to physiological properties have brought us to consider these trophoblastins as a new class of type I IFN termed IFN-τ (Charpigny et al., 1988a,b; Martal et al., 1988; Imakawa et al., 1989; Kleeman et al., 1990; Roberts et al., 1991; Charlier et al., 1991; Martal et al., 1991a). The trophoblastins obtained by conceptus culture exhibit antiviral properties characteristic to IFNs (Pontzer et al., 1988; Martal et al., 1988; Martal and Chêne, 1992). As opposed to most IFNs, the antiviral activity of IFN-τ is not species specific since oTP is active on bovine, porcine, human, murine, canine cells challenged with VSV (Vesicular Stomatitis Virus). Similarly a strong antiviral activity can be detected in COS cells medium after oTP cDNA transfection (Charlier et al., 1989).

Both IFN-τ and IFN-ω polypeptide chains are composed of 172 amino acids (aa) whereas IFN-α and IFN-ß contain 166 aa. IFN-τ show two disulfide bridges (1-99, 29-139), a consensus sequence characteristic of IFNs (139-146: Cys-Ala-Trp-Glu-Ile-Val-Arg-Val) and a signal peptide of 23 aa. IFN-τ also share the consensus nucleotide sequence characteristic to inflammatory mediators, TTATTTAT: IFNs, TNF (Tumor Necrosis Factor) and Interleukins.

2. Expression of trophoblastins. Six oTP isoforms can be produced by only one embryo (Charpigny et al., 1988a,b; Martal et al., 1988; Fillion et al., 1990; Reinaud et al., unpublished data). This result is consistent with the finding of several oTP cDNA and oTP genes (Kleeman et al., 1990; Roberts et al., 1991; Nephew et al., 1993). Several functional oTP genes can therefore be simultaneously expressed by a single conceptus.

The experiments of intra-uterine injections of trophoblast homogenates or vesicles first showed that trophoblastin, an antiluteolytic protein is produced during a short period of time by the conceptus from day 9 to 22 (Rowson and Moor, 1967; Martal et al., 1979, 1987; Heyman et al., 1984). These results have been confirmed by Northern blotting of oTP mRNA (Hansen et al., 1988; Charlier et al., 1989) and by culture experiments of ovine conceptuses (Godkin et al., 1982). oTP and its mRNA have been specifically located into the extra-embryonic trophectoderm by immunohistological techniques (Godkin et al., 1984b), in situ hybridization and immunohistofluorescence (Farin et al., 1989; Guillomot et al., 1990). Surprisingly, that last method revealed oTP expression into not cultivated blastocysts from day 10 of pregnancy (Guillomot et al., 1990) suggesting that oTP expression could be derepressed in vitro before this stage. The induction of IFN-τ expression seems to be genetically determined because bTP mRNA transcripts have been found by RT-PCR in bovine blastocysts cultured in vitro after in vitro oocyte maturation and in vitro fertilization (Hernandez-Ledezma et al., 1992). However, the TP expression level might be modulated in vivo by the maternal uterine environment. Following this hypothesis the regulation of the IFN-τ expression needs further investigation. The maximum of oTP secretion is observed on day 14 of gestation and this level is decreasing by day 16. The arrest of oTP synthesis has been found during the implantation of the trophoblast on maternal caruncles (Guillomot et al., 1990) and begins on day 15 of pregnancy in sheep. Several growth factors or cytokines are involved in the positive control of the oTP secretion such as IGF_I and IGF_{II} (Ko et al., 1991) or GM-CSF (Nephew et al., 1993). Indeed, these factors are produced by both trophoblast and endometrium during the ovine peri-implantation period (Chêne et al., 1991, and unpublished data). In addition, it is known that other cytokines as LIF (Leukemia Inhibitory Factor) and IL_6 (Interleukin 6) are capable of inducing a trans nuclear factor, IRF_1 (Interferon Regulatory Factor-1) which increases the IFN-α expression (Abdollahi et al., 1991).

II. Other trophoblast IFNs in non-ruminants

1. The pig trophoblast constitutively produces IFNs (Cross and Roberts, 1989; Mirando et al., 1990a; La Bonnardière et al., 1991). Like in ruminants, antiviral activity is detected in pig uterine flushings and in pig conceptus culture media during the peri-implantation period (day 12 to 20) with a maximum on days 14-16 of pregnancy. After day 20 of gestation, IFN activity is almost undetectable either in the uterine flushings, or in the conceptus culture media. The comparison of trophoblast IFN production between the very prolific Meishan breed (about 15 piglets by litter) and Large White breed (about 10.5 piglets by litter) showed a slight but significant precocity for Meishan. The antiviral activity produced by pig conceptuses on day 15 is much lower than that produced by ruminants: about 10^5 IU per pig conceptus *versus* 10^8 IU per ovine conceptus at the same stage. In addition, the nature of trophoblast IFN in pig is very different from that in ruminants. The porcine conceptus secretes IFN-γ (Lefèvre et al., 1990) the encoding nucleotide sequence of which is strictly identical to that of porcine lymphocyte IFN-γ (146 aa and several introns). Only a single IFN-γ gene was observed. The antiviral activity produced by pig trophoblast is not completly neutralized by an anti-r.p IFN-γ (recombinant porcine IFN-γ) serum. It is also immunoneutralized by an anti-human leukocyte IFN serum (La Bonnardière et al., 1991). Although expressed much lower than pig trophoblast IFN-γ, this other pig trophoblast IFN corresponds to a new family of type I IFN, different from IFN-α, ß, ω and τ (Lefèvre et al., 1991). The amino acid sequence deduced from cDNA revealed a very short polypeptide (149 aa) as compared to type I IFN which shares an identity between 27 and 42% with IFN-α, ß, ω, τ. The temporary name spI IFN (for short porcine type I IFN) has been recently proposed by Lefèvre (personal communication). The co-expression of type I and II IFNs in the pig trophectoderm therefore suggests a unique tissue-specific regulation of these IFN genes. Like the other type I IFN, the spI IFN gene is unspliced.

Since the maintenance of the pig pregnancy corpus luteum is considered to be due to an oestrogen-dependent mechanism of a bi-directional change in endocrine to exocrine uterine $PGF_{2\alpha}$ secretions (Bazer and Thatcher, 1977; Bazer, 1989), the porcine trophoblast IFNs (γ and spI) do not seem to be involved in the control of the luteal function as is the case with ruminant trophoblast IFNs. Their possible physiological effects are not clear, since they might play a role in the mechanisms of embryo non-rejection by the dam or in the control of the embryogenesis (La Bonnardière and Martal, 1991).

2. The equine trophoblast produces low antiviral activity (10^3 IU/conceptus or 100 times lower than in pig and 10 000 times lower than in ruminants: bovine, ovine, caprine) (Zouari et al., 1991a). It is worth underlining that this antiviral activity is observed during a very short period: 14 to 18 days of pregnancy. This antiviral activity is also present in concentrated uterine flushings. This antiviral activity can be seroneutralized with a serum against oTP. This protein bears an apparent molecular weight of 20 kDa and has been located by immunohistofluorescence in the extra-embryonic trophectoderm like pig and ruminant trophoblast IFNs. On the other hand, this yet unidentified equine IFN only appears during the pre-implantatory phase (implantation occurs on Day 30 in horse). The kinetic of its expression corresponds exactly to the organogenesis from inner cell mass: on Day 15 the totipotent cells of embryonic disc can be observed. So can the main organs such as the heart which starts beating on Day 17.

3. The rabbit endometrium exhibits an antivral activity (Zouari et al., 1991b). Paradoxically, the uterine secretion of not only the pregnant rabbit but also the pseudopregnant one shows an antiviral activity on MDBK cells infected by VSV. This activity is immunoneutralized by sera against natural oTP. However, the molecular nature of this activity has still to be determined. Furthermore there is no immunological cross-reaction with rabbit IFN-ω (Charlier et al., 1993). Although the concentration of the rabbit antiviral activity secreted in uterine flushings and endometrial fibroblast cultures is particularly low as compared to ruminant and pig trophoblast interferons, it is regularly expressed on Days 6 and 7 of pregnancy and pseudopregnancy. As opposed to trophoblast IFNs previously described in other species, an antiviral activity was exceptionnally observed in the rabbit conceptus, in our experimental conditions. Thus, the ubiquitous property of trophoblast IFN in early pregnancy is not established in every placental mammals.

4. In the human species, the existence of placental IFN throughout pregnancy has been demonstrated by Lebon et al. (1982, 1985), Duc-Goiran et al. (1985), Taguchi et al. (1985), Bocci et al. (1985), Chard et al. (1986). IFN-α and IFN-ß are present in the foetal tissues. IFN-α was detectable at very low concentrations in the blood (Shiozawa et al., 1988) and tissues (Khan et al., 1989) of normal subjects. The highest circulating levels are measured in young adults (Shiozawa et al., 1986, 1989) but much higher concentrations have been determined in the foetal-placental unit: foetal blood, foetal organs, placenta, membranes, amniotic fluid and decidua (Chard et al., 1986; Chard and Iles, 1992). Nevertheless, these placental IFN-α and IFN-ß are expressed much later and at a much lower concentration than the ruminant and pig embryonic IFNs.

However several human genomic clones have recently been shown to exhibit a high structural identity with ovine trophoblastin (about 85%) (Whaley et al., 1991), but the physiological chronology of their expression during pregnancy is still unknown.

5. In the mouse trophoblast, the existence of embryonic IFN is controversial (Cross and al., 1990; Baker and Nieder, 1990; Perez et al., 1991; Roberts et al., 1992a,b) although there is no doubt later in pregnancy about the presence of placental IFN (Fowler et al., 1980; Weislow et al., 1983; Barlow et al., 1984; Yamada et al., 1985).

Finally, the presence of IFNs during the peri-implantation period seems to be confirmed in several species but their physiological importance and/or molecular nature need to be determined.

B. PARACRINE ANTILUTEOLYTIC FUNCTION OF TROPHOBLASTINS IN RUMINANTS

I. Local antiluteolytic effect of conceptus

In sheep, Moor and Rowson (1966) were the first to demonstrate that the conceptus (embryo and membranes) inhibits the uterine cyclic luteolytic activity after transfer into the uterine horn ipsilateral or contralateral to corpus luteum. On the contrary, when one uterine horn is ligated, the embryo can inhibit the cyclic luteolysis only if it is transfered into the uterine horn adjacent to the corpus luteum. Thus the antiluteolytic embryonic factor acts by a local mechanism and not by general circulation as opposed to human Chorionic Gonadotropin. The authors have also demonstrated that embryo transfer into cyclic recipient ewes inhibits luteolysis only when performed before day 12 of oestrous cycle. Daily uterine infusions of ovine 14-16-day-old trophoblast homogenates to recipient cyclic ewes maintain luteal function over one month whereas infusions of 21-23-day-old do not (Rowson and Moor, 1967; Martal et al., 1979). The embryonic antiluteolytic signal is a protein because it is inactivated by heat and by protease. It is synthesized at least from day 12 to 22 and it was been first termed trophoblastin (Martal et al., 1979), then oTP-1 (ovine Trophoblastic Protein-1) (Godkin et al., 1984a,b), oTP-B (ovine Trophoblastic Protein B) (Martal et al., 1984), oTP (Trophoblastin) (Charlier et al., 1989) and at last IFN-τ.

II. Antiluteolytic effect of natural purified oTP

Daily uterine infusions of partly purified natural oTP into recipient cyclic ewes can delay the luteolysis for several days (Godkin et al., 1984a; Fincher et al., 1986; Vallet et al., 1988). Similar results with bTP were obtained in cattle (Knickerbocker et al., 1986; Helmer et al., 1987; Thatcher et al., 1989). oTP specifically binds to the endometrial receptors (Godkin et al., 1984b) and competes with recombinant bovine IFN-α (r.b.IFNα) (Stewart et al., 1987). oTP binds to both high and low affinity endometrial receptors whereas r.b.IFNα binds only to the high affinity ones (Hansen et al., 1989). Uterine infusions of oTP considerably reduce the secretion of the main stable $PGF_{2\alpha}$metabolite (PGFM: 15-keto-13-14 dihydro $PGF_{2\alpha}$) even after injections of oestradiol or oxytocin (Finsher et al., 1986; Vallet et al., 1988) the luteolytic activity of which has been established. In these experimental conditions, highly enriched purified oTP maintained luteal function as uterine infusions of total proteins of ruminant conceptus culture media did. Nevertheless, it was difficult to know from these experiments whether some contaminant of purified natural oTP or some isoforms play an additional important function in the maintenance of the cyclic corpus luteum.

III. Antiluteolytic effect of recombinant bovine IFN-α (r.b.IFN-α)

Although structurally distinct (~ 35%) from IFN-τ but very similar in *in vitro* biological assays, commercially available (Ciba-Geigy) recombinant bovine IFN-α (r.b.IFN-α) was used in experiments with cyclic ewes and cows. Uterine infusions of high concentrations (2 mg daily per ewe) r.b.IFN-α could maintain the cyclic corpus luteum but only for several days (Plante et al., 1988; Stewart et al., 1989; Flint et al., 1991; Schaule-Francis et al., 1991; Parkinson et al., 1992).

When r.b.IFN-α (kind gift from Ciba-Geigy) and recombinant oTP are injected in strictly similar experimental conditions (350 µg twice daily from days 10 to 17, by uterine route) in cyclic recipient ewes, recombinant bovine IFN-α is far less efficient on the maintenance of the corpus luteum than r.oTP. It appears therefore that IFN-α has to be introduced in the uterus in considerably higher quantities compared to natural trophoblast IFN in order to achieve an extension of the luteal function (Roberts et al., 1992b).

IV. Antiluteolytic effect of recombinant trophoblastin (r.oTP)

From the oTP cDNA (Charlier et al., 1989) we obtained high amounts of r.oTP thanks to the kind help of Transgène Society (Strasbourg, France). After trying several expression systems: baculovirus-insect cells system (Cerruti et al., 1991), *E. Coli* and yeast (Degryse et al., 1992), the latter has been chosen because of the addition of a dipeptide spacer that significantly improved the secretion of ovine trophoblastin. Large amounts of r.oTP (5-10 mg/l) are obtained by yeast fermentation. oTP-cDNA was fused to the pre-pro nucleotide sequence encoding the pre-pro peptide of the yeast α-factor precursor (KEX 2 gene product). The cleavage site for the processing enzyme (KEX 2p) was at the Lys-Arg↑Cys_1 bond liberating mature oTP. To improve the r.oTP secretion, Degryse et al. (1992) performed an extension of the signal peptidase cleavage site in the pre-oTP fusion in the plasmid pTG 7908. The presence of the Ala-Pro extension of r.oTP was observed by N-terminal sequencing of purified r.oTP by HPLC. The specific activity ($1-2.5 \times 10^8$ IU/mg) of the Ala-Pro-r.oTP is similar to natural oTP and the recombinant molecule retains its biological activity *in vivo* (Martal et al., 1990). No significant differences have been

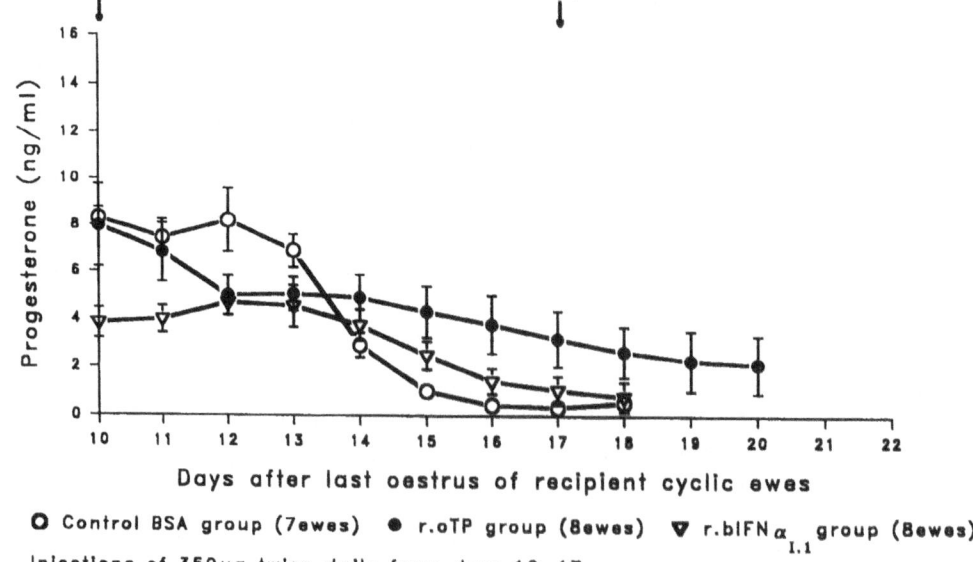

Figure 1. Comparative effects of r.oTP and r.b.IFN-α on the maintenance of the corpus luteum after intra-uterine injections into recipient cyclic ewes.

observed between r.oTP and natural oTP in many assays: SDS-polyacrylamide electrophoresis, radioimmunoassays, inhibition of *in vitro* $PGF_{2\alpha}$ secretion by endometrial epithelial cells, cellular immunological assays. New conditions of semi-preparative purification of yeast culture media have been defined and r.oTP preparations have been obtained with a purity degree over 98%. Uterine influsion of r.oTP (350 µg twice daily per cyclic ewe from days 10 to 17) was able to extend the luteal function for about one month in 60% of recipient animals (Martal et al., 1990). Recombinant ovine trophoblastin appeared as efficient as trophoblast homogenates or vesicles (Martal and Chêne, 1992) to increase the duration of the progesterone luteal secretion. Since r.oTP is produced by yeasts from only one cDNA, other isoforms are not absolutely necessary to the paracrine antiluteolytic function of the conceptus. Nevertheless, one has to inject quantities equivalent to the sum of the isoforms in order to produce the same result as in natural conditions. This observation leads to the conclusion that the isoforms help the paracrine antiluteolytic function. These results also show that no other proteins than oTP need to be considered to explain the luteolytic inhibition mechanisms in early pregnancy.

Finally, it has been demonstrated that intramuscular injections of r.b.IFN-α could improve both the embryonic survival and the prolificity of treated pregnant ewes (Nephew et al., 1990). This suggest that r.oTP which is more efficient than r.b.IFN-α in the luteolysis inhibition assay, could be useful in the new reproductive biotechnologies.

V. Mechanisms of *in vitro* antiluteolytic action of trophoblastin

In endometrial cell culture, oTP reduces $PGF_{2\alpha}$ and PGE_2 secretion (Salamonsen et al., 1988, 1991). Basal levels of $PGF_{2\alpha}$ secreted by uterine epithelial cells are 100 times higher than those of stromal cells. After a 12-24h pre-treatment, oTP reduced by about 50% the secretion of $PGF_{2\alpha}$ by epithelial cells and by around 30% the secretion by stromal fibroblasts. When added *in vitro* before and along with oxytocin, oTP can inhibit the $PGF_{2\alpha}$ secretion induced by oxytocin (Charpigny et al., 1991). We know that oxytocin, once bound to its receptor, stimulates the prostaglandin synthesis by phosphoinositides hydrolysis (Flint et al., 1986). The resulting triphosphate inositol (IP_3) provokes intracellular calcium mobilization which activates phospholipase A_2 and then, from hydrolysed phospholipids, the formation of arachidonic acid, the prostaglandins precursor. Despite of oxytocin, oTP inhibits phosphatidylinotisols cycle (Bazer, 1989; Vallet and Bazer, 1989; Mirando et al., 1990b). oTP also blocks the $PGF_{2\alpha}$ synthesis induced by a protein kinase C activator (phorbol ester, PMA) (Charpigny et al., 1991). oTP could foster the synthesis and/or the activation of an endometrial prostaglandin synthetase inhibitor (EPSI) (Gross et al., 1988). oTP could also be responsible for the induction and/or activation of an other potential endometrial inhibitor involved in the phospholipase A_2 (PLA_2) activity reduction, as suggested by the comparison of this PLA_2 activity in pregnant ewes (D12-D14) and cyclic ewes (D12-D14) endometrium (Tamby et al., 1991).

VI. Mechanism of *in vivo* oTP antiluteolytic paracrine action: inhibition of endometrial $PGF_{2\alpha}$ pulsatile secretion

1. **$PGF_{2\alpha}$ pulsatile secretion.** Endometrial $PGF_{2\alpha}$ is the hormone responsible for luteolysis in ruminants (Mc Cracken et al., 1981; Schramm et al., 1983). It crosses by counter-current from the utero-ovarian vein to the ovarian artery (Mc Cracken et al., 1972). $PGF_{2\alpha}$ secretion starts on Day 12 of the oestrous cycle and requires at least 5 successive pulses to induce complete luteolysis. In early pregnancy those $PGF_{2\alpha}$ peaks are inhibited (Thorburn et al., 1973; Nett et al., 1976; Zarco et al., 1988). During luteolysis, the simultaneous increases in $PGF_{2\alpha}$ and oxytocin result from reciprocal interactions between luteal oxytocin and uterine $PGF_{2\alpha}$ (Flint and Sheldrick, 1986). The plasma oxytocin

concentrations are almost identical in pregnant and cyclic ewes between Day 13 and Day 16 (Hooper et al., 1986). On the opposite, the concentrations of endometrial receptors to oxytocin are high in luteolysis and low in early pregnancy (Mc Cracken, 1980; Sheldrick and Flint, 1985; Flint and Sheldrick, 1986).

2. Endocrine control of endometrial receptors to oxytocin (R_{OT}). The results above underlined the importance of oxytocin receptors in regulation of $PGF_{2\alpha}$ pulsatile secretion (Flint et al., 1990). In ruminants, the endometrial concentrations of R_{OT} are controlled by oestrogens and progesterone during luteolysis. Oestrogens could allow the increase in R_{OT} only after progesterone has fallen (Mc Cracken, 1981; Zhang et al., 1992). A sequential treatment of progesterone (10 days)-oestrogens (2 days) and progesterone (5-12 days) showed that the progesterone level particularly increases the endometrial R_{OT} concentrations (Vallet et al., 1990) instead of reducing it, like in the Mc Cracken's classical hypothesis of R_{OT} regulation (1980). In addition, with a continuous treatment of progesterone (P_4) and oestradiol (E_2) during 9 days followed by only E_2 treatment (3 days), the endometrial R_{OT} increase, but not if P_4 and E_2 are continuous during 12 days (Zhang et al, 1992). We can no more assume that the absence of oestrogens is responsible for the reduction of endometrial R_{OT} in early pregnancy. In cyclic ewes, the endometrial R_{OT} concentration is strongly reduced by uterine infusion of total proteins of conceptus culture media or of r.b.IFNα (1 mg/horn) (Vallet and Lamming, 1991), suggesting that trophoblastins could inhibit the R_{OT} synthesis.

3. Inhibition of $PGF_{2\alpha}$ pulsatility by r.oTP. The main stable $PGF_{2\alpha}$ metabolite (PGFM: 15-keto-13-14-dihydro-$PGF_{2\alpha}$) was measured by radioimmunoassay in blood samples taken from the jugular vein, hourly round the clock, from days 12 to 18 of oestrous cycle of recipient ewes. In this way PGFM pulsatile episodes can be determined as Zarco et al. (1988) first showed. Silastic catheters were surgically introduced into the uterine horns (Martal et al., 1979) and corpora lutea marked with India ink. Control cyclic recipient ewes received, by uterine route, bovine serum albumin at the same dose (350 µg twice daily from days 10 to 17) as cyclic ewes treated with recombinant trophoblastin. As shown in Figure 2b, r.oTP is able to inhibit $PGF_{2\alpha}$ pulsatility in recipient cyclic ewes exhibiting a maintenance of luteal progesterone secretion (Assal et al., unpublished data). Trophoblastin could act in two ways: by decreasing uterine $PGF_{2\alpha}$ synthesis and possibly by reducing endometrial oxytocin receptors.

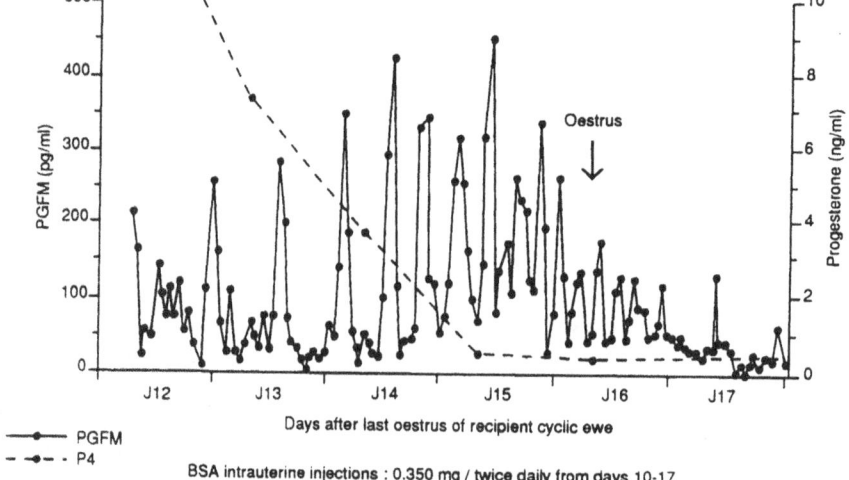

Figure 2a. Comparative patterns of PGFM and P_4 of control recipient cyclic ewe (9463). BSA intrauterine injections: 350 µg/twice daily from days 10-17.

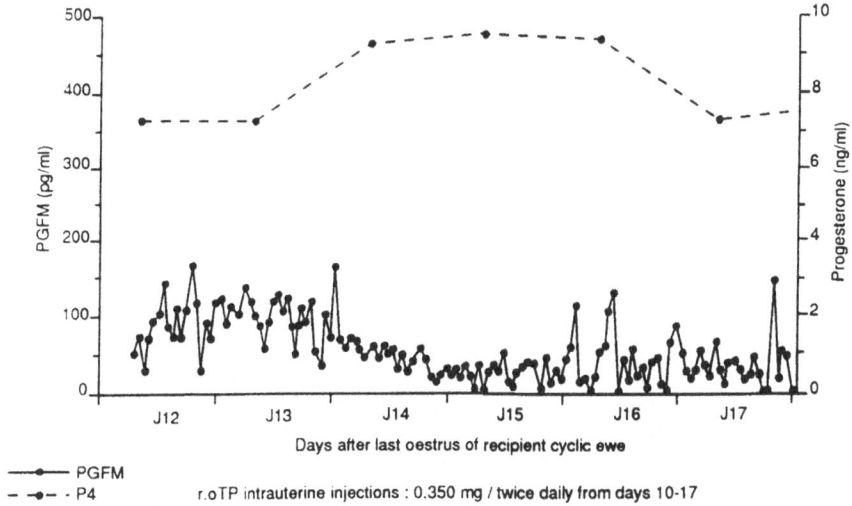

Days after last oestrus of recipient cyclic ewe

——•—— PGFM
– –•– – P4   r.oTP intrauterine injections : 0.350 mg / twice daily from days 10-17

Figure 2b. Comparative patterns of PGFM and P_4 of recipient cyclic ewe treated by r.oTP (7582). r.oTP intrauterine injections: 350 µg/twice daily from days 10-17.

Consequently, in ruminants, trophoblast interferons constitute the main embryonic signal of maternal recognition of pregnancy. We can consider with Chard (1991) and othersthat IFN-τ acts like an actual reproductive hormone although it remains confined into the conceptus-endometrium interface and does not circulate in blood.

C. IMMUNO-ENDOCRINE FUNCTION OF TROPHOBLASTIN IN THE MECHANISMS OF IMMUNE TOLERANCE OF THE CONCEPTUS

I. Necessity of local immunoregulatory mechanisms for maintaining the embryonic semi-allograft

The non-rejection of the conceptus presents an immunological paradox because the fetus is produced by the mating of genetically dissimilar individuals resulting in an allograft which is highly successful.

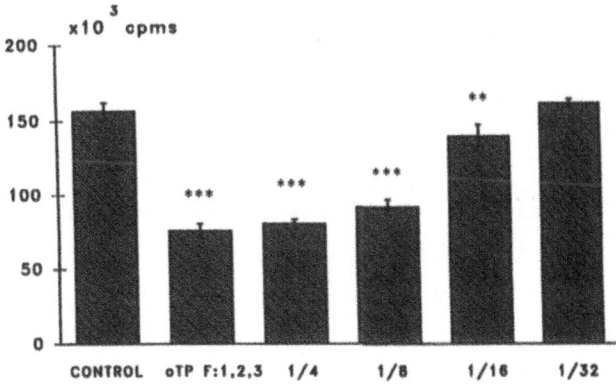

Figure 3. Inhibitory effect of oTP isoforms 1, 2, 3 on PHA driven proliferation of human lymphocytes. *** $P < 0.001$; ** $P < 0.05$; oTP isoforms were obtained after DEAE HPLC chromatography in according with Charpigny et al. (1988).

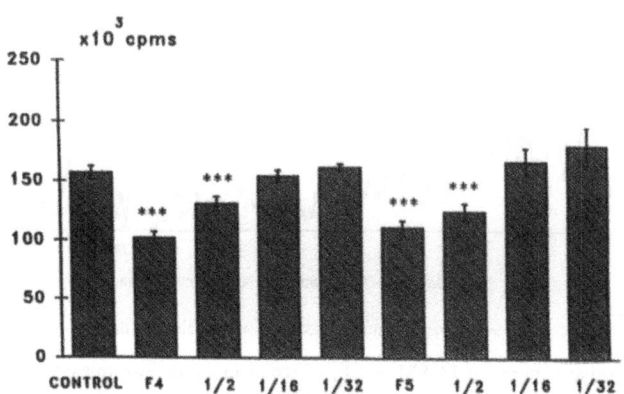

Figure 4. Inhibitory effect of oTP isoforms 4, 5 on PHA driven proliferation of human lymphocytes. *** P < 0.001; oTP isoforms were obtained after DEAE HPLC chromatography in according with Charpigny et al. (1988).

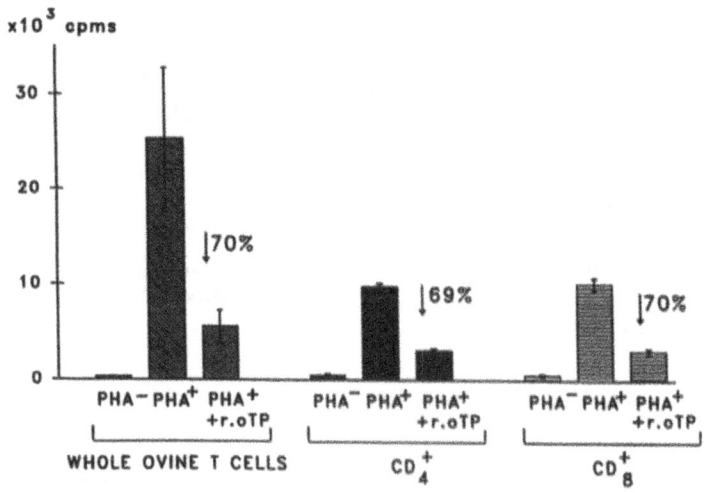

Figure 5. Inhibitory effect of r.oTP on ovine T lymphocyte proliferation assessed on purified CD_4^+ and CD_8^+ PHA blasts selected by panning 24h after mitogen stimulation and placed in r.oTP-containing medium. 5.10^6 (C_3H/He) mouse lymphocytes per ml + 5.10^6 (Balb/c) mouse lymphocytes per ml. oTP and r.oTP: 3 µg/ml.

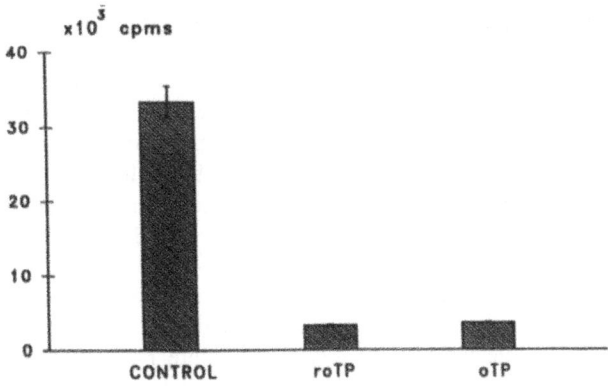

Figure 6. Inhibitory effect of r.oTP and oTP on murine mixed lymphocyte reaction.

At the early implantation stage, the conceptus is resistant to Natural Killer cells (NK) and to Cytotoxic T Lymphocyte (CTL)-mediated lysis (Croy and Rossant, 1987) in mice. However, freshly isolated trophoblast cells could be the target of LAKCs (lymphokine-activated Killer cells) or CTLs endowed with some LAKCs properties after growth *in vitro* (Drake and Head, 1988, 1989). Moreover TNF (Tumor Necrosis Factor) released by NK, CTL, and others, could trigger the uterine smooth muscle motility (Warner and Libby, 1989). Delayed type hypersensitivity (DTH)-like phenomenon to embryonic or paternal antigens could also lead to an inflammatory reaction or abnormal uterine contractions resulting in early pregnancy wastage. Thus, local immunosuppressive mechanisms appear necessary for the embryo to avoid its own rejection at the implantation eventhough MHC (Major Histocompatibility Complex) antigens are few and transitory for some of them (Chaouat, 1978).

II. The IFN-τ inhibits the helper (CD_4^+) and cytotoxic (CD_8^+) T lymphocyte proliferation

Five natural ovine trophoblastin isoforms exhibit antiproliferative properties on the lymphocyte blastogenesis: they inhibit *in vitro* PHA (Phytohemagglutinin A)-induced proliferation of heterologous (murine and human) lymphocytes, as shown in Figures 3 and 4 (Fillion et al., 1991) and homologous (ovine) ones as shown in Figure 5 (Assal-Meliani et al., 1991, 1992). After isolation of CD_4^+/CD_8^+ T lymphocytes by using monoclonal antibodies against CD_4^+/CD_8^+ murine, human (Fillion et al., 1991) and ovine determinants (Assal et al., 1991, 1992; Martal et al., 1991b) oTP and r.oTP show a cytostatic activity on CD_4^+ lymphocytes suggesting that IFN-τ is immunosuppressive on helper T lymphocytes across species barriers. On the contrary, this ovine trophoblast IFN acts in different ways on PHA-induced proliferation of heterologous CD_8^+ T lymphocytes: for example it does not block the murine IL_2 dependent CTL-L_2 but inhibits human ones (Fillion et al., 1991) and ovine CD_8 T cells, in figure 5 (Assal et al., 1993a; ovine monoclonal antibodies against CD_4^+ and CD_8^+ were kindly gifted by Dr. Mackay of Basel Institute for Immunology). In addition in these various immunological assays, oTP and r.oTP does not modify the lymphocyte viability consistently with the physiological nature of these non-viral induced trophoblast interferons, suggesting no cytotoxic effect, in own experimental conditions (Assal-Meliani et al., 1993).

Finally, oTP can inhibit PHA-induced proliferation of helper T lymphocytes (CD_4^+) across species barriers (mice, human, sheep) and particularly its homologous CTL (CD_8^+).

III. The IFN-τ inhibits *in vitro* and *in vivo* allogenic reactions (MLR and GVHR)

Ovine trophoblastin exhibits a cytostatic activity in the *in vitro* allogenic type cytolytic reactions: oTP and r.oTP strongly inhibited cellular lysis obtained in the Mixed Lymphocyte Reactions (MLR) between two lymphocyte populations obtained from different mice strains (Balb/c and C_3H) (Assal-Meliani et al., 1993). In these *in vitro* uni- or bi-directional MLR, CTLs are produced by the allogeneic lymphocytes and are able to provoke the lysis of the other population. In presence of IFN-τ, the lysis is strongly inhibited (Figure 6).

Similarly, *in vivo*, in the local graft versus host reaction, r.oTP highly inhibited the immune rejection reaction of allogenic splenic lymphocytes (Balb/c) versus host lymph node cells (Balb/b x B_6) injected into the foot pads (Figure 7). Recombinant trophoblastin inhibited both the reactional lymph nodes and the cellular proliferation of lymph nodes (Assal-Meliani et al., 1993). Consequently, IFN-τ are able to inhibit the local GVH across strong species barriers.

IV. The IFN-τ slightly stimulates NK proliferation

Natural purified oTP and r.oTP stimulated NK activation (mice K462 target cell line) as did IFN-α (much more) and IFN-γ (still more). In spite of this apparent paradoxal property which seems to favour conceptus rejection, the immunotrophism theory (Wegman, 1987; Wegman and Gill, 1991) considered that NK could increase, by a weak TNF (Tumor Necrosis Factor) secretion, the GM-CSF or CSF_1 production by macrophages, for instance. Then, these cytokines could increase trophoblast growth. This source of cytokine secretion added to the natural production by trophoblast and/or endometrium could favour implantation success. With this new integrative hypothesis (Wegman et al., 1993; Chaouat et al., 1993), we could better understand how scid/scid, bg/bg mice which are deficient in B, T and NK cells, can exhibit normal implantation and fertility (Croy and Chapeau, 1990). Wegman et al. (1993) distinghished on the one hand, negative cytokines which have deleterious effect such as TNF, IL_2 (Interleukin 2) and IFN-γ (at least in mice) and on the other hand, positive cytokines favouring the trophoblast growth, such as GM-CSF, CSF_1, IL_3, IL_6, IL_{10}... especially in mice and human (Athanassakis et al., 1987; Pollard et al., 1987; Armstrong et al., 1989; Wegman et al., 1989; Chaouat, 1990a,b; Mathialagan et al., 1992; Chaouat et al., 1993). We could add also the LIF (Leukemia Inhibitor Factor) (Fry

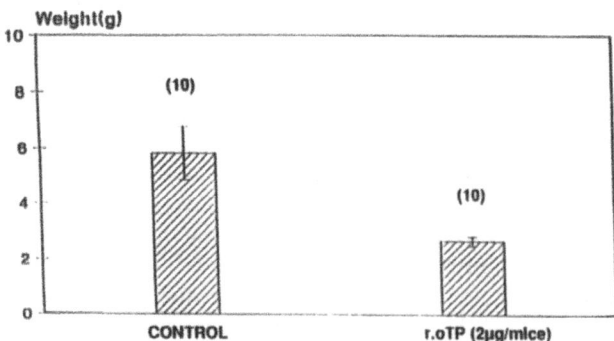

Figure 7. Effect of r.oTP on graft versus host reaction. Mean popliteal lymph node (PLN) stimulation index as estimated from PLN weight on Balb/c spleen cells injected into footpads of (Balb/b x B_6) F1 mice recipients. PLN were harvested on day 5. ( ) Number of animals. r.oTP: 2 µg/ml.

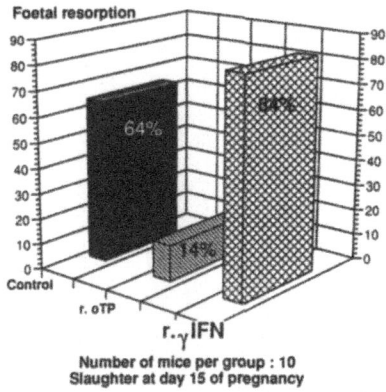

Figure 8. Foetal resorption in mice ♀ CBA/J x ♂ DBA/2 treated by recombinant ovine trophoblastin i.p. given on Days 5 and 8.

et al., 1992; Stewart et al., 1992) and the IFN-τ and possibly other trophoblast IFNs. Moreover some Growth Factors such as $TGF_{\beta 2}$ are produced by Natural Suppressive cells (NS) (Clark et al., 1991). In sheep, some of these aforementioned cytokines or growth factors are actually secreted by trophoblast and/or endometrium (Chêne et al., 1991 and unpublished data).

V. The IFN-τ highly improves the embryonic semi-allograft in the immunological murine CBA/J x DBA/2 model (Assal-Meliani et al., unpublished data)

In DBA/2 mated-CBA/J females, we usually observed 50 to 70% spontaneous fetal resorptions which would be mediated by cell of the NK lineage (Asialo GM1+) (Chaouat et al., 1990a,b, 1991). After ovine recombinant trophoblastin treatment on Days 5 and 8 of pregnancy, these mice exhibited a 80% decrease in fetal resorption observed on Day 15 (Fig. 8). On the contrary, r.h.IFN-γ treatment on Days 5 and 8 of pregnancy resulted in 85% of embryonic wastage in this NK mediated DBA/J x DBA/2 system according to Chaouat et al. (1991). The stage chosen for r.oTP treatment is determinant since when administered on Days 5 and 8 or only on Day 5, this IFN-τ decreased the fetal abortion but increased it after treatment on Days 8 and 10 (Figure 9). On the opposite, recombinant murine IFN-α or r.h.IFN-γ led to a strong fetal resorption at any stage (Figure 9). The embryonic mortality CBA/J x DBA/2 model allowed to demonstrate for the first time such a great difference between IFN-τ and IFN-α in their biological properties.

In fact, when we further analyze in Figure 10 the r.oTP fetal resorption decrease, we can observe that the number of implantation sites doubles (49 *vs* 23) and the number of living foetuses increases three times (42 *vs* 14). In other words, embryonic mortality rate actually tripled as soon as implantation started and IFN-τ therefore improved embryonic survival and consequently the fetal one.

Recent results obtained with Wegman's collaboration (Chaouat et al., 1993, unpublished data) showed that IFN-τ stimulated particularly the decidua and placental IL_{10} production in the r.oTP treated CBA/J x DBA/2 mice. In addition, IL_{10} treatment to the DBA/2 mated-CBA/J females from Day 5 of pregnancy, could reduce strongly foetal

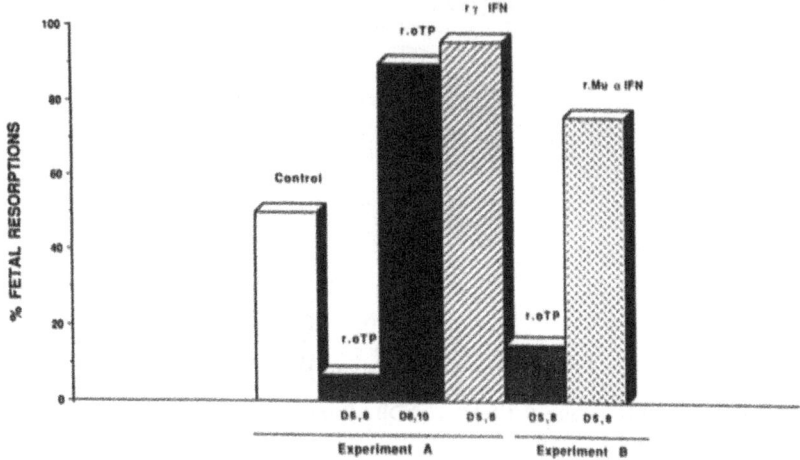

Figure 9. Fetal resorptions in mice ♀ CBA/J x ♂ DBA/2 treated by r.oTP given on Days 5 and 8, or 8 and 10, by r.h.IFN-γ and by r.MuIFN-α.

resorption like r.oTP treatment. Consequently, the immunosuppressive effect of IFN-τ seems to be partly mediated by the endometrial and placental IL_{10} production in this CBA/J x DBA/2 murine model.

Obviously, the different experiments reported here in mice do not prove the immunoregulatory properties of trophoblastin in ruminants. Nevertheless, the oTP immunological activity in early pregnancy in rodents, strongly suggests a possible implication of IFN-τ in the natural immunosuppressive mechanisms in the conceptus tolerance by its dam.

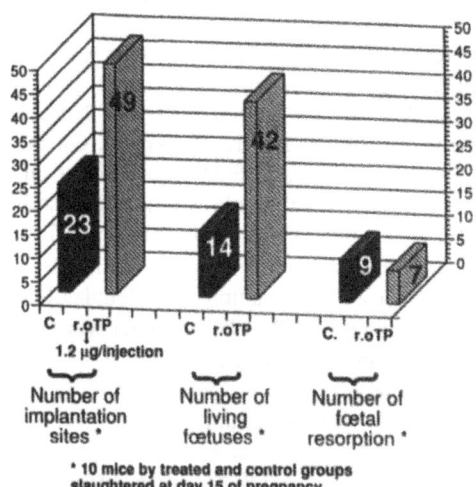

Figure 10. Increase in foetal survival in mice ♀ CBA/J x ♂ DBA/2 treated by recombinant ovine trophoblastin i.p. given on Days 5 and 8.

208

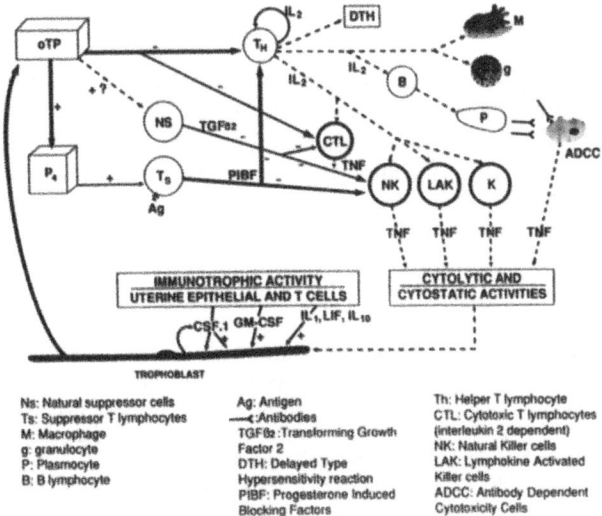

Ns: Natural suppressor cells
Ts: Suppressor T lymphocytes
M: Macrophage
g: granulocyte
P: Plasmocyte
B: B lymphocyte

Ag: Antigen
—⊀ :Antibodies
TGFβ₂ :Transforming Growth
Factor 2
DTH: Delayed Type
Hypersensitivity reaction
PIBF: Progesterone Induced
Blocking Factors

Th: Helper T lymphocyte
CTL: Cytotoxic T lymphocytes
(interleukin 2 dependent)
NK: Natural Killer cells
LAK: Lymphokine Activated
Killer cells
ADCC: Antibody Dependent
Cytotoxicity Cells

Figure 11. Working hypothesis of the immuno-endocrine function of ovine Trophoblastin (oTP) in the mechanisms of embryo immunotolerance. NS: Natural suppressor cells; Ts: Suppressor T lymphocytes; M: macrophage; g: granulocyte; P: plasmocyte; B: B lymphocyte; Ag: antigen; —⊀ : antibodies; $TGF_{\beta2}$/ Transforming Growth Factor 2; DTH: Delayed Type Hypersensitivity reaction; PIBF: Progesterone Induced Blocking Factors; Th: Helper T lymphocyte; CTL: Cytotoxic T lymphocytes (interleukin 2 dependent); NK: Natural Killer cells; LAK: Lymphocyte Activated Killer cells; ADCC: Antibody Dependent Cytotoxicity Cells.

VI. Working hypothesis of possible immuno-endocrine function of IFN-τ (Fig. 11)

1. oTP directly acts through inhibition of the ovine helper T lymphocyte (CD_4^+) proliferation, probably on TH_1 population. In addition, oTP could stimulate TH_2 population because it increases in murine trophoblast or endometrium IL_{10} production. Helper TH_1 lymphocytes (CD_4^+) play a key role in activating most lymphocyte IL_2 secretion i.e. blastogenesis of B lymphocytes, suppressor T lymphocytes (TS), cytolytic lymphocytes (CTL), Killer cells (NK, K, LAK) and Delayed type-like hypersensitivity reaction (DTH). So, in presence of IFN-τ, CTL and Killer cells would not proliferate and TNF secretion would remain very low, thus inhibiting their cytolytic and cytostatic activities which favour conceptus rejection.

2. oTP, at least in sheep, and other trophoblast substances such as $TGF_{\beta2/\beta1}$ (both present in ovine endometrium and trophoblast) are able to inhibit the CTL (CD_8^+) and Killer cells proliferation. It seems, at least in mice, that oTP could inhibit Natural Suppressive (NS) cells.

3. oTP is indirectly involved in the immunoregulatory mechanisms by its antiluteolytic activity allowing the maintenance of the luteal progesterone secretion. In fact, progesterone (P_4) exhibits its own immunosuppressive activity. In mice especially, progesterone in presence of antigens, here trophoblast antigens, is able to stimulate the secretion, by the suppressor T lymphocytes, of blocking factors such as PIBF (Progesterone Induced Blocking Factor) (Szekeres-Bartho et al., 1985) which have the ability to inhibit Killer cells activity.

4. oTP might favour trophoblast growth by slightly activating NK cells which produce cytokines. Similarly, trophoblast, endometrium and immune cells produce positive cytokines: GM-CSF, CSF_1, IL_1, IL_4, IL_6, IL_{10}...), according to Wegman et al. (1993).

Acknowledgments

We are grateful to Miss M.E. Martal for the translation of this text in English and Mrs M.E. Marmillod for typing the manuscript. Dr. E. Degryse and Transgene Society (Strasbourg, France) are gratefully acknowledged for recombinant oIFN-τ and Dr. Martinod and Ciba-Geigy for kind gift of recombinant bovine IFN-α. This work was supported by grants for Ministère de la Recherche et de l'Espace.

REFERENCES

ABDOLLAHI, A., LORD, K.A., HOFFMAN-LIEBERMANN, B., and LIEBERMANN, D.A., 1991, Interferon regulatory factor-1 is a myeloid differentiation primary response gene induced by interleukin-6 and leukemia inhibitory factor-role in growth inhibition, *Cell Growth Diff.* 2:401-407

ARMSTRONG, D.T., and CHAOUAT, G., 1989, Effects of lymphokines and immune complexes on murine placental cell growth *in vitro*, *Biol. Reprod.* 40:466-474.

ASSAL-MELIANI, A., CHARPIGNY, G., MARTAL, J., and CHAOUAT, G., 1991, Natural and recombinant oTP1 trophoblastin has immunoregulatory activities. Ann. Meeting of the Intern. Society for Interferon Research, Nice, *J. IFN Res.* 11, Suppl. 1:S102.

ASSAL-MELIANI, A., CHARPIGNY, G., REINAUD, P., MARTAL, J., and CHAOUAT, G., 1993, Recombinant ovine trophoblastin inhibits ovine and murine lymphocyte proliferation. J. Reprod. Immunol. (in press).

ASSAL-MELIANI, A., KINSKY, R., CHARPIGNY, G., REINAUD, P., MARTAL, J., CHAOUAT, G., 1992, Recombinant ovine trophoblastin inhibits ovine and murine lymphocyte proliferation and popliteal lymph node (GVH) assays, 5th International congress of reproductive immunology, Rome (Italie): 73.

ATHANASSAKIS, I., BLEACKLEY, R.C., PAETKAU, V., GUILBERT, L., BARR, P.J., WEGMANN, T.G., 1987, The immunostimulatory effect of T cells and T cell lymphokines on murine fetally derived-placental cells, *J. Immunol.* 138:37-44.

BAKER, D.J., and NIEDER, G.L., 1990, Interferon activity is not detected in blastocyst secretions and does not induce decidualization in mice, *J. Reprod. Fertil.* 88:307-313.

BARLOW, D.P., RANDLE, B.J., and BURKE, D.C., 1984. Interferon synthesis in the early post-implantation mouse embryo, *Differentiation* 27:229-235.

BAUMBACH, G.A., DUBY, R.T., and GODKIN, J.D., 1990, N-glycosylated and unglycosylated forms of caprine trophoblast protein-1 are secreted by peri-implantation goat conceptuses, *Biochem. Biophys. Res. Comm.* 172:16-21.

BAZER, F.W., 1989, Establishment of pregnancy in sheep and pigs, *Reprod. Fertil. Dev.* 1:237-242.

BAZER, F.W., and THATCHER, W.W., 1977, Theory of maternal recognition of pregnancy in swine based on estrogen controlled endocrine versus exocrine secretion of prostaglandin F-2α by the uterine endometrium, *Prostaglandins* 14:397-400.

BOCCI, V., PAULESCU, L., and RICCI, M.G., 1985. The physiological interferon response, IV: Production of interferon by the perfused human placenta at term, *Proc. Soc. Exp. Biol. Med.* 180:137-143.

CERUTTI, M., HUE, D., CHARLIER, M., L'HARIDON, R., PERNOLLET, J.C., DEVAUCHELLE, G., and GAYE, P., 1991, Expression of a biologically active ovine trophoblastic interferon (oTP) using a Baculovirus expression system, *Bioch. Biophys. Res. Commun.* 181:443-448.

CHAOUAT, G., 1978, Aspects immunologiques de l'ovo-implantation, *in:* "L'implantation de l'oeuf", pp. 231-238, F. du Mesnil du Buisson, A. Psychoyos, and K. Thomas, eds, Masson, Paris.

CHAOUAT, G., ASSAL-MELIANI, A., MARTAL, J., RAGHUPATHY, R., HUI, L., and WEGMANN, T.G., 1993, IL10 corrects resorptions in the CBA/J x DBA/2 murine abortion model and is induceable by tau interferons, (in press).

CHAOUAT, G., MENU, E., CLARK, D.A., DY, M., MINKOWSKI, M., and WEGMANN, T.G., 1990a, Control of fetal survival in CBA x DBA/2 mice by lymphokine therapy, *J. Reprod. Fert.* 89:447-458.

CHAOUAT, G., MENU, E., and KINSKY, R., 1990b. Animal models of the fetal allograft. *Immun. All. Clin. North America*, 10, 13-25.

CHAOUAT, G., MENU, E., LELAIDIER, C., DELAGE, G., MOREAU, J.F., ASSAL-MELIANI, A., DJIAN, V., ROPERT, S., DAVID, F., WEGMANN, T.G., HUI, L., RAGHUPATHY, R., MARTAL, J., and FRYDMAN, R., 1993, Cytokines and immuno endocrine network as determinants of early pregnancy success or failure and in parturition, *VIIIth World congress on in vitro fertilization and alternate assisted reproduction*, Kyoto, Japan.

CHAOUAT, G., MENU, E., SZEKERES-BARTHO, J., REBUT-BONNETON, C., BUSTANY, P., KINSKI, R., DY, M., MINKOWSKI, M., CLARK, D.A., and WEGMANN, T.G., 1991, Immunological and endocrinological factors that contribute to successful pregnancy, *in:* "Molecular and cellular immunobiology of the maternal fetal interface", T.G. Wegmann, T.J. Gill, and E. Nisbet-Brown, eds, Oxford University Press, New York.

CHARD, T., 1991, Interferon-α is a reproductive hormone, *J. Endocr.* 131:337-338.

CHARD, T., CRAIG, P.H., MENABAWAY, M., and LEE, C., 1986, Alpha interferon in human pregnancy, *Br. J. Obstet. Gynaecol.* 93:1145-1149.

CHARD, T., and ILES, R., 1992, Interferon as a fetoplacental signal in pregnancy, *Troph. Res.* 6:55-72, *in:* "Placental signals autocrine and paracrine control of pregnancy", L. Cédard and A. Firth, eds, University of Rochester Press.

CHARLIER, M., HUE, D., BOISNARD, M., MARTAL, J., and GAYE, P., 1991, Cloning and structural analysis of two distinct families of ovine interferon-α genes encoding functional class II and trophoblast (oTP) α-interferons, *Mol. Cell. Endocrinol.* 86:161-171.

CHARLIER, M., HUE, D., MARTAL, J., and GAYE, P., 1989, Cloning and expression of cDNA encoding ovine trophoblastin: Its identity with a class-II alpha interferon, *Gene* 77:341-348.

CHARLIER, M., L'HARIDON, R., BOISNARD, M., MARTAL, J., and GAYE, P., 1993, Cloning and structural analysis of four genes encoding interferon omega in rabbit, *J. IFN Res.* (in press).

CHARPIGNY, G., REINAUD, P., HUET, J.C., GUILLOMOT, M., CHARLIER, M., PERNOLLET, J.C., and MARTAL, J., 1988a, High homology between a trophoblastic protein (trophoblastin) isolated from ovine embryo and interferons, *FEBS Lett.* 228:12-16.

CHARPIGNY, G., REINAUD, P., LA BONNARDIERE, C., GUILLOMOT, M., HUET, J.C., PERNOLET, J.C., and MARTAL, J., 1988b, Evidence for antiviral properties of three purified isoforms of oTPB. In Proc. Int. Workshop on maternal recognition of pregnancy and maintenance of the corpus luteum, Jerusalem (Israël) (abstr.):72.

CHARPIGNY, G., REINAUD, P., TAMBY, J.P., MARTAL, J., 1991, Ovine embryonic interferon blocks prostaglandin synthesis induced by activators of protein kinase C: evidence for a dual inhibitory mechanisms. Ann. meeting of the intern, *J. IFN Res.* 11, suppl. 1: Abstr. S166.

CHENE, N., HAMROUCHE, N., PUISSANT, C., GRENOUILLET, C., and MARTAL, J., 1991, Distribution of growth factors (IGF-I, IGF-II, TGF beta 1 and TGF beta 2) mRNAs in the ovine conceptus during periimplantation, *J. Reprod. Fert.* suppl. serie 8:abstr. 107.

CLARK, D.A., HEAD, J.R., DRAKE, B.L., FULOP, G., BRIERLEY, J., MANUEL, J., BANWATT, D., and CHAOUAT, G., 1991, Role of a factor related to transforming growth factor Beta-2 in successful pregnancy, *in:* "Molecular and cellular immunobiology of the maternal fetal interface", pp. 294-311, T.G. Wegmann, T.J. Gill and E. Nisbet-Brown, eds, Oxford University Press, New-York.

CROSS, J.C., FARIN, C.E., SHARIF, S.F., and ROBERTS, R.M., 1990, Characterization of the antiviral activity constitutively produced by murine conceptuses: Absence of placental mRNAs for interferon alpha and beta, *Mol. Reprod. Dev.* 26:122-128.

CROSS, J.C., and ROBERTS, R.M., 1989, Porcine conceptuses secrete an interferon during the pre-attachment period of early pregnancy, *Biol. Reprod.* 40, 1109-1118.

CROY, B.A., and CHAPEAU, C., 1990, Evaluation of the pregnancy immunotrophism hypothesis by assessment of the reproductive performance of young adult mice of genotype scid/scid bg/bg, *J. Reprod. Fert.* 88:231-239.

CROY, B.A., and ROSSANT, J., 1987, Mouse embryonic cells become susceptible to CTL mediated lysis after midgestation, *Cell Immunol.* 104:355-365.

DEGRYSE, E., DIETRICH, M., NGUYEN, M., ACHSTETTER, T., CHARLIER, M., CHARPIGNY, G., GAYE, P., and MARTAL, J., 1992, Addition of a dipeptide spacer significantly improves secretion of ovine trophoblast interferon in yeast, *Gene* 118:47-53.

DE MAYER, E., and DE MAYER-GUIGNARD, J., 1988, Interferon and other regulatory cytokines, Wyleg, New York.

DRAKE, B.L., and HEAD, J.R., 1988, Murine trophoblast cells are susceptible to Lymphokine Activated Killer (LAK) cell lysis, *Am. J. Reprod. Immunol.* 16:114.

DRAKE, B.L., and HEAD, J.R., 1989, Murine trophoblast cells can be killed by allospecific cytotoxic T lymphocytes generated in Gibco OPTI MEM medium, *J. Reprod. Immunol.* 15:71-77.

DUC-GOIRAN, P., ROBERT-GALLIOT, B., LOPEZ, J., and CHANY, C., 1985. Unusual apparently constitutive interferons and antagonists in human placental blood, *Proc. Natl. Acad. Sci. USA* 82:5010-5014.

FARIN, C.E., IMAKAWA, K., and ROBERTS, R.M., 1989, *In situ* localization of mRNA for the interferon, ovine trophoblast protein-1, during early embryonic development of the sheep, *Mol. Endocr.* 3:1099-1107.

FILLION, C., CHAOUAT, G., CHARPIGNY, G., REINAUD, P., and MARTAL, J., 1990, Immunoregulatory effects of trophoblastin (oTP): all 5 isoforms are immunosuppressive of PHA driven lymphocyte proliferation, *J. Reprod. Immunol.* 19:237-249.

FINCHER, K.B., BAZER, F.W., HANSEN, P.J., THATCHER, W.W., and ROBERTS, R.M., 1986, Ovine conceptus secretory proteins suppress induction of uterine prostaglandin F-2α release by oestradiol and oxytocin, *J. Reprod. Fert.* 76:425-433.

FLINT, A.P.F., PARKINSON, T.J., STEWART, H.J., VALLET, J.L., and LAMMING, G.E., 1991, Molecular biology of trophoblast interferons and studies of their effects *in vivo. J. Reprod. Fert.* suppl. 43:13-25.

FLINT, A.P.F., and SHELDRICK, E.L., 1986, Ovarian oxytocin and the maternal recognition of pregnancy, *J. Reprod. Fert.* 76:831-839.

FLINT, A.P.F., SHELDRICK, E.L., McCANN, T.J., and JONES, D.S.C., 1990, Luteal oxytocin: characteristics and control of synchronous episodes of oxytocin and $PGF_{2\alpha}$ secretion at luteolytis in ruminants, *Dom. Anim. Endocr.* 7:111-124.

FOWLER, A.K., REED, C.D., and GIRON, D.J., 1980, Identification of an interferon in murine placentas, *Nature* 286:266-267.

FRY, R.C., BATT, P.A., FAIRCLOUGH, R.J., and PARR, R.A., 1992, Human leukemia inhibitory factor improves the viability of cultured ovine embryos, *Biol. Reprod.* 46:470-474.

GODKIN, J.D., BAZER, F.W., MOFFATT, J., SESSIONS, F., and ROBERTS, R.M., 1982, Purification and properties of a major, low component weight protein released by the trophoblast of sheep blastocysts at day 13-21, *J. Reprod. Fert.* 65:141-150.

GODKIN, J.D., BAZER, F.W., and ROBERTS, R.M., 1984a, Ovine trophoblast protein 1, an early secreted blastocyst protein binds specifically to uterine endometrium and affects protein synthesis, *Endocrinology* 114:120-130.

GODKIN, J.D., BAZER, F.W., THATCHER, W.W., and ROBERTS, M., 1984b, Proteins released by cultured day 15-16 conceptuses prolong luteal maintenance when introduced into uterine lumen of cyclic ewes, *J. Reprod. Fert.* 71:57-64.

GROSS, T.S., THATCHER, W.W., HANSEN, P.J., JOHNSON, J.W., and HELMER, S.D., 1988, Presence of an intracellular endometrial inhibitor of prostaglandin synthesis during early pregnancy in the cow, *Prostaglandins* 35:359-378.

GUILLOMOT, M., MICHEL, C., GAYE, P., CHARLIER, M., TROJAN, J., and MARTAL, J., 1990, Cellular localization of an embryonic interferon, ovine trophoblastin and its m-RNA in sheep embryos during early pregnancy, *Biol. Cell.* 68:205-211.

HANSEN, T.R., IMAKAWA, K., POLITES, H.G., MAROTTI, K.R., ANTHONY, R.V., and ROBERTS, R.M., 1988, Interferon RNA of embryonic origin is expressed transiently during early pregnancy in the ewe, *J. Biol. Chem.* 263:12801-12804.

HANSEN, T.R., KAZEMI, M., KEISLER, D.H., MALATHY, P.V., IMAKAWA, K., ROBERTS, R.M., 1989, Complex binding of the embryonic interferon, ovine trophoblast protein-1, to endometrial receptors, *J. Interferon Res.* 9:215-225.

HELMER, S.D., HANSEN, P.J., ANTHONY, R.V., THATCHER, W.W., BAZER, F.W., and ROBERTS, R.M., 1987, Identification of bovine trophoblast protein-1, a secretory protein immunologically related to ovine trophoblast protein-1, *J. Reprod. Fert.* 79:83-91.

HERNANDEZ-LEDEZMA, J.J., SIKES, J.D., MURPHY, C.N., WATSON, A.J., SCHULTZ, G.A., and ROBERTS, R.M., 1992, Expression of bovine trophoblast interferon in conceptuses derived by *in vitro* techniques, *Biol. Reprod.* 47:374-380.

HEYMAN, Y., CAMOUS, S., FEVRE, J., MEZIOU, W., and MARTAL, J., 1984, Maintenance of corpus luteum after uterine transfer of trophoblastic vesicles in cyclic cows and ewes, *J. Reprod. Fert.* 70:533-540.

HOOPER, S.B., WATKINS, W.B., and THORBURN, G.D., 1986, Oxytocin, oxytocin-associated neurophysin, and prostaglandin F-2α concentrations in the utero-ovarian vein of pregnant and nonpregnant sheep, *Endocrinology* 119:2590-2597.

IMAKAWA, K., ANTHONY, R.V., KAZEMI, M., MAROTTI, K.R., POLITES, H.G., and ROBERTS, R.M., 1987, Interferon-like sequence of ovine trophoblast protein secreted by embryonic trophectoderm, *Nature* 330:377-379.

IMAKAWA, K., HANSEN, T.R., MALATHY, P.C., ANTHONY, R.V., POLITES, H.G., MAROTTI, K.R., and ROBERTS, R.M., 1989, Molecular cloning and characterization of complementary deoxyribonucleic acids corresponding to bovine trophoblast protein-1 and bovine interferon-α_{II}, *Molec. Endocrinol.* 3:127-139.

ISAACS, A., and LINDENMANN, J., 1957, Virus interference. I. The interferon. *Proc. Royal. Soc. Lond.* B 147:258-267.

KHAN, N.U.D., PULFORD, K.A.F., FARQUHARSON, M.A., HOWATSON, A., STEWART, C., JACKSON, R., McNICOL, A.M., and FOULIS, A.K., 1989, The distribution of immunoreactive interferon-alpha in normal tissues, *Immunology* 66:201-206.

KLEMANN, S.W., IMAKAWA, K., and ROBERTS, R.M., 1990, Sequence variability among ovine trophoblast interferon cDNA, *Nucleic Acids Res.* 18:22.

KNICKERBOCKER, J.J., THATCHER, W.W., BAZER, F.W., BARRON, D.H., and ROBERTS, R.M., 1986, Inhibition of uterine prostaglandin $F_2\alpha$ production by bovine conceptus secretory proteins, *Prostaglandins* 31:777-793.

KO, Y., LEE, C.Y., OTT, T.L., DAVIS, M.A., SIMMEN, R.C.M., BAZER, F.W., SIMMEN, F.A., 1991, Insulin-like growth factors in sheep uterine fluids: concentrations and relationship to ovine Trophoblast Protein-1 production during early pregnancy, *Biol. Reprod.* 45:135-142.

LA BONNARDIERE, C., and MARTAL, J., 1991, Les interférons embryonnaires, *Le Point vétérinaire* 23: 55-61.

LA BONNARDIERE, C., MARTINAT-BOTTE, F., TERQUI, M., LEFEVRE, F., ZOUARI, K., MARTAL, J., and BAZER, F.W., 1991, Production of two species of interferon by Large White and Meishan pig conceptuses during the peri-attachment period, *J. Reprod. Fert.* 91:469-478.

LEBON, P., DAFFOS, F., CHECOURY, A., GRANGEOT-KEROS, L., FORESTIER, F., and TOUBLANC, J.E., 1985, Presence of an acid-labile alpha-interferon in sera from fetuses and children with congenital rubella, *J. Clin. Microbiol.* 21:775-778.

LEBON, P., GIRARD, S., THEPOT, F., and CHANY, C., 1982. The presence of α-interferon in human amniotic fluid, *J. Gen. Virol.* 59:393-396.

LEFEVRE, F., 1989, Le système interféron : structure, biologie, applications, *Ann. Rech. Vét.* 20:17-38.

LEFEVRE, F., BOULAY, V., and LA BONNARDIERE, C., 1991, A new type one interferon with unusual regulatory elements. *J. IFN Res.* 11, Suppl. 1:Abstr. 14.

LEFEVRE, F., MARTINAT-BOTTE, F., GUILLOMOT, M., ZOUARI, K., CHARLEY, B., and LA BONNARDIERE, C., 1990, Interferon-gamma gene and protein are spontaneously expressed by the trophectoderm early in gestation. *Eur. J. Immunol.* 20:2485-2490.

MARTAL, J., ASSAL, A., ASSAL, E., CHARLIER, M., CHENE, N., CHARPIGNY, G., GUILLOMOT, M., and CHAOUAT, G., 1991a, Characterization, antiluteolytic and immunosuppressive effects of ovine trophoblastin (oTP, α-interferon of class "III"), *in:* "Cellular and Molecular Biology of the materno-fetal relationship", G. CHAOUAT, and J. MOWBRAY, eds, Colloque INSERM/John Libbey Eurotext Ltd, 212:217-324.

MARTAL, J., ASSAL, N.E., ASSAL, A., CHARPIGNY, G., REINAUD, P., and CHAOUAT, G., 1991b, Involvment of trophoblast interferons in the control progesterone-dependant uterine environment and prevention of rejection of the conceptus as an allograft, *J. Reprod. Fert.* abstr. series No. 8.

MARTAL, J., CHARLIER, M., CAMOUS, S., FEVRE, J., and HEYMAN, Y., 1984, Origin of embryonic signals allowing the establishment of pregnancy corpus luteum in ruminants, 10th Int. Congress on Animal Reproduction and Artificial Insemination, Urbana-Champaign, USA, 111, 509:3.

MARTAL, J., CHARLIER, M., CHARPIGNY, G., CAMOUS, S., CHENE, N., REINAUD, P., SADE, S., and GUILLOMOT, M., 1987, Interference of trophoblastin in ruminant embryonic mortality. A review, *Livest. Prod. Sci.* 17:193-210.

MARTAL, J., CHARPIGNY, G., REINAUD, P., HUET, J.C., GUILLOMOT, M., ZOUARI, K., PERNOLLET, J.C., and LA BONNARDIERE, C., 1988, Embryonic signals and corpus luteum: Why three isoforms of trophoblastin can be considered as interferons α of class II? *J. Reprod. Fert.* 2:Abstr. 11.

MARTAL, J. and CHENE, N., 1992, Functions of embryonic interferons and of the main serum proteins specific for pregnancy, *Troph. Res.* 6:73-122, *in:* "Placental signals autocrine and paracrine control of pregnancy", L. Cédard and A. Firth, eds, University of Rochester Press.

MARTAL, J., DEGRYSE, E., CHARPIGNY, G., ASSAL, N., REINAUD, P., CHARLIER, M., GAYE, P., and LECOCQ, J.P., 1990, Evidence for extended maintenance of the corpus luteum by uterine infusion of a recombinant trophoblast α-interferon (trophoblastin) in sheep, *J. Endocrinol.* 127:R5-R8.

MARTAL, J., LACROIX, M.C., LOUDES, C., SAUNIER, M., and WINTENBERGER-TORRES, S., 1979, Trophoblastin, an antiluteolytic protein present in early pregnancy in sheep. *J. Reprod. Fert.* 56:63-73.

MATHIALAGAN, N., BIXBY, J.A., and ROBERTS, R.M., 1992, Expression of interleukin-6 in porcine, ovine, and bovine preimplantation conceptuses, *Mol. Reprod. Dev.* 32:324-330.

McCRACKEN, J.A., 1980, Hormone receptor control of prostaglandin F-2_α secretion by the ovine uterus, *Adv. Prostaglandin Thromboxane Res.* 8:1329-1344.

McCRACKEN, J.A., CARLSON, J.C., GLEW, M.E., GODING, J.R., BAIRD, D.T., GREEN, K., and SAMUELSSON, B., 1972, Prostaglandin F-2α identified as a luteolytic hormone in sheep. *Nature* 238:129-134.

Mc CRACKEN, J.A., SCHRAMM, W., BARCIKOWSKI, B., and WILSON, L., 1981, Identification of PGF-2α as a uterine luteolytic hormone and the hormonal control of its synthesis, *Acta. Vet. Scand.* suppl. 77:71-88.

MIRANDO, M.A., HARNEY, J.P., BEERS, S., PONTZER, C.H., TORRES, B.A., JOHNSON, H.M., and BAZER, F.W., 1990a, Onset of secretion of proteins with antiviral activity by pig conceptuses, *J. Reprod. Fert.* 88:197-203.

MIRANDO, M.A., OTT, T.L., VALLET, J.L., DAVIS, M., and BAZER, F.W., 1990b, Oxytocin-stimulated inositol phosphate turnover in endometrium of ewes is influenced by stage of the estrus cycle, pregnancy and intrauterine infusion of ovine secretory proteins, *Biol. Reprod.* 42:98-105.

MOOR, R.M., and ROWSON, L.E.A., 1966, The corpus luteum of the sheep: functional relationship between the embryo and the corpus luteum, *J. Endocr.* 34:233-239.

NEPHEW, K.P., McCLURE, K.E., DAY, M.L., XIE, S., ROBERTS, R.M., and POPE, W.F., 1990, Effects of intra-muscular administration of recombinant bovine interferon-Alpha$_1$1 during the period of maternal recognition of pregnancy, *J. anim. Sci.* 68:2766-2770.

NEPHEW, K.P., WHALEY, A.E., CHRISTENSON, R.K., and IMAKAWA, K., 1993, Differential expression of distinct mRNAs for ovine Trophoblast Protein-1 and related sheep type I interferons, *Biol. Reprod.* 48:768-778.

NETT, T.M., STAIGMILLER, R.B., AKBAR, A.M., DIEKMAN, M.A., ELLINWOOD, W.E., and NISWENDER, G.D., 1976, Secretion of prostaglandin F-2$_α$ in cycling and pregnant ewes, *J. Anim. Sci.* 42:876-880.

PARKINSON, T.J., LAMMING, G.E., FLINT, A.P.F., and JENNER, L.J., 1992, Administration of recombinant bovine interferon-α$_I$ at the time of maternal recognition of pregnancy inhibits prostaglandin F$_{2α}$ secretion and causes luteal maintenance in cyclic ewes, *J. Reprod. Fert.* 94:489-500.

PEREZ, A., RIEGO, E., MARTINEZ, R., CASTRO, F.O., LLEONART, R., and DE LA FUENTE, J., 1991, Presence of mRNAs homologous to IFN-α, but not to IFN-ß in preimplantation mouse embryos, *J. IFN Res.* 11, suppl. 1:Abst. 62.

PLANTE, C., HANSEN, P.J., and THATCHER, W.W., 1988. Prolongation of luteal lifespan in cows by intrauterine infusion of recombinant bovine alpha-interferon. *Endocrinology* 122:2342-2344.

POLLARD, J.W., BARTOCCI, A., ARCECI, R., ORLOFSKY, A., LADNER, M.B., and STANLEY, E.R., 1987, *Nature* 330:484-486.

PONTZER, C.H., TORRES, R.A., VALLET, J.L., BAZER, F.W., and JOHNSON, H.M., 1988, Antiviral activity of the pregnancy recognition hormone ovine trophoblast protein-1, *Bioch. Biophys. Res. Commun.* 152:801-807.

ROBERTS, R.M., CROSS, J.C., and LEAMAN, D.W., 1992b, Interferons as hormones of pregnancy, *Endocr. Rev.* 13:432-452.

ROBERTS, R.M., KLEMANN, S.W., LEAMAN, D.W., BIXBY, J.A., CROSS, J.C., FARIN, C.E.E, IMAKAWA, K., and HANSEN, T.R., 1991, The polypeptides and genes for ovine and bovine trophoblast protein-1, *J. Reprod. Fertil.* suppl. 43:3-12.

ROBERTS, R.M., LEAMAN, D.W., and CROSS, J.C., 1992a, Role of interferons in maternal recognition of pregnancy in ruminants, *Proc. Soc. Exp. Biol. Med.* 200:7-18.

ROWSON, L.E.A., MOOR, R.M., 1967, The influence of embryonic tissue homogenate infused into the uterus on life-span of the corpus luteum in the sheep, *J. Reprod. Fert.* 13:511-516.

SALAMONSEN, L.A., CHERNY, P.A., and FINDLAY, J.K., 1991, *In vitro* studies of the effects of interferons on endometrial metabolism in sheep, *J. Reprod. Fert.* Suppl. 43:27-38.

SALAMONSEN, L.A., STUCHBERY, S.J., O'GRADY, C.M., GODKIN, J.D., and FINDLAY, J.K., 1988, Interferon α mimics effects of ovine trophoblast protein-1 on prostaglandin and protein secretion by ovine endometrial cells *in vitro*, *J. Endocr.* 117:R1-R4.

SCHAULE-FRANCIS, T.K., FARIN, P.W., CROSS, J.C., KEISLER, D., and ROBERTS, R.M., 1991, Effect of injected bovine interferon-α$_I$1 on oestrous cycle length and pregnancy success in sheep, *J. Reprod. Fert.* 91:347-356.

SCHRAMM, W., BOVIARD, L., GLEW, M.E., SCHRAMM, G., and McCRACKEN, J.A., 1983, Corpus luteum regression induced by ultra-low pulses of prostaglandin F-2α. *Prostaglandins* 26:347-363.

SHELDRICK, E.L., and FLINT, A.P.F., 1985, Endocrine control of uterine oxytocin receptors in the ewe, *J. Reprod. Fert.* 77:523-529.

SHIOZAWA, S., CHIHARA, K., SHIOZAWA, K., FUJITA, T., IKEGAMI, H., KOYAMA, S., and KURIMOTO, M., 1986, A sensitive radioimmunoassay for alpha-interferon circulating alpha-interferon-like substance in the plasma of healthy individuals and rheumatoid arthritis patients, *Clin. Exper. Immunol.* 66:77-87.

SHIOZAWA, S., SHIOZAWA, K., SHIMIZU, S., TANAKA, Y., MORIMOTO, M., YOSHIHARA, R., and FUJITA, T., 1989, Age distribution of circulating alpha-interferon, *Experientia* 45:764-765.

STEWART, H.J., FLINT, A.P.F., LAMMING, G.E., McCANN, S.H.E., and PARKINSON, T.J., 1989, Antiluteolytic effects of blastocyst-secreted interferon investigated *in vitro* and *in vivo* in the sheep, *J. Reprod. Fert.* suppl. 37:127-138.

STEWART, C.L., KASPAR, P., BRULET, L.J., BHATT, H., GADI, I., KONTGEN, F., ABBONDANZO, S.J., 1992, Blastocyst implantation depends on maternal expression of leukaemia inhibitory factor, *Nature* 359:76-79.

STEWART, H.J., McCANN, S.H.E., BARKER, P.J., LEE, K.E., LAMMING, G.E., and FLINT, A.P.F., 1987, Interferon sequence homology and receptor binding activity of ovine trophoblast antiluteolytic protein, *J. Endocr.* 115:R13-R15.

STEWART, H.J., McCANN, S.H.E., and FLINT, A.P.F., 1990, Structure of an interferon-α_{II} gene expressed in the bovine conceptus early in gestation, *J. Mol. Endocrinol.* 2:65-70.

STEWART, H.J., Mc CANN, S.H.E., NORTHROP, A.J., LAMMING, G.E., and FLINT, A.P.F., 1989b, Sheep antiluteolytic interferon cDNA sequence and analysis of mRNA levels, *Mol. Endocrinol.* 2:65-70.

SZEKERES BARTHO, J., KILAR, F., FALKAY, G., CSERNU, V., TOROK, A., and PACSA, A.S., 1985, Progesterone-treated lymphocytes release a substance inhibiting cytotoxicity and prostaglandin synthesis. *Am. J. Reprod. Immunol.* 9:15-24.

TAGUCHI, F., KAJIOKA, J., and SHIMADA, N., 1985, Presence of interferon and antibodies to BK virus in amniotic fluid of normal pregnant women, *Acta Virol.* 29:299-304.

TAMBY, J.P., CHARPIGNY, G., REINAUD, P., and MARTAL, J., 1991, Evidence for inhibition of phospholipase A_2 activity in endometrium by embryonic interferon (ovine Trophoblast Protein, oTP) in sheep, *J. IFN Res.* 11, Suppl. 1:Abstr. S169.

THATCHER, W.W., HANSEN, P.J., GROSS, T.S., HELMER, S.D., PLANTE, C., and BAZER, F.W., 1989, Antiluteolytic effects of bovine trophoblast protein-1, *J. Reprod. Fert.* suppl. 37:91-99.

THORBURN, G.D., COX, R.I., CURRIE, W.B., RESTALL, B.J., and SCHNEIDER, W., 1973, Prostaglandin F and progesterone concentrations in the utero-ovarian venous plasma of the ewe during the oestrous cycle and early pregnancy, *J. Reprod. Fert.* suppl. 18:151-158.

VALLET, J.L., and BAZER, F.W., 1989, Effect of ovine trophoblast protein-I, oestrogen and progesterone on oxytocin-induced phosphatidylinositol turnover in endometrium of sheep, *J. Reprod. Fert.* 87:755-761.

VALLET, J.L., BAZER, F.W., FLISS, M.F.V., and THATCHER, W.W., 1988, Effect of ovine conceptus secretory proteins and purified ovine trophoblast protein-1 on interoestrous interval and plasma concentrations of prostaglandins $F_2\alpha$ and E and of 13,14-dihydro-15-prostaglandin $F_2\alpha$ in cyclic ewes, *J. Reprod. Fert.* 84:493-504.

VALLET, J.L., and LAMMING, G.E., 1991, Ovine conceptus secretory proteins and bovine recombinant interferon $\alpha_1$1 decrease endometrial oxytocin receptor concentration in cyclic and progesterone-treated ovariectomized ewes, *J. Endocr.* 131:475-482.

VALLET, J.L., LAMMING, G.E., and BATTEN, M., 1990, Control of endometrial oxytocin receptor and uterine response to oxytocin by progesterone and oestradiol in the ewe, *J. Reprod. Fert.* 90:625-634.

WARNER, S.J.C., and LIBBY, P., 1989, Human vascular smooth muscle. Target for and source of tumor necrosis factor, *J. Immunol.* 142:100-110.

WHALEY, A.E., CARROLL, R.S., and IMAKAWA, K., 1991, Cloning and analysis of a gene encoding ovine interferon α-II, *Gene* 106:281-282.

WHALEY, A.E., CARROLL, R.S., NEPHEW, K.P., IMAKAWA, K., 1991, Molecular cloning of unique interferons from human placenta, *J. IFN Res.* 11, suppl. 1:Abstr. 69.

WEGMANN, T.G., 1987, Placental immunotrophism: maternal T cells enhance placental growth and function, *Am. J. Reprod. Immunol.* 15:67-70.

WEGMANN, T.G., ATHANASSAKIS, I., GUILBERT, L., BRANCH, D., DY, M., MENU, E., and CHAOUAT, G., 1989, The role of M-CSF and GM-CSF in fostering placental growth, fetal growth, and fetal survival, *Transplantation Proc.* 21.1.P1, 89:566-569.

WEGMANN, T.G., and Gill, T.J. III, 1991, The molecular and cellular nature of maternal-fetal immune signaling: An overview, *in:* "Molecular and cellular immunobiology of the maternal fetal interface", T.G. WEGMANN, T.J. GILL, and E. Nisbet-Brown, Oxford University Press, New York.

WEGMANN, T.G., LIN, H., GUILBERT, L., and MOSMANN, T.R., 1993, Bidirectional cytokine interactions in the maternal-fetal relationship: successful pregnancy is a TH2 phenomenon, *Immun. Today* (in press).

WEISLOW, O.S., KISER, R., ALLEN, P.T., and FOWLER, A.K., 1983, Partial purification of a placental interferon with atypical characteristics, *J. Interferon Res.* 3:291-298.

YAMADA, K., SHIMIZU, Y., OKAMURA, K., KUMAGAI, K., and SUZUKI, M., 1985, Study of interferon production during pregnancy in mice and antiviral activity in the placenta, *Am. J. Obstet. Gynecol.* 153:335-341.

ZARCO, L., STABENFELDT, G.H., QUIRKE, J.F., KINDAHL, H., BRADFORD, G.E., 1988, Modification of prostaglandin F-2α synthesis and release in the ewe during the establishment of pregnancy, *J. Reprod. Fert.* 83:527-536.

ZHANG, J., WESTON, P.G., HIXON, J.E., 1992, Role of progesterone and oestradiol in the regulation of uterine oxytocin receptors in ewes, *J. Reprod. Fert.* 94:395-404.

ZOUARI, K., BEZARD, J., REINAUD, P., GUILLOMOT, M., PALMER, E., and MARTAL, J., 1991a, Evidence for the presence of equine trophoblastic interferons during early pregnancy, *J. IFN Res.* 11, suppl. 1: Abstr. S134.

ZOUARI, K., REINAUD, P., LA BONNARDIERE, C., and MARTAL, J., 1991b, Uterine interferon during early pregnancy and pseudopregnancy in rabbits, *J. IFN Res.* 11, Suppl. 1:Abstr. S135.

THE DECIDUAL HORMONES AND THEIR ROLE IN PREGNANCY RECOGNITION

Geula Gibori

Department of Physiology and Biophysics
University of Illinois College of Medicine
Chicago, IL 60680

INTRODUCTION

How does the mother recognize the presence of the blastocyst and adjust to provide the right milieu for the embryo to implant, to grow, and to remain in a safe uterine environment until able to survive outside the womb? The answer to this question, as discussed by Dr. Mori and Kanzaki in another chapter, appears to differ in different species. However one common denominator to all species is the necessity to maintain adequate levels of ovarian steroids in the maternal circulation and to rapidly change the uterine environment so that contact between mother and fetus becomes possible without damage to the mother or rejection of the fetus.

An important component for such necessary maternal adaptation in the rat is the decidual tissue. Its role in maintaining the endocrine and the intrauterine milieu for a successful pregnancy has recently attracted significant attention. Most interestingly, it has become clear that different subpopulations of decidual cells play different roles in either sustaining steroid levels in the circulation or in providing an intrauterine environment necessary to prevent extensive tissue damage during trophoblast invasion. The rat decidual tissue is a striking feature of the maternal response to pregnancy and can be induced by either the implanting blastocyst or by artificial stimuli. An interesting feature in the formation of this tissue is that decidualization, i.e. the growth and differentiation of the endometrial stromal cells, occurs differently in different regions of the uterus. Decidualization begins in the antimesometrial region where the blastocyst implants. Extensive stromal endometrial cell proliferation gives rise to large polyploid and tightly packed cells. Approximately two days later, decidualization occurs in the mesometrial region close to where blood vessels gain access to the uterus. The mesometrial cells are very different from the antimesometrial cells. For a reason not yet known, they are smaller, rarely polyploid, and loosely packed. It is in the mesometrial decidual cells that the trophoblast invades to establish the definitive hemochorial placenta! (Gibori et al., 1987; Gu et al., 1992a; and Gibori, 1993).

Differential Role of Mesometrial and Antimesometrial Cells in Pregnancy

Our investigations on the role of the decidual tissue in pregnancy have revealed that the antimesometrial and the mesometrial decidual cells forming the rat decidua have different functions in pregnancy. The antimesometrial cells form an endocrine tissue which secretes hormones that profoundly affect ovarian secretion of steroids. The decidual cells forming the mesometrial tissue have no apparent endocrine activities but express rather abundantly α_2 macroglobulin, a protease inhibitor which may play an important role during placentation.

What has facilitated these studies is that the antimesometrial tissue is dark and can be visually separated from the lighter mesometrial decidua by dissection. We also took advantage of the difference in cell size and succeeded at separating the two decidual cell populations by elutriation (Gibori et al, 1987, Gu et al, 1992b, Gibori, 1993). The antimesometrial cells are not only significantly larger than the mesometrial cells but they also exhibit striking morphological differences. In culture, the large antimesometrial cells have a syncytial-like appearance, they are polynucleated, contain numerous lipid droplets, and secrete hormones that profoundly affect ovarian secretion of steroids. In contrast, the small mesometrial cells remain less differentiated in culture, they are sometimes binucleated, have few lipid droplets and have no endocrine activity. Yet these subpopulations of decidual cells abundantly secrete protease inhibitor(s) and are responsive to the hormone secreted by the antimesometrial cells (Gu et al., 1992a; Gibori, 1993; Gu et al., 1992b).

What led to investigate the possibility that decidual cells have endocrine activity was the earlier findings which indicated that the presence of the decidual tissue in the uterus markedly affects the production of progesterone by the corpus luteum (Gibori et al., 1974; Basuray and Gibori, 1980; Gibori et al., 1981; Basuray et al., 1983; Gibori et al., 1984; Jayatilak et al.,1984) and that of estradiol by the follicle (Gibori et al., 1985). The antimesometrial decidual cells express and secrete hormones which sustain luteal production of progesterone and concomitantly reduce the ability of the follicle to aromatize androgen and to secrete estradiol. The decidual hormone that has luteotropic activity was found to be a PRL-like protein that migrates on SDS-PAGE at an apparent Mr of 28,000. On gel chromatography, it runs as a 24 KDa protein. It binds to PRL receptors on luteal cell membranes and acts, as does PRL, on different parameters involved in progesterone production. This decidual luteotropin was shown to be expressed solely by antimesometrial cells (Jayatilak et al., 1985; Herz et al., 1985; Jayatilak et al., 1989). Recently, a decidual PRL-related protein was cloned and sequenced (Roby et al., 1993). The clone encodes for a 28 KDa protein. The deduced amino acid sequence and biochemical characteristics of the decidual protein are consistent with its inclusion in the PRL family. This decidual PRL-related protein and PRL contain similarly positioned cysteine residues suggesting a correspondence in folding and thus biological activity. Indeed, the recombinant decidual hormone, expressed in CHO cells, binds with high affinity to PRL receptors on luteal cell membranes (Gibori and Soares, unpublished). It is, however, not yet clear whether the decidual luteotropin is a single protein or whether it is comprised of more than one 28 KDa PRL-related polypeptide. Croze et al. (1990) have shown that another PRL-related protein, termed PLP-β, is expressed in both trophoblast and decidua, and they suggested that PLP-β may be a candidate for the decidual luteotropin. However, PLP-β gene expression in the decidual tissue is very low when compared to the decidual PRL-related hormone indicating that the message detected in the decidua with PLP-β cDNA may be due to cross hybridization due to homology between the proteins.

Analysis of tissue specific expression with the decidual PRL-related protein cDNA demonstrated that this hormone is highly expressed in the decidua. A weak signal is also detected in the trophoblast, however, it is not clear whether this is due to decidual tissue contamination or to cross hybridization with other PRL-like molecules expressed in the trophoblast. Interestingly, this message is not detected in the nondecidualized uteri nor is it present in the small mesometrial cells indicating that it is an excellent marker for both decidualization and for the specific antimesometrial cell subpopulation of the decidua (Gu et al., 1993).

The presence of the decidual tissue in the uterus causes a remarkable reduction in both ovarian secretion of estradiol and follicular aromatase activity (Gibori et al., 1985). This led us to hypothesize that the decidual tissue secretes another molecule that depresses follicular aromatase. Since the aromatase gene in the follicle is regulated by FSH, we examined the possibility that the decidual tissue produces a signal that inhibits FSH secretion, and thus follicular aromatase activity. Because a inhibin mRNA is not detectable in the decidua, the expression of the follistatin gene, another FSH inhibitor, was investigated (Kaiser et al., 1990). In situ hybridization and Northern analysis indicated that follistatin gene expression is confined to the antimesometrial decidual cells. No follistatin message could be detected in the trophoblast nor in the mesometrial decidual cells (Gu, Jayatilak, and Gibori, 1992a; Gibori, 1993; Kaiser et al., 1990). Because of the large mass of the antimesometrial decidua which can reach more than one gram in weight and the abundance of the follistatin message, the total content of follistatin mRNA in the decidua is impressive. Follistatin has been shown to affect estradiol production by the follicle not only by down regulating FSH secretion but also by acting directly on the follicle (Xiao et al., 1990; Mather et al., 1992). Our recent evidence suggests that follistatin action may be direct on the follicle rather than being due to a decrease in FSH secretion; however, rigorous experimentation is still needed to demonstrate without a doubt that the decrease in follicular aromatase by the decidual tissue is due solely to decidual follistatin. If this is revealed to be true, then decidual follistatin may be of great physiological significance for the maintenance of pregnancy. Indeed whether low levels of estradiol, together with progesterone, is a prerequisite for the normal development of the decidual tissue, elevated levels of estradiol between days 6-12 can cause decidual tissue collapse and abortion (Rothchild et al., 1940; Yochim and DeFeo, 1963). It is of interest that follistatin is expressed in the decidua specifically from day 6 to day 12. Follistatin may thus play an important role by reducing ovarian secretion of estadiol to levels compatible with the normal maintenance of pregnancy.

Whereas the mesometrial cells secrete neither follistatin nor decidual luteotropin, they possess receptors for PRL (Jayatilak and Gibori, 1986; Gibori et al., 1987; Gu et al., 1992a; Gibori, 1993) to which decidual luteotropin binds. Originally, it was demonstrated that rat decidual membrane has high affinity binding sites for PRL and that decidual luteotropin competes in a dose-related manner with PRL. Whereas decidual luteotropin was almost totally confined to the antimesometrial cells, twice as many PRL receptors were found in the mesometrial than in the antimesometrial cells (Jayatilak and Gibori, 1986; Gibori et al., 1987). Recently, the expression of the short and long form of the PRL receptor message was examined in the decidual tissue. The results revealed that the antimesometrial decidua express the message for the long form of the receptor whereas no message for the short form could be detected. In contrast, the mesometrial decidua expressed more abundantly both short and long forms of the PRL receptor (Gu Y, Clarke D, Linzer D and Gibori G, unpublished).

The presence of receptors for decidual luteotropin in the mesometrial cells led us to investigate the possibility that this hormone secreted by the antimesometrial cells may act by a paracrine mechanism to control protein secretion, by the neighboring mesometrial cells. We were specifically interested by a_2

macroglobulin, since it is the major protein secreted by the mesometrial cells and since a_2 macroglobulin gene was shown to be regulated in the ovary by prolactin (Gaddy-Kurten and Richards, 1989). Results of this investigation (Gibori et al., 1987; Jayatilak et al., 1989) revealed a clear and marked up-regulation of a_2 macroglobulin expression in the mesometrial cells by both PRL and decidual luteotropin, indicating that this decidual hormone has not only an endocrine action on the ovary but also influences the secretion of the a_2 macroglobulin by the decidual tissue itself.

What may be the physiological role of a_2 macroglobulin expression specifically in the mesometrial cells and that of its up-regulation by decidual luteotropin? a_2 macroglobulin binds a host of growth factors and prevents their action (Huang et al., 1988; Chu et al., 1991). This may be the reason why the mesometrial decidual cells remain small and relatively undifferentiated in relation to the antimesometrial cells. Another potentially more important role of a_2 macroglobulin is that it is a potent proteinase inhibitor (Barrett and Starkey, 1973; Sottrup-Jensen, 1989). a_2 macroglobulin binds to a wide range of proteinases including metalloproteinase and prevents them from destroying the extracellular matrix. The fact that this potent inhibitor is secreted specifically by the mesometrial tissue, the site of trophoblast invasion may be one important reason why the mesometrial decidua can limit the tissue damage during trophoblast invasion. By up-regulating a_2 macroglobulin secretion, decidual luteotropin appears to play an important role in assuring that enough of this protease inhibitor is available when most needed during placentation.

In summary, the antimesometrial and mesometrial cells forming the decidua play different roles in pregnancy. The mesometrial cells secrete a protease inhibitor that may play an important role during placentation whereas the antimesometrial cells produce hormones that act by both endocrine and paracrine mechanisms to allow the existance of the right milieu for not only the recognition of pregnancy but also its normal progress.

Acknowledgments

The studies reported in this manuscript were supported by HD12356.

REFERENCES

Barrett AJ, Starkey PM, 1973. The interaction of a_2-macroglobulin with proteinase. Biochem. J. 133:709-724.

Basuray R, Gibori G, 1980. Luteotropic action of the decidual tissue in the pregnant rat. Biol. Reprod. 23:507-512.

Basuray R, Jaffe RC, Gibori G, 1983. Role of decidual luteotropin and prolactin in the control of luteal cell receptors for estradiol. Biol. Reprod. 28:551-556.

Chu CT, Rubenstein DS, Engheld JJ, Pizzo SV, 1991. Mechanism of insulin incorporation into a_2 macroglobulin: implication for the study of peptide and growth factor binding. Biochemistry 30:1551-1560.

Croze F, Kennedy TG, Schroeder IC, Friesen HG, 1990. Expression of rat prolactin-like protein-B in deciduoma of pseudopregnant rat and in decidua during early pregnancy. Endocrinology 127:2665-   .

Gaddy-Kurten D, Richards JS, 1989. Regulation of a_2 macroglobulin by luteinizing hormone and prolactin during cell differentiation in the rat ovary. Mol. Endocrinol. 5:1280-1291.

Gibori G, 1993. Decidual hormones in pregnancy. Endocrinology 130: Abstract #20.

Gibori G, Basuray R, McReynolds B, 1981. Luteotropic role of the decidual tissue in the rat: dependency on intraluteal estradiol. Endocrinology 108:2060-2066.

Gibori G, Jayatilak PG, Khan I, Rigby B, Puryear T, Nelson S, Herz Z, 1987. Decidual luteotropin secretion and action: Its role in pregnancy maintenance in the rat. In: Mahesh VB, Dhindsa DS, Anderson E, Kalra SP (eds.), Regulation of Ovarian and Testicular Function. Plenum Press, New York, pp. 379-397.

Gibori G, Kalison B, Basuray R, Rao MC, Hunzicker-Dunn M, 1984. Endocrine role of the of the decidual tissue: Decidual luteotropin regulation of luteal adenylyl cyclase, cyclase activity, luteinizing hormone receptors, and steroidogenesis. Endocrinology 115:1157-1163.

Gibori G, Kalison B, Warshaw, ML, Basuray R. Glaser LA, 1985. Differential action of decidual luteotropin on luteal and follicular production of testosterone and estradiol. Endocrinology 116:1784-1791.

Gibori G, Rothchild I, Pepe GJ, Morishige WK, Lam P, 1974. Luteotropic action of the decidual tissue in the rat. Endocrinology 95:1113-1118.

Gu Y, Jayatilak PG and Gibori G, 1992a. Cell specific expression and binding of decidual luteotropin in the rat decidua. Biol. Reprod. 467. Suppl. 1:140, Abstract #43.

Gu Y, Jayatilak PG, Parmer TG, Gauldie J, Fey GH and Gibori G, 1992b. a_2-Macroglobulin expression in the mesometrial decidua and its regulation by decidual luteotropin and prolactin. Endocrinology, 131:1321-1328.

Gu Y, Soares MJ and Gibori G, 1993. Rat decidual PRL-related protein gene: Cell specific expression and regulation. Biology of Reproduction 48: Abstract #478.

Herz Z, Khan I, Jayatilak PG, Gibori G, 1985. Evidence for the synthesis and secretion of decidual luteotropin: A prolactin-like hormone produced by rat decidual cells. Endocrinology 118:2203-2209.

Huang SS, O'Grady P, Huang JS, 1988. Human transforming growth factor β - a_2-macroglobulin complex is a latent form of transforming growth factor β. J. Biol. Chem. 264:7210-7216.

Jayatilak PG and Gibori G, 1986. Ontogeny of prolactin receptors in rat decidual tissue: Binding by locally produced prolactin-like hormone. J. Endocrinol. 110:115-121.

Jayatilak PG, Glaser LA, Basuray R, Kelly PA, Gibori G, 1985. Identification and partial characterization of a prolactin-like hormone produced by the rat decidual tissue. Proc. Natl. Acad. Sci. USA 82:217-221.

Jayatilak PG, Glaser LA, Warshaw ML, Herz Z, Gruber JR, Gibori G, 1984. Relationship between luteinizing hormone and decidual luteotropin in the maintenance of luteal steroidogenesis. Biol. Reprod. 31:556-564.

Jayatilak PG, Puryear TK, Herz Z, Fazleabas, Gibori G, 1989. Protein secretion by mesometrial and antimesometrial rat decidual tissue: Evidence for differential gene expression. Endocrinology 125:659-666.

Kaiser M, Gibori G and Mayo KE, 1990. The rat follistatin gene is highly expressed in decidual tissue. Endocrinology 126:2768-2770.

Mather JP, Woodruff TK and Krummen LA, 1992. Paracrine regulation of reproductive function by inhibin and activin. Proc. Soc. Exptl. Biol. Med. 201:1-15.

Roby KF, Deb S, Gibori G, Sziper C, Levan G, Kwok SCM and Soares JM, 1993. Decidual prolactin related proteins: Identification, molecular cloning and characterization. J. Biol. Chem. 268:3136-3142.

Rothchild I, Meyer RK, Spielman MA, 1940. A quantitative study of estrogen-progesterone interaction in the formation of placentomata in the castrate rat. Am. J. Physiol. 128:213-217.

Sottrup-Jensen L, 1989. a_2 macroglobulins: Structure, shape, and mechanism of proteinase complex formation. J. Biol. Chem. 264:11539-11542.

Xiao S, Findlay JK and Robertson DM, 1990. The effect of bovine activin and follicle-stimulating hormone (FSH) suppressing protein/follistatin on FSH-induced differentiation of rat granulosa cells in vitro. Mol. Cell Endocrinol. 69:1-8.

Yochim JM, DeFeo VJ, 1963. Hormonal control of the onset, magnitude and duration of uterine sensitivity in the rat by steroid hormones of the ovary. Endocrinology 72:317-325.

EMBRYO-ENDOMETRIAL INTERACTIONS
Angiogenic Growth Factor Expression and Human Implantation

Stephen K. Smith

Department of Obstetrics and Gynaecology
University of Cambridge
Rosie Maternity Hospital
Cambridge CB2 2SW
U.K.

Introduction

A key event in the process of implantation is the development of a blood supply in both the maternal and fetal environments. In primates, this is achieved on the maternal side by the growth of spiral arterioles during the menstrual cycle, which eventually provide the blood supply to the placenta. In the placenta, the initial blood compartment consists of lacunae but defined fetal blood vessels become recognised as the villous system develops within the placenta.

The technique of cyclical menstrual shedding evolved in primates demands a profound period of angiogenic activity to repair the spiral arterioles after menstruation. Angiogenesis is not a common event in most organs of the body where endothelial cells undergo slow turnover. Thus the endometrium (and ovary) are unique in their need to retain a cyclical angiogenic activity. Whilst the preparation of endometrium for implantation has in the past been defined primarily in terms of ovarian steroids, it is now clear that the principal agents controlling angiogenesis are local peptides which may be modulated by or interact with these steroids (Folkman and Klagsbrun, 1987).

Angiogenic Growth Factor Expression in Endometrium

Many agents exert angiogenic properties but these can be divided into four categories (Klagsbrun and D'Amore, 1991). Some growth factors such as epidermal growth factor (EGF) and transforming growth factor-α (TGFα) have wide ranging effects on cells of ectodermal, mesenchymal and endothelial lineage. Others including the heparin binding fibroblast growth factors have profound angiogenic activities but retain a predominant action on cells derived from the mesenchyme. A third group of growth factors, vascular endothelial growth factor (VEGF) and platelet derived endothelial cell growth factor (PDEGF), have a much narrower target being restricted to endothelial cells. Finally, various cytokines and colony stimulating growth factors modulate angiogenesis but probably by indirect rather than direct actions on endothelial cells.

a) EGF and TGFα

EGF and TGFα mRNA is present in low copy number in human endometrium but RT-PCR demonstrates the presence of mRNA encoding both growth factors in this tissue (Haining et al, 1991a). Provisional results suggest that levels of mRNA are highest in the early proliferative phase of the cycle when vascular repair is proceeding rapidly in the desquamated endometrial surface. Immunohistochemistry indicates the greatest intensity of staining in the basal parts of the endometrial glands in the late proliferative phase of the cycle. The clarity of staining is lost during the luteal phase of the cycle when the intensity of staining becomes reduced (Haining et al, 1991b)

b) Acidic and basic FGF

Messenger RNA encoding both growth factors is present in human endometrium throughout both stages of the menstrual cycle (Ferriani et al, 1993a). Interestingly, the greatest intensity of staining is found in the surface and glandular epithelium. This is most intense in the luteal phase of the cycle. The significance of these findings are difficult to determine. FGFs do not contain a secretory signal sequence and are presumably released from the cells at the time of tissue destruction. Their role at menstruation would be easy to envisage with the destruction and cell death that arises in endometrium at this time. They cannot however function in a similar way to which they do in wound sites in other parts of the body as endometrial repair does result in fibrosis. When it does, it produces Asherman's syndrome and infertility. The putative role for FGFs in the development of the spiral arterioles is probably restricted to expression in the vascular smooth muscle cells of the arterioles themselves.

The part that steroids play in the regulation of FGF expression is less clear. Presta (1988) demonstrated that estradiol stimulated bFGF synthesis from human carcinoma cell lines HEC, yet Ferriani et al (1993a) were unable to identify mRNA encoding bFGF in similar cells.

c) VEGF

VEGF is a dimeric peptide (Tischer et al, 1989) of which there are four splice variants in human endometrium encoding peptides of 189, 165, 145 and 121 amino acids (Charnock-Jones et al, 1993). In situ hybridization suggests that the temporal and spatial expression of VEGF varies throughout the menstrual cycle. In the proliferative phases of the cycle, hybridization is present in both the glandular and stromal compartments of the endometrium.

Within the stroma no clearly defined groups of cells can be found expressing the peptide. However, in the luteal phase of the cycle, the site of expression changes dramatically, hybridization occuring strongly only in the glands with absent expression in the stroma. The intensity of hybridization reaches its peak at the time of menstruation (Charnock-Jones et al, 1993). VEGF is a secreted peptide but it is not clear if the glandular cells have an apical or basal preference in the secretion of VEGF. Clearly, if basal, it would promote angiogenesis in the endometrium, whilst if apical it could influence the pre-implantation embryo.

Estradiol increases levels of mRNA encoding VEGF in HEC-1A cells but studies on the effects of steroids on primary cell cultures or endometrial explants have not yet been reported. The increased intensity of hybridization in glandular cells of secretory endometrium suggest that progesterone may increase expression in estradiol primed endometrium.

d) Cytokines

Probably the best characterised is CSF-1, which is present in human epithelial and endothelial cells in endometrium at implantation (Arceci et al, 1989). Its presence in the endothelial cells could suggest a role in endothelial cell growth or differentiation. Stem cell factor, SCF, is also expressed in human endometrium but is more widely expressed being found in both epithelial and stromal compartments.

Angiogenic Growth Factor Expression at the Embryo-Endometrial Interface

a) EGF

EGF staining is intense in surface epithelium of gestational decidua and is also present in the syncytiotrophoblast but not cytotrophoblast in first trimester pregnancy (Hofmann et al, 1991). This pattern of staining is retained throughout pregnancy. The function of EGF in regulating angiogenesis is not clear but as it is a secreted peptide a putative role is possible.

b) Acidic and basic FGF

Immunohistochemical staining for a and bFGF is clearly present in the cytotrophoblast of first trimester human placentae (Ferriani et al, 1993b). In addition, extravillous trophoblast passing into the decidua stains for both growth factors intensely. The depth of invasion of the cells does not affect the intensity of staining. This pattern of staining alters as pregnancy progresses such that by mid-gestation, a and bFGF immunoreactivity is found in fetal vascular smooth muscle in tertiary stem villae and in syncytiotrophoblast. It is not clear how the FGFs influence angiogenesis if they are not secreted. They would provide a source of angiogenic growth factors if the cells expressing them were injured or underwent apoptosis. It is possible that in vessels, the vascular smooth muscle cells are continuously undergoing cell death and depositing the FGFs into the extracellular matrix where they are bound to heparin. Proteolytic enzymes then regulate the release of the FGFs from heparin sulphate proteoglycans (Baird and Ling, 1987).

c) VEGF

In situ hybridization demonstrates mRNA encoding VEGF in syncytiotrophoblast of early human pregnancy (Sharkey et al, 1993a). The intensity of hybridization declines as pregnancy progresses. Of particular interest is the finding that intense hybridization arises in specific cells accumulated at the placental site. Subsequent serial staining of these cells has identified them as macrophages. Two of the splice variants, 165 and 121 are expressed by cellular macrophages and peripheral monocytes and although the in situ hybridization is not splice specific, presumably these cells are expressing these splice variants. Both of these variants are secreted peptides with profound angiogenic activity. This raises the prospect that maternal macrophages, attracted to the implanting placentae may facilitate the development of maternal blood vessels supplying the placenta. There would appear to be no need for this peptide to be transported to the fetal compartment. In the terminal villae, intense hybridization is present in cells identified as Hofbauer cells, i.e. fetal macrophages. Thus both the maternal and fetal circulations may have angiogenesis regulated by VEGF expressed by cells of the immune system.

In rodents, the immune system has been shown to be a crucial mediator of trophoblast growth (Robertson et al, 1992) though this has not been confirmed in humans. The above observations raise a further interactive system at the embryo-endometrial interface in which the immune systems of the mother and fetus regulate angiogenesis.

d) Cytokines

The previous observations concerning the expression and function of a variety of cytokines at the implantation site suggests the possibility of paracrine loops regulating trophoblast growth and invasion (Pollard, 1990), which in the light of the above findings, raises the prospect that trophoblast may also mediate the immune system which in turn regulates angiogenesis. Various splice variants of SCF are expressed by human trophoblast (Sharkey et al, 1992). Several of these variants contain the products of exons which permit the extracellular processing of the peptide and thus its release. Recently, we have demonstrated that the macrophages expressing VEGF, also express the receptor for SCF, which is the product of the oncogene c-kit (Sharkey et al, 1993b). Furthermore, the cells which most obviously express the receptor are found in the implantation site at the limit of the invasion of the extravillous trophoblast cells, expressing VEGF. FACS sorting of these cells has allowed us to further characterise the macrophages expressing c-kit as a subset of precursor macrophages. This suggests that SCF facilitates the development of macrophages which express VEGF. Cytokines would in this way regulate expression of VEGF and thus control the process of angiogenesis. Interestingly, in the fetal villae, Hofbauer cells stain with antibody against c-kit indicating a similar regulatory system in the fetus.

These observations are of particular importance with the recent finding that LIF, another cytokine, is obligatory for implantation in mice (Stewart et al, 1992). Recent studies in our laboratory indicate that LIF mRNA also shows cyclical variation of expression in human endometrium and thus may be required for implantation. It's site of action is yet to be determined but if macrophages express the receptor for LIF it opens a further route by which cytokines could regulate angiogenesis.

Conclusion

Implantation in humans is dependent on the cyclical shedding of endometrium and its rapid repair. Profound angiogenesis arises in the process of this repair and the endometrium expresses a range of growth factors known to promote new vessel formation. It is likely that they are, at least in part, regulated by ovarian steroids. The role of angiogenic growth factors in the very earliest development of the human embryo is still unclear but it is evident in mice that they play a crucial role. Trophoblast expresses a wide range of angiogenic growth factors whose temporal and spatial expression alter during pregnancy. This expression is not limited to the villous core as extravillous trophoblast also express angiogenic growth factors. Of particular interest is the finding that macrophages in the implantation site have high levels of expression for angiogenic growth factors. In addition trophoblast expresses cytokines whose receptors are present on the macrophages suggesting a paracrine loop whereby angiogenic activity of cells derived from the immune system could be regulated by the trophoblast. This system appears to be in place both in the maternal and fetal circulation of the embryo-endometrial interface. Further work is needed to establish if abnormalities in this system give rise to infertility or abnormal placentation leading to serious obstetrical disorders such as intrauterine growth retardation and catastrophic haemorrhage.

References

Arceci, A.J., Shanahan, F., Stanley, E.R., Pollard, J.W.,1989, Temporal expression and location of colony-stimulating factor 1 (CSF-1) and its receptor in the female reproductive tract are consistent with CSF-1-regulated placental development, Proc. Nat. Acad. Sci. USA, 86:8818-8822.

Baird, A., Ling, N.,1987, Fibroblast growth factors are present in the extracellular matrix produced by endothelial cells in vitro: implications for a role of heparinase-like enzymes in the neovascular response, Biochem. Biophys. Res. Comm. 142:428-435.

Charnock-Jones, D.S., Sharkey, A.M., Rajput-Williams, J., Burch, D., Schofield, J.P., Fountain, S.A., Boocock, C.A., Smith, S.K., 1993, Identification and localization of alternately spliced mRNAs for vascular endothelial growth factor in human uterus and steroid regulation in endometrial carcinoma cell lines, Biol. Reprod. 48: 1120-1128.

Ferriani, R.A., Charnock-Jones, D.S., Prentice, A., Thomas, E.J., Smith, S.K., 1993a, Immunohistochemical localisation of acidic and basic fibroblast growth factors in normal human endometrium and endometriosis and the detection of their mRNA by PCR, Human Reprod. 8:11-16.

Ferriani, R.A., Ahmed, A., Sharkey, A., Smith, S.K., 1993b, Immunolocalization of acidic and basic fibroblast growth factor (FGF) in human placenta and basic FGF-stimulated mitogenesis and phospholipase C activiation in trophoblast cell line JEG-3, Placenta (submitted).

Folkman, J., Klagsbrun, M., 1987, A family of angiogenic peptides, Nature 329:671-672.

Haining, R.E.B., Schofield, J.P., Jones, D.S.C., Rajput-Williams, J., Smith, S.K., 1991a, Identification of mRNA for epidermal growth factor and transforming growth factor a present in low copy number in human endometrium and decidua using reverse transcriptase-polymerase chain reaction, J. Mol. Endocrinol. 6:207-214.

Haining, R.E.B., Cameron, I.T., van Papendorp, C., Davenport, A.P., Prentice, A., Thomas, E.J., Smith, S.K., 1991b, Epidermal growth factor in human endometrium: proliferative effects in culture and immunocytochemical localization in normal and endometriotic tissues, Human Reprod. 6:1200-1205.

Hofman, G.E., Scott, R.T., Bergh, P.A., Deligdisch, L., 1991, Immunohistochemical localization of epidermal growth factor in human endometrium, decidua and placenta, J. Clin. Endocrinol. Metab. 73:882-887.

Klagsbrun, M., D'Amore, P.A., 1991, Regulators of angiogenesis, Ann. Rev. Physiol. 53:217-239.

Pollard, J.W., 1990, Regulation of polypeptide growth factor synthesis and growth factor-related gene expression in the rat and mouse uterus before and after implantation, J. Reprod. Fertil. 88:721-731.

Presta, M., 1988, Sex hormones modulate the synthesis of basic fibroblast growth factor in human endometrial adenocarcinoma cells: implications for the neovascularization of normal and neoplastic endometrium, J. Cell. Physiol. 137:593-597.

Robertson, S.A., Brannstrom, M., Seamark, R.F.,1992, Cytokines in rodent reproduction and the cytokine-endocrine interaction, Curr. Opinion Immunol. 4:585-590.

Sharkey, A.M., Jones, D.S.C., Brown, K.D., Smith, S.K., 1992, Expression of messenger RNA for kit-ligand in human placenta: localisation by in situ hybridisation and identification of alternatively spliced variants, Mol. Endocrinol. 6:1235-1242.

Sharkey, A.M., Charnock-Jones, D.S., Boocock, C., Brown, K.D., Smith, S.K., 1993a, Expression of mRNA for vascular endothelial cell growth factor in human placenta, J. Reprod. Fertil. in press.

Sharkey, A.M., Jokhi, P.P., King, A., Loke, Y.W., Brown, K.D., Smith, S.K., 1993b, Expression of c-kit and kit ligand at the human maternofetal interface, Development, submitted.

Stewart, C.L., Kaspar, P., Brunet, L.J., Bhatt, H., Gadi, I., Kontgen, F., Abbondanzo, S.J., 1992, Blastocyst implantation depends on maternal expression of leukaemia inhibitory factor, Nature, 359:76-79.

Tischer, E., Gospodarowicz, D., Mitchell, R., Silva, M., Schilling, J., Lau, K., Crisp, T., Fiddes, J.C., Abraham, J.A., 1989, Vascular endothelial growth factor: a new member of the platelet-derived growth factor gene family, Biochem. Biophys. Res. Comm. 165:1198-1206.

INSULIN-LIKE GROWTH FACTORS AND THEIR BINDING
PROTEINS IN THE ENDOMETRIUM

Liam J. Murphy

Departments. of Medicine and Physiology
University of Manitoba, Winnipeg
Manitoba R3E 0W3 Canada

INTRODUCTION

In mammals and rodents the endometrium undergoes cyclical changes in response to the fluctuations in the circulating levels of ovarian steroid hormones. These changes serve to prime the endometrium for embryo implantation. In rodents, blastocyst attachment to the hormone-sensitized luminal epithelium initiates a process termed decidualization [Psychoyos, 1973]. The blastocyst stimulates a regional differentiation of the underlying stromal cells to form the decidua. The decidua eventually completely surrounds the developing embryo and forms the interface with the trophoblasts of the developing placenta. In humans, blastocyst implantation is not necessary for decidualization and pre-decidual changes can be seen in the stroma in the late secretory phase of the menstrual cycle [Padykula, 1991]. After implantation of the blastocyst the process of decidualization continues with further proliferation and differentiation of stromal cells. This differentiation is accompanied by expression of a variety of biochemical markers such as prolactin, insulin-like growth factor binding protein-1, transforming growth factor-a, adenosine deaminase, metallothionein, desmin and decidual luteotrophin [Croze, et al., 1990; Han et al., 1987; Hong et al., 1991; De et al., 1989; Glasser and Julian, 1986; Gibori et al., 1984]. It is now clearly established that estrogen, progesterone, growth hormone and thyroxine are all required for decidualization of the rodent endometrium [Kennedy and Doktorcik, 1988]. There is also evidence that each of these hormones can enhance IGF-I expression at least in some tissues [Murphy and Friesen, 1988; Murphy et al.,1988a; Norstedt et al.,1989; Wolf et al.,1989]. Although some changes consistent with decidualization can be achieved when human stromal cells are cultured under appropriate conditions, it has not been possible to completely reproduce this process of decidualization in the stromal cells *in vitro* [Tabanelli et al., 1992]. Suitable *in vitro* models to address the underlying molecular mechanisms involved in decidualization still remain to be established. The failure to demonstrate *in vitro*, the same steroid hormone responsiveness in endometrial cells that is observed *in vivo*, suggests that cell-cell interactions and cell-extracellular matrix interactions are likely to be important in the processes of decidualization and placentation just as they are important in the response of the endometrium to steroid hormones.

Table 1. Growth factors and cytokines which are expressed in the endometrium

Growth factor or cytokine	Regulation by steroids[1]	Cell type.	Involved in decidualization[2]
IGF-I	+	Stromal cells, macrophages	+
IGF-II	+	Stromal cells, macrophages	?
EGF	+	Epithelial cells	?
TGF-α	+	Stromal/epithelial cells	+
TGF-β	+	Stromal/epithelial cells	+
bFGF	+	?	?
PDGF	?	Macrophages, vascular cells	?
CSF-1	+	Epithelial cells	+
TNF-α	?	Macrophages + others	?
INF-γ	?	Lymphocytes + others	?
INF-α	?	?	?

[1] The + sign indicates some published data demonstrating an effect of estrogen or progesterone on expression while the ? sign indicates no available data.

[2] The + sign indicates some published data demonstrating uterine expression during decidualization while the ? sign indicates no available data.

A wide variety of mechanisms have been described which mediate cell-cell interactions. In addition to paracrine or juxtacrine growth factors, cell membrane bound and secreted proteolytic enzymes may be involved in the release of growth modulators from the extracellular matrix and uterine cells may communicate directly through gap junctions [Garfield et al., 1977]. A variety of growth factors and cytokines are expressed in the endometrium and

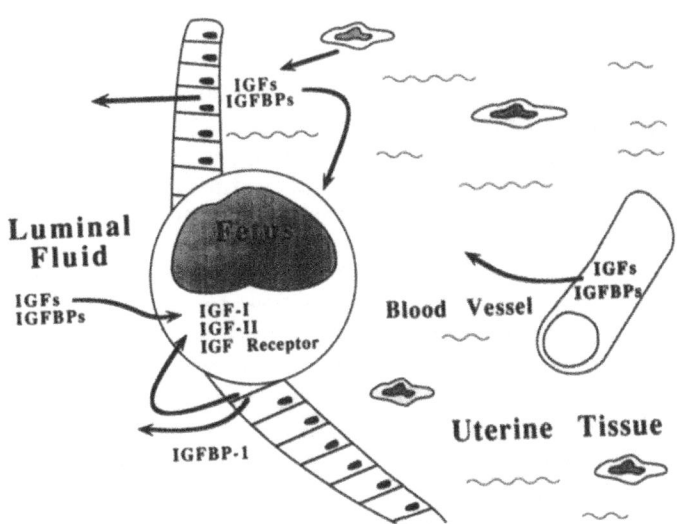

Figure 1. Expression of the IGF system during implantation and decidualization. IGF-I, IGF-II and IGFBPs are present in the luminal fluid and are able to interact with receptors on the embryo. The decidualized stromal cells and epithelial cells are also able to express these components of the IGF system. In addition to blood and tissue-derived IGFs and IGFBPs some of these factors are also expressed in the developing embryo.

may have some role in the endometrial response to steroid hormones and the processes of decidualization and placentation (Table 1). A more detailed discussion of expression of each of the different classes of growth factors and cytokines in the uterus can be found in other recent reviews and will not be discussed here [Brigstock et al., 1989; Tabibzadeh, 1991]. This review will concentrate on the insulin-like growth factors which represent one group of growth factors which are thought to be important in physiological responses of the uterus to steroid hormones, in the process of decidualization and in early fetal development. I will initially discuss the regulation of expression of the various components of the insulin-like growth factor system in the non-pregnant and pregnant uterus and conclude with some discussion of the changes which occur during decidualization and placentation. In addition, I will review the role of the insulin-like growth factor system in embryogenesis and early fetal development. It is important to note that early mammalian development unlike avian and amphibian embryogenesis occurs in close proximity to maternal tissue. Not only are maternal factors important in implantation, placentation and fetal development, recent studies in the mouse suggest that the maternal influence can be demonstrated prior to implantation. This maternal effect is likely to be mediated via nutrients, growth factors and cytokines which are present in the luminal fluid to which the pre-implantation embryo is exposed. Furthermore, as the embryo's own genome is activated, the embryo produces a variety of growth factors and cytokines which could potentially have effects on the adjacent maternal tissue. Thus, it is important to consider the entire feto-maternal unit when considering the role of the IGF system in endometrial physiology (Fig. 1).

BIOLOGICAL ACTIONS OF THE INSULIN-LIKE GROWTH FACTORS

The insulin-like growth factors were originally thought to be specific mediators of growth hormone however it is now apparent that they have a general role in many growth responses including some responses which are clearly growth hormone independent. The latter group includes the uterotrophic response to estrogen and renal regeneration [Murphy et al.,1988a; Fagin and Melmed, 1987]. Like other growth factors, the name belies the functional importance. The IGFs clearly have functions other than simply stimulation of cellular proliferation These include synergistic actions with pituitary hormones in steroid and thyroid hormone production [Lowe, 1991]. To date, two IGFs have been identified, IGF-I and II. They are both small peptides which have structural and functional similarities to proinsulin and are probably part of a larger gene family. A number of insulin and IGF-related peptides have been identified in lower species, suggesting that other as yet unidentified members of this gene family may exist [Smit et al., 1988; Iwami et al., 1989; Nagamatsu et al., 1991]. Relaxin, a polypeptide hormone of some importance in the female reproductive tract, was initially thought to be a distant member of this gene family however more recent critical analysis of the protein sequence does not support this concept.

In recent times there has been considerable emphasis on paracrine and autocrine actions of the IGFs, however it is important to realize that the IGFs also function in an endocrine fashion. A significant, but as yet undetermined proportion of uterine IGF-I and IGF-II is derived from the blood stream and possibly from blood derived mononucleated cells such as macrophages. In this regard it is also important to appreciate that sex steroids have rapid and profound effects on uterine blood flow [Szego and Roberts, 1953]. Thus, in addition to any effect estrogen may have on growth factor expression in uterine tissue it is likely to enhance delivery of blood derived growth factors to this organ. Estrogen may also enhance the uterine expression of chemotaxic factors which could result in migration into the endometrium of cells which secrete growth factors and cytokines [Lee et al., 1989]. There is enhanced vascular permeability at the site of implantation which could facilitate delivery of blood-derived growth factors and nutrients to the developing embryo.

REGULATION OF IGF EXPRESSION IN THE UTERUS

The IGF-I and IGF-II mRNAs have been demonstrated in uterine tissue of the rat, mouse, pig, cow as well the human uterus [Murphy et al., 1988a; Hoppener et al., 1988; Ogasawara et al., 1989; Geisert et al., 1991]. In both the rat and the human uterus IGF-I and IGF-II mRNAs are localized to cells of mesenchymal origin such as the smooth muscle cells and the stromal cells. The luminal and glandular epithelial cells demonstrate little expression of IGF-I or II [Ghahary et al.,1990]. In studies of IGF-I expression in the rat uterus using the *in situ* hybridization technique, IGF-I mRNA was been localized to the peri-glandular stromal cells [Ghahary et al.,1990]. In most species so far examined immunoreactive IGF-I and II can be detected in uterine luminal fluid and uterine extracts although it still remains to be determined what proportion of this is derived from local uterine synthesis compared to blood-derived IGF-I. Truncated forms of IGF-I with enhanced bioactivity were originally isolated from porcine uterine tissue as a uterine-derived cell growth factor [Ogasawara et al., 1989]. The enhanced *in vitro* and *in vivo* biological activity of this truncated IGF-I appears to be largely due to the reduced affinity of the truncated IGF-I for the IGF binding proteins [Ballard et al., 1987]. Although the truncation of the amino terminal three residues of IGF-I may eventually prove to be an artifact of the extraction procedure, it may also be due to a specifically regulated enzymatic activation which could have physiological importance. A variant form of IGF-II resulting from alternate splicing of the mRNA may also be present in the uterus however the bioactivity of this IGF-II variant has not been determined and the physiological role of the variant in the uterus is not clear [Yeh et al., 1991].

In the rodent uterus both IGF-I and IGF-II mRNAs increase following estrogen administration [Murphy et al., 1988a]. The estrogen-induced uterine IGF-I response is pituitary-independent and does not require continuing protein synthesis [Murphy and Luo, 1989]. This response to estrogen is not generalized but rather restricted to a select few tissues and cell types. A similar induction of IGF-I mRNA by estrogen can be demonstrated in certain cell types in ovarian and skeletal tissue [Hernandez et al., 1989; Ernst and Rodan, 1991]. In most other tissues estrogen has no effect or even reduces IGF-I mRNA abundance [Murphy and Friesen, 1988]. There is also an increase in immunoreactive IGF-I in the uterus after estrogen administration to ovariectomized rats [Murphy et al.,1988a]. This increase in extractable IGF-I is likely to reflect both increased blood-derived IGF-I and increased local synthesis of IGF-I. Estrogen can also up-regulate IGF-I expression in the porcine endometrium [Simmen et al., 1990] and IGF-I expression may also be estrogen dependent in the human endometrium although the evidence for this is indirect [Rein et al., 1990]. The changes in uterine IGF-I expression observed in response to the physiological changes in circulating estradiol during the reproductive cycle are less marked than the estrogen response in ovariectomized or immature rodents [Murphy et al., 1988b]. This may be partly due to the fact that the majority of the IGF-I mRNA is located in the myometrium, a layer of the uterus which shows little proliferative response to the cyclic variations in estrogen levels in the mature animal. More detailed examination of the stromal component of the endometrium using *in situ* hybridization may reveal more dramatic changes. Increased levels of IGF-I and IGF-II protein can also be demonstrated in bovine uterine washings from proestrus compared to other stages of the cycle [Geisert et al., 1991]. IGF-I transcripts are most abundant in human endometrium from the late proliferative stage of the menstrual cycle while IGF-II transcripts are most abundant in the early proliferative phase of the menstrual cycle [Boehm et al., 1991].

Progesterone is also important in sensitizing the uterus for blastocyst attachment, implantation and decidualization. The major effects of progesterone in the uterus are thought to be inhibition of estrogen-induced proliferation and induction of differentiation. This may be a rather simplistic view of progesterone action since this hormone can stimulate early response genes such as *fos* and *myc* in some endometrial cell types and a proliferative effect of progestins on stromal can be demonstrated *in vitro* [Huet-Hudson et al., 1989; Holinka and Gurpide, 1992]. Studies in primate endometrium using the thymidine labeling techniques and studies of endometrial cancer cells *in vitro* support the concept that at least under certain circumstances endometrial epithelial cells are growth stimulated by progestins [Padykula, 1991]. The role of progesterone in regulation of uterine IGF-I expression is not clear. Progesterone can enhance the estrogen-induced increase in uterine IGF-I mRNA levels in the immature rat but has no effect on uterine IGF-I expression in hypophysectomized rats [Norstedt et al., 1989; Carlsson and Billig, 1991]. In the human endometrium both IGF-I and IGF-II transcript abundance decreases after ovulation. This would be consistent with a negative effect of progesterone on IGF expression however other explanations are possible. An equally confusing set of data has been obtained in the porcine uterus where progesterone treatment appears to have opposite effects depending upon estrogen pretreatment, sexual maturity and timing of injection [Simmen et al., 1990].

The other hormones which have been clearly demonstrated to be necessary for decidualization in the rodent are growth hormone and thyroid hormone. These hormones also have effects on IGF expression. Growth hormone enhances IGF-I expression in most if not all tissues including the uterus [Murphy et al., 1988a]. Thyroid hormones enhance the effects of growth hormone on IGF-I expression in a variety of tissues however convincing proof of this synergy in the uterus has not been reported [Wolf et al.,1989].

EXPRESSION OF THE IGF RECEPTORS IN THE UTERUS

Specific receptors for IGF-I and IGF-II have been demonstrated in uterine tissue. The type 1 or IGF-I receptor is a heterotetrameric complex which is structurally and functionally very similar to the insulin receptor and appears to share some common signal transduction mechanisms. IGF-II can interact with the IGF-I receptor but also interacts with a specific receptor, the type 2 receptor which is a monomeric protein of 260 kDa. Cloning of the type 2 receptor revealed that it was identical to the mannose-6-phosphate receptor. Both the type 1 and type 2 receptors are expressed in uterine tissue. The affinity, capacity and biochemical properties of these receptors in uterine tissue are similar to that reported for other tissues [Ghahary and Murphy, 1989; Chandrasekhar et al., 1991]. Administration of estradiol to immature and ovariectomized rats increases the uterine IGF-I receptor number but has no significant effect on receptor affinity [Ghahary and Murphy, 1989]. Uterine IGF-I receptor number show a variation throughout the estrous cycle with maximal uterine ^{125}I-IGF-I binding capacity observed during proestrus and the lowest binding in diestrus. This contrasts with uterine IGF-I mRNA levels where the highest levels are seen late in diestrus and early in proestrus [Murphy et al.,1988b]. These receptors are demonstrable in stromal and epithelial cell from the rat uterus [Ghahary and Murphy, 1989]. It is not clear whether the effect of estrogen on uterine IGF receptors is direct or secondary to up-regulation of IGF-I expression or other estrogen induced proteins. Estrogen has no effect on IGF-I receptors in primary cultures of normal endometrial stromal and glandular tissue. In contrast, progesterone up-regulated IGF-I receptors in endometrial glandular tissue [Reynolds et al., 1990]. Expression of the mannose-6-phosphate/IGF-II receptor has also been demonstrated in the rat uterus but there does not appear to be any significant variation throughout the estrous cycle [Chandrasekhar et al., 1991].

EXPRESSION OF THE IGF BINDING PROTEINS IN THE UTERUS

Additional components of the IGF system which are expressed in endometrial tissue are the IGF binding proteins (IGFBPs). These high affinity binding proteins are present in most biological fluids and tissue extracts. Six IGF binding proteins (IGFBPs) have now been identified and cDNA clones for each of these IGFBPs are now available. They show differences in tissue expression and regulation suggesting that they may have very different functional roles. At least four distinct IGFBPs can be detected in rat uterine extracts and uterine luminal fluid. Transcripts for IGFBP-1,3,4 and 5 are easily detected in mRNA from rat uteri [Molnar and Murphy, 1993]. IGFBP-6 mRNA is also present in very low abundance. The relative abundance of the IGFBPs detected by ligand blotting in luminal fluid is similar to that observed in serum and no consistent changes in the relative abundance of IGFBPs is apparent throughout the estrous cycle [Murphy and Ghahary, 1990; Molnar and Murphy, 1993].

The role of the IGFBPs in the uterine physiology has yet to be determined. IGFBP-1 was the first of the binding proteins to be characterized and although it appears to be only a relatively minor component of the IGF binding capacity in the circulation, it is present in high concentrations in amniotic fluid and is abundantly expressed in both primate and rodent endometrium during decidualization [Drop et al., 1979; Croze et al.,1990]. In the rat, immunohistochemistry and *in situ* hybridization localizes IGFBP-1 expression to the uterine luminal and glandular epithelial tissue [Croze et al.,1990]. This contrasts with the pattern of expression seen in the primate where IGFBP-1 is localized predominantly to the stromal cells [Fazleabas et al., 1989]. Regulation of IGFBP-1 expression also appears to be different in the rodent and primate. In the immature rat, estrogen administration results in a marked decrease in uterine IGFBP-1 mRNA abundance [Murphy, 1991] and in the mature cycling rat the highest levels of IGFBP-1 mRNA are seen in diestrus and lowest levels are seen in proestrus [Murphy, 1991]. The changes in IGFBP-1 mRNA level during the estrous cycle are the inverse of the changes seen in uterine IGF-I expression and IGF-I receptors. Since at least some isoforms of IGFBP-1 can actually inhibit IGF-I action, the attenuation of expression of this binding protein in concert with an increased expression of IGF-I and IGF-I receptors in response to estrogen would enhance the bioavailability and functional activity of uterine IGF-I [Elgin et al., 1987; Ritvos et al.,1988]. Isoforms of IGFBP-1 which differ in the degree of phosphorylation have been detected in amniotic fluid and culture medium of endometrial cells and may account for the differential effects of inhibition and stimulation of IGF-I action reported in some assay systems [Frost and Tseng, 1991]. The phosphorylated isoform of IGFBP-1 has a 5 fold higher affinity for IGF-I than does the non-phosphorylated form and would be expected to have a greater capacity to inhibit IGF-I action. The nature of the kinases involved and their regulation has yet to be determined, however this may represent yet another step where steroid hormones may indirectly modulate IGF action in the endometrium.

In the primate endometrium progesterone appears to be the major regulator of IGFBP-1 expression. In human endometrial tissue IGFBP-1 levels are highest in the follicular phase of the menstrual cycle [Waites et al., 1988] and expression of IGFBP-1 by cultured endometrial stromal cells *in vitro* can be induced by progestins [Bell et al., 1991]. Progesterone may have some role in regulating expression of IGFBP-1 in the rodent since endometrial IGFBP-1 expression in ovariectomized rats increases in response to a combination of progesterone and estrogen [Croze et al.,1990]. The effects of other hormones on IGFBP-1 expression in the endometrium has not been examined in detail although relaxin has been reported to enhance IGFBP-1 expression by cultured human endometrial stromal cells [Bell et al., 1991]

IGFBP-3 is one of the most abundant binding proteins present in uterine luminal fluid [Murphy, 1991]. Uterine IGFBP-3 mRNA is down-regulated by estrogen in ovariectomized rats and shows a similar variation in abundance during the estrous cycle to that seen for IGFBP-1 [Molnar and Murphy, 1993]. The cellular localization of IGFBP-3 in the uterus has not been determined. The regulation of expression of the other binding proteins which are expressed in the rodent uterus has not been investigated.

A variety of enzymes which could potentially be involved in the regulation of IGF action are expressed in the endometrium. As stated above the majority of IGF-I extracted from the porcine uterus is the biologically more active des 1-3 variant [Ogasawara et al., 1989]. This truncated variant retains receptor binding activity but has a very much reduced affinity for the IGFBPs. The enzyme or enzymes which catalyze the cleavage of the amino terminal three residues from mature IGF-I may be yet another potential step where IGF-I action can be regulated. In addition, proteolytic enzymes which destroy the IGF binding capacity of the IGFBPs have also been described. For example plasmin is able to dissociate IGFs from their binding proteins [Campbell et al., 1992]. Both plasminogen activators and inhibitors are expressed in the endometrium and the regulation is under sex hormone control [Casslen et al., 1986].

THE IGF SYSTEM DURING IMPLANTATION AND DECIDUALIZATION

In the rodent, differentiation of hormonally sensitized stromal cells to decidual cells occurs in response to the blastocyst. A similar response is seen in the sensitized endometrium with artificial stimuli indicating that this process does not require any specific factors from the blastocyst. However the decidualization of stromal cells requires the interaction of a number of hormonal and local factors to bring about both proliferation and differentiation of the stromal cells [Psychoyos, 1973]. The relative abundance of the IGFs and their binding proteins, particularly IGFBP-1 in the reproductive tract suggests that these factors may have some role in decidualization. The expression of IGF-I and IGFBP-1 has been examined in the controlled setting of induction of decidualization in the hypophysect-omized, ovariectomized rat [Croze et al.,1990]. In this setting there is a gradual decline in endometrial IGF-I expression with a reciprocal increase in IGFBP-1 expression. These changes were restricted to the endometrium with little change occurring in the underlying myometrium.

In the decidualized endometrium the spatial localization of IGF-I and IGFBP-1 expression suggests an important relationship between these two proteins. In situ hybridization localizes IGFBP-1 mRNA to the luminal and glandular epithelium with little hybridization seen in the adjacent stromal tissue (Fig. 2). Immunohistochemistry also confirms this specific localization of the IGFBP-1 mRNA in the rodent uterus. This contrasts with the primate

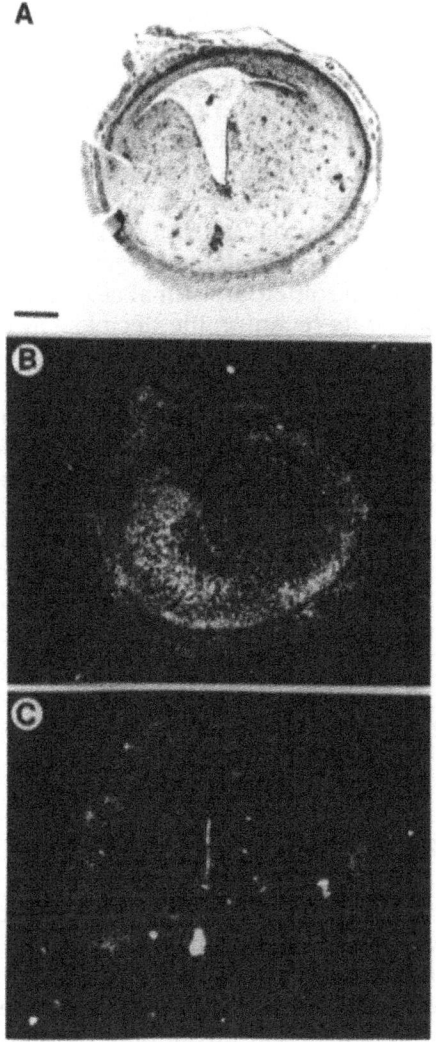

Figure 2. Localization of IGF-I and IGFBP-1 mRNA in the rodent decidualized endometrium. Panel A shows a stained cross-section of a decidualized rat uterus. In panel B the pattern of hybridization obtained with an IGF-I antisense riboprobe is shown while panel C shows the pattern of hybridization obtained with an IGFBP-1 antisense riboprobe. Reproduced from Croze et al., 1990 with permission.

decidualized endometrium where IGFBP-1 expression is predominantly but not exclusively localized to the stroma [Fazleabas et al., 1989]. In the rodent decidualized endometrium IGF-I mRNA is more abundant in anti-mesometrial stromal tissue and is particularly abundant in the stromal cells immediately adjacent to the luminal and glandular epithelia (Fig. 2). There is a gradation of IGF-I expression in the stromal cells with cells more distant from the epithelial cells demonstrating less hybridization than peri-epithelial stromal cells.

Although in the rat both growth hormone and thyroid hormone replacement is required for decidualization neither of these hormones have significant effects on uterine IGFBP-1 or IGF-I expression during the process of decidualization [Croze et al., 1990]. The major determinant of IGF-I expression appeared to be estradiol as is the case in other rodent models where uterine IGF-I expression has been examined. The decline in the abundance of IGF-I mRNA was observed after the introduction of progesterone possibly reflecting an ability of progesterone to antagonize this action of estradiol. In contrast IGFBP-1 expression was not seen until after the administration of progesterone [Croze et al., 1990]. It is of interest that in the human uterus, IGFBP-1 is not detected in endometrium during the proliferative phase of the menstrual cycle whereas secretion of IGFBP-1 is easily detected late in the secretory phase when cellular proliferation has diminished [Waites et al., 1988]. Since in the human endometrial stromal cell cultures IGFBP-1 is able to inhibit binding of IGF-I to its receptor, the decrease in IGF-I expression and concomitant increase in IGFBP-1 expression may well be part of the mechanism whereby cells are switched from proliferation to differentiation. Decidualization of the uterine stromal tissue involves two separate processes, cellular proliferation and differentiation. There is *in vitro* evidence using a variety of cell lines which suggest that these two processes are often mutually exclusive. Indeed, it is often necessary to arrest growth by serum depletion, to induce the differentiated phenotype and differentiating agents such as retinoic acid and phorbol esters are often growth inhibitory. Estrogen is clearly necessary to induce decidualization. The estrogen induced local synthesis of IGF-I appears to enhance the mitogen response since proliferation of stromal cells is stimulated by IGF-I [Ghahary et al., 1990]. Induction of IGFBP-1 expression, possibly in response to progesterone, is necessary to inhibit the mitogenic activity of IGF-I and allow for the differentiation of the stromal tissue.

THE IGF SYSTEM IN THE DEVELOPING EMBRYO

Although murine embryos can develop in defined medium devoid of serum or added growth factors, development appears to lag behind embryos that develop *in vivo* [Bowman and Mclaren, 1970; Harlow and Quinn, 1982]. This suggests that even at this very early stage of development that the maternal microenvironment of the developing embryo contributes factors which favor embryo development. The same may also be true in other species. The relative contribution of the various constituents of the oviduct and uterine luminal fluid to embryo development has not been determined. However IGF-I and IGF-II which are likely to be present in oviduct fluid and have been demonstrated in uterine luminal fluid, have significant effects on protein, RNA and DNA synthesis in the pre-implantation mouse embryo [Harvey and Kaye, 1988; Rao et al., 1990; Harvey and Kaye, 1992]. Physiological concentrations of IGF-I when added to 2-cell mouse culture result in an increase in the number of cells in the inner cell mass compared to control embryos but no effect on trophoectoderm cell proliferation [Harvey and Kaye, 1992]. The concentration of the IGF required to demonstrate these effects are in the same order as that which would be anticipated to be present in the oviduct and uterine luminal fluid. It is possible therefore that the developmental delay observed when embryos are cultured *in vitro* is due in part to lack of exposure to maternal growth factors derived from the biological fluids to which the pre-implantation embryo is exposed.

Implantation in the rodent is of the interstitial type, thus the developing embryo is in close proximity to maternal decidual cells immediately after implantation. Of relevance to the discussion here is the fact that conditioned medium from decidualized stromal cells enhances growth of pre-implantation mouse embryos [Liu et al., 1991]. As discussed above decidualized stromal cells produce both IGF-I and IGFBPs. However, it has yet to be determined that these growth factors are the factors responsible for the growth stimulating effects of decidual cell conditioned medium. In both the rat and man IGFBP-1 appears to be particularly abundant in the decidual tissue. The presence of IGFBP-1 in high concentrations at the maternal-embryo interface suggests that it might have some role in stabilizing and or transporting IGFs to the developing embryo.

In addition to these extra-embryonic components of the IGF system, expression of various components of this system can be detected in the developing embryo. Activation of the embryonic genome occurs sometime after the 2-cell stage with the mRNA present before this stage being predominantly of maternal origin. Using reverse transcriptase polymerase chain reaction technique it has been possible to demonstrate IGF-II receptor transcripts in the 2-cell mouse embryo stage, and IGF-I receptor and insulin receptor transcripts at the 8-cell stage [Rappolee et al., 1990]. Insulin and IGF-I can not be detected until day 8, approximately 4 days after implantation [Heyner et al., 1989]. Thus, both the IGF receptors and their ligands are expressed early in embryogenesis with IGF-II and its receptor being expressed very early in development. If these mRNA are translated into protein it would be consistent with the production of IGF-II as early as the two-cell stage and IGF-I at day 8. The functional significance of expression of these growth factors at this very early stage of embryogenesis is not clear however recent experiments using homologous recombination to knock-out IGF-II expression has demonstrated the importance of IGF-II in fetal growth [DeChiara et al., 1990]. It is significant that in rodents unlike man, IGF-II levels are low in post-natal life and thus the maternal IGF-II concentrations that the pre-implantation embryo is exposed to are quite low in the rodent compared to man.

There are as yet no reports of IGFBP expression in the preimplantation embryo. Systematic examination of IGFBP expression in post-implantation rodent embryo has been reported for IGFBP-2. This binding protein is expressed as early as embryonic day 11 in the rat [Wood et al.,1990]. Although it is most likely that the IGFs and their binding proteins produced by the embryo function as local paracrine or autocrine modulators within the embryo it is possible that these proteins produced by the embryo interact with maternal tissues.

CONCLUSIONS

Various components of the insulin-like growth factor system are expressed in the uterus and expression appears to be under steroid hormone control. During decidualization and at the time of implantation there is abundant expression of IGF-1 and IGFBP-1 in both the rodent and primate endometrium. In addition at the time of implantation or shortly therefore IGF-II, IGF-I and IGFBP expression can be detected in the developing embryo. The data reported to date suggest that the insulin-like growth factor system could be potentially important in decidualization, implantation and placentation however definite experiments are still required to determine the exact role of the IGFs in these processes.

ACKNOWLEDGMENTS

The work described in this review was supported by grants from the Medical Research Council of Canada and the National Cancer Institute of Canada. L.J.M is the recipient of an endowed research professorship in metabolic diseases.

REFERENCES

Ballard, F.J., Francis, G.L., Ross, M., Bagley, C.J., May, B., and Wallace, J.C., 1987, Natural and synthetic forms of insulin-like growth factor-I (IGF-I) and the potent derivative, destripeptide IGF-I: biological activities and receptor binding, *Biochem. Biophys. Res. Commun.* 149:398-404.

Bell, S.C., Jackson, J.A., Ashmore, J., Zhu, H.H., and Tseng, L., 1991, Regulation of insulin-like growth factor binding protein-1 synthesis and secretion by progestin and relaxin in long term cultures of human endometrial stromal cells, *J. Clin. Endocrinol .Metab .* 72:1014-1024.

Boehm, K.D., Daimon, M., Gorodeski, I.G., Sheean, L.A., Utian, W.H., and Ilan, J., 1990, Expression of insulin-like and platelet-derived growth factor genes in human uterine tissues, *Mol . Reprod. Dev.* 27: 93-101.

Bowman, P., and McLaren, A., 1970, Cleavage rate of mouse embryos in vivo and in vitro, *J. Embryol. Expl. Morphol.* 24:203-207.

Brigstock, D.R., Heap, R.B., and Brown, K.D., 1989, Polypeptide growth factors in uterine tissues and secretions, *J. Reprod. Fert.* 85: 747-758.

Campbell, P.G., Novak, J.F., Yanosick, T.B., and McMaster, J.H., 1992, Involvement of the plasmin system in dissociation of the insulin-like growth factor-binding protein complex, *Endocrinology* 130:1401-1412.

Carlsson, B., and Billig, H., 1991, Insulin-like growth factor-I gene expression during development and estrous cycle in the rat uterus, *Mol. Cell. Endocrinol.* 77:175-180.

Casslen, B., Andersson, A., Nilsson, I.N., and Astedt, B., 1986, Hormonal regulation of the release of plasminogen activators and of a specific activator inhibitor from endometrial tissue in culture, *Proc. Soc. Exp. Biol. Med.* 182:419-424.

Chandrasekhar, Y., Narayan, S., Singh, P., and Nagamani, M., 1991, Receptors for insulin-like growth factor II in the rat uterus: characterization and variation throughout the estrous cycle. *Acta Endocrinol. (Copenh.)* 124:434-41.

Croze, F., Kennedy, T.G., Schroedter, I.C., Friesen, H.G., and Murphy, L.J., 1990, Expression of insulin-like growth factor-I and insulin-like growth factor binding protein-1 in the rat uterus during decidualization, *Endocrinology* 127:1995-2001.

De, S.K., McMaster, M.T., Dey, S.K., and Andrews, G.K., 1989, Cell-specific metallothionein gene expression in mouse decidua and placentae, *Development* 107:6111-621.

DeChiara, T., Efstratiadis, A., Robertson, E.A., 1990, Growth-deficient phenotype in heterozygous mice carrying an insulin-like growth factor II gene disruption by targeting, *Nature* 345:78-80.

Drop, S.P.S., Valiquette, G., Guyda, H., Corvol, H.J., Corvol, M.T., and Posner, B.I., 1979, Partial purification and characterization of a binding protein for insulin-like activity (ILAs) in human amniotic fluid: a possible inhibitor of insulin-like activity, *Acta. Endocrinol ..Copenh)*. 90:505-518.

Elgin, R.G., Busby, W.H., and Clemmons, D.R., 1987, An insulin-like growth factor (IGF) binding protein which enhances the biological response to IGF-I, *Proc. Natl. Acad. Sci. U.S.A.* 84:3254-9.

Ernst, M., and Rodan, G.A., 1991, Estradiol regulation of insulin-like growth factor-I expression in osteoblastic cells: Evidence for transcriptional control, *Mol. Endocrinol.* 5:1081-1089.

Fagin, J.A., and Melmed, S., 1987, Relative increase in insulin-like growth factor I messenger ribonucleic acid levels in compensatory renal hypertrophy, *Endocrinology* 120:718-724.

Fazleabas, A.T., Jaffe, R.C., Verhage, H.G., Waites, G., and Bell, S.C., 1989. An insulin-like growth factor binding protein in the baboon (papio anubis) endometrium: synthesis, immunocytochemical localization and hormonal regulation, *Endocrinology* 124:2321-2329.

Frost, R,A, and Tseng, L., 1991, Insulin-like growth factor-binding protein-1 is phosphorylated by cultured human endometrial stromal cells and multiple protein kinases in vitro, *J. Biol. Chem.* 266:18082-18088.

Garfield, R.E., Sims, S., and Daniel, E.E. 1977, Gap junctions: their presence and necessity in myometrium during parturition, *Science* 198: 958-960.

Geisert, R.D., Lee, C-Y., Simmen, F.A., Zavy, M.T., Fliss, A.E., Bazer, F.W., and Simmen, R.C.M., 1991, Expression of messenger RNAs encoding insulin-like growth factor-I -II and insulin-like growth factor binding protein-2 in bovine endometrium during the estrous cycle and early pregnancy, *Biol. Reprod.* 45:975-983.

Ghahary, A., Chakrabarti, S., Murphy, L.J., 1990, Localization of the sites of synthesis and action of insulin-like growth factor-I in the rat uterus, *Mol. Endocrinol.* 4:191-7.

Ghahary, A., and Murphy, L.J., 1989, Regulation of uterine insulin-like growth factor receptors by estrogen and variation throughout the estrous cycle, *Endocrinology* 125: 597-604.

Gibori, G., Kalison, B., Basuray, R., Rao, M.C., and Hunzicker-Dunn, M., 1983, Endocrine role of decidual tissue: decidual luteotropin regulation of luteal adenyl cyclase activity, luteinizing hormone receptors and steroidogenesis, *Endocrinology* 115: 1157-1163.

Glasser, S.R. and Julian, J.A., 1986, Intermediate filament protein as a marker for uterine stromal cell differentiation, *Biol. Reprod.* 35,463-474.

Han, V. K. M., Hunter, E. S., Pratt, R. M., Zendegui, J. G., and Lee, D. C., 1987, Expression of rat transforming growth factor alpha mRNA during development occurs predominantly in the maternal decidua, *Mol. Cell.Biol.*, 7, 2335, 1987.

Harlow, G.M., and Quinn, P., 1982, Development of preimplantation mouse embryo in vivo and in vitro, *Aust. J. Biol. Sci.* 35:187-193

Harvey, M.B., and Kaye, P.L., 1988, Insulin stimulates protein synthesis in compacted mouse embryos, *Endocrinology* 123:1182-1183.

Harvey, M.B., and Kaye, P.L., 1992, Insulin-like growth factor-I stimulates growth of mouse preimplanation embryo in vitro, *Mol. Reprod. Develop.* 31:195-199.

Hernandez, E.R., Roberts, C.T., LeRoith, D., and Adashi, E.Y., 1989, Rat ovarian insulin-like growth factor-I (IGF-I) gene expression is granulosa cell selective: 5'-untranslated mRNA variants representation and hormonal regulation, *Endocrinology* 89:572

Heyner, S., Smith, R.M., and Schultz, G.A., 1989, Temporally regulated expression of insulin and insulin-like growth factors and their receptors in early mammalian development, *Bioessays* 11:171-176.

Holinka, C.F., and Gurpide, E., 1992, Growth-promoting effects of progesterone in a human endometrial cancer cell line (Ishikawa-var I), *J. Steroid Biochem. Molec. Biol.* 43:635-641.

Hong, L., Mulholland, J., Chinsky, J.M., Kudsen, T.B., Kellems, R.E., and Glasser, S.R., 1991, Developmental expression of adenosine deaminase during decidualization in the rat uterus, *Biol .Reprod.* 44:83-93.

Hoppener, J.W.M., Mosselman, S., Roholl, P.J.M., Lambrechts, C., and Slebos, R.J.C., de Pagter-Holthuzen, P., Lips, C.J.M., Jansz, H.S., and Sussenbach, J.S., 1988, Expression of insulin-like growth factor-I and -II genes in human smooth muscle tumours, *EMBO J.* 7:1379-1385.

Huet-Hudson, Y. M., Andrews, G. K., and Dey, S. K., 1989, Cell type-specific localization of c-myc protein in the mouse uterus: modulation by steroid hormones and analysis of the periimplantation period, *Endocrinology*, 125, 1683-9.

Iwami, M., Kawakami, A., Ishizaki, H., Takahashi, S.Y., Adachi, T., Suzuki, Y., Nagasawa, H., and Suzuki, A., 1989, Cloning of a gene encoding Bombyxin, an insulin-like brain secretory peptide of the silkmoth bombyx mori with prothoracicotropic activity, *Develop. Growth and Differ.* 31:31-37.

Kennedy, T.G., and Doktorcik, P.E., 1988, Uterine decidualization in hypophysectomized-ovariectomized rats: effects of pituitary hormones, *Biol. Reprod.* 39:318-23.

Lee, Y.H., Howe, R.S., Sha, S., Teuscher, C., Sheehan, D.M., and Lyttle, C.R. 1989, Estrogen regulation of an eosinophil chemotactic factor in the immature rat uterus. *Endocrinology* 125:3022-8.

Liu, H.C., Tseng, L., and Rosenwaks, Z., 1991, Communication between embryo and endometrium via secretory proteins, *Proc. 39th Ann. Meeting Society for Gynecological Investigation.* San Antonio, Abst. 165

Lowe, W.L., 1991, Biological actions of the insulin-like growth factors, in:: "Insulin-Like Growth Factors: Molecular and Cellular Aspects", D. LeRoith, ed., CRC Press, Cleveland. pp49-85.

Molnar, P., and Murphy, L.J., 1993, Effects of estrogen on uterine IGFBP expression and regulation throughout the estrous cycle. *Proc. 75th Ann, Meeting Endocrine Society* Las Vagas, Abst.

Murphy, L.J., 1991, The uterine insulin-like growth factor system, in:: "Modern Concepts of Insulin-Like Growth Factors," E.M. Spencer, ed., Elsevier, New York pp 275-285.

Murphy, L.J., and Friesen, H.G., 1988, Differential effects of estrogen and growth hormone on uterine and hepatic insulin-like growth factor-I gene expression in the ovariectomized, hypophysectomized rat, *Endocrinology* 122:325-332.

Murphy, L.J., and Ghahary, A., 1990, Uterine insulin-like growth factor-I: regulation of expression and its role in estrogen-induced uterine proliferation,*Endocr. Rev.* 11, 443-453.

Murphy, L.J., and Luo, J.M., 1989, Effects of cycloheximide on hepatic and uterine insulin-like growth factor-I mRNA, *Mole Cell Endocrinol* 64:81-86.

Murphy, L.J., Murphy, L.C., and Friesen, H.G., 1988a, Estrogen induces insulin-like growth factor-I expression in the rat uterus, *Mol .Endocrinol* . 1: 445-450.

Murphy, L.J., Murphy, L.C., and Friesen, H.G., 1988b, A role for the insulin-like growth factors as estromedins in the rat uterus, *Trans .Ass. Amer. Physic* . 99:204-211.

Nagamatsu, S., Chan, S.J., Falkmer, S., and Steiner, D.F., 1991, Evolution of the insulin gene superfamily. Sequence of a preproinsulin-like growth factor cDNA from the atlantic hagfish, *J. Biol. Chem.* 266:2397-2402.

Norstedt, G., Levinovitz, A., and Eriksson, H., 1989, Regulation of uterine insulin-like growth factor-I mRNA and insulin-like growth factor-II mRNA by estrogen in the rat, *Acta Endocrinol. (Copenh.)* 120: 466-471.

Ogasawara, M., Karey, K.P., Marquardt, H., and Sirbasku, D.A., 1989, Identification and purification of truncated insulin-like growth factor I from porcine tissue. Evidence for high biological potency. *Biochemistry* 28:2710-2721.

Padykula, H.A., 1991, Regeneration in the primate uterus: the role of stem cells. *Ann. N.Y. Acad. Sci* . 622:47-56.

Psychoyos, A., 1973, Endocrine control of egg implantation, *in:* "Handbook of Physiology, Sect. 7. Vol II Part 2." R.O. Greep, E.B. Astwood and S.R. Geiger, eds., American Physiological Society, Bethesda, pp 187-215.

Rao, L.V., Wilarczuk, M.L., Heyner, S., 1990, Functional roles of insulin and insulinlike growth factors in preimplantation mouse embryo development, *In vitro Cell. Dev. Biol.* 26:1043-1048.

Rappolee, D.A., Sturm, K., Schultz, G.A., Pedersen, R.A., and Werb, Z., 1990, The expression of growth factor ligands and receptors in preimplatation mouse embryos, in: "UCLA Symposia on Molecular and Cellular Biology, New Series: Early Embryo Development and Paracrine Relationships," S. Heyer and L. Wiley, eds., Wiley-Liss, New York, pp11-25.

Rein, M.S., Friedman, A.J., Pandian, M.R., and Heffner, L.J., 1990, The secretion of insulin-like growth factors I and II by explant cultures of fibroids and myometrium from women treated with a gonadotropin-releasing hormone agonist, *Obstet. Gynecol.* 76:388-94

Reynolds, R.K., Talavera, F., Roberts, J.A., Hopkins, M.P., and Menon, K.M.J., 1990, Regulation of epidermal growth factor and insulin-like growth factor-I receptors by estradiol and progesterone in normal and neoplastic endometrial cell cultures, *Gynecol. Oncol.* 38:396-406.

Ritvos, O., Ranta, T., Jalkanen, J., Suikkari, A.M., Voutilainen, R., Bohn, H., and Rutanen, E.M., 1988, Insulin-like growth factor (IGF) binding protein from human decidua inhibits the binding and biological action of IGF-I in cultured choricarcinoma cells, *Endocrinology* 122:2150-8.

Simmen, R.C.M., Simmen, F.A., Hogif, A., Farmer, S.J., and Bazer, F.W., 1990, Hormonal regulation of insulin-like growth factor gene expression in the pig uterus, *Endocrinology* 127:2166-2174.

Smit, A.B., Vreugdenhil, E., Ebberink, R.H.M., Geraerts, W.P.M., Klootwijk, J., Joosse, J., 1988, Growth-controlling molluscan neurons produce the precursor of an insulin-related peptide, *Nature* 331:535-538.

Szego, C.M., and Roberts, S., 1953, Steroid action and interaction in uterine metabolism. *Recent Prog. Horm. Res.* 8:419-32.

Tabanelli, S., Tang, B., and Gurpide, E., 1992, In vitro decidualization of human endometrial stromal cells. *J. Steroid Biochem. Molec. Biol.* 42:337-344.

Tabibzadeh, S., 1991, Human endometrium: an active site of cytokine production and action. *Endocrine Rev.* 12, 272-290.

Waites, G.T., James, R.F.L., and Bell, S.C., 1988, Immunohistological localization of human endometrial secretory proteins "pregnancy-associated endometrial secretory 1-globulin" (1-PEG), an insulin-like growth factor binding protein, during the menstrual cycle, *J. Clin. Endocrinol. Metab.* 67:1100 -1104.

Wolf, M., Ingbar, S.H., and Moses, A.C., 1989, Thyroid hormones and growth hormone interact to regulate insulin-like growth factor-I messenger ribonucleic acid and circulating levels in the rat, *Endocrinology* 125:2905-2914.

Wood, T.L., Brown, A.L., Rechler, M.M., and Pintar, J.E., 1990, The expression pattern of an insulin-like growth factor (IGF)-binding protein gene is distinct from IGF-II in the midgestational rat embryo, *Mol. Endocrinol.* 4:1257-1263.

Yeh, J., Danehy, F.T., Osathanondh, R., Villa-Komaroff, L.. 1991, mRNAs for insulin-like growth factor-II (IGF-II) and variant IGF-II are co-expressed in human fetal ovary and uterus. *Mol. Cell. Endocrinol.* 80:75-82

INSULIN-LIKE GROWTH FACTOR 2 (*IGF2*) EXPRESSION AT THE EMBRYONIC/MATERNAL BOUNDARY

Tomas J. Ekström, Lars Holmgren, Anna Glaser and Rolf Ohlsson

Department of Drug Dependence Research
Karolinska Hospital, P.O. Box 60 500
S-104 01 Stockholm, Sweden

Introduction

IGF-II plays an important but not pivotal role during normal development in the mouse. By inactivating the endogenous *Igf2* gene, DeChiara et al could show that the absence of IGF-II resulted in perfectly healthy and symmetrical offspring although these were 40% smaller than normal littermates (DeChiara et al.1990). In humans, similar evidence is lacking although genetic data indicate that an overactive *IGF2* results in organ hyperplasia as manifested in the Beckwith-Wiedemann syndrome BWS(Junien, 1992). It is interesting in this context that the *IGF2* expression patterns during early human development show striking similarities with the organomegaly typical of the BWS (Hedborg et al.1993).

The *IGF2* is first activated in the trophoblastic lineage of both mouse and man (Ohlsson et al, 1989, Lee et al.1990). Of all embryonic and extra-embryonic cell types investigated, the highest levels of *IGF2* transcripts are found in proliferative cytotrophoblasts of early human placenta (Ohlsson et al, 1989a, Ohlsson et al.1989b) while both type 1 and 2 IGF-receptor (*IGFR*) mRNA is found more generally in cytotrophoblasts and mesenchymal stroma of placental villi (Hedborg et al.1993, Ohlsson et al.1989b). Since IGF-II has been shown to stimulate DNA synthesis in primary cultures of first trimester placenta (Ohlsson et al 1989b), it is reasonable to assume that IGF-II functions as a local-acting mitogen during the early phase of human prenatal development.

The IGF-II ligand can be complexed to a variety of IGF-binding protein (IGFBP) genes (Baxter, 1988). Such IGFBP-complexed IGF-II differs from the free IGF-II ligand in various ways. It has been suggested that IGFBPs protect against hypoglyceamia by preventing IGFs from binding to the insulin receptor, for example (Zapf et al, 1979). In addition, the formation of IGF-IGFBP complexes has been shown to prolong the half-life of the growth factor (9) while also serving as a protein-carrier in the plasma (Baxter & Martin, 1989a). So far six IGF-binding proteins (IGFBP) have been identified and characterized (Baxter, 1988). In this report we are concerned only with the IGFBP-1-3. IGFBP-1 was cloned and sequenced from cDNA libraries of the human HepG2 hepatoma cell line (Lee et al, 1988), human placenta (Brinkman et al, 1988) and human decidua (Brewer et al, 1988). The gene product is also abundant in amniotic fluid from which it was originally purified (Baxter et al, 1987). The IGFBP-2 structure was deduced from cDNA of human fetal liver (Binkert et al,

1989), rat BRL-3A cells (Brown et al, 1989) and rat adult liver (Margot et al, 1989). The affinity of IGFBP-1 is higher for IGF-I than for IGF-II while the opposite is seen for IGFBP-2. The IGFBP-3 clone was isolated from human liver cDNA (Wood et al, 1988). This protein binds IGF-I and IGF-II with equal affinities and is growth hormone-dependent (Baxter & Martin, 1989b).

IGF and the fetal-maternal boundary

The boundary between the placenta and the decidua limits the developing potentials of the embryo proper. In order to gain some insight to this issue, we have studied the distribution of active genes encoding IGF-l, IGF-II, the type 1 IGF-receptor and IGFBP-1, -2 and -3 using Northern blot and *in situ* hybridization analysis. The *IGF2* and the *IGF1R* genes are coexpressed in both the placenta and decidua (Glaser et al, 1992). In the Jeg-3 choriocarcinoma cell line, the levels of *IGF2* and *IGF1R* transcripts are similar to that found in normal cytotrophoblasts from term placenta (Glaser et al, 1992). As was mentioned above, the IGFBPs are expected to modulate the actions of the IGF-II ligand and its receptor. Therefore, the expression of IGFBP-1, -2 and -3 genes was investigated by Northern blot. (Table I, Glaser et al, 1992, Wang et al, 1988). Placenta and purified trophoblasts show high expression of the IGFBP-3 gene. IGFBP-1 transcripts, on the other hand, could be detected only at very low levels, a fact that may be due to decidual contamination since the decidua expresses high levels of IGFBP-1 mRNA. Other investigators have suggested that the binding protein that is primarily responsible for IGF-II modulation is IGFBP-2, since they

TABLE 1. Cell type-specific expression patterns [a]

	IGF1	IGF2	IGFBP1	IGFBP2	IGFBP3	IGF1R
Decidua						
Gland/surface epithelia.	+++	—	—	+	+++	+++
Mesenchymal stroma.	—	++	+++	+++	++	+/++
Placenta						
Extravillous cytotrophoblasts.	—[b]	+++	—	—	+++	+/++
Villous cytotrophoblasts.	—[b]	+	—	—	+	+/++
Syncytio trophoblasts	+++[b]	—	—	—	—	—
Mesenchymal stroma.	—	+	—	—/+	+	+/++

— = not detected
+++ = highest signal observed, indicates no relation in expression levels of different genes' expression.
[a]/ in situ hybridization analysis of first trimester decidua and placenta. For more details see Glaser et al 1992.
[b]/ see Wang et al, 1988.

found it to be expressed mainly at junctional surfaces between the fetal and maternal compartments in the rat and human (Zhou & Bondy, 1992).

More detailed information of the fine tuned spatial expression patterns can be obtained by *in situ* hybridization analysis. Using this technique, we have found that the high abundance of *IGF2* transcripts in cytotrophoblasts originates from the highly proliferative extravillous cytotrophoblasts, Table I, (Hedborg et al, 1993, Ohlsson et al, 1989a, Ohlsson et al, 1989b,Glaser et al, 1992), while the *IGFIR* transcripts are found more uniformly in cytotrophoblastic cells (Hedborg et al, ,Ohlsson et al, 1989b). The IGFBP-3 gene was also found to be expressed primarily in proliferative cytotrophoblasts together with *IGF2* (Glaser et al, 1992). This result shows an exception to the common finding that IGFBP-3 gene expression is primarily associated with more adult cell types (Baxter & Martin, 1989a). The seemingly coordinated spatial expression of the *IGF2* and IGFBP-3 genes prevails throughout placental development. Even at term, when the majority of the remaining cytotrophoblasts have invaded the decidua basalis of the maternal lining, the IGFBP-3 and *IGF2* genes are coexpressed in these cells, despite the fact that the decidua itself is expressing all three IGFBP genes (Glaser et al, 1992).

In the decidua, the *IGF2* expression is found primarily in a subset of mesenchymal stroma cells and is virtually absent in epithelial cells, (Table I, Glaser et al, 1992). This is in contrast to the *IGFI* gene which is expressed at detectable levels in the glandular and surface epithelia. While the *IGFI* receptor transcripts can be found in a subpopulation of stroma cells *IGFIR* is expressed at a higher level in the epithelial cells, (Table I, Glaser et al, 1992). It is therefore reasonable to assume that IGF-II plays a role primarily in the microenvironment within the mesenchymal stroma, while IGF-l plays a role in the proliferation of the epithelial cell layers. We conclude that IGF-II and IGF-l are expressed and are, in all likelihood, used in cell type-specific fashions within the decidua.

The potential modulation of the IGF ligand function was investigated by analysing the expression patterns of the IGFBP genes. The IGFBP-1 gene is expressed at high levels in a subset of mesenchymal stroma cells but not in epithelial cells. IGF-II function may therefore be regulated, either positively or negatively by IGFBP-1 in a paracrine manner in glandular epithelia but in a short-range paracrine or autocrine mode within the mesenchymal stroma. IGFBP-2 transcripts are evenly distributed in the mesenchymal stroma cells which may reflect a function as a general modulator of IGF-II action. IGFBP-3 gene expression is not restricted to the mesechymal stroma cells but can also be found in epithelial cell layers. Assuming that the spatial distribution of mRNA also reflects the pattern of protein production, as has been shown for the *IGF2* gene in the placenta (Ohlsson et al, 1989a), we suggest that the IGFBP-1 and -2 act in parallel to regulate the *IGF2* function in the decidua. Conversely, the IGFBP-3 might play a role in the autocrine/short-range paracrine action of IGF-l for the epithelial cell layers.

Promoter usage in the *IGF2* gene

With the knowledge of *IGF2* expression in placental and decidual cells at hand, we turned to more specific aspects of transcriptional regulation. Five different transcripts from at least four promoters (P1 - P4) are found to be involved in IGF-II production (vanDijk et al, 1991,). The P1 promoter (exon 1) appears to be active exclusively in the postnatal liver (Sussenbach et al, 1992) and choroid plexus/leptomeninges (Ohlsson et al, 1993a), whereas the P2, P3 and P4 promoters (exons 4, 5 and 6, respectively) are active during prenatal development and in some postnatal tissues. The usage of these promoters appears to be coordinated during human embryogenesis (Ohlsson et al, 1993a). We have investigated the specific promoter usage in the placenta and decidua in an effort to learn how the promoter usage reflects the growth condition of the tissues. By using exon specific RNA probes (Figure 1)

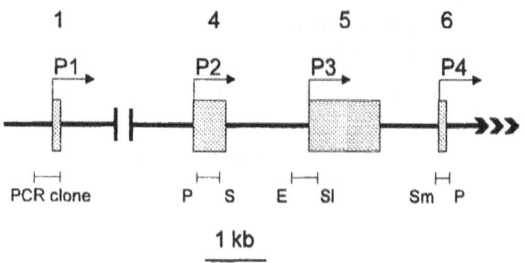

Figure 1. Exon organization of the human *IGF2* gene. Denoted are the clones used as probes in Northern blot hybridizations.

it has been possible to acurrately investigate the *IGF2* promoter usage with Northern blot hybridizations. The probes used for this purpose have been verified by RNase protection analysis (Ohlsson et al, 1993a).

Total cellular RNA extracted from embryo/fetus and the corresponding placenta and decidua were analyzed with these probes. Only the placenta and fetus express substantial levels of exon 4-containing transcripts indicating that the P2 promoter is not important for production of IGF-II in the decidua (Figure 2). The P3 promoter was used to a high extent in all three tissues while P4-driven transcription was much lower in the decidua. We conclude that the decidua (an adult tissue) uses at least two of the so-called fetal promoters. Furthermore, P3 is used in decidua to an equally high extent as in the embryonic and extraembryonic tissues.

IGF2 is parentally imprinted

The finding that the decidua is expressing *IGF2* and also apparently using more or less similar transcriptional machinery to the fetus, made us look into other possible differences of fetal and maternal regulation of this gene.

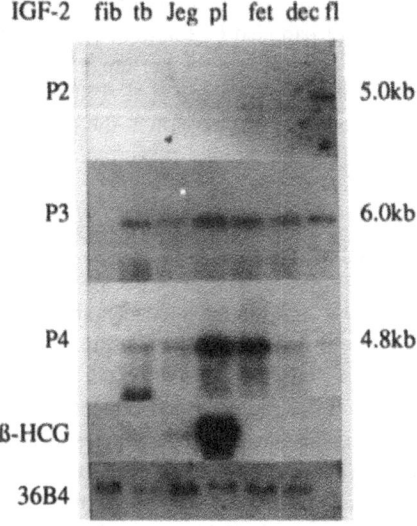

Figure 2. Northern blot analysis. Exon-specific probes were used to probe total RNA from: fibroblasts (fib), trophoblasts (tb), Jeg-3 cell line (Jeg), placenta (pl), fetus (fet), decidua (dec), fetal liver (fl). 36B4 was used as a control for RNA input. b-HCG was used to probe for trophoblast contamination within the decidua.

The phenomenon of parental imprinting means that the sex of the parent but not the sex of the offspring determines the transcriptional activity of the inherited alleles (Surani, 1991). An asymmetric allelic expression of imprinted growth factor and related genes could be one of the mechanisms regulating the interplay between fetal and maternal tissues. In the mouse, the *Igf2r* (Barlow et al, 1991) and *Igf2* (DeChiara et al, 1991) are expressed from the opposite parental alleles. We have shown recently that the *IGF2* gene in human is imprinted in the same way as in mouse, i.e. it is expressed from only the paternally-inherited allele (Ohlsson et al, 1993b). Preliminary results show that the decidua also expresses a parentally imprinted *IGF2* (Ekström et al, 1993). Since the active *IGF2* allele of the decidua was different from the active *IGF2* allele used in the placenta (Ekström et al, 1993), we can further conclude that *IGF2* expression within the mesenchymal stroma of the decidua does not constitute a contamination of *IGF2*-positive trophoblasts.

Parental imprinting and the decidua: Possible evolutionary implications

According to the theory by Haig & Graham (Haig & Graham, 1991), the conflict between maternal and paternal reproductive interests provides an evolutionary pressure for developing and maintaining parental imprinting. This means that the mother is striving to give birth to a litter small in size in order to increase the chances of her own as well as the offspring's survival, while the father's interest is to propagate his own genome via a small number of large and strong offspring. Haig predicted that genes important for the growth of the fetus (by providing a well-functioning placenta) should be expressed from the paternally-inherited allele only. In addition, genes antagonistic to this function should be expressed from only the maternally-inherited allele. The reciprocal imprinting of *Igf2* and *Igf2r* in the mouse seemed to support this notion since the IGF-II receptor is perceived as a scavenger for IGF-II (Haig & Graham, 1991). An inherent contradiction of the proposal is, however, the conflict between actively expressed genes inherited from the maternal grandfather and the father, respectively. It is therefore in the interest of the maternal grandfather to perpetuate his genes through his daughter which, according to the hypothesis would be in conflict with the interests of the father of the F2 generation. Our observation that human decidua is likely to express an imprinted *IGF2*, may be central to this issue. What is the role of the decidua in this scenario? Although this tissue in all likelihood adapts to and nurtures the developing conceptus, it will also provide the physical confines of the growing conceptus (Starkey, 1993). A clear role for the decidua in support of either parent is therefore not readily apparent. Were it to support primarily the maternal interests, the generality of the hypothesis would be challenged, since the interests of the maternal grandfather and the father would be antagonistic with respect to the embryo/fetus (here representing the F2 generation). Were the decidua to operate primarily in the interests of the growth of the fetus and hence be beneficial to the reproductive interests of the father, the genome of the maternal grandfather would cooperate with that of the father against the interests of the grandfather's own offspring (his daughter). Either of these possibilites is not easily reconcilable with the hypothesis of Haig and Graham.

Our experiments have shown that several components involved in the regulation of *IGF2* function are expressed at both sides of the fetal-maternal boundary. In addition, only one parental *IGF2* allele is expressed in both decidual and placental cells. The imprinted *IGF2* in the decidua is not reconcilable with the prevalent hypotheses of the roles of parental imprinting. The fetal-maternal boundary may constitute an evolutionary hot-spot which may require parental imprinting or provide new roles for the phenomenon of parental imprinting.

Abbreviations: **IGF-II**, insulin-like growth factor 2 polypeptide: *IGF2*, human insulin-like growth factor 2 gene: *Igf2*, murine insulin-like growth factor 2 gene: **IGFBP**, insulin-like growth factor binding protein.

Acknowledgements: Dr. Gail Adam is appreciated for reviewing the manuscript. This work was supported by the Swedish Cancer Society and the Swedish Natural Science Research Council to T.J.E. and R.O.

References

Barlow, D.P., Stöger, R., Herrmann, B.G., Saito, K., and Schweifer, N. 1991. *The mouse insulin-like growth factor type-2 receptor is imprinted and closely linked to the Tme locus.* Nature. **349**: 84-87.

Baxter, R.C., Martin, J.L., and Wood, M.H. 1987. *Two immunoreactive binding proteins for insulin-like growth factors in human amniotic fluid: relationship to fetal maturity.* J.Clin.Endocrinol.Metab. **65**: 423-431.

Baxter, R.C. 1988. *The insulin-like growth factors and their binding proteins.* Comp.Biochem.Physiol. **91B**: 229-235.

Baxter, R.C. and Martin, J.L. 1989a. *Binding proteins for the insulin-like growth factors: structure, regulation and function.* Progr.Growth Fact.Res. **1**: 49-68.

Baxter, R.C. and Martin, J.L. 1989b. *Structure of the Mr 140,000 growth hormone-dependent insulin-like growth factor binding protein complex: determination by reconstitution and affinity-labeling.* Proc.Natl.Acad.Sci.USA. **86**: 6898-6902.

Binkert, C., Landwehr, J., Mary, J.L., Schwander, J., and Heinrich, G. 1989. *Cloning, sequence analysis and expression of a cDNA encoding a novel insulin-like growth factor binding protein (IGFBP-2).* EMBO J. **8**: 2497-2502.

Brewer, M., Stetler, G., Squires, C., Thompson, R., Busby, W., and Clemmons, D. 1988. *Cloning, characterization, and expression of a human insulin-like growth factor binding protein.* Biochem. Biophys.Res.Commun. **152**: 1289-1297.

Brinkman, A., Groffen, C., Kortleve, D.J., van., G., Kessel, A., and Drop, S.L. 1988. *Isolation and characterization of a cDNA encoding the low molecular weight insulin-like growth factor binding protein (IBP-1).* EMBO J. **7**: 2417-2423.

Brown, A., *et al.* 1989. *Nucleotide sequence and expression of a cDNA clone encoding a fetal rat binding protein for insulin-like growth factors.* J.Biol.Chem. **264**: 5148-5154.

DeChiara, T.M., Efstratiadis, A., and Robertson, E.J. 1990. *A growth-deficiency phenotype in heterozygous mice carrying an insulin-like growth factor II gene disrupted by targeting.* Nature. **345**: 78-80.

DeChiara, T.M., Robertson, E.J., and Efstratiadis, A. 1991. *Parental imprinting of the mouse insulin-like growth factor II gene.* Cell. **64**: 849-859.

Ekström, T.J., Rutanen, E.-M., Nyström, A., Holmgren, L., Flam, F., and Ohlsson, R. 1993. *Mono-allelic expression of IGF2 at the human feto-maternal boundary: possible evolutionary implications.* Submitted.

Glaser, A., Luthman, H., Stern, I. and Ohlsson, R. 1992. *Spatial distribution of active genes implicated in the regulation of insulin-like growth factor stimulatory loops in human decidual and placental tissue of first-trimester pregnancy.* Mol.Reprod.Develop. **33**: 7-15.

Haig, D. and Graham, C. 1991. *Genomic imprinting and the strange case of the insulin-like growth factor II receptor.* Cell. **64**: 1045-1046.

Hedborg, F., Holmgren, L., Sandstedt, B., and Ohlsson, R. 1993. *Cell type-specific IGF2 expression during early human development: implications for the Beckwith-Wiedemann syndrome.* Submitted.

Junien, C. 1992. *Beckwith-Wiedemann Syndrome, tumorigenesis and imprinting.* Curr.Op.Genet.Develop. **2**: 103-123.

Lee, Y.-L., Hintz, R., James, P., Lee, P., Shively, J., and Powell, D. 1988. *Insulin-like growth factor (IGF) binding protein complementary deoxyribonucleic acid from human HepG2 hepatoma cells: Predicted protein sequence suggests an IGF binding domain different from those of the IGF-I and IGF-II receptors.* Mol.Endocrinol. **2**: 404-411.

Lee,J.E., Pintar,J. and Efstratiadis,A. 1990. *Pattern of the insulin-like growth factor II gene expression during early mouse embryogenesis.* Development **110**: 151-159.

Margot, J.B., Binkert, C., Mary, J.L., Landwehr, J., Heinrich, G., and Schwander, J. 1989. *A low molecular weight insulin-like growth factor binding protein from rat: cDNA cloning and tissue distribution of its   messenger RNA.* Mol.Endocrinol. **3**: 1053-1060.

Ohlsson,R., Larson,E., Nilsson,O., Wahlström,T. and Sundström,P. 1989a. *Blastocyst implantation precedes induction of insulin-like growth factor II gene expression in human trophoblasts.* Development **106**: 555-559.

Ohlsson, R., Holmgren, L., Glaser, A., Szpecht, A., and Pfeifer-Ohlsson, S. 1989b. *Insulin-like growth factor 2 and short-range stimulatory loops in control of human placental growth.* EMBO J. **8**: 1993-1999.

Ohlsson, R., Hedborg, F., Holmgren, L., Walsh, C., and Ekström, T.J. 1993a. Glaser et al, 1992. *Overlapping patterns of IGF2 and H19 expression during human development: biallelic IGF2 expression correlates with lack of H19 expression.* Submitted.

Ohlsson, R., Nyström, A., Pfeifer-Ohlsson, S., Töhönen, V., Hedborg, F., Schofield, P., Flam, F., and Ekström, T.J. 1993b. *IGF2 is parentally imprinted during human embryogenesis and in the Beckwith-Wiedemann syndrome.* Nature Genet. **4**: 94-97.

Starkey, P.M. 1993. *The decidua and factors controlling placentation.* in The Human Placenta, (Redman CWG., Sargent IL, Starkey PM, eds.) Blackwell Scientific Publications, Oxford, pp. 362-413.

Surani, M.A.H. 1991. *Genomic imprinting: developmental significance and molecular mechanisms.* Curr.Op.Genet.Develop. **1**: 241-246.

Sussenbach, J.S., Steenbergh, P.H., and Holthuizen, P. 1992. *Structure and expression of the human insulin-like growth factor genes.* Growth Regul. **2**: 1-9.

van Dijk, M.A., van Schaik, F.M.A., Bootsma, H.J., Holthuizen, P., and Sussenbach, J.S. 1991. *Initial characterization of the four promoters of the human insulin-like growth factor II gene.* Mol Cell Endocrinol. **81**: 81-94.

Wang, C-Y., Daimon, M., Shen, S-J., Engelmann, G.L., and Ilan, J. 1988. *Insulin-like growth factor-1 messenger ribonucleic acid in the developing human placenta and in term placenta of diabetics.* Mol.Endocrinol. **2**: 217-229.

Wood, W.I., et al. 1988. *Cloning and expression of the growth hormone-dependent insulin-like growth factor-binding protein.* Mol.Endocrinol. **2**: 1176-1185.

Zapf, J., Hauri, C., Waldvogel, M., and Froesch, E.R. 1986. *Acute metabolic effects and half-lives of intravenously administered insulin-like growth factors I and II in normal and hypophysectomized rats.* J.Clin.Invest. 77: 1768-1775.

Zapf, J., Schoenle, E., Jagers, E., Sand, I., and Froesch, E.R. 1979. *Inihibition of the action of non-suppressible insulin-like activity on isolated rat fat cells by binding to its carrier protein.* J.Clin.Invest. **63**: 1077-1084.

Zhou, J. and Bondy, C. 1992. *Insulin-like growth factor-II and its binding proteins in placental development.* Endocrinology. **131**: 1230-40.

COLONY STIMULATING FACTOR-1 IN THE FEMALE REPRODUCTIVE TRACT DURING THE PRE- AND PERI- IMPLANTATION PERIODS

Jeffrey W. Pollard

Department of Developmental and Molecular Biology
  and Obstetrics and Gynecology
Albert Einstein College of Medicine
1300 Morris Park Avenue
Bronx, NY 10461

INTRODUCTION

Viviparous reproduction in mammals demands co-ordination between maternal physiology and embryonic development. This includes the synchronization between pre-implantation embryonic development and the uterine preparation for implantation, the coordinated development of maternal and fetal cells during the formation of the placenta, the requirements of parturition and the development of the mammary gland in preparation for lactation. Paradoxically, in mice, the uterine preparation for implantation, pre-implantation embryonic development and decidualization can proceed independently of one another. Thus, fertilized oocytes can be cultured to the blastocyst-stage in defined medium, whereafter they can be sucessfully transferred to pseudopregnant recipients to complete embryonic development. Similarly, non-pregnant ovariectomized mice can be made pseudopregnant by administration of the appropriate regimens of steroid hormones and to be induced to undergo decidualization by an inert stimulus such as arachis oil applied into the uterine lumen in the absence of an embryo. However, embryonic development *in vitro* is slower than observed *in vivo* and such development is restricted to a few inbred mouse strains. In addition, in rodents it is essential to maintain synchrony between pre-implantation

Endocrinology of Embryo-Endometrium Interactions
Edited by S.R. Glasser *et al.*, Plenum Press, New York, 1994

embryonic development and the uterine preparation for pregnancy *in vivo* so that the embryo is ready to implant during the relatively small window of uterine receptivity (Brinster, 1963; Finn, 1980; Finn and Martin, 1972; Harlow and Quinn, 1982).

The uterine preparation for implantation in the mouse is strictly under the regulation of progesterone (P_4) and estrogen. Ovarian estrogen synthesized in a cyclical manner causes uterine cell proliferation which is largely restricted to the lumenal and glandular epithelium. In the mouse, if copulation occurs, the corporea lutea are maintained and P_4 is synthesized. P_4 suppresses uterine epithelial cell proliferation and prepares the stromal cells to respond to nidatory estrogen with a wave of cell division between day 3 and 4 of pregnancy. Although the kinetics of stromal cell proliferation have been well documented the responsive cell types have not been delineated. The nidatory estrogen is absolutely required for implantation and following the round of stromal cell proliferation, there is a relatively short receptive period during which the blastocyst can attach to the uterine epithelium and implant (Clarke and Sutherland, 1990; Finn, 1980; Finn and Martin, 1972; Martin and Finn, 1969; Martin and Finn, 1971). At implantation, luminal epithelial cells die by apoptosis (Welsh and Enders, 1993), and there is a transformation of the underlying stromal cells into decidual cells. Decidualization, which is strictly dependent on the presence of P_4, includes a remarkable proliferative response and differentiation of uterine stromal cells with the consequence that the decidua entirely surrounds the invading embryo (Das and Martin, 1978; Finn, 1980). The cellular composition of the uterus also changes at this time since some stromal cells undergo morphogenic transformation and other cells, such as macrophages, are excluded from the decidual area (Pollard *et al.*, 1991a; Tachi and Tachi, 1989).

Recently, the uterine epithelium has been shown to be a major site of cytokine synthesis. These cytokines are directed to cells both within the uterus and the embryo. Among the growth factors synthesized by the uterine epithelium are epidermal growth factor (EGF), transforming growth factor (TGF) α and TGF ß1, tumor necrosis factor α (TNF), leukemia inhibitory factor (LIF), colony stimulating factor-1 (CSF-1) and granulocyte-macrophage colony stimulating factor (GM-CSF). The synthesis of some of these have been directly shown to be influenced by estradiol-17ß (E_2) and P_4, whilst others are synthesized in a temporal sequence that suggests sex steroid hormone control. In the hormone-induced uterine cell proliferation and differentiation described above it is becoming apparent that such growth factors play a role in mediating the local responses to these hormones. In addition, it seems likely that growth factors are involved in the synchronization of these uterine events with pre-implantation embryonic development (Brigstock, 1991; Pollard, 1990, 1991; Pollard *et al.*, 1991b; Tabibzadeh, 1991b).

This review will focus on uterine growth factors that have both embryonic cells and uterine macrophages as their targets. We will particularly emphasize our work on CSF-1 and use this study as an example to illustrate the multifunctional roles that growth factors can have during pregnancy.

COLONY STIMULATING FACTOR-1 AND ITS RECEPTOR DURING PREGNANCY

Colony stimulating factor-1 was originally described as a circulating growth factor which regulates the proliferation, differentiation and viability of mononuclear phagocytes. CSF-1 is also chemotactic for macrophages and can stimulate the expression of a wide range of secreted products, including growth factors, biologically active lipids and proteases (Roth and Stanley, 1992). In the majority of cases, CSF-1 is synthesized as a homodimeric transmembrane precursor which is cleaved in secretory vesicles, following modification by O- and N-linked glycosylation and the addition of chondroitin sulphate side chains, to give a secreted molecule. Further extracellular processing results in a variety of molecular species, the smallest of which lacks the proteoglycan moiety but is nevertheless fully biologically active in *in vitro* and *in vivo* assays. In addition to the secreted form, an alternative mRNA species lacking exon 6 and therefore, the coding region for the secretory vesicle processing site, directs the synthesis of a biologically active transmembrane form (Kawasaki and Ladner, 1990; Price *et al.*, 1992; Roth and Stanley, 1992). CSF-1 appears to act exclusively through a single high affinity transmembrane tyrosine kinase receptor encoded by the c-*fms* proto-oncogene. This receptor, upon ligand binding, dimerizes, resulting in trans-phosphorylation of tyrosine residues located in the intracellular domain of the receptor and the activation of the intrinsic receptor-tyrosine kinase activity required for the propagation of the signal transduction pathway (Roth and Stanley, 1992; Sherr, 1990, 1991; Sherr and Stanley, 1990).

In the early 1970's the mouse pregnant uterus was demonstrated to be a rich source of hematopoietic colony stimulating activity (Bradley *et al.*, 1971; Rosendaal, 1975). Most of this activity is CSF-1 with the highest concentration of uterine CSF-1 being found late in pregnancy. However, five-fold elevated concentrations over non-pregnant uteri can be detected at the time of implantation (Bartocci *et al.*, 1986; Sanford *et al.*, 1992; Wood *et al.*, 1992). Uterine CSF-1 is largely, if not entirely, synthesized by the uterine epithelium. Both E_2 and P_4 given alone to ovariectomized mice stimulate uterine CSF-1 synthesis and these hormones act synergistically to further increase uterine CSF-1 concentrations to those concentrations found just before implantation in pregnant mice. Consistent with these observations, elevated levels of CSF-1 and CSF-1 mRNA can be detected at proestrus and in the uterine epithelium of pregnant mice as soon as P_4 is synthesized (Arceci *et al.*, 1989; Bartocci *et al.*, 1986; Hunt *et al.*, 1993; Pollard *et al.*, 1987; Sanford *et al.*, 1992; Wood *et al.*, 1992). CSF-1 mRNA however, has not been detected in mouse pre-implantation embryos (Arceci *et al.*, 1992a).

Our studies on uterine CSF-1 mRNA show that although the major species is an alternatively spliced 2.3 kb mRNA, usually found at lower relative abundance in other cell types, there are also a number of other CSF-1 mRNA synthesized. These include a 4.6 kb mRNA which has an identical coding region to the 2.3 kb mRNA but because of alternative splicing event lacks the untranslated exon 9 and contains the long exon 10 (Kawasaki and Ladner, 1990; Pollard et al., 1987). In addition, there are two mRNA species, a ~1.6 kb and an ~3.0 kb CSF-1 mRNA that both lack most of exon 6 because of the use of an alternative splice-acceptor sequence, but differ in the use of the untranslated exons 9 and 10 (Pampfer et al., 1991b). The 2.3 and 4.6 kb species encode the secreted, often proteoglycanated, forms of CSF-1, the 1.6 and 3.0-kb species encode the cell surface form of CSF-1 (Kawasaki and Ladner, 1990; Rettenmier and Roussel, 1988). All these species of CSF-1 mRNA have been detected in day 5 and day 14 pregnant uteri (F.C. Chuan and J.W. Pollard, unpublished data).

After implantation, in mice the uterine epithelium is destroyed within the implantation site, but CSF-1 is continually synthesized by the lumenal and glandular epithelium of the inter-implantation sites. In these cells, there is a progressive elevation of CSF-1 mRNA to reach a peak between day 14-16 of pregnancy (Arceci et al., 1989; Pollard et al., 1987; Regenstreif and Rossant, 1989). The elevation in CSF-1 concentration and increased CSF-1 mRNA levels can be reproduced in hormonally treated mice induced to undergo decidual stimulation by arachis oil. The further elevation of CSF-1 production over that of P_4/E_2 treated mice requires a decidual stimuli suggesting additional controls over and above the basic sex-steroid hormonal regulation (Pollard et al., 1987). Similarly, after day 15 of pregnancy there is a decline in CSF-1 mRNA concentrations, suggesting negative regulatory stimuli (Arceci et al., 1989).

In humans, a rather similar pattern of CSF-1 synthesis is observed during pregnancy, although the magnitude of change is not as great as that detected in the mouse (Daiter et al., 1992). CSF-1 mRNA and protein is found in non-pregnant uterine glandular epithelial cells (Pampfer et al., 1991b). The concentration of CSF-1 mRNA varies through the menstrual cycle, with high levels being detected at the proliferative and mid-to-late luteal phases (Pampfer et al., 1991b), matched by similar changes in protein concentrations (Kauma et al., 1991). This cyclical nature of the CSF-1 mRNA expression suggests that the hormonal regulation is similar between mice and humans. Elevated concentrations over non-pregnant levels of both CSF-1 and CSF-1 mRNA are detected in the endometrium from first trimester pregnancies (Daiter et al., 1992; Kauma et al., 1991), but these concentrations decline in the second and third trimester (Daiter et al., 1992). Immunohistochemical staining for CSF-1 shows the uterine glandular epithelium is a major source of this growth factor (Daiter et al., 1992). However, in contrast to the mouse, the placenta is a site of CSF-1 synthesis in the human. This placental synthesis may compensate for the decline in uterine epithelial synthesis caused by the relative compaction and destruction of epithelial cells during human pregnancy (Daiter et al., 1992; Kauma et al., 1991). Analysis of CSF-1 mRNA in the uterus and placenta shows a dominant ~4.0 kb mRNA whose concentration is elevated in first trimester endometrium (Daiter et al., 1992; Kauma et al., 1991). Unexpectedly, in uterine epithelium isolated from non-pregnant uteri approximately 40% of the CSF-1 mRNA is a ~3.0 kb mRNA which encodes a cell surface form of CSF-1 (Pampfer et al., 1991b).

All cells of the mononuclear phagocytic family have been demonstrated to express the CSF-1R. Human and mouse uterine macrophages are unlikely to be an exception to this rule and this has been shown for the Hofbauer cells of the human placenta (Jokhi *et al.*, 1993, and our unpublished results). In addition, in the female reproductive tract, CSF-1R mRNA is detected in non-macrophage cells of both maternal and fetal origin. Before fertilization CSF-1R mRNA is expressed in mouse oocytes and this mRNA is rapidly degraded following fertilization, to be resynthesized by zygotic transcription at the late two-cell stage. Thereafter, CSF-1 mRNA is expressed in the embryo through to implantation (Arceci *et al.*, 1992a). Implantation triggers the immediate expression of CSF-1R mRNA in cells of the sub-epithelial uterine stroma (Arceci *et al.*, 1992b). As decidualization progresses, expression is markedly enhanced in cells that form the primary decidual zone. Thereafter, expression declines in the decidual capularis but is maintained in the decidual basalis (Arceci *et al.*, 1989, 1992b). The decidual expression of the CSF-1R is unlikely to be in macrophages since these decidual cells do not display the diagnostic mononuclear phagocytic-specific cell surface marker, F4/80 (Pollard *et al.*, 1991a). Decidual cells in humans also express CSF-1R (Pampfer *et al.*, 1992). In fetal, extra-embryonic tissues of the mouse very high levels of expression of CSF-1R mRNA is observed in secondary polar and mural giant trophoblasts with substantial expression in the diploid cells of the ectoplacental cone. As the placenta matures, the giant trophoblastic layer has the highest expression with diminishing expression in the spongiotrophoblastic and labyrinthine layers, respectively (Arceci *et al.*, 1989, 1992b; Regenstreif and Rossant, 1989). A similar distribution occurs in humans with high levels of CSF-1R expression in syncytial and intermediate trophoblasts with somewhat lower levels in cytotrophoblasts (Jokhi *et al.*, 1993; Pampfer *et al.*, 1992). In humans, alternative use of an upstream exon results in use of a promoter in trophoblasts which is different from that used in monocytes, implying differential regulation (Pampfer *et al.*, 1992; Visvader and Verma, 1989).

OTHER UTERINE SYNTHESIZED CYTOKINES THAT AFFECT MACROPHAGE FUNCTION

GM-CSF can also promote the proliferation, differentiation and survival of macrophages (Metcalf, 1991). However, it is not the primary regulator of such functions for most macrophage populations *in vivo* (Wiktor-Jedrzejczak *et al.*, 1990, 1991). It acts through a hetero-dimeric receptor: one subunit (ß) of which is common to the receptors for IL-3 and IL-5 in humans and for the IL-5 receptor in mice. The other ligand binding subunit (α) is unique. This receptor is a member of the hematopoietin family of receptors (Boulay and Paul, 1992). GM-CSF is also synthesized by the uterine epithelium on day 1 of pregnancy, but only if copulation with intact males has occurred. Thus GM-CSF is synthesized in the uterus before the embryo has arrived and is presumably directed towards the uterine macrophage population. GM-CSF is also synthesized after implantation by decidual and uterine epithelial cells and by the placenta (Crainie *et al.*, 1990;

Kanzaki *et al.*, 1991; Robertson *et al.*, 1992; Robertson and Seamark, 1990). There have been no reports yet on GM-CSF receptor expression in mice, but receptors have been detected in human trophoblasts and placental membranes (Loke *et al.*, 1992; Scheffler *et al.*, 1990).

TNFα is a multifunctional cytokine that is both produced by and acts upon macrophages. It has a wide range of pleiotropic effects on many cell types. Two receptors, p60 and p80, of high and low affinity, belonging to the nerve growth factor-related receptor family, have been isolated for TNFα. These receptors are expressed in a wide range of cell types, although in varying proportions (Vassalli, 1992). TNFα is synthesized by oviductal and uterine cells during the pre-implantation period. In the mouse uterus, TNFα is synthesized by the uterine lumenal and glandular epithelium during the pre-implantation period (Hunt *et al.*, 1993). TNFα is also detected in rat uterine epithelial cells during the estrous cycle at varying concentrations and studies with ovariectomized rats and mice reconstituted with female sex steroid hormones indicate that the level of TNFα mRNA is endocrine regulated (De *et al.*, 1992; Hunt, 1993; Yelavarthi *et al.*, 1991). TNFα is also detected in the myometrium at low levels and in myometrial macrophages at high levels. After implantation, TNFα continues to be expressed by the uterine epithelium as well as in the primary decidua. In the fetal extra-embryonic tissues giant, polar trophoblasts are major sites of synthesis (Hunt *et al.*, 1993).

In human endometrium, TNFα expression is lacking in the basalis region, but stromal cells in the functionalis region are positive. During human pregnancy, in a manner reminiscent of the mouse, uterine epithelial cells synthesize TNFα as do large decidual cells. Syncytial trophoblasts also synthesize TNFα. TNFα receptors are detected in human trophoblasts (Chen *et al.*, 1991; Hunt, 1993; Tabibzadeh, 1991a).

CSF-1 AND EMBRYONIC DEVELOPMENT

In both humans and mice, the CSF-1 receptor is expressed, in addition to macrophages, in cells of the trophectodermal lineage and the maternal decidual. These data suggest paracrine roles for uterine synthesized CSF-1 in the development and differentiation of those cells that will ultimately constitute the placenta. In addition, CSF-1R but not CSF-1 mRNA has been detected in mouse pre-implantation embryos. CSF-1 is synthesized in the oviduct and at increasing levels by the uterus from day 3 of pregnancy, coincident with the blastocyst entering the uterus. This indicates that the pre-embryo is continuously exposed to CSF-1.

To explore the role of CSF-1 in development we have used the osteopetrotic (*op/op*) mouse mutant which is completely deficient in CSF-1 owing to a recessive, inactivating mutation in the CSF-1 gene (Wiktor-Jedrzejczak *et al.*, 1990; Yoshida *et al.*, 1990). These *op/op* mice have very compromised fertility, especially apparent in homozygous mutant matings. Consistent with the complex patterns of CSF-1R expression, deficits in fertility in the *op/op* mice are manifest at all stages of pregnancy. Firstly, the frequency of successful matings is reduced: of those female *op/op* mice that mate, only approximately fifty percent become pregnant. Secondly, of those mice that do become pregnant, there are fewer implantations and greater losses are sustained throughout gestation. Thirdly, greater than 90% of *op/op* mothers cannot feed their young, although the young can survive if transferred to normal foster mothers. These data confirm an important role for CSF-1 during pregnancy. However, despite the high level of expression of the CSF-1R in cells that will ultimately constitute the placenta, placental weights and the placental expression of A4311 (a spongiotrophoblast marker), c-*kit* and CSF-1R mRNAs are the same in *op/op* and +/*op* mothers (Pollard *et al.*, 1991a). This indicates that, although CSF-1 can stimulate proliferation of placental cells in culture (Athanassakis *et al.*, 1987), it cannot be the major mitogen for trophoblasts *in vivo*.

Reconstitution experiments of *op/op* mice with CSF-1 have shown that CSF-1 cannot restore fertility if it is administered beginning at the time of mating and continuing through pregnancy (Wiktor-Jedrzejczak *et al.*, 1991). Indeed, there are reports that CSF-1 administered during the first five days of pregnancy is abortifacient, suggesting that inappropriate doses of CSF-1 may be harmful (Tartakovsky *et al.*, 1991). However, CSF-1 administered from birth to *op/op* females has no embryonic toxicity and significantly improves the success of mating. This improvement is only observed if the male used in matings has CSF-1 (+/*op*) or is an *op/op* treated with CSF-1 (Pollard, J.W., Chisholm, O., Dominguez, M.G. and Stanley, E.R., unpublished results). This data, coupled with the observation that heterozygous males have greater fertility with *op/op* females than *op/op* males despite the latter exhibiting almost normal fertility with +/*op* females, indicates that *op/op* female fertility can be restored by mating with a CSF-1-stimulated male. This suggests male seminal fluid supplies CSF-1 or CSF-1-induced uterine factors which compensate for maternally produced CSF-1 (Pollard *et al.*, 1991a). Such a compensating uterine factor could be GM-CSF, whose uterine epithelial synthesis requires copulation with an intact male (Robertson and Seamark, 1990). The litter sizes of *op/op* mothers however, are not significantly increased by CSF-1 given from birth (Pollard, J.W., Chisholm, O., Dominguez, M.G. and Stanley, E.R., unpublished results), consistent with the failure of the systemically administered growth factor to mimic the extremely high local concentration of uterine CSF-1.

Despite these observations with *op/op* mice which have established an important role for CSF-1 in pregnancy, the functions for CSF-1 still have to be elucidated. Similarly, roles for TNFα and GM-CSF remain to be defined. All three growth factors interact with trophoblasts and have been shown to modestly stimulate the production of human chorionic gonadotrophins by human tropho-blasts in culture (Li *et al.*, 1992; Saito *et al.*, 1991). But whether the stimulation of hormone production is specific or due to a general stimulation of metabolism, maintenance of cell viability or induction of differentiation, remains to be determined. GM-CSF stimulates DNA synthesis in mouse and human tropho-blasts and CSF-1 does the same in mixed placental cell cultures (Athanassakis *et* al., 1987; Armstrong and Chaouat, 1989; Loke *et al.*, 1992). TNFα inhibits the proliferation of rat trophoblastic cell lines (Hunt, 1993) and inhibits the development of 2-cell embryos to the blastocyst stage (Hill *et al.*, 1987). GM-CSF inhibits the growth of 2-cell embryos in culture but stimulates post-morula stage embryonic development (Hill *et al.*, 1987; Robertson *et al.*, 1990). CSF-1 enhances embryonic development from the 2-cell to the blastocyst stage and stimulates trophoblastic outgrowth in an implantation model (Haimovici and Anderson, 1993; Pampfer *et al.*, 1991a). Such data suggests an interplay between these three cytokines in regulating pre-implantation embryonic development *in utero* and in trophoblast function after implantation.

CYTOKINE REGULATION OF UTERINE MACROPHAGE POPULATIONS

Macrophages represent a major uterine cell type accounting for 10-15% of cells in the uterine stroma of non-pregnant mice. In the virgin mouse they are distributed throughout the stroma (De *et al.*, 1991). At day 1 of pregnancy the numbers increase 2-fold and they are recruited adjacent to the uterine epithelium. After day 1 they disperse again through the endometrium. Just before implantation, the numbers of stromal macrophages increases about 4-fold (De *et al.*, 1991). In the myometrium, macrophages are in the connective tissue and closely associated with the serous epithelium. After implantation, the tissue distribution is relatively unchanged in the myometrium and non-decidualized endometrial stroma, but macrophages are almost completely excluded from the decidualized stroma (De and Wood, 1991; Pollard *et al.*, 1991a; Tachi and Tachi, 1989). In the metrial gland, a unique structure in rodents formed at the mesometrial triangle, macrophages represent 10-20% of the cells. Overall, macrophage numbers continue to increase until day 9 of pregnancy (De and Wood, 1991). In human endometrium, macrophages are also abundant and, unlike the mouse, macrophages also represent a significant portion of the decidua, being adjacent to the invading trophoblasts (Hunt and Pollard, 1992).

The relative absence of uterine macrophages in the osteopetrotic mouse, has demonstrated that this cell population is primarily regulated by CSF-1 (Pollard *et al.*, 1991a). Systemic restoration of circulating CSF-1 concentrations from day 3 of life in homozygous mutant (*op/op*) mice corrects the deficiency in uterine macrophages (Pollard, J.W., Chisholm, O., Dominguez, M.G. and Stanley, E.R., unpublished observations). In normal mice, during the pre-implantation period of pregnancy, there is a temporal relationship between elevated CSF-1 concentrations and increases in uterine macrophage numbers (De *et al.*, 1993). These data suggest that uterine-synthesized CSF-1 may be the primary regulator of uterine macrophage number during the pre-implantation period. Consistent with this suggestion is the increase in uterine macrophage number observed in ovariectomized mice treated with intra-luminal uterine injections of CSF-1 (Wood *et al.*, 1992). However, at day 1 of pregnancy, macrophages are aligned with the epithelium whilst, at day 5 they are distributed throughout the stroma despite the observation that the uterine epithelium is the major source of CSF-1 at both times. This suggests that both the number and location of uterine macrophages may be also be influenced by GM-CSF whose uterine epithelial synthesis is elevated at day 1 of pregnancy.

CSF-1 regulates macrophage cell proliferation and viability and it is also chemoattractive for monocytes. The increased macrophage number in the uterus during pregnancy could be due to one or all of these effects. It is tempting to speculate that part of the stromal cell proliferation in response to nidatory estrogen could be through the CSF-1 induced proliferation of mononuclear phagocytes. CSF-1 also regulates macrophage cell spreading *in vitro* (Boocock *et al.*, 1989) and this is also clearly a function for CSF-1 in the uterus since in the *op/op* mouse those macrophages that are present are not highly dendritic. Although the consequences of this failure of macrophages spreading is unknown, it is interesting to note that adherence affects macrophage gene expression (Haskill *et al.*, 1988).

In the post-implantation *op/op* mouse uterus, abnormally rounded F4/80 positive cells appear in the metrial gland at day 7-8. By day 10, however, these cells have disappeared and there continues to be a paucity of macrophages in the mutant mice (Pollard *et al.*, 1991a). These data indicate that during pregnancy, CSF-1 is the primary regulator of the uterine macrophage population, influencing their location and viability, but that at the peak of decidualization, another growth factor is synthesized that transiently recruits macrophages in the uterus. Such a growth factor could be GM-CSF or TNFα whose expression, at least in the latter case, is unaffected by the absence of CSF-1 (Hunt *et al.*, 1993a). During the latter half of pregnancy, CSF-1 concentrations continue to logarithmically increase without concurrent increases in macrophage number (Bartocci *et al.*, 1986; De *et al.*, 1993), suggesting that this dramatic elevation in CSF-1 concentration is concerned with the regulation of placental function through CSF-1 receptors expressed on decidual cells and trophoblasts.

There have been many suggestions for functions of macrophages during pregnancy ranging from their producing immunosuppressive molecules which protect the fetal allograft from rejection, to their producing cytokines that positively influence both embryonic and uterine development, to immunological roles protecting against infection, to roles in uterine remodelling (Hunt and Pollard, 1992). However, important functions for uterine macrophages *in vivo* have not been definitively shown. The relative failure of pregnancy in *op/op* mice may, in part, be determined by the deficiency of uterine macrophages.

CONCLUSIONS

This paper is concerned principally with CSF-1, but also with GM-CSF and TNF-α. All these growth factors influence not only macrophage function and development, but also affect embryonic cells. For all three growth factors, either direct or circumstantial evidence suggests that their uterine epithelial synthesis is modulated by the female sex steroid hormones, suggesting that they are local mediators of sex steroid hormone action. Conclusive evidence, derived from the *op/op* mouse demonstrates that this is the case for CSF-1.

Figure 1 shows a scheme for the multifunctional action of these cytokines during the pre-implantation period. The elevation of uterine concentrations of these cytokines in response to the P_4 and E_2 synthesized after copulation results in increased numbers and activity of stromal macrophages. These macrophages can synthesize cytokines (e.g. TNFα, TGFα, and IFNα) in response to these growth factors as well as other biologically potent molecules such as prostaglandins and plasminogen activator. Such molecules will, in turn, be targeted to other uterine cells, influencing their function and development. In addition, the embryo is exposed to all these growth factors. We have suggested that uterine cytokines synthesized in carefully orchestrated patterns in response to the changing hormonal milieu may be involved in regulating the synchrony between embryonic development and the uterine preparation for implantation. The increased failure over wild-type mice of plugged *op/op* females to progress to pregnancy could be due to the relative failure of this synchrony. However, the exact defect in *op/op* mice fertility has yet to be determined.

Undoubtedly, there is a complex interplay between cytokines within the uterus. Null mutations in cytokine genes will illuminate the roles for these

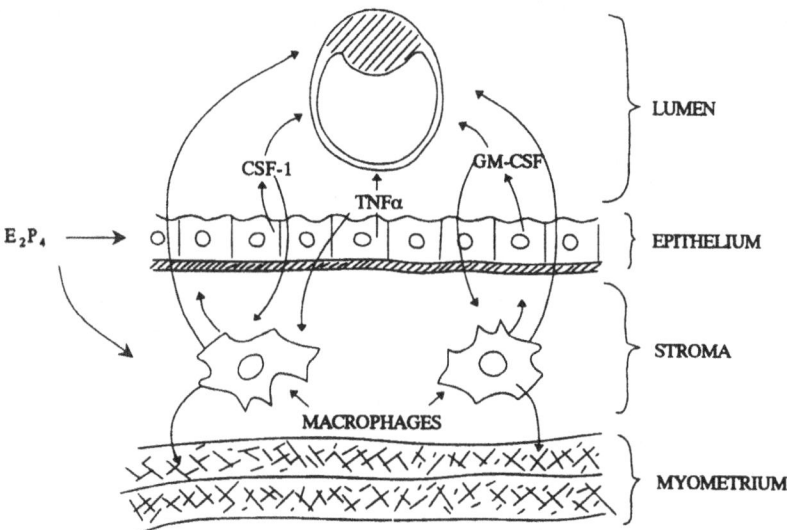

Fig. 1. Schematic diagram of the proposed actions of macrophage-reactive cytokines in the uterus during the pre-implantation period. CSF-1, TNFα and GM-CSF synthesized in the uterine epithelium under the regulation of E_2 and P_4 stimulate stromal macrophages to synthesize cytokines that influence both uterine and embryonic cellular functions. In addition, CSF-1, GM-CSF and TNFα interact directly with the developing blastocyst co-ordinating its development with the uterine receptivity for implantation.

molecules in reproduction. For example, a major role for uterine synthesized LIF in regulating implantation has been demonstrated by the creation of a null mutation by homologous recombination and ES cell technology (Stewart *et al.*, 1992, see this volume). However, a critical role for TGFα has been made unlikely by the observations of complete fertility in TGFα-lacking null mice (Luetteke *et al.*, 1993). In this fashion the availability of mutant mice will allow the unraveling of the cytokine interactions operating during both the pre- and peri-implantation periods.

ACKNOWLEDGEMENTS

This manuscript was written whilst the author's work was supported by grants from the NIH (HD27322) and ACS (DB-28). J.W.P. is a Monique Weill-Caulier Scholar. I would also like to thank all my many colleagues who have contributed to this research.

REFERENCES

Athanassakis, I., Bleackley, C., Paetkau, V., Guilbert, L., Barr, P.J. and Wegmann, T.G., The immunostimulatory effect of T cells and T-cell lymphokines on murine fetally derived placental cells, *J. Immunol.* 138:37-44 (1987).

Arceci, R.J., Shanahan, F., Stanley, E.R. and Pollard, J.W., Temporal expression and location of colony-stimulating factor 1 (CSF-1) and its receptor in the female reproductive tract are consistent with CSF-1-regulated placental development, *Proc. Natl. Acad. Sci. USA* 86:8818-8822 (1989).

Arceci, R.J., Pampfer, S. and Pollard, J.W., Expression patterns of CSF-1/c-*fms* and SF/c-*kit* mRNA during early pre-implantation mouse development, *Dev. Biol.* 151:1-8 (1992a).

Arceci, R.J., Pampfer, S. and Pollard, J.W., Role and expression of colony-stimulating factor-1 and steel factor receptors and their ligands during pregnancy in the mouse, *Reprod. Fertil. Dev.* 4:619-632 (1992b).

Armstrong, D.T. and Chaouat, G., Effects of lymphokines and immune complexes on murine placental cell growth in vitro, *Biol. Reprod.* 40:466-474 (1989).

Bartocci, A., Pollard, J.W. and Stanley, E.R., Regulation of colony-stimulating factor 1 during pregnancy, *J. Exp. Med.* 164:956-961 (1986).

Boocock, C.A., Jones, G.E., Stanley, E.R. and Pollard, J.W., Colony stimulating factor-1 induces rapid behavioral responses in the mouse macrophage cell-line, BAC1:2F5, *J. Cell Sci.* 93:447-456 (1989).

Boulay, J-L. and Paul, W.E., The interleukin-4-related lymphokines and their binding to hematopoietin receptors, *J. Biol. Chem.* 267:20525-20528 (1992).

Bradley, T.R., Stanley, E.R. and Sumner, M.A., Factors from mouse tissues stimulating colony growth of mouse bone marrow cells *in vitro*, *Aust. J. Exp. Biol. Med. Sci.* 49:595-603 (1971).

Brigstock, D.R., Growth factors in the uterus: steroidal regulation and biological actions, in: Bailliere's Clinical Endocrinology and Metabolism, Bailliere Tindall, (1991), pp. 791-808.

Brinster, R.L., A method for *in vitro* cultivation of mouse ova from two-cell to blastocyst, *Exp. Cell Res.* 32:205-208 (1963).

Chen, H-L., Yang, Y., Hu, X-L., Yelavarthi, K.K., Fishback, J.L. and Hunt, J.S., Tumor necrosis factor-alpha mRNA and protein are present in human placental and uterine cells at early and late stages of gestation, *Am. J. Pathol.* 139:327-335 (1991).

Clarke, C.L. and Sutherland, R.L., Progestin regulation of cellular proliferation, *Endocrine Reviews* 11:266-301 (1990).

Crainie, M., Guilbert, L.J. and Wegmann, T., Expression of novel cytokine transcripts in murine placenta, *Biol. Reprod.* 43:999-1005 (1990).

Daiter, E., Pampfer, S., Yeung, Y.G., Barad, D., Stanley, E.R. and Pollard, J.W., Expression of colony stimulating factor-1 in the human uterus and placenta, *J. Clin. Endocr. Metab.* 74:850-858 (1992).

Das, R.M. and Martin, L., Uterine DNA synthesis and cell proliferation during early decidualization induced by oil in mice, *J. Reprod. Fert.* 53:125-128 (1978).

De, M., Choudhuri, R. and Wood, G.W., Determination of the number and distribution of macrophages, lymphocytes, and granulocytes in the mouse uterus from mating through implantation, *J. Leuk. Biol.* 50:252-262 (1991).

De, M. and Wood, G.W., Analysis of the number and distribution of macrophages, lymphocytes, and granulocytes in the mouse uterus from implantation through parturition, *J. Leuk. Biol.* 50:381-392 (1991).

De, M., Sanford, T.R. and Wood, G.W., Interleukin-1, interleukin-6, and tumor necrosis factor α are produced in the mouse uterus during the estrous cycle and are induced by estrogen and progesterone, *Dev. Biol.* 151:297-305 (1992).

De, M., Sanford, T. and Wood, G.W., Relationship between macrophage colony-stimulating factor production by uterine epithelial cells and accumulation and distribution of macrophages in the uterus of pregnant mice, *J. Leuk. Biol.* 53:240-248 (1993).

Finn, C.A., The endometrium during implantation, in: The endometrium, F.A. Kimball, ed., Spectrum Publications, Inc., London (1980), pp. 43-53.

Finn, C.A. and Martin, L., Endocrine control of the timing of endometerial sensitivity to a decidual stimulus, *Biol. Reprod.* 7:82-86 (1972).

Haimovici, F. and Anderson, D.J., Effects of growth factors and growth factor-extracellular matrix interactions on mouse trophoblast outgrowth in vitro, *Biol. Reprod.* 49:124-130 (1993).

Harlow, G.M. and Quinn, P., Development of preimplantation mouse embryos *in vivo* and *in vitro*, *Aust. J. Biol. Sci.* 35:187-193 (1982).

Haskill, S., Johnson, C., Eierman, D., Becker, S. and Warren, K., Adherence induces selective mRNA expression of monocyte mediators and proto-oncogenes, *J. Immunol.* 140:1690-1694 (1988).

Hill, J.A., Haimovici, F. and Anderson, D.J., Products of activated lymphocytes and macrophages inhibit mouse embryo development *in vitro*, *J. Immunol.* 139:2250-2259 (1987).

Hunt, J.S. and Pollard, J.W., Macrophages in the uterus and placenta, *Curr. Prog. Microbiol. Immunol.* 181:39-63 (1992).

Hunt, J.S., Endocrine regulation of tumor necrosis factor-α, *Reprod. Fertil. Dev.* (1993). (In Press)

Hunt, J.S., Chen, H-L., Hu, X.L. and Pollard, J.W., Normal distribution of tumor necrosis factor-α messenger ribonucleic acid and protein in the uteri, placentas and embryos of osteopetrotic (*op/op*) mice lacking colony stimulating factor-1, *Biol. Reprod.* (1993). (In Press)

Jokhi, P.P., Chumbly, G., King, A., Gardner, L. and Loke, Y.W., Expression of the colony stimulating factor-1 receptor (c-fms product) by cells at the human uteroplacental interface, *Laboratory Investigation* 68:308-320 (1993).

Kanzaki, H., Crainie, M., Lin, H., Yui, J., Guilbert, L.J., Mori, T. and Wegmann, T.G., The *in situ* expression of granulocyte-macrophage colony-stimulating factor (GM-CSF) mRNA at the maternal-fetal interface, *Growth Factors* 5:69-74 (1991).

Kauma, S.W., Aukerman, S.L., Eierman, D. and Turner, T., Colony-stimulating factor-1 and c-*fms* expression in human endometrial tissues and placenta during the menstrual cycle and early pregnancy, *J. Clin. Endocr. Metab.* 73:746-751 (1991).

Kawasaki, E.S. and Ladner, M.B., Molecular biology of macrophage colony stimulating factor, in: Colony Stimulating factors: Molecular and Cellular Biology, T.M. Dexter, J.M. Garland and N.G. Testa, eds., Marcel Dekker, Inc., New York (1990), pp. 155-176.

Li, Y., Matsuzaki, N., Masuhiro, K., Kameda, T., Taniguchi, T., Saji, F., Yone, K. and Tanizawa, O., Trophoblast-derived tumor necrosis factor-α induces release of human chorionic gonadotrophin using interleukin-6 (IL-6) and IL-6-receptor-dependent system in the normal human trophoblasts, *J. Clin. Endocr. Metab.* 74:184-191 (1992).

Loke, Y.W., King, A., Gardner, L. and Carter, N.P., Evidence for the expression of granulocyte-macrophage colony-stimulating factor receptors by human first trimester extravillous trophoblast and its response to this cytokine, *J. Reprod. Immunol.* 22:33-45 (1992).

Luetteke, N.C., Qiu, T.H., Peiffer, R.L., Oliver, P., Smithies, O. and Lee, D.C., TGFα deficiency results in hair follicle and eye abnormalities in targeted and waved-1 mice, *Cell* 73:263-278 (1993).

Martin, L. and Finn, C.A., Hormone secretion during early pregnancy in the mouse, *J. Endocr.* 45:57-65 (1969).

Martin, L. and Finn, C.A., Oestrogen-gestagen interactions on mitosis in target tissues, in: Basic Actions of Sex Steroids on Target Organs, S.Karger, Basel, Switzerland (1971), pp. 172-188.

Metcalf, D., Control of granulocytes and macrophages: molecular, cellular and clinical aspects, *Science* 254:529-533 (1991).

Pampfer, S., Arceci, R.J. and Pollard, J.W., Role of colony stimulating factor-1 (CSF-1) and other lympho-hematopoietic growth factors in mouse preimplantation development, *BioEssays* 13:535-540 (1991a).

Pampfer, S., Tabibzadeh, S., Chuan, F.C. and Pollard, J.W., Expression of colony stimulating factor-1 (CSF-1) mRNA in human endometrial glands during the menstrual cycle: Molecular cloning of a novel transcript that predicts a cell surface form of CSF-1, *Mol Endo.* 5:1931-1938 (1991b).

Pampfer, S., Daiter, E., Barad, D. and Pollard, J.W., Expression of the colony-stimulating factor-1 receptor (c-*fms* proto-oncogene product) in the human uterus and placenta, *Biol. Reprod.* 46:48-57 (1992).

Pollard, J.W., Bartocci, A., Arceci, R., Orlofsky, A., Ladner, M.B. and Stanley, E.R., Apparent role of the macrophage growth factor, CSF-1, in placental development, *Nature* 330:484-486 (1987).

Pollard, J.W., Regulation of polypeptide growth factor synthesis and growth factor-related gene expression in the rat and mouse uterus before and after implantation, *J. Reprod. Fert.* 88:721-731 (1990).

Pollard, J.W., Lymphohematopoietic cytokines in the female reproductive tract, *Current Opinion in Immunology* 3:772-777 (1991).

Pollard, J.W., Hunt, J.W., Wiktor-Jedrzejczak, W. and Stanley, E.R., A pregnancy defect in the osteopetrotic (*op/op*) mouse demonstrates the requirement for CSF-1 in female fertility, *Dev. Biol.* 148:273-283 (1991a).

Pollard, J.W., Pampfer, S. and Arceci, R.J., Class III tyrosine kinase receptors at the maternal-fetal interface, in: Cellular and Molecular Biology of the Materno-Fetal Relationship, G. Chauoat and J. Mowbray, eds., INSERM/John Libbey Eurotext Ltd., (1991b), pp. 81-89.

Price, L.K.H., Choi, H.U., Rosenberg, L. and Stanley, E.R., The predominant form of secreted colony stimulating factor-1 is a proteoglycan, *J. Biol. Chem.* 267:2190-2199 (1992).

Regenstreif, L.J. and Rossant, J., Expression of the c-*fms* proto-oncogene and of the cytokine, CSF-1, during mouse embryogenesis, *Dev. Biol.* 133:284-294 (1989).

Rettenmier, C.W. and Roussel, M.F., Biosynthesis of macrophage colony-stimulating factor (CSF-1): Differential processing of CSF-1 precursors suggests alternative mechanisms for stimulating CSF-1 receptors, *Mol. Cell. Biol.* 8:5026-5034 (1988).

Robertson, S.A., Lavranos, T.C. and Seamark, R.F., *In vitro* models of the maternal-fetal interface, in: The Molecular and Cellular Immunobiology of the Maternal-Fetal Interface, T.G. Wegmann, E. Nisbet-Brown and T.J. Gill III, eds., Oxford University Press, New York (1990), pp. 191-206.

Robertson, S.A. and Seamark, R.F., Granulocyte macrophage colony stimulating factor (GM-CSF) in the murine reproductive tract: stimulation by seminal factors, *Reprod. Fertil. Dev.* 2:359-368 (1990).

Robertson, S.A., Mayrhofer, G. and Seamark, R.F., Uterine epithelial cells synthesize granulocyte-macrophage colony-stimulating factor and inter-leukin-6 in pregnant and nonpregnant mice, *Biol. Reprod.* 46:1069-1079 (1992).

Rosendaal, M., Colony-stimulating factor (CSF) in the uterus of the pregnant mouse, *J. Cell Sci.* 19:411-423 (1975).

Roth, P. and Stanley, E.R., The biology of CSF-1 and its receptor, *Curr. Topics Microbiol. Immunol.* 181:141-147 (1992).

Saito, S., Saito, M., Motoyoshi, K. and Ichijo, M., Enhanced effects of human macrophage colony-stimulating factor on the secretion of human chorionic gonadotrophin by human chorionic villous cells and tPA30-1 cells, *Biochem. Biophys. Res. Comm.* 178:1099-1104 (1991).

Sanford, T.R., De, M. and Wood, G.W., Expression of colony-stimulating factors and inflammatory cytokines in the uterus of CD1 mice during days 1 to 3 of pregnancy, *J. Reprod. Fert.* 94:213-220 (1992).

Scheffler, J.E., Fleissner, L.C., Seelig, G.F., Nagabhushan, T.L. and Trotta, P.P., Characterization of the human granulocyte-macrophage colony-stimulating factor receptor from placenta, in: Hematopoiesis, Alan R. Liss, Inc., NY (1990), pp. 107-115.

Sherr, C.J., Colony stimulating factor-1 receptor, *Blood* 75:1-12 (1990).

Sherr, C.J. and Stanley, E.R., Colony-stimulating factor 1/macrophage colony-stimulating-factor, in: Handbook of Experimental Pharmacology: Peptide Growth Factors and their Receptors, M.B. Sporn and A.B. Roberts, eds., Springer-Verlag, Heidelberg, New York (1990), pp. 667-698.

Sherr, C.J., Mitogenic response to colony-stimulating factor 1, *Trends Genet.* 7:398-402 (1991).

Stewart, C.L., Kaspar, P., Brunet, L.J., Bhatt, H., Gadi, I., Köntgen, F. and Abbondanzo, S.J., Blastocyst implantation depends on maternal expression of leukaemia inhibitory factor, *Nature* 359:76-79 (1992).

Tabibzadeh, S., Ubiquitous expression of TNF-α/cachectin immunoreactivity in human endometrium, *Am. J. Rep. Immunol.* 26:1-4 (1991a).

Tabibzadeh, S., Human endometrium: An active site of cytokine production and action, *Endocrine Reviews* 12:272-290 (1991b).

Tachi, C. and Tachi, S., Role of macrophages in the maternal recognition of pregnancy, *J. Reprod. Fert.* 37:63-68 (1989).

Tartakovsky, B., Goldstein, O. and Brosh, N., Colony-stimulating factor-1 blocks early pregnancy in mice, *Biol. Reprod.* 44:906-912 (1991).

Vassalli, P., The pathophysiology of tumor necrosis factors, *Ann. Rev. Immunol.* 10:411-452 (1992).

Visvader, J. and Verma, I.M., Differential transcription of exon 1 of the human c-*fms* gene in placental trophoblasts and monocytes, *Mol. Cell. Biol.* 9:1336-1341 (1989).

Welsh, A.O. and Enders, A.C., Chorioallantoic placenta formation in the rat. III. Granulated cells invade the uterine luminal epithelium at the time of epithelial cell death, *Biol. Reprod.* 49:38-57 (1993).

Wiktor-Jedrzejczak, W., Bartocci, A., Ferrante, A.W.,Jr., Ahmed-Ansari, A., Sell, K.W., Pollard, J.W. and Stanley, E.R., Total absence of colony-stimulating factor 1 in the macrophage-deficient osteopetrotic (*op/op*) mouse, *Proc. Natl. Acad. Sci. USA* 87:4828-4832 (1990).

Wiktor-Jedrzejczak, W., Urbanowska, E., Aukerman, S.L., Pollard, J.W., Stanley, E.R., Ralph, P., Ansari, A.A., Sell, K.W. and Szperl, M., Correction by CSF-1 of defects in the osteopetrotic *op/op* mouse suggests local, developmental, and humoral requirements for this growth factor, *Exp. Hematol.* 19:1049-1054 (1991).

Wood, G.W., De, M., Sanford, T. and Choudhuri, R., Macrophage colony stimulating factor controls macrophage recruitment to the cycling mouse uterus, *Dev. Biol.* 152:336-343 (1992).

Yelavarthi, K.K., Chen, H-L., Yang, Y., Cowley Jr., B.D., Fishback, J.L. and Hunt, J.S., Tumor necrosis factor-α mRNA and protein in rat uterine and placental cells, *J. Immunol.* 146:3840-3848 (1991).

Yoshida, H., Hayashi, S.I., Kunisada, T., Ogawa, M., Nishikawa, S., Okamura, H., Sudo, T., Shultz, L.D. and Nishikawa, S-I., The murine mutation osteopetrosis in the coding region of the macrophage colony stimulating factor gene, *Nature* 345:442-444 (1990).

LEUKAEMIA INHIBITORY FACTOR AND THE
REGULATION OF BLASTOCYST IMPLANTATION

Colin L. Stewart

Roche Institute of Molecular Biology
Roche Research Center
Nutley, NJ 07110

Embryos, following activation of their developmental program at fertilization usually progress to adulthood as a continuous series of proliferative and differentiative events. In the mammalian embryo this is also the usual course. However they have to pass one critical point, namely implantation, the stage at which the embryo assumes a more intimate physical association with the maternal uterine tissues.

The extent of physical contact or invasion of the uterus, by the embryonic trophoblast varies between species and all are dependent on a close interaction with the uterine tissues for their continued development (reviewed Renfree, 1984). Many mammalian species have exploited this point of transition at which the autonomously developing embryo becomes dependent on the maternal environment as a means to regulating the timing of birth, principally to coincide this event with an optimal availability of nutrients (Renfree, 1978; Sandell, 1990). Thus, embryogenesis in many mammalian species is characterised by an arrest of development at implantation and this is referred to as delayed implantation or embryonic diapause.

Two forms of delayed implantation have been described; facultative and obligate. Facultative, or lactational induced delay was first described by Fernand Lataste approximately 100 years ago. He recognized that the gestational period for a litter of rats was longer in females that were already suckling a previous litter than in a non-lactating female. He also correctly assumed that the increase in the length of gestation was due to implantational delay (Lataste,1891). Since his discovery, lactational inhibition of implantation has also been shown to occur in other rodents, as well as in many marsupials. In rodents, embryonic diapause evolved possibly as a means to maximize their reproductive efficiency in their relatively short life span (Renfree, 1978).

Obligate implantational delay occurs in many of the larger mammalian species, such as Roe deer, seals and bears. In these species delay usually occurs as a means to slowing embryonic development after conception, so that birth of the offspring occurs at an optimal time of year, when nutritional resources are at their most abundant. (Renfree, 1978; Sandell, 1990).

Since Lataste's discovery, delayed implantation has been of particular interest because of its unique form of growth control. The majority of studies in attempting to understand how this event is physiologically regulated have been performed on rodents, particularly rats and mice undergoing lactational delay. The conclusions from these studies have been that the initiation of delay is maternally induced. Much subsequent research has centered on trying to identify which factors or molecules regulate or synchronise the receptivity of the uterus with blastocyst development. A variety, ranging from the steroid hormones to histamine and prostaglandins have been proposed to having a critical role. However, none has been shown to have a conclusive role (reviewed, Glasser *et. al.*, 1991). It is however clear that following ovulation the uterus is initially primed in response to the steroid hormone, progesterone. A subsequent pulse of estrogen makes the uterus receptive for blastocyst attachment, which if this occurs, results in implantation and the induction of the decidual response (Psychoyos, 1973).

Recently it has become apparent that another class of physiological regulators, namely cytokines, also have a central role in regulating maternal-fetal interactions (Pollard 1991; Pampfer *et al.*, 1991). This became apparent with the demonstration that the hemopoietic factor colony stimulating factor-1 (CSF-1) was abundantly expressed in the pregnant uterus and its receptor, the proto-oncogene *c-fms* was present on the adjacent trophoblast (Pollard *et al.*, 1987; Arceci *et al.*, 1989; Regenstreif and Rossant 1989). That this ligand and its receptor functions in regulating fetal-maternal interactions was revealed by the discovery of a naturally occuring null mutation in the CSF-1 gene, the *op/op* or osteopetrotic mutant (Yoshida *et al.*, 1990). Females, homozygous for this mutation exhibited a reduced frequency of implantation when mated to heterozygous males and failed to produce any viable offspring when crossed to fertile homozygous males (Pollard *et al.*, 1991 and Pollard; this volume).

Following these observations, another cytokine, Leukemia Inhibitory Factor (LIF) has been found to have a crucial role in regulating maternal-fetal interactions and as will be reviewed in this chapter it may be the factor that synchronises uterine receptivity with the regulation of blastocyst attachment.

Leukaemia inhibitory factor, also referred to as DIA (differentiating inhibitory activity) or CDF (cholinergic differentiating factor) was originally characterised and cloned because it induced the differentiation of one myeloid leukaemia cell line, M1 (Gearing *et al.*, 1987; Gough *et al.*, 1988; Hilton *et al.*, 1988). By inducing the differentiation of this cell line, the differentiated derivatives eventually ceased to proliferate. Although LIF receptors appear to be widely distributed among somatic cells, LIF's biological effects to date, appear to be mainly restricted to stimulating megakaryocyte proliferation (Hilton *et al.*, 1991; Metcalf *et al.*, 1991) and to sustaining hemopoietic stem cells (Escary *et al.*, 1993). Since its identification, LIF has been shown to have multiple effects in different physiological systems, (Hilton, 1992) such as inducing the acute phase response proteins in hepatocytes (Bauman and Wong, 1989), regulating bone resorbtion (Abe *et al.*, 1986) and adipocyte formation (Mori *et al.*, 1989), as well as influencing neuronal phenotype (Yamomori *et al.*, 1989) and neuronal survival (Murphy *et al.*, 1992). However, one of its most striking characteristics is its requirement for inhibiting the differentiation of embryonic stem cells derived from mouse embryos (Evans and Kaufman, 1981; Bradley *et al.*, 1984; Wagner *et al.*, 1985).

Embryonic stem (ES) cells are pluripotent cells isolated from the inner cell mass (ICM) of cultured mouse blastocysts (Evans and Kaufman, 1981). They have attracted much interest, both as a means of studying the molecular basis of early mouse development and as a route to genetically manipulating the genome of mice (Koller and Smithies, 1992). Their growth *in vitro* is dependent on the presence of a "feeder" layer of mouse embryonic fibroblasts. If the fibroblasts are removed the stem cells differentiate and eventually cease to proliferate (Figs. 1A and B). The fibroblasts synthesize and secrete LIF and with the availability of purified recombinant protein, it was demonstrated that LIF inhibited ES cell differentiation without inhibiting their proliferation (Figs. 1C and D) (Williams *et al.*, 1988; Smith *et al.*, 1988). These

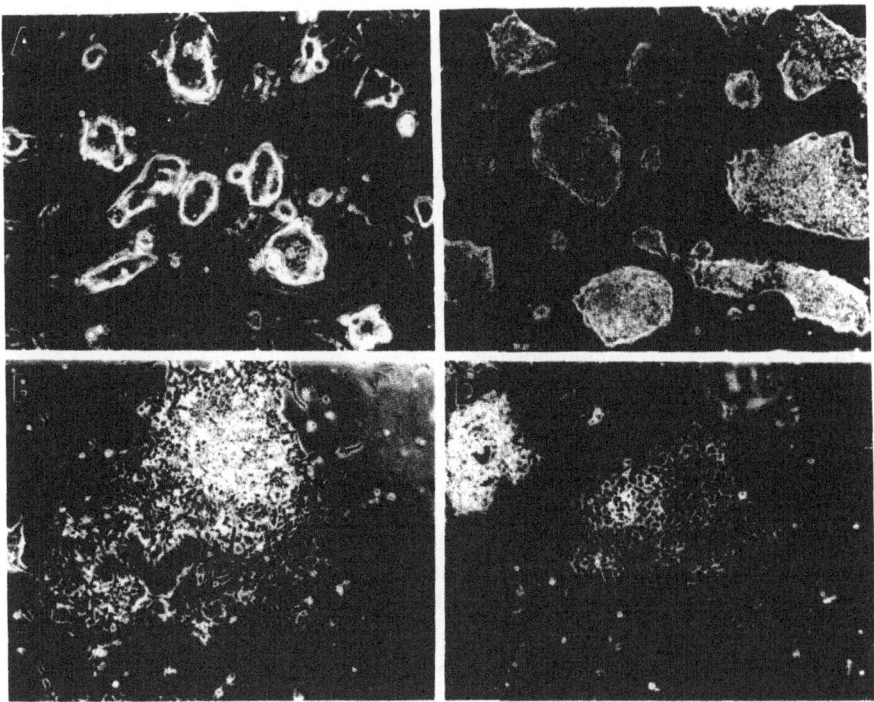

Figure 1. (A) Colonies of ES cells growing on a feeder layer of fibroblasts.(B) In the absence of feeders the ES colonies differentiate.(C) Colonies of undifferentiated ES cells growing in the absence of a feeder layer, but in medium supplemented with LIF.(D) In the absence of LIF the ES cells differentiate.

observations were at the time, of particular interest and significance, since LIF was the first protein identified that influenced the differentation of cells derived from the early mouse embryo and suggested that it might have a role in regulating their development.

The first step in analyzing LIF's potential *in vivo* embryonic role, was to determine when and where it was expressed. LIF transcripts were found in a wide variety of different tissues, both in adult and embryonic stages, (Bhatt *et al*, 1991) including the preimplantation embryo (Conquet and Brulet, 1990; Murray *et al.*, 1990). Overall, the levels were very low, although stronger signals were seen in newborn skin and in the small intestine. Higher levels were detected in the uterus, and a detailed analysis revealed 2 distinct peaks of expression: the first at oestrous, concurrent with ovulation (Shen and Leder, 1992), and the second on the 4th and 5th days of pregnancy (Fig. 2) (Bhatt *et al.*, 1991). *In situ* hybridization showed that the LIF expression in the pregnant uterus was confined to the endometrial glands, which in the endometrium, synthesize and secrete proteins into the uterine lumen (Fig. 3). On the 5th day following implantation and decidua formation, the glands degenerate and stop transcribing LIF (Bhatt *et al.*, 1991). The location of LIF expression in the estrous uterus has not yet been determined.

What was intriguing about these results, especially regarding the second peak of expression on the 4th-5th days, was the coincidence in timing between arrival of the embryo, as a blastocyst in the uterine lumen, with the onset of LIF expression (Fig. 4). This suggested that LIF expression could be induced by some signal originating from the embryo or alternatively,

LIF expression may occur independent of the embryo and be under maternal control. To distinguish between these two possibilities, LIF expression was analyzed in pseudopregnant females and in those undergoing delayed implantation.

In mice, copulation is essential for physiologically stimulating the female into preparing for pregnancy. Thus, if a female mates with a vasectomized male, changes are initiated in her that are associated with the onset of pregnancy, although there are no fertilized eggs present to produce viable embryos. LIF expression in the uteri of these pseudopregnants paralleled that observed in normal pregnant females, showing that LIF expression on the 4th-5th days of pregnancy was not dependent on the presence of viable embryos.

In a futher series of experiments, LIF expression was analysed in females undergoing delayed implantation. Under normal circumstances, this can occur when a female mouse gives birth to a litter of pups. Almost immediately afterwards she ovulates and the eggs, if fertilized, develop to the blastocyst stage, but do not implant. The embryos lose the surrounding *zona pellucida*, elongate and cell proliferation eventually ceases (Given,1988). They can remain in this state for many days. This situation occurs in response to the suckling stimulus of the first litter of pups that prevent the implantational (nidatory) pulse of estrogen secretion occuring. If the suckling of the first litter is interrupted, either by loss or weaning of the pups, the delayed blastocysts are activated and implant within 18–24 hours and resume normal development. Delayed implantation can also be experimentally induced in mice by ovariectomising the pregnant female on the 3rd day of pregnancy. Providing that relatively high circulating levels of progesterone are maintained, the embryos will enter into a delayed state once they form blastocysts. This can be maintained for days, even weeks, because the pulse of nidatory estrogen, produced by the ovaries between the 3rd and 4th days of pregnancy, cannot occur. If a single injection of oestrogen is given, implantation occurs followed by development of the embryo.

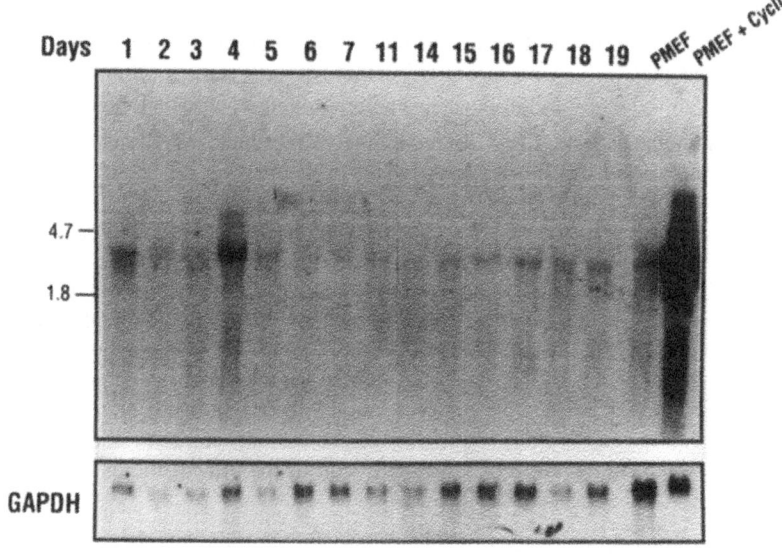

Figure 2. LIF expression in the mouse uterus during pregnancy. LIF is transcribed as a 4.2 kb mRNA. Days of pregnancy (1–19) are at the top. Total RNA prepared from primary mouse embryonic fibroblasts (PMEFs) with and without cyclohexamide (+ cyclo) was used as a positive control. Note the increased LIF expression on the 4th day of pregnancy.

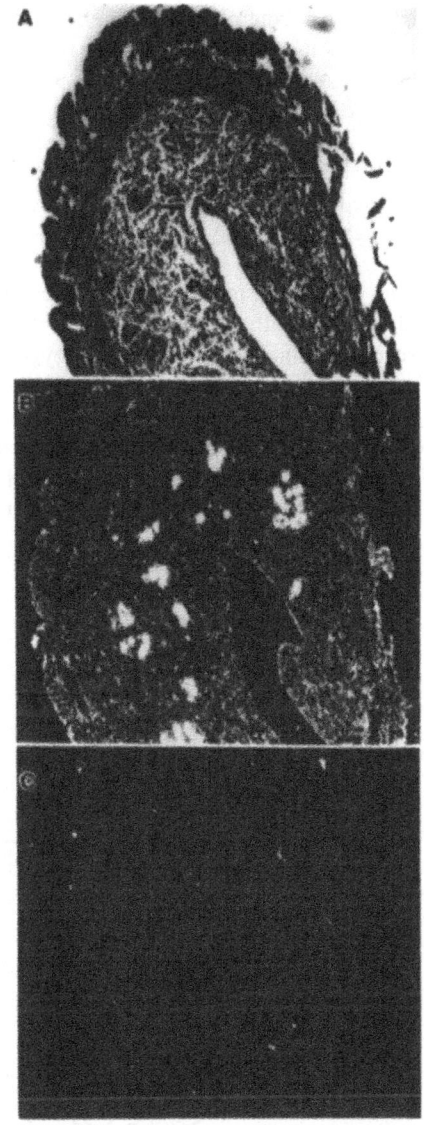

Figure 3. Localization of LIF transcripts to the endometrial glands in the 4th day uterus.(A) The glands are arrowed in the stained section.(B) Antisense LIF probes are located over the glands.(C) Sense control.

In females, undergoing either naturally or experimentally induced delay, no LIF expression was detected in the uteri containing viable blastocysts. However, as soon as delay was interrupted, by either removing the suckling pups, or by giving a single injection of oestrogen, LIF expression was detected within 18–24 hours. These results showed that LIF expression was under maternal control and that expression always preceeded implantation (Bhatt *et al.*, 1991). Furthermore, since the 2 peaks of LIF expression coincided with elevated levels of oestrogen, or could be induced following a single injection of estrogen, suggested that LIF expression might be under the direct control of this hormone.

Direct evidence for LIF regulating implantation came with the derivation of mice lacking a functional LIF gene. This was achieved by targeted mutagenesis using homologous recombination to disrupt the LIF gene. ES clones were screened for homologous recombinants and those carrying a mutated LIF gene were injected into blastocysts to derive chimeras. Their offspring, carrying the mutated LIF gene, were identified and intercrossed to produce homozygotes with both genes mutated which were identified at the expected frequency of 25% following heterozygous matings. The homozygotes were overtly normal, except that they were 25–30% smaller in body size.

When the homozygotes were test-bred, the males were found to be fertile, while the females, despite repeated matings with homozygous or wild type males, never became pregnant. However, a closer examination of the female homozygotes, following mating to male homozygotes, revealed that they were fertile in that overtly normal blastocysts and in the appropriate numbers (6–8) could be recovered from their uteri on the 4th day of pregnancy. Further analysis of homozygous females on the 7th day of pregnancy, revealed that they also contained blastocysts (Fig.5). However, there was no indication that implantation occurred or had even been initiated. Rather, the blastocysts morphologically resembled those undergoing

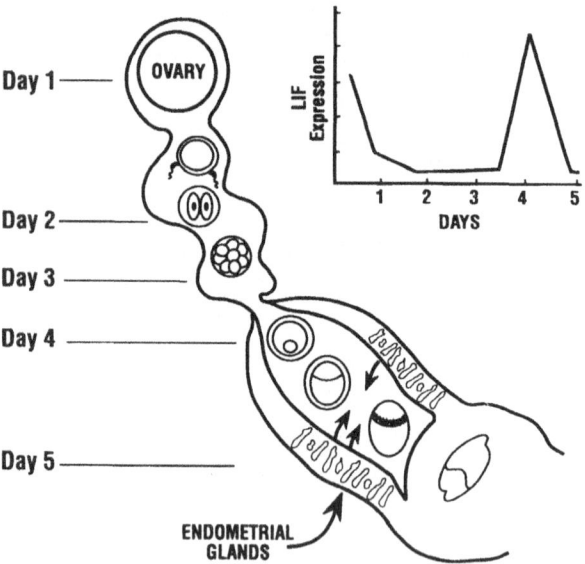

Figure 4. Diagram summarizing the pattern of LIF expression in the uterus during the first 5 days of pregnancy. The stages of embryonic development are shown with LIF expression from endometrial glands occurring on the 4th day. By the 5th day, implantation has occurred.

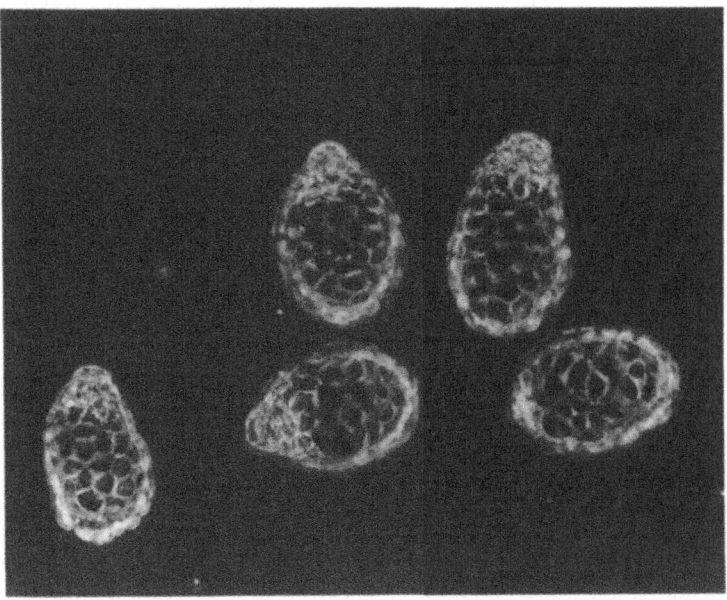

Figure 5. 5 blastocysts recovered from a 7 day pregnant LIF deficient homozygous female that had been mated to a LIF deficient male. The blastocysts should have implanted 2 days previously. They have the typical morphology of embryos undergoing delayed implantation (i.e. elongated with no *zona pellucida*).

delayed implantation with loss of the *zona pellucida* with the embryos assuming an elongated shape. By transplanting the embryos to wild type pseudopregnant recipients, it was shown that these blastocysts, that had developed in a completely LIF deficient environment, were able to develop normally to term if placed in an environment in which LIF was present. Thus failure to implant was maternal in origin and not due to abnormalities associated with the blastocysts (Stewart *et al.*, 1992).

Exactly how LIF regulates implantation is at present not understood. It could be acting on the uterus, preparing it to be receptive to the blastocyst. or, it could be acting on the blastocyst alone, or both. Currently, we favor the last possibility for the following reasons. We have evidence that LIF receptors are expressed in the uterus, although their exact cellular distribution is not yet clear. Furthermore, it is well established that the proliferative and differentiative response of the uterus associated with implantation (the decidual response) can be induced experimentally in normal mice by subjecting the uterus at the appropriate time (i.e. 4th-5th day of pregnancy) to a stimulus of a non-specific nature (e.g. the injection of a small amount of paraffin oil into the uterine lumen). Attempts to induce this in the LIF deficient females (mated to vasectomized males) failed to produce a decidual response, whereas all heterozygous controls responded (C.L.Stewart, *in preparation*). This indicates that the LIF deficient uterus is unresponsive to implantational stimuli. However, we also believe that LIF can directly affect the embryo. The evidence for this is that there are LIF receptors on the trophectoderm (S.Heyner and C.L. Stewart, *in preparation*). Furthermore, expression of LIF, under the control of a heterologous constitutive promoter in transgenic preimplantation embryos results in abnormal development and failure to form a blastocyst (L.Brunet and C.L Stewart, submitted). Thus, the embryo both expresses LIF receptors and is responsive to LIF.

The blastocysts in the LIF deficient females resemble those in delay, which in normal mice is associated with an absence of LIF expression in the uterus. We also have preliminary evidence that these delayed blastocysts, when exposed to LIF, activate the expression of certain genes. Nevertheless, the results do show that the transient burst of LIF expression that precedes implantation is essential for this event to occur. Furthermore, the results from these studies indicate that LIF is not essential for overt differentiation of the preimplantation mouse embryo. Rather, its function may have more to do with controlling proliferation in the pre-implantation/peri-implantation embryo. As to whether LIF may have a similar role in other mammalian species we have evidence that it is more abundantly expressed in the uterine endometrium in women during the secretory or post ovulatory phase of the menstrual cycle, than in the pre-ovulatory phase. (Abbondanzo *et al., in preparation*) Therefore, LIF may be of general use in regulating implantation in mammals.

In conclusion, studies that were initiated to determine which factors regulated the differentiation of ES cells *in vitro* have resulted in the identification of a previously cloned and characterised protein, LIF, as being responsible for inhibiting the differentiation and maintaining the proliferation of these cells. Subsequent *in vivo* studies resulted in the rather surprising discovery that one of this protein's most important functions is to regulate implantation of the embryo by both priming the uterus and possibly regulating cell proliferation in the blastocyst. Since LIF may interact with and affect both the uterus and blastocyst, it is an effective means to synchronizing the interaction between these two tissues, so ensuring that the changes in maternal physiology that are associated with pregnancy and implantation are co-ordinated with development of the embryo.

The task now is to determine what are the uterine cellular targets for LIF and what activities in these cells LIF is regulating. In addressing these questions a molecular understanding of how implantation is regulated should be attainable.

Acknowledgements

I would like to thank all my colleagues in my laboratory who participated in this work and Sharon Perry for help in preparation of this manuscript.

References

Abe, E., Tanaka, H., Ishimi, Y., Miyaura, C., Hayashi, T., Ngasawa, H., Tomida, M., Yamaguchi, Y., Hozumi, M., and Suda, T., 1986, Differentiation inducing factor purified from conditioned medium of mitogen treated spleen cultures stimulates bone resorption, *Proc.Natl.Acad.Sci.U.S.A.*, 83: 5958–5962,

Arceci, R.J., Shanahan, F., Stanley, E.R., and Pollard, J.W., 1989, Temporal expression and location of colony stimulating factor-1 (CSF-1) and its receptor in the female reproductive tract are consistent with CSF-1 regulated placental development, *Proc.Natl.Acad.Sci.U.S.A.*, 86: 8818–8822,

Baumann, H., and Wong, G.G., 1989, Hepatocyte stimulating factor III shares structural and functional identity with leukaemia inhibitory factor, *J. Immunol.*, 143: 1163–1167,

Bhatt, H., Brunet, L.J., and Stewart, C.L., 1991, Uterine expression of leukemia inhibitory factor coincides with the onset of blastocyst implantation, *Proc.Nat.Acad.Sci.U.S.A.*, 88: 11408–11412,

Bradley, A., Kaufman, M.H., Evans, M.J. and Robertson, E.J., 1984, Formation of germ line chimaeras from embryo-derived cell lines, *Nature*, 309: 255–256,

Conquet, F and Brulet, P., 1990, Developmental expression of myeloid leukemia inhibitory factor gene in preimplantation blastocysts and in extraembryonic tissue of mouse embryos, *Mol.Cell.Biol.*, 10: 3801–3805,

Escary, J-L., Perreau, J., Dumenil, D., Ezine, S., and Brulet, P., 1993, Maintenance of haematopoietic stem cells and thymocyte stimulation require leukemia inhibitory factor, *Nature*, 363: 361–364,

Evans, M.J., and Kaufman, M.H., 1981, Establishment in culture of pluripotential cells from mouse embryos, *Nature*, 292: 154–156,

Gearing, D.P. Gough, N.M., King, J.A., Hilton, D.J., Nicola, N.A., Simpson, R.J., Nice, E.C., Kelso, A., and Metcalf, D., 1987, Molecular cloning and expression of cDNA encoding a murine myeloid leukaemia inhibitory factor (LIF). *EMBO.J.*, 6: 3995–4001,

Given, R.L., 1988, DNA synthesis in the mouse blastocyst during the beginning of delayed implantation, *J.exp.Zool*, 248: 365–370,

Glasser, S.R., Mulholland, J., Mani, S.K., Julian, J., Munir, M.I., Lampelo, S., and Soares, M. J., 1991, Blastocyst-endometrial relationships: reciprocal interactions between uterine epithelial and stromal cells and blastocysts, *Trophoblast.Res.*, 5: 229–280,

Gough, N.M., Gearing, D.P., King, J.A., Willson, T.A, . Hilton, D.J,. Nicola, N.A, and Metcalf, D., 1988, Molecular cloning and expression of the human homologue of the murine gene encoding myeloid leukemia-inhibitory factor, *Proc.Natl.Acad.Sci.U.S.A.* 85: 2623–2627,

Hilton, D.J., 1992 LIF: lots of interesting functions, *Trends.BiochemSci.*, 17: 72–76,

Hilton, D.J., Nicola, N.A.and Metcalf, D., 1988, Specific binding of murine leukemia inhibitory factor to normal and leukemic monocytic cells, *Proc.Natl.Acad.Sci.U.S.A.*, 85: 5971–5975,

Hilton, D.J., Nicola, N.A. and Metcalf, D., 1991, Distribution and comparison of receptors for leukemia inhibitory factor on murine hemopoietic and hepatic cells, *J.Cell.Physiol.*, 146: 207–215,

Koller, B.H., and Smithies, O., 1992, Altering genes in animals by gene targeting, *Ann.Rev.Immunol.*, 10: 705–730,

Lataste, F., 1891, De la variation de duree la gestation chez les mammifières et des circonstances qui déterminent ces variations, *Mem.Soc.Biol.*, 43: 21–31,

Metcalf, D., Hilton, D., and. Nicola, N.A.1991, Leukemia inhibitory factor can potentiate murine megakaryocyte production in vitro *Blood.*, 77: 2150–2153,

Mori, M., Yamaguchi, K., and Abe, K., 1989, Purification of a lipoprotein lipase-inhibiting protein produced by a melanoma cell line associated with cancer cachexia. *Biochem.Biophys.Res.Commun.*, 160: 1085–1092,

Murphy, M., Reid, K., Hilton, D.J., Bartlett, P.F., 1992, Generation of sensory neurons is stimulated by leukemia inhibitory factor.*Proc.Natl.Acad.Sci.U.S.A.*, 88: 3498–3501,

Murray, R., Lee, F., and Chiu, C.P., 1990, The genes for leukemia inhibitory factor and interleukin-6 are expressed in mouse blastocysts prior to the onset of hemopoiesis, *Mol.Cell.Biol.*, 10: 4953–4956,

Pampfer, S., Arceci, R.J., and Pollard J.W., 1991, Role of colony stimulating factor-1 (CSF-1) and other lympho-hematopoietic growth factors in mouse pre-implantation development. *Bioessays*,13: 535–540,

Pollard, J.W., Bartocci, A., Arceci, R., Orlofsky, A., Ladner, M.B., and Stanley, E.R., 1987, Apparent role of the macrophage growth factor, CSF-1, in placental development. *Nature*, 330: 484–486,

Pollard, J.W., Hunt, J.S., Wiktor-Jedrzejczak, W., and Stanley E.R., 1991, A pregnancy defect in the osteopetrotic (op/op) mouse demonstrates the requirement for CSF-1 in female fertility. *Dev.Biol*, 148: 273–283,

Pollard, J.W., 1991, Lymphohematopoietic cytokines in the female reproductive tract. *Curr.Opin.Immunol.*, 3: 772–777,

Psychoyos, A., 1973, Endocrine control of egg implantation, *in* "Handbook of physiology; Endocrinology," pp187–215 R.O. Greep and E.B. Astwood, eds., Vol 2, Williams and Wilkins, Baltimore.

Regenstreif, L.J. and Rossant, J., 1989, Expression of the *c-fms* proto-oncogene and the cytokine CSF-1, during mouse embryogenesis, *Dev.Biol*, 133: 284–294,

Renfree, M.B., 1978, Embryonic diapause in mammals-A developmental strategy, *in* "Dormancy and developmental arrest: experimental analysis in plants and animals "pp. 1–45, M.E. Clutter, ed., Academic press, New York.

Renfree, M.B., 1984, Implantation and placentation, *in* "Reproduction In Mammals, Vol. 2: Embryonic and Fetal Development" pp 26–69. C.R.Austin and R.V.Short, eds Cambridge University Press Cambridge.

Sandell, M., 1990, The evolution of seasonal delayed implantation, *Quart.Rev.Biol.*,65:23–42,

Shen, M.M., and Leder, P., 1992, Leukemia inhibitory factor is expressed by the preimplantation uterus and selectively blocks primitive ectoderm formation in vitro. *Proc.Natl.Acad.Sci.U.S.A.* 89: 8240–8244,

Smith, A.G., Heath, J.K., Donaldson, D.D., Wong, G.G., Moreau, J., Stahl, M., and Rogers., 1988, Inhibition of pluripotential embryonic stem cell differentiation by purified polypeptides, *Nature*, 336: 688–690,

Stewart, C.L., Kaspar, P., Brunet, L.J., Bhatt, H., Gadi, I., Köntgen, F. and Abbondanzo, S.J., 1992, Blastocyst implantation depends on maternal expression of leukaemia inhibitory factor, *Nature*, 388: 76–79,

Wagner, E.F., Keller, G., Gilboa, E., Ruther, U. and Stewart, C., 1985, Gene transfer into murine stem cells and mice using retroviral vectors, Cold Spring Harbor 50th Anniversary Symposium, 691–700,

Williams, R.L., Hilton, D.J., Pease, S., Willson, T.A., Stewart, C.L., Gearing, D.P., Wagner, E.F., Metcalf, D., Nicola, N.A., and Gough, N.M., 1988, Myeloid leukaemia inhibitory factor maintains the developmental potential of embryonic stem cells. *Nature*, 336: 684 -686,

Yamamori, T., Fukada, K., Aebersold, R., Korsching, S., Fann, M.J., and Patterson, P.H., 1989, The cholinergic neuronal differentiation factor from heart cells is identical to leukemia inhibitory factor, *Science*, 246: 1412–1416,

Yoshida, H., Hayashi, S-I., Kunisada, T., Ogawa, M., Nishikawah, H., Sudo, T., Shultz, L.D.,and Nishikawa, S-I, 1990, The murine mutation osteopetrosis is in the coding region of the macrophage colony stimulating factor gene, *Nature*, 345: 442–444,

CARBOHYDRATES AND IMPLANTATION OF THE MAMMALIAN EMBRYO

Susan J. Kimber

Department of Cell and Structural Biology
University of Manchester
Oxford Rd
Manchester M13 9PT
UK

INTRODUCTION

Initially, the mammalian embryo moves freely through the reproductive tract as it undergoes cell division and the first morphogenetic and differentiative steps of its development. After several days a cavity forms in the embryo and it develops into the blastocyst containing an inner clump of cells, the inner cell mass, from which all the tissues of the fetus and some of the extraembryonic membranes are derived. The outer layer surrounding the cavity and inner cell mass, the trophectoderm epithelium, will form the trophoblast of the placenta. It is responsible for attachment of the embryo to the wall of the uterine lumen and the process of implantation.

In mice the 'free' period lasts only 4.5 days while in other mammals it lasts longer, for instance possibly 6-7 days in the human and about 14-18 days in the pig. In all placental mammals the majority of prenatal life takes place with more intimate contact between fetus and mother. After moving into the uterus the blastocyst develops close contact with the epithelium lining the uterine cavity. The degree of penetration of the maternal tissue by the conceptus varies from none (superficial) to interstitial implantation in which the embryo becomes surrounded by uterine connective tissue. In species such as the mouse and the human, in which implantation is invasive, the trophectoderm is activated at the time of attachment (Dickson 1963; Holmes & Lindenberg 1988) and has the potential to invade. The cells undergo a transition from an epithelial to a highly invasive phenotype, infiltrate the epithelium and move into the underlying stroma. However, before there is any evidence of firm association between embryo and luminal epithelium the presence of the embryo triggers changes in the underlying stroma which are particularly pronounced in rats and mice. The stroma cells undergo a process of differentiation, or decidualisation, with the production and reorganisation of extracellular matrix components (Wewer et al 1986; Glasser et al 1987).

In order for implantation to occur and the vital placental organ of gas and nutritive exchange to develop, a number of preparatory changes must occur on both the maternal and embryonic side. These maternal and embryonic events must occur in parallel and in a well ordered manner. Differentiation (and physiology) must reach an appropriate stage on both the maternal and fetal sides for the tissues to cooperate to form the complex placental architecture. The influence of progesterone (produced by cells of the corpus luteum of the evacuated ovarian follicle) and estrogen in preparing the uterus to receive the blastocyst is well documented (Psychoyos 1976; 1986). In the endometrium progesterone induces changes which convert it to a `sensitized`state prepared to respond to estrogen. The uterine epithelial lining does not provide a suitable surface for implantation in most species until a surge of estrogen from the ovary (McCormack & Greenwald 1974; Gidley Baird 1981) or from the conceptus (Dickman et al 1976; Gadsby et al 1980) has initiated further events. Following this surge the uterus is briefly receptive to implantation of an appropriately differentiated blastocyst. However this receptivity is transient, occurring in mice and rats between about 12 and 36 h after the nidatory surge of estrogen, and thus lasting only about 24h (Psychoyos 1973). The uterus then becomes refractory to implantation, even if increased estrogen is given, but continued progesterone is essential for maintenance of the refractory state in ovariectomised animals (Mulholland & Leroy 1988; Meyers 1970). Two points, pertinent to what follows, have frequently been noted regarding the transient receptivity of the uterus. Firstly, self-adhesiveness of the apical surfaces of epithelia lining lumina of organs or ducts is not a normal situation or lumina would not remain patent (Denker 1990). Yet this occurs in trophoblast-endometrial epithelial interaction (Schlafke & Enders 1975). Secondly the activated blastocyst, removed from the uterine horns, is catholic in what it will adhere to. Indeed there are few surfaces to which it will not attach <u>in vitro</u> (Sherman 1978; Morris & Potter 1990), or ectopically <u>in vivo</u> (Kirby,1965).

APPOSITION AND ATTACHMENT

Implantation has commonly been divided into three phases: apposition, adhesion and in those species where it occurs, invasion. The embryo ceases motion, coming into close contact with the apical surface of the luminal epithelium. The initial attachment is critical, being a prerequisite for the subsequent species-specific pattern of trophoblast penetration of the endometrial epithelium and infiltration of the underlying stroma, leading to the full establishment of the placenta. In the mouse the embryo attaches and starts to implant on day 5 of pregnancy. Initial attachment occurs by the abembryonic trophoblast, away from the inner cell mass (Kirby et al 1967; Kaufman, 1983), whilst development of the chorio-allantoic placenta occurs from the embryonic pole. Even this initial apposition phase may have a number of components. The closing down of the lumen towards the end of the preimplantation period brings the embryo into close proximity with the endometrial surface, perhaps aided by the muscular activity of the uterine myometrium. However, apposition must be more than transient: it must lead to a firmer adhesive interaction from which the trophectoderm can penetrate the epithelium.

CHANGES IN ENDOMETRIAL EPITHELIAL GLYCOCONJUGATES DURING THE PRE-IMPLANTATION PERIOD

Fundamental changes in the expression of glycoconjugates on the endometrial epithelial cell surface lining the uterine lumen were predicted from ultrastructural studies which revealed the thinning of the glycocalyx around the time of implantation (Enders & Schlafke 1974; Chavez & Anderson, 1985) in parallel with a decrease in surface charge (Hewitt et al 1979; Morris & Potter 1984). There is also a fundamental alteration in the

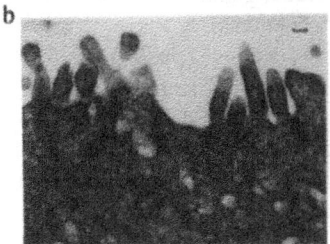
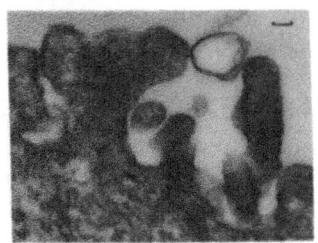

Fig. 1. Electron micrographs of luminal endometrial epithelium of the mouse. a) the cuboidal nature of the cells on day 5 of pregnancy; b) detail of the apical surface on day 4 showing microvilli; and c) day 5 showing bulbous processes. Scale bars a) 0.5μm; b, c) 0.1μm. (R. Waterhouse unpublished micrographs).

morphology of the cells during the preimplantation period: whereas on days 1-2 of pregnancy the cells are several times higher than they are broad, by day 4 they are almost cuboidal (Fig 1). The polarity seen in the early preimplantation period (e.g. basal nuclei; Psychoyos, 1973) becomes less pronounced towards implantation. At the cell surface the density of apical microvilli decreases and these are replaced by bulbous protrusions (Fig 1b,c; Enders & Schlafke 1974). In the implantation chamber itself the luminal cells adjacent to the trophoblast undergo cell rounding and apoptotic cell death at the start of implantation (El Shershaby & Hinchcliffe 1975; Parr et al 1987). Changes also occur in the interactions between luminal epithelial cells, for instance the cell adhesion molecule C-CAM is down regulated during the preimplantation period (Svalander et al 1990). Alterations have been demonstrated in the nature of epithelial gap junctional connexins around the time of implantation in the rabbit and in the implantation chamber in the rat (Winterhager et al 1988,1991 and this volume).

Ovarian steroids have been reported to trigger both quantitative and qualitative alterations in endometrial epithelial glycoproteins in several species. Estrogen can stimulate the synthesis of specific proteins (Kuivanen & De Sombre 1985; Teng et al 1986; Anderson et al 1986) but during implantation the role of nidatory estrogen appears primarily to restrict progesterone stimulated protein synthesis in the uterus (Glasser & McCormack 1979; Mulholland & Leroy 1989). Estrogen has been shown to stimulate glycosylation in the absence of progesterone (Dutt et al 1986; Carson et al 1987a). In uterine tissue slices estrogen, with or without progesterone, preferentially stimulates synthesis of branched lactosaminoglycans, oligosaccharide chains containing repeating Galß1-GlcNAc units (Dutt et al 1988). The majority are inhibited by Tunicamycin, indicating that they are N-linked, and of a broad range of sizes. There is a small O-linked component and in polarised cultures most apical neuraminidase-insensitive carbohydrates which bound the lectin WGA (affinity for GlcNAc residues) appeared to be O-linked (Valdizan et al 1992).

Some years ago we screened the uterine epithelium in early pregnancy with a panel of monoclonal antibodies (mAbs) against lactosaminoglycan related oligosaccharides. A number of mAbs which recognise histo-blood group antigens, such as H-, Le[a], Le[b], Le[x] and Le[y], reacted specifically with the uterine epithelium of glands and lumen (Kimber et al 1988). These antigens are mono or di-fucosylated terminal structures carried on linear or branched chains composed of the repeat disaccharide (Galß-GlcNAc)n in either 1-3 or 1-4 linkage. In rodents they are tissue rather than blood group antigens (Oriol et al 1986). Most of the carbohydrates which we found to be specific markers of endometrial epithelial cells stained both glandular and luminal epithelium during the pre-implantation period, with a steady immunofluorescence intensity which decreased slightly towards the time of implantation. Frequently not all cells in an individual gland stained and some glands were entirely negative when the majority were brightly stained.

Expression of the H-type 1 blood group antigen (Fucα1-2Galß1-3GlcNAcß1-) recognised by mAb 667/9E9 changed during the peri-implantation period. Staining was intense on all cells up to day 3 of pregnancy. However between day 4 and day 5 of pregnancy patches of unstained cells appeared. By the day following implantation this determinant could rarely be detected on the epithelium. The staining pattern indicated that the epitope was present at the cell surface and in the cytoplasm. It was detected in glandular and luminal secretions on day 2-3 but absent from the lumen on day 5 of pregnancy. Staining was abolished in the presence of the free pentasaccharide LNF-1 (Lacto-N-fucopentaose 1: Fucα1-2Galß1-3GlcNAcß1-3Galß1-4Glc). In contrast to the endometrial epithelium, neither the trophectoderm nor the ICM reacted with mAb 667/9E9 at any time (Lindenberg et al 1988).

ROLE OF CARBOHYDRATES IN EMBRYO ATTACHMENT

Carbohydrate based cell-cell interactions have previously been suggested during fertilisation and embryonic development (see Zalik 1991; Kimber, 1988, 1990; Wassarman 1992). In order to investigate whether the change in distribution of the H-type 1 epitope was functionally connected with the period of implantation we used an in vitro approach. Luminal epithelial cells were cultured to confluence in tissue culture wells and blastocysts added in the presence of different concentrations of oligosaccharides. Attachment of the

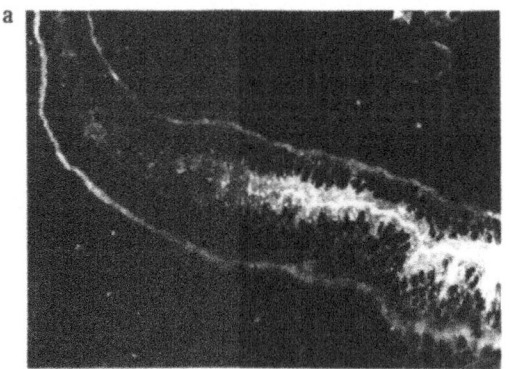
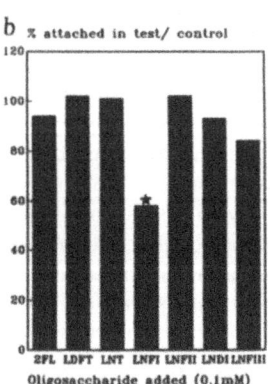

Fig. 2. a) Staining of endometrial epithelial cells on day 5 of pregnancy with mAb 667/9E9.b) Effect of oligosaccharides (0.1mM) on blastocyst attachment to endometrial epithelial cells. Total number of blastocysts tested shown in brackets below. 2FL, 2-fucosyl lactose (97); LDFT, Lactodifucotetraose (74); LNT, Lacto-N-tetraose (70); LNFI, Lacto-N-fucopentaose I (95); LNFII, Lacto-N-fucopentaose II (86); LNDI, Lacto-N-difucohexaose (81); LNFIII, Lacto-N-fucopentaose III (86). Asterisk, significantly different from control at p<0.001 by the Mantel-Haenszel chi[2] test.

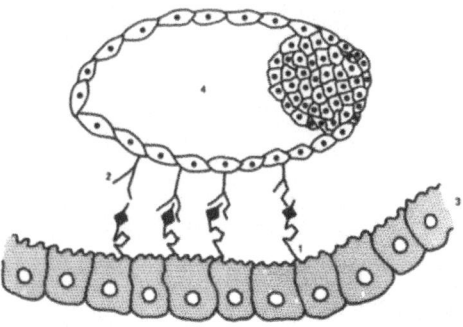

Fig. 3. Model illustrating attachment of the murine blastocyst (4) via binding of a receptor (2) on the abembryonic trophectoderm to a glycoproteins carrying the H-type-1 determinant (1) on the luminal epithelium (3). From Lindenberg et al 1990 with permission.

embryos to the epithelial cells was scored by tandem motion between the endometrial epithelial layer and blastocysts when the dish was tapped gently. In a blind study we investigated 7 different coded Gal-GlcNAc based oligosaccharides (at concentrations between 0.1 and 5mM) including those recognised by mAbs which bound to the endometrial epithelium. Only the LNF-1 pentasaccharide was able to inhibit attachment (Fig 2). At 0.1mM the percentage of blastocysts attaching was reduced to 59% of the controls. At maximum we could inhibit attachment of 75% of blastocysts using 5mM LNF-1 but even at high levels of sugar we could never abolish attachment of all blastocysts. Inhibition was similarly incomplete when we used mAb 667/9E9 (attachment of 53% inhibited) or a neoglycoprotein consisting of the pentasaccharide LNF-1 conjugated to BSA or human serum albumen (HSA; 10-20 sugars per molecule). The neoglycoprotein at 0.04mM inhibited attachment by 65%, again incompletely, but neither LNFIII-HSA or LNT-HSA had a significant effect. The simplest hypothesis to explain these results is that free LNF-1 (as pentasaccharide or neoglycoprotein) binds competitively to a lectin-like molecule on the surface of the trophectoderm. This receptor is thus prevented from interacting with a glycoconjugate on the luminal epithelium carrying the same or a very similar sugar structure (Model, Fig. 3). The antibody is also assumed to inhibit competitively by binding the endometrial carbohydrate. This hypothesis presupposes that the epithelial cells grown in culture express the carbohydrate epitope. We demonstrated this using mAb 667/9E9. However by no means all of the cells stained and although the cell surface staining was punctate, as in vivo, the staining density was quite low. This suggested that embryos might use other mechanisms to attach to H-type-1-free patches of cells or gaps which may appear in the epithelial layer. Zona pellucida-free blastocysts attach both to plastic and extracellular matrix (ECM) substrates of various kinds including complex ECM material (Glass et al 1983; Welsh & Enders 1989; Wordinger et al 1991), various collagens, fibronectin and laminin etc (referenced below) and extracellular matrix components are deposited by murine endometrial epithelial cells in culture (Waterhouse & Kimber unpublished). The trophoblast in these circumstances may attach by its basal surface (Morris & Potter 1990) rather than the apical surface by which attachment occurs in vivo (Schlafke & Enders 1975). This difference could account for discrepancies between observations made on material in utero and those on interaction of trophoblast with acellular substrates or non-polarised cell layers. However, during the peri-implantation period the composition of apical and basal surfaces may not be dissimilar and may change rapidly (Morris & Potter 1990; Denker 1990; Carson et al 1990; Glasser & Mulholland 1993).

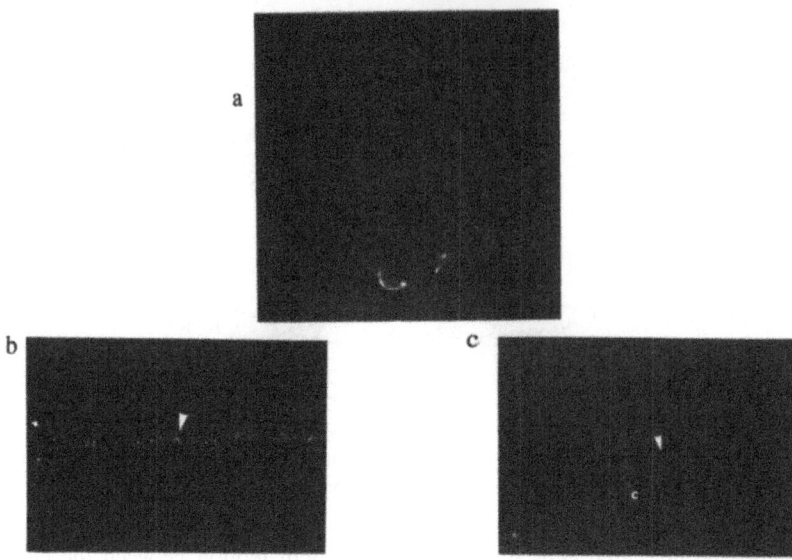

Fig. 4. a) staining of cultured hatched blastocyst with LNF-1 neoglycoprotein showing binding to abembryonic trophectoderm. b) H-type-1 epitope expression on a section of a polarised endometrial epithelial monolayer cultured for 8 days. c) control for b). C, Cellagen; arrow apical surface.

Recently we have grown luminal endometrial epithelial cells on suspended cross linked collagen membranes (Cellagen, ICN; Kimber et al 1993) in a modification of the method developed by Glasser et al (1988) for rat endometrial epithelium. The cells retain their polarity on Cellagen having a cuboidal appearance with abundant apical microvilli. When confluent monolayers of these cells were stained with mAb 667/9E9 the proportion of cells stained and the intensity of staining were greater (Fig. 4b). We co-cultured embryos with endometrial epithelial cells on such membranes using a basal hormonal milieu mimicking that found during the nidatory surge of estrogen. Under these conditions blastocyst attachment was delayed by about 20 h in the presence of LNF-1 but eventually blastocysts attach in equal numbers to those seen in control cultures without LNF-1 or in the presence of the control sugar Lacto-N-fucopentaose III (R. Waterhouse and S.Kimber in preparation). This suggests that in this static model system in which the blastocysts remain in contact with a patch of epithelial cells under gravity, LNF-1 carrying glycoconjugates may be the first mediators of attachment. However if the trophoblast and epithelial surfaces remain in contact further mechanisms of adhesion may come into play. These sequential mechanisms contribute to the normal blastocyst attachment-adhesion-invasion cascade. Additionally, one would predict redundancy in the molecules involved in an event as important to the species as implantation.

In order for our simple hypothesis of initial attachment to hold we needed to demonstrate that there was an H-type-1-specific receptor on the blastocyst. We examined binding to mouse embryos of fluorescein isothiocyanate (FITC) conjugated neoglycoproteins carrying 10-20 LNF-1 or control sugars. Embryos were denuded of their zonae and stained, either after flushing from the reproductive tract, or following culture from the two cell stage (Lindenberg et al 1990). All these experiments were undertaken in the presence of 0.02% azide to prevent uptake of neoglycoproteins into the cells. No embryos bound LNF-1-BSA-FITC conjugates at any stage up to day 4 of pregnancy. However specific staining of 65% of flushed embryos and 16% of cultured embryos was observed on the morning of day 5 of pregnancy. By the morning of equivalent day 6 of pregnancy 76% of cultured embryos were stained (Fig. 4a). Staining could be inhibited partially by LNF-1 but almost completely by the unlabelled multi-ligand LNF-1-HSA (7% hatched blastocysts weakly stained) suggesting multiple low affinity binding. This has been demonstrated for other cell adhesion systems involving carbohydrates (Stowell & Lee 1978; Baenziger & Maynard 1980; Fenderson et al 1984). Interestingly the appearance of the receptor correlated with hatching of blastocysts from the zona pellucida and expression occurred specifically on the abembryonic mural trophectoderm which first contacts the luminal epithelial surface in vivo. Our findings have recently been confirmed by Yamagata et Yamazaki (1991) who also found an abembryonic distribution for the neoglycoprotein binding receptor.

The staining pattern with mAb 667/9E9 suggested that the H-type 1 determinant is carried both on a membrane protein of luminal cells and on secreted component(s). This latter was also indicated by its presence in uterine fluid early in pregnancy. Clearly, new expression of H-type-1 epitope on the luminal epithelial cell surface does not occur in the receptive period. However the balance between the secreted and cell surface molecules carrying LNF-1 may be crucial in determining whether trophoblast cells adhere to the epithelium or not. The secreted component decreases to a low level between days 4 and 5 of pregnancy and both secreted and cell-surface components disappear from the endometrial epithelium after implantation (day 6-7). The earlier disappearance of this epitope from the uterine fluid would avoid competition for the trophectoderm receptor between the secreted and endometrial cell surface molecules carrying the determinant. However the presence of molecules containing the epitope in the luminal fluid up to day 4 would also fit in with the inability of implantation-competent blastocysts to adhere in a pre-receptive uterus (Psychoyos, 1973) although this cannot be the only factor involved. Since the trophectoderm receptor appears between day 4 and 5 of pregnancy the presence of moderate levels of secreted H-type-1 containing molecules in the uterine fluid on day 4 may in fact delay attachment of advanced, receptor-positive embryos until day 5. On the other hand the disappearance of H-type-1 epitope from the cell surface by day 6 of pregnancy could contribute to the refractory period. By definition in this period, >36h following the nidatory estrogen surge, the embryo can no longer attach.

In a preliminary study glycolipids were extracted from the uterine epithelium and separated by thin layer chromatography (TLC), but staining with mAb 667/9E9 was negative. In contrast, the antibody bound glycolipid in lipid extracts from other tissues and mAbs recognizing other sugar determinants reacted with endometrial epithelial glycolipids (P. Pahlsson, S. Lindenberg & Kimber unpublished data). This suggested that the H-type 1 antigen is not present in the uterine glycolipid fraction. However extraction of frozen sections of mouse uterus with a chloroform: methanol 2:1, but not methanol alone, abolished antibody binding to the epithelial cell surface (Kimber et al 1988). Thus the cell surface molecules localized by immunocytochemistry, although not lipids themselves, may contain hydrophobic domains embedded in the membrane, possibly in close association with lipid. Staining of the endometrial oligosaccharide is insensitive to trypsin but sensitive to pronase, suggesting that the determinant is carried on a trypsin-insensitive protein core.

We have partially characterised the murine endometrial epithelial glycoproteins carrying the H-type 1 determinant   (A.Cook & S.Kimber in preparation).   The cell bound glycoproteins are transmembrane molecules which require non ionic-detergent for extraction. By immunoblotting with mAb 667/9E9 we identified a broad band between 120 and 180 kD separated by SDS PAGE under reducing conditions (Fig. 5). More recent evidence suggests the reactive material may consist of a single glycoprotein in a number of glycoforms generally running as about 120-130kD. The band is present in extracts of endometrial epithelial cells from cycling mature fertile females and immature superovulated mice on day 2 of pregnancy, confirming immunocytochemical data. Immuno-reactive material was not found after implantation or following birth of the young during suckling. Preliminary evidence suggests that luminal fluid contains species within the same MW range, so the transmembrane and secreted molecules may be similar.

Uterine glycoproteins involved in implantation would be expected to be controlled by ovarian steroids. We examined the control of expression of the LNF-1 epitope and other epithelium-specific saccharides by ovariectomising mice on day 3 of pregnancy and treating with ovarian steroids to mimic various phases of the estrous cycle or early pregnancy. These studies indicated that appearance of LNF-1 on the endometrial epithelium required the presence of estrogen (Kimber & Lindenberg 1990). Originally it appeared that progesterone priming followed by a single dose of estrogen with progesterone (P+E) had a greater stimulatory effect than estrogen alone. However more recent studies suggest that progesterone does not enhance the effect of estrogen. In the uteri of mice given progesterone followed by single dose of P +10ng E, the epitope was expressed in patches but it was absent after 100ng E (Kimber et al 1993b). When we stained sections of uterine epithelium from mice carefully selected for stage in the estrous cycle it was clear that moderate amounts of the epitope were present at all stages except early proestrus just before the ovulatory estrogen peak (White & Kimber 1993). We found little change in expression of Le^x during the estrous cycle but Babiarz and Hathaway (1988) reported subtle hormone-dependent modulation of this antigen: In ovariectomised mice Le^x expression on the glands required both estrogen and progesterone treatment but on the luminal epithelium it was also present also after estrogen alone. We also found P+E treatment to subtly enhanced Le^x expression on the luminal epithelium (Kimber & Lindenberg 1990). Sialylated linear Galß1-4Galß1-3GlcNAc chains, also required the presence of P+E for epithelial expression (Babiarz and Hathaway 1988).

To assess whether the presence of the embryo influenced expression of the H-type-1 epitope we compared uterine expression in female mice mated with fertile males with that in females mated with either vasectomised males or those of strain T145 which are homozygous sterile. Sterile matings lead to the condition of pseudopregnancy. We could not detect consistent differences of any magnitude in the expression of H-type 1 between pseudopregnant and pregnant mice. There were subtle differences in expression of some other carbohydrate epitopes but whether these have significance in terms of reproductive function is unclear.

Ovarian steroidal control of the H-type-1 epitope might indicate turnover of entire glycoproteins in response to the hormonal milieu. However a more energy-efficient mechanism of regulating expression of carbohydrate epitopes would be hormonal regulation of the glycosylation of stable proteins. We have investigated the hormonal

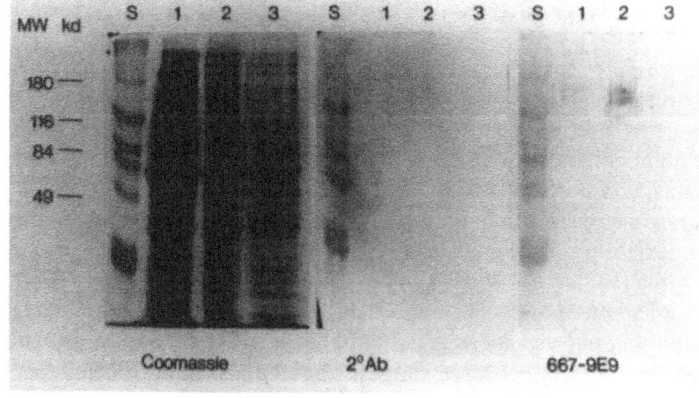

Fig 5. Immunoblot showing staining of proteins from luminal epithelium a) Coomassie stained 7.5% gel b) staining with second antibody c) staining with mAb 667/9E9.Lane S, MW markers; Lane 1, Hepes buffer extract; Lane 2, 0.1% NP-40 extract; Lane 3, 300mM KI extract.

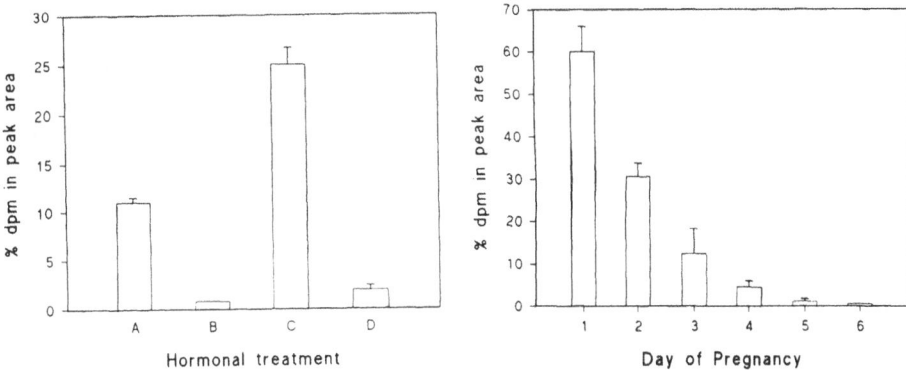

Fig 6. Uterine luminal epithelial $\alpha(1\text{-}2)$fucosyltransferase (α1-2FT) activity (%dpm bound in peak areas on chromatogram) a) after ovariectomy and hormone replacement, b) during the first six days of pregnancy. $\alpha(1\text{-}2)$FT was assayed using $GDP[_{14}C]$fucose as donor and phenylgalactoside as specific acceptor for α1-2fucosylation. Products were separated by paper chromatography. In a) all mice were given 3 daily injections of estradiol benzoate and left for two days before treating as follows: A, No hormone, 4 daily injections of vehicle only; B, 4 daily injections of 500µg progesterone; C, 4 daily injections of 100ng estradiol benzoate; D, 3 daily injections of 500µg progesterone followed by a final injection of progesterone + 10ng estradiol. Animals were killed 18h after the final injection.

control of synthesis of the H-type 1 epitope. The enzyme catalysing addition of the terminal fucose, in α1-2 linkage to galactose is a specific α1-2fucosyltransferase. We therefore assessed α1-2fucosyltransferase activity in the luminal epithelium during the pre- and peri-implantation period, the estrous cycle and in steroid supplemented ovariectomised mice. Enzyme activity was high on days 1 and 2 of pregnancy but decreased rapidly through the preimplantation period becoming undetectable by day 4 (Fig. 6). Similarly changes occurred during the estrous cycle, with high activity being demonstrated at estrus but much lower activity at other periods in the cycle. In ovariectomised mice treated with E and, or, P the α1-2fucosyltransferase enzyme(s) was stimulated by E and strongly inhibited by P (White & Kimber 1993). This agrees with our previous immunocytochemical studies suggesting that predominant control of expression of the H-type-1 epitope is through regulation of fucosylation by ovarian steroids.

Recently carbohydrate ligands have been identified for the selectin family of cell adhesion molecules which carry terminal lectin-like domains (Springer 1990; McEver 1991; Lasky 1991 1992). Members of this family, of which three have been identified so far, include L-selectin, the lymphocyte homing receptor which binds to high endothelial venules. All three, E-, L-, and P- selectin are involved in mediating initial adhesion between circulating leukocytes and the vascular endothelium prior to extravasation at sites of inflammation. E-selectin, the endothelial-leukocyte cell adhesion molecule binds sialyl Le^x as a major ligand (Phillips et al 1990; Tyrrell et al 1991) although it may bind to other structures. Contributory evidence that this carbohydrate functions as a ligand for E-selectin was the demonstration that transfection of a non-myeloid cell line, lacking $\alpha(1\text{-}3)$fucosyltransferase, with the human gene for this enzyme conferred E-selectin-dependent cell-cell adhesion on these cells (Lowe et al 1990). Thus, control of cell adhesion dependent on selectin-carbohydrate interactions might also occur at the level of expression of fucosyltransferase activity. Furthermore, E-Selectin dependent cell adhesion of colon carcinoma cells to activated endothelial cells has been demonstrated to be inhibited by an antibody to H-type-1 epitope (Dejana et al 1992) suggesting that $\alpha(1-2)$ fucosylation may be an important control point in modulating selectin based adhesion during metastasis. Indeed the similarities between the selectin-carbohydrate mediated initial neutrophil-endothelial interaction and the carbohydrate based mechanism of initial adhesion between the blastocyst and the endometrial epithelium is striking (Kimber et al 1993a,b). Perhaps H-type-1 mediated adhesion brings blastocysts to a halt in the same way that the rolling motion caused by low affinity selectin-sugar interactions with endothelial cells stops neutrophils. The latter allows other stronger adhesion mechanisms, specifically integrin-immunoglobulin superfamily interactions, to take effect.

COMPONENTS OF THE 'IMPLANTATION ADHESION CASCADE'

Although we suggest that H-type-1 bearing carbohydrates may be responsible for initial embryo attachment, it is clear that other molecules are involved. A sequence of interactions takes place as the trophoblast adheres to and then displaces and penetrates the epithelium and moves into the stroma. Most studies have examined attachment of blastocysts to a culture dish and the subsequent outgrowth of trophoblast cells in two dimensions. This is considered to mimic trophoblast invasion to some extent. Care must be taken in interpreting these experiments since blastocysts will attach to molecules which they are unlikely to contact in vivo. In vivo, the timing of appearance of receptors and ligands must be carefully regulated to give an orderly sequence of trophoblast interactions with uterine cells and extracellular molecules.

Attachment and outgrowth on fibronectin can be inhibited by RGD containing peptides or heparin and this appears to rely on interaction with peptides of fibronectin rather than its oligosaccharides (Armant et al 1986a,b; Farach et al 1987). However Sutherland et al (1988) reported inhibition of trophoblast outgrowth (but not attachment) by RGD tripeptide on both fibronectin and laminin. Attachment on laminin requires the protein but not the oligosaccharide moiety. Outgrowth of trophoblast utilises the E8 fragment but is unaffected by heparin, even though E8 contains the heparin binding domain (Armant 1991). Interaction of trophoblast galactosyltransferase with laminin oligosaccharides also promotes outgrowth of trophoblast in vitro (Romagnano & Babiarz 1990). Laminin has been identified on the apical surface of the trophectoderm in the pre-hatched blastocyst in vitro and attachment stage blastocysts in vivo, suggesting that basal and apical surfaces may not be entirely distinct at this time (Dziadek & Timpl 1985; Carson et al 1992). So laminin could interact with ligands on the endometrial epithelium and embryo in the early phase of attachment. Carson's group suggest that heparan sulphate proteoglycan (HSPG) may be involved in initial attachment: HSPG can be detected immunologically on the outer surface of the blastocyst before implantation (Dziadek et al 1985; Carson et al 1992). Staining with an antibody to perlecan, the basement membrane form of HSPG increased as embryos became adhesive and was present over the entire blastocyst surface after hatching in vivo (Carson et al 1992). In the luminal epithelium Potter & Morris (1992) observed basal and lateral staining at met-, di- and pro-estrus but only basal staining at estrus with antibodies to syndecan (a cell-surface HSPG). During early pregnancy staining decreased between days 4 and 5. Apical staining, marginally above background, appeared on day 5. Staining for HSPG at the luminal surface was also reported in mice of undefined endocrine status (Tang et al 1987). In keeping with these findings, Morris et al (1988) showed that turnover of HSPG is stimulated by estrogen. The apical appearance of HSPG in the peri-implantation period is difficult to reconcile with the reported reduction in negative apical surface charge (Hewitt et al 1979; Morris & Potter 1984) unless there is a large reduction in other negatively charged components. Thrombospondin, another ECM glycoprotein, is also present on the cell surface of the early embryo and peri-implantation blastocyst. Trophoblast migration on thrombospondin is inhibited by specific antibodies (O'Shea et al 1990). However its heparin binding domain is an ineffective inhibiter.

Collagens have been implicated in supporting attachment and outgrowth (Jenkinson & Wilson 1973; Sherman et al 1980; Carson et al 1988; 1990). For instance, attachment and outgrowth of day 5 blastocysts was reported to occur much more rapidly on collagen types II and VI than on types I, III, IV and V. Inhibition studies indicated an initial RGD dependent mechanism of outgrowth which was overcome with time. However, in the rat, collagen VI, present in undecidualised stroma, disappeared on formation of deciduoma (Mulholland et al 1992), suggesting interaction of trophoblast with collagen VI may not be important in vivo. It is likely that interaction with fibronectin and collagen IV comes into play once the trophoblast has started to penetrate the epithelium.

Reorganisation and degradation of the characteristic glycoproteins and GAGs of the epithelial basement membrane must be an important component in trophoblast movement through the basement membrane and into the stroma (Kubo et al 1981; Glass et al 1983; Welsh & Enders 1989; Behrendtsen et al 1992; Lala & Graham 1990). Both the luminal epithelium (Welsh & Enders 1989) and decidual cells (Blankenship & Given 1992) may aid penetration through the epithelial basement membrane by their degradative activity.

Once the decidual response has been initiated, stromal cells produce increased amounts of laminin, entactin, fibronectin, type IV collagen and HSPG and organise it into a basement membrane-like layer around themselves (Grinnell et al 1982; Wewer et al 1986; Glasser et al 1987; Aplin et al 1988). Furthermore, the type and quantity of fibronectin mRNA expressed changes as particular spliced variants become concentrated anti-mesometrially (Rider et al 1992). Embryos can also attach and outgrow on hyaluronate - HA (Carson et al 1987b). HA synthesis by stromal cells was reported to increase as they decidualise in vitro (Carson et al 1987b) but localisation with a specific probe shows HA is not present adjacent to the luminal epithelium, especially anti-mesometrially where blastocyst attachment occurs (Brown & Papaioannou 1992). Chondroitin sulphate proteoglycans, a major class of proteoglycan synthesised by endometrial stroma cells in vitro in the mouse (Jacobs & Carson 1991) have been reported to inhibit outgrowth on fibronectin and collagen (Carson et al 1988). The presence of adhesive and inhibitory molecules may facilitate cell movement which requires firm attachment and detachment from the substrate. The blastocyst is capable of interacting with a large number of the macromolecules present in the epithelium and decidualised stroma, suggesting extremely complex interactions in vivo.

EMBRYONIC CELL SURFACE CARBOHYDRATE EXPRESSION DURING THE PERI-IMPLANTATION PERIOD

It is known that once the blastocyst has hatched from the zona pellucida it is extremely sticky and can attach indiscriminately to many surfaces (Holmes & Dickson 1973; Holmes & Lindenberg 1988). Indeed, blastocysts which have hatched in vitro tend to stick avidly to one another. Early experiments revealed that embryos flushed on day 5 of pregnancy, or reactivated to implant after delay, bound colloidal iron or alcian blue indicators of negatively charged carbohydrate (Holmes & Dickson 1973; Nilsson et al 1980). In ovariectomised, steroid-treated mice embryonic staining correlated with estrogen treatment of the dams. Lectin binding suggested changes in cell surface glycoconjugates at the time of implantation (Chavez & Enders 1981; Chavez 1986 1990; reviewed Kimber 1990). However results from studies using lectins have been conflicting, preventing definite conclusions about the role of carbohydrates in trophectodermal adhesiveness.

The advent of the monoclonal antibody technique (Kohler & Milstein 1975) has greatly enlarged our knowledge of changing cell surface expression during preimplantation development (Feizi, 1985 1988; Richa & Solter 1986; Kimber 1990; Fenderson et al 1990). Although antigens have been identified we frequently remain ignorant about the molecules to which they are attached. This is particularly true for carbohydrate epitopes: changes in their expression correlate in many cases with developmental transitions in the pre-implantation embryo but we have few clues as to their role. However, there is evidence that cell surface glycoconjugates function in cellular interactions in the embryo (Bird & Kimber 1984; Fenderson et al 1984; Bayna et al 1988; Rastan et al 1985).

Antigens have been identified, using mAbs, which change at, or just before, implantation. For instance, the Forssman glycolipid has long been known to appear first on the cell surface of the mouse embryo at the late morula to early blastocyst stage. It is present on the trophectoderm of the expanded blastocyst but decreases thereafter, although remaining on the ICM lineage (Stinnakre et al 1981; Willison et al 1982). Antibodies to human blood group related antigens have proved particularly helpful in studying cell-surface changes in embryonic development, as they have for studying the endometrial epithelium. Modifying terminal structures would be an economical way of changing the surface composition presented to the external environment. The expression of such antigens has been reviewed (Richa & Solter 1986; Fenderson et al 1990; Kimber 1990) and only observations relevant to implantation will be outlined here. Certain epitopes appear while others disappear as the trophoblast differentiates. For instance the SSEA-1 determinant (Le^x) is first expressed at the 8-cell stage, and may be involved in stabilisation of cell-cell interactions in the embryo at that time (Bird & Kimber 1984; Fenderson et al 1984). It diminishes on the trophectoderm of expanded blastocyst stage embryos but is retained on the inner cell mass (Solter et al 1978; Fox et al 1981; Wood et al 1992). In contrast, the Le^y antigen appears at the 8-16 cell stage in embryos flushed from the uterus, but not on embryos flushed earlier from the oviduct and cultured in vitro (Fenderson et al 1986; Kimber 1990). Apparently synthesis of the antigen is triggered by, or the antigen is absorbed from, the uterine fluid. The latter may be more likely as glycosylated molecules have been reported to be absorbed from the female tract (e.g Gaunt 1985: Kapur & Johnson 1988; St Jaques et al 1992) emphasising the potential influence of reproductive tract fluid on the embryo. However, embryos themselves synthesise molecules carrying these epitopes later in development (Gaunt 1985; Fenderson et al 1986) and some of these may be involved in implantation (see below).

What might be the nature of a trophectodermal receptor which binds to carbohydrate on the endometrial epithelium? In similar systems, such as plant lectin-sugar interaction or the binding of carbohydrate by selectins, heterophilic interaction occurs between a protein domain and the carbohydrate. S-type and C-type lectins have been identified in animal cells and may be involved in cell-cell interactions. S-type lectins have mainly been found to be galactose-binding molecules (Drickamer 1988). C-type but not S-type lectins are calcium dependent, a feature exhibited, albeit weakly, by binding of LNF-1-neoglycoproteins to the blastocyst. This suggests that the receptor for H-type-1 carrying glycoproteins might be a C-type lectin. We are currently testing this. An alternative to protein-carbohydrate interaction has been proposed by Fenderson and colleagues (1991).

They incorporated Le^x or Le^y into liposomes composed of ^{14}C-cholesterol and examined binding of these liposomes to glycolipid derivitised substrates (Eggens et al 1989; Fenderson et al 1991). Le^x-liposomes bound Le^x-glycolipid but Le^y liposomes bound H-type-1 and H-type-2 glycolipids but not others in a calcium- and density-dependent manner. Thus, binding of H-type-1 epitopes of endometrial glycoproteins could occur by a carbohydrate-carbohydrate interaction involving Le^y present on the blastocyst surface (see above). Future research will allow us to distinguish between these possibilities.

The initial attachment phase of implantation appears morphologically similar between species, so our data on the murine embryo may act as a pointer to universal mechanisms. Fucosylated components have been detected in the endometrial epithelium of other mammals, such as equids (Whyte & Allen 1985) and humans (Bychkov & Toto 1986; Damjanov & Lee 1986; Inoue et al 1987,1990; Garin-Chesa & Rettig 1989; Aplin, 1991; Ravn et al 1992). Immunocytochemical examination of the endometrial epithelium of the rat, cow and human revealed the presence of oligosaccharide antigens similar to the H-type-1 determinant involved in embryonic attachment in the mouse. Rat endometrial epithelial cells express the H-type-1 epitope under the control of ovarian steroids (Kimber & Glasser in preparation). In the human molecule(s) carrying this epitope appear to be masked by sialic acid. They increase in the luteal phase at the expected time for implantation (S. Lindenberg & S.Kimber unpublished). Thus the H-type 1 epitope may prove a useful adjunct in assessment of uterine biopsies (Aplin 1991). If similar molecules are involved in human implantation they may be clinically important in the diagnosis of infertility.

ACKNOWLEDGEMENTS

Thanks are due to members of my laboratory for data yet unpublished. Different areas of the research outlined in this article were supported by the Medical Research Council UK, Nuffield Foundation, BioCarb Inc Lund Sweden, World Health Organisation, Victoria University of Manchester and the Science and Engineering Research Council UK to which I am extremely grateful.

REFERENCES

Anderson, T.L. Olson, G.E. and Hoffman, L.H. (1986) Stage-specific alterations in the apical membrane glycoproteins of endometrial epithelial cells related to implantation in rabbits. Biol. Reprod. 34, 701-720.

Aplin, J.D. (1991) Glycans as biochemical markers of human endometrial secretory differentiation. J. Reprod. Fert. 91, 525-541.

Aplin, J.D. Charlton, A.K., and Ayad, S. (1988) An immunohistochemical study of endometrial extracellular matrix during the menstrual cycle and first trimester of pregnancy. Cell Tiss. Res., 253, 235-240.

Armant, R.D. (1991) Cell interactions with laminin and its proteolytic fragments during outgrowth of primary mouse trophoblast cells. Biol. Reprod. 45, 664-672.

Armant, D.R., Kaplan, H. A. and Lennarz, W.J. (1986a) Fibronectin and laminin promote in vitro attachment and outgrowth of mouse blastocysts Dev. Biol. 116, 519-523.

Armant, D.R., Kaplan, H. A. Mover, H. and Lennarz, W.J. (1986b) The effect of hexapeptides on attachment and outgrowth of mouse blastocysts cultured in vitro: Evidence for the involvement of the cell recognition tripeptide Arg-Gly-Asp. Proc Natl. Acad. Sci. 83, 6751-6755.

Babiarz, B.B. and Hathaway, H.J. (1988) Hormonal control of the expression of antibody-defined lactosaminoglycans in the mouse uterus. Biol. Reprod. 39, 699-706.

Baenziger, J.U. and Maynard, Y. (1980) Human hepatic lectins. J. Biol. Chem. 255, 4607-4613.

Bayna, E. M., Shaper, J.H. and Shur, B.D. (1988) Temporally specific involvement of cell surface ß1,4galactosyl- transferase during mouse embryo morula compaction. Cell 53, 145-157.

Behrendtsen, O. Alexander, C.M. and Werb, Z. (1992) Metalloproteinases mediate extracellular matrix degradation by cells from mouse blastocyst outgrowths. Development 114, 447-456.

Bird, J.M. and Kimber, S.J. (1984) Oligosaccharide containing fucose linked α(1-3) and α(1-4) cause decompaction of mouse morulae. Devl. Biol. 87, 267-276.

Blankenship, T.N. and Given, R. L. (1992) Penetration of the uterine basement membrane during blastocyst implantation in the mouse. Anat. Rec. 233, 196-204.

Brown, J.G. and Papaioannou, V.E. (1992) Distribution of hyaluronan in the mouse endometrium during periimplantation period. Differentiation 52, 61-68.

Bychkov, V. and Toto, P.D. (1986) Lectin binding to normal human endometrium. Gynecol. Obstet., Invest. 22, 29-33.

Carson, D.D., Dutt, A. and Tang, J-P. (1987b) Glycoconjugate synthesis during early pregnancy: hyaluronate synthesis and function. Dev. Biol. 120, 228-235.

Carson, D.D., Tang, J-P and Gay, S. (1988) Collagen supports embryo attachment and outgrowth in vitro: effects of the Arg-Gly-Asp sequence. Dev. Biol. 127, 368-375.

Carson, D.D., Tang, J-P. and Hu G. (1987a) Estrogen influences dolichyl phosphate distribution among glycolipid pools in mouse uteri. Biochem 26, 1598-1606.

Carson, D.D. Tang, J-Y. and Julian, J. (1992) Heparan sulfate proteoglycan (Perlecan) expression by mouse embryos during acquisition of attachment competence. Dev. Biol. 155, 97-106.

Carson, D.D., Wilson, O.F. and Dutt, A. (1990) Glycoconjugate expression and interactions at the cell surface of mouse uterine epithelial cells and periimplantation stage embryos. In `Trophoblast invasion and endometrial receptivity` (eds H-W. Denker & J.D. Aplin). Trophoblast Res. 4, 211-241.

Chavez, D.J. (1986) Cell surface of mouse blastocysts at the trophectoderm-uterine interface during the adhesive phase of implantation. Am. J. Anat. 176, 153-158.

Chavez, D. (1990) Possible involvement of D-galactose in the implantation process. In `Trophoblast invasion and endometrial receptivity` (eds H-W. Denker & J.D. Aplin). Trophoblast Res. 4, 259-272.

Chavez, D.J. and Anderson, T.L. (1985) The glycocalyx of the mouse uterine luminal epithelium during estrus; early pregnancy; the peri-implantation period and delayed implantation. Biol. Reprod. 32, 1135-1142.

Chavez, D.J. and Enders, A.C. (1981) Temporal changes in lectin binding of peri-implantation mouse blastocysts. Dev. Biol. 87, 267-276.

Damjanov, I. and Lee M.-C. (1986) Pregnancy-related changes in murine and human endometrium by differential binding of fluoresceinated lectins. In `Pregnancy Proteins in Animals' (ed. J. Hau) Walter de Gruyter & Co Berlin p178-183.

Dejana, E., Martin-Padura, I. Lauri, D., Bernasconi, S., Bani, M.R., Garofalo, A., Giavazzi, R. Magnani, J., Mantovani, A. and Menard, S. (1992) Endothelial leukocyte adhesion molecule-1-dependent adhesion of colon carcinoma cells to vascular endothelium is inhibited by an antibody to lewis fucosylated type 1 carbohydrate chain. Lab.Invest. 66, 324-330.

Denker, H.-W. (1990) Trophoblast-endometrial interactions at embryo implantation: a cell biological paradox. In `Trophoblast invasion and endometrial receptivity` (H-W. Denker & J.D. Aplin eds). Trophoblast Res. 4, 3-29.

Dickmann, Z., Dey, S.K. and SenGupta, J. (1976) A new concept: Control of early pregnancy by steroid hormones originating in the preimplantation embryo. Vitam. Horm. 34, 215-242.

Dickson, A.D. (1963) Trophoblast giant cell transformation of mouse blastocysts. J. Reprod Fert., 6, 465-466.

Drickamer, K. (1988) Two distinct classes of carbohydrate-recognition domains in animal lectins. J. Biol. Chem. 263, 9557-9560.

Dutt, A, Tang, J.-P., Welply, J.K. and Carson, D.D. (1986) Regulation of N-Linked glycoprotein assembly in uteri by steroid hormones. Endocrinology 118, 661-673.

Dutt, A. Tang, J-P. and Carson, D.D. (1988) Estrogen preferentially stimulates lactosaminoglycan-containing oligosaccharide synthesis in mouse uteri. J. Biol. Chem, 263, 2270-2279.

Dziadek, M. abd Timpl, R. (1985) Expression of nidogen and laminin in basement membranes during mouse embryogenesis and in teratocarcinoma cells. Dev. Biol. 111, 372-382.

Dziadek, M., Fujiwara, S., Paulsson, M. and Timpl, R. (1985) Immunological characterization of basement membrane types of heparan sulfate proteoglycan. EMBO J. 4, 905-912.

Eggens, I., Fenderson, B., Toyokuni, T., Dean, B., Stroud, M. and Hakomori, S-I. (1989) Specific interaction between Le^x and Le^x determinants. J. Biol. Chem. 264, 9476-9484.

El Shershaby, A.M. and Hinchcliffe, J.R. (1975) Epithelial autolysis during implantation of the mouse blastocyst: an ultrastructural study. J. Embryol. exp. Morphol. 33, 1067-1080.

Enders A.C. and Schlafke, S. (1974) Surface coats of the mouse blastocyst and uterus during the preimplantation period. Anat. Rec. 180, 137-150.

Farach, M.C., Tang, J.P. Decker, G.L. and Carson, D.D. (1987) Heparin/heparan sulfate is involved in attachment and spreading of mouse embryos in culture. Dev. Biol. 123, 401-410.

Fenderson, B.A., Holmes, E.H., Fukushi, Y. and Hakomori, S. (1986) Coordinate expression of X and Y haptens during murine embryogenesis. Dev. Biol. 114. 12-21.

Fenderson, B.A., Zehavi U. and Hakomori S-I. (1984) A multivalent lacto-N-fucopentaose III conjugate decompacts preimplantation mouse embryos while the free oligosaccharide is ineffective. J. Exp. Med., 160: 1591-1596.

Fenderson, B.A., Eddy, E.M. and Hakomori, S. (1990) Glycoconjugate expression during embryogenesis and its biological significance Bioessays 12. 173-179.

Fenderson, B., Kojiima, N., Stroud, M., Zhu, Z. and Hakomori, S. (1991) Specific interaction between Le^y and H as a possible basis for trophectoderm-endometrium recognition during implantation. Glycoconjugate J. 8 177a.

Feizi, T. (1985) Demonstration by monoclonal antibodies that carbohydrate structures of glycoproteins and glycolipids are onco-developmental antigens. Nature. 314. 53-57.

Feizi, T. (1988) Oligosaccharides in molecular recognition. Biochem. Soc. Trans. 16(6) 930-934.

Fox, N.W., Damjanov, I., Martinez-Hernandez, A., Knowles, B.B. and Solter, D. (1981) Immunohistochemical localization of the early embryonic antigen (SSEA-1) in postimplantation mammlian embryos and fetal and adult tissues Dev. Biol. 83, 391-398.

Gadsby, J.E., Heap, R.B. and Burton, R.D. (1980) Oestrogen production by blastocyst and early embryonic tissue of various species. J.Reprod. Fertil. 60, 409-417.

Garin-Chesa P. and Rettig W.J. (1989) Immunohistochemical analysis of LNT, NeuAc2-3LNT and Le^x carbohydrate antigens in human tumours and normal tissues. Am J. Pathol, 1315-1327.

Gaunt, S.J. In vivo and in vitro cultured mouse preimplantation embryos differ in their display of teratocarcinoma cell surface antigen: possible binding of an oviduct factor. J. Embryol. exp. Morph. 88, 55-69.

Gidley-Baird, A.A. (1981) Endocrine control of implantation in rats and mice. J. Reprod. Fert. 29, 97-109.

Glass, R.H. Aggeler, J. Spindle, A. Pedersen, R.A. and Werb, Z. (1983) Degradation of mouse trophoblast outgrowths: a model for implantation. J.Cell Biol. 96, 1108-1116.

Glasser, S.R. Lampelo, S. Munir, M.I. and Julian, J. (1987) Expression of desmin, laminin and fibronectin during in situ differentiation (decidualisation) of rat uterine stromal cells. Differentiation 35, 132-142.

Glasser, S.R. and McCormack, S.A.(1979) Estrogen modulated uterine gene transcription in relation to decidualization. Endocrinol. 104, 1112-1118.

Glasser, S.R. Julian, J. Decker, G.L. Tang, J-P. and Carson, D.D. (1988) Development of morphological and functional polarity in primary cultures of immature rat uterine epithelial cells. J.Cell Biol. 107, 2409-2423.

Grinnell, F., Head, J.R. and Hoffpauir, J. (1982) Fibronectin and cell shape in vivo: studies on the endometrium during pregnancy. J. Cell Biol. 94, 597-606.

Hewitt K., Beer, A.E. and Grinnell, F. (1979) Disappearance of anionic sites from the surface of the rat endometrial epithelium at the time of implantation. Biol. Reprod. 21, 691-707.

Holmes, P.V. and Dickson, A.D. (1973) Estrogen induced surface coat and enzyme changes in implanting mouse blastocysts. J. Embryol. exp. Morph. 29, 639-645.

Holmes P.V. and Lindenberg, S. (1988) Behaviour of mouse and human trophoblast cells during adhesion to and penetration of the endometrial epithelium. In `Eukaryote Cell Recognition: concepts and model systems. (eds Chapman, G.P., Ainsworth, C.C. Chatham, C.J.) Camb. Univ. Press p225-237.

Inoue M., Nakayama, M. and Tanizawa, O. (1990) Altered expression of Lewis blood group and related antigens in fetal, normal adult and malignant tissues of the uterine endometrium. Virchows Arch. A Pathol. Anat. 416, 221-228.

Inoue, M., Sasagawa, T., Saito, J., Shimizu, H., Ueda, G., Tanizawa, O. and Nakayama, M. (1987) Expression of blood group A,B, H, Lewis-a and Lewis-b in fetal normal and malignant tissues of the uterine endometrium. Cancer, 60. 2985-2993.

Jacobs, A.L. and Carson, D.D. (1991) Proteoglycan synthesis and metabolism by mouse uterine stroma cultured in vitro. J. Biol. Chem 266, 15464-15473.

Jenkinson E.J. & Wilson,I.B. (1973) In vitro studies on the control of trophoblast outgrowth in the mouse. J,Embryol. Exp. Morphol. 30, 21-30.

Kapur, R.P. and Johnson, L.V. (1988) Ultrastructural evidence that specialised regions of the murine oviduct contribute a glycoprotein to the extracellular matrix of mouse oocytes. Anat. Rec. 221, 720-729.

Kaufman, M. (1983) The origin properties and fate of trophoblast in the mouse. In Biology of Trophoblast (Loke,C. & Whyte,J. eds) 23-67.

Kimber, S.J. (1988) The role of fucosylated glycoconjugates in cell-cell interactions of the mammalian pre-implantation embryo. In `Eukaryote Cell Recognition: concepts and model systems. (Chapman, G.P., Ainsworth, C.C. Chatham, C.J. eds) Camb. Univ. Press p194-224.

Kimber, S.J. (1990) Glycoconjugates and cell surface interactions in pre- and peri-implantation development. Int. Rev. Cytol. 120, 53-163.

Kimber S.J. and Lindenberg S. (1990) Hormonal control of carbohydrate determinants involved in implantation. J. Reprod. Fertil 89, 13-21.

Kimber, S.J. Lindenberg, S. and Lundblad, A.(1988) Distribution of some Galβ1-3(4)GlcNAc related carbohydrate antigens on the mouse uterine epithelium in relation to the peri-implantation period. J. Reprod Immunol. 12, 297-313.

Kimber, S., White, S., Cook, A. and Illingworth, I. (1993) The initiation of implantation. Parallels between attachment of the embryo and neutrophil-endothelial interaction? In `Gametesand Embryo Quality' (Ed L. Mastrioianni, Jr) Parthenon publ., Carnforth, in press.

Kimber, S.J. Waterhouse, R. and Lindenberg, S. (1993a) In vitro models for implantation. Serono Symposium on Preimplantation Development (Ed. B. Bavister) Springer Verlag, pp 244-263.

Kirby, D.R.S. (1965) The invasiveness of the trophoblast. In The early conceptus, normal and abnormal. (W. Park ed) University of St Andrew's Press, Edinburgh, pp 60-73

Kirby, D.R.S., Potts, D.M. and Wilson, I.B. (1967) On the orientation of the implanting blastocyst. J.Embryol. exp. Morphol. 17, 527-523.

Kohler, G. and Milstein, C. (1975) Continuous cultures of fused cells secreting antibody of predefined specificity. Nature 256, 495-497.

Kubo, H. Spindle, A. and Pedersen, R.A. (1981) Inhibition of blastocyst attachment and outgrowth by protease inhibitors. J. Exp. Zool. 216, 445-451.

Kuivanen, P.C. and De Sombre, E.R. (1985) The effects of sequential administration of 17ß-estradiol on the synthesis and secretion of specific proteins in the immature rat uterus. Biochem 22, 439-451.

Lala, P.K. and Graham, C.H. (1990) Mechanisms of trophoblast adhesiveness and their control: the role of proteases and protease inhibitors. Cancer Metastasis Reviews 9, 369-379.

Lasky, L.A. (1991) Lectin cell adhesion molecules (LECCAMS): a new family of cell adhesion proteins involved with inflammation. J. Cell. Biochem. 45, 139-146.

Lasky, L.A. (1992) Selectins: Interpreters of cell-specific carbohydrate information during inflammation. Science 258, 964-969.

Lindenberg, S., Kimber S.J. and Kallin, E. (1990) Carbohydrate binding properties of mouse embryos. J. Reprod. Fert., 89, 431-439.

Lindenberg, S., Sundberg, K., Kimber, S.J. and Lundblad, A. (1988) The milk oligosaccharide, lacto-N-fucopentaose I inhibits attachment of mouse blastocysts on endometrial monolayers. J. Reprod. Fertil. 83, 149-158.

Lowe, J.B. Stoolman, L.M., Nair, R.P. Larsen, R.D. Berhend. T.L. and Marks, R.M. (1990) ELAM-1-dependent cell adhesion to vascular endothelium determined be a transfected human fucosyltransferase gene. Cell 63, 475-484.

McCormack, J.T. and Greenwald, G.S. (1974) Progesterone and estradiol-17β concentrations in the peripheral plasma during pregnancy in the mouse. J. Endocrinol. 62, 101-107.

McEver, R.P. (1991) Selectins: novel receptors that mediate leukocyte adhesion during inflammation. Thrombosis and Haemostasis 65, 223-228.

Meyers, K.P. (1970) Hormonal requirements for the maintenance of estradiol-induced inhibition of uterine sensitivity in the ovariectomised rat. J. Endocrinol. 46, 341-346.

Morris, J.E. and Potter, S.W. (1984) A comparison of developmental changes in surface charge in mouse blastocyst and epithelium using DEAE beads and dextran sulphate in vitro. Dev. Biol. 103, 190-199.

Morris, J.E. and Potter, S.W. (1990) An in vitro model for studying interactions between the mouse trophoblast and uterine epithelial cells. In `Trophoblast invasion and endometrial receptivity` (H-W. Denker and J.D. Aplin eds). Trophoblast Res. 4, 51-69.

Morris, J.E., Potter, S.W. and Gaza-Bulseco, G. (1988) Estradiol induces an accumulation of free heparan sulfate glycosaminoglycan chains in uterine epithelium. Endocrinol. 122, 242-253.

Mulholland, J. and Leroy, F. (1988) Protein and mRNA synthesis in the peri-implantation rat endometrium. In `Blastocyst implantation' (K.Yoshinaga, ed.) Serono Symposium USA. Adams Pub. Group. Boston USA, p31-38.

Mulholland, J., Aplin, J.D. Ayad, S. Hong, L. and Glasser, S.R. (1992) Loss of collagen type VI from rat endometrial stroma during decidualisation. Biol. Reprod. 46, 1136-1143.

O'Shea, K.S. Liu, L.-H.J., Kinnunen, L.H. & Dixit, V.M. (1990) Role of thrombospondin in the early development of the mouse embryo. J. Cell Biol. 111, 2713-2723.

Oriol, R., Le Pendu, J. and Mollicone, R. (1986) Genetics of ABO, H, Lewis, X and related antigens. Vox Sang 51, 161-171.

Parr, E.L., Tung, H.N. and Parr, M.B. (1987) Apoptosis as the mode of uterine epithelial cell death during embryo implantation in mice and rats. Biol. Reprod. 36, 211-225.

Phillips, M.L., Nudelman, E., Gaeta, F.C.A., Perez, M.P., Singhal, A.K. Hakomori, S.-I. and Paulson, J.C. (1990) ELAM-1 mediated cell adhesion by recognition of a carbohydrate ligand, sialyl-Le^x. Science 250, 1130-1132.

Psychoyos, A. (1973) Hormonal control of ovoimplantation. Vitam. Horm. 31, 201-256.

Psychoyos, A. (1976) Hormonal control of uterine receptivity for nidation. J. Reprod. Fertil. 25, 17-28.

Psychoyos, A. (1986) Uterine receptivity for nidation. Annals NY Acad. Sci. USA 476, 36-42.

Rastan, S., Thorpe,S.J. Scudder, P., Brown, S., Gooi, H.C. and Feizi, (1985). Cell interactions of preimplantation embryos: evidence for involvement of the poly-N-acetyllactosamine series. J. Embryol. Exp. Morph. 87, 115-128.

Ravn, V. Teglbjaerg, C.S., Mandel, U. and Dabelsteen, E. (1992) The distribution of type-2 chain histoblood group antigens in normal cycling human endometrium. Cell & Tissue Res. 270, 425-433.

Richa, J. and Solter, D. (1986) Role of cell surface molecules in early mammalian development. In: "Experimental approaches to mammalian embryonic development. (eds J. Rossant, R. Pederson) Camb. Univ. Press p 293-320.

Rider, V. Carlone, D.L. Witrock, D. Cai, C. and Oliver, N. (1992) Uterine fibronectin mRNA content and localisation are modulated during implantation. Devel. Dynamics 195, 1-14.

Romagno, L. and Babiarz, B. (1990) The role of murine cell surface galactosyltransferase in trophoblast: laminin interactions in vitro. Dev. Biol. 141, 254-261.

Sclafke, S. and Enders, A. (1975) Cellular basis of interaction between trophoblast and uterus at implantation. Biol. Reprod. 12, 41-65.

Sherman, M.I. (1978) Implantation of mouse embryos in vitro. In `Methods in mammalian reproduction.' (J.C. Daniel Jr. ed) Academic press pp 247-272.

Sherman, M.I., Gay, R., Gay, S. and Miller, E.J. (1980) Association of collagen with preimplantation and peri-implantation mouse embryos. Dev. Biol. 74, 470-478.

Solter, D. and Knowles, B.B. (1978) Monoclonal antibody defining a stage specific mouse embryonic antigen (SSEA-1).Proc Natl. Acad. Sci. USA 75, 5565-5569.

Springer, T.A. (1990) The sensation and regulation of interactions with the extracellular environment - the cell biology of lymphocyte adhesion receptors. Ann. Rev. Cell Biol. 6. 359-402.

St Jaques, S. Malette, B. Chevalier, S. Roberts, K.D. and Bleau, G. (1992) The zona pellucida binds the mature form of an oviductal glycoprotein (Oviductin). J. Exp. Zool. 262, 97-104.

Stinnakre, M.G. Evans, M.J. Willison, K.R. and Stern, P.L. (1981) Expression of Forssman antigen in the postimplantation mouse embryo. J. Embryol. exp. Morph. 61, 117-131.

Stowell, C.P. and Lee, Y.C. (1978) The binding of D-glucosyl-neoglycoproteins to the hepatic asialoglycoprotein receptor. J. Biol. Chem. 253, 6107-6110.

Sutherland, A.E. Calarco, P.G. and Damsky, C.H. (1988) Expression and function of cell surface extracellular matrix receptors in mouse blastocyst attachment and outgrowth. J.Cell Biol. 106, 1331-1348.

Svalander, P.C. Odin, P. Nilsson, B.O. and Obrink, B. (1990) Expression of cellCAM-105 in the apical surface of rat uterine epithelium is controlled by ovarian steroid hormones. J. Reprod. Fertil. 88, 213-221.

Tang, J.-P., Julian, J., Glasser, S.R. and Carson, D.D. (1987) Heparan sulfate proteoglycan synthesis and metabolism by mouse uterine epithelial cells in vitro. J. Biol. Chem. 262, 12832-12842.

Teng, T.T., Walker, M.P. Bhattacharyya, S.N., Klapper, D.G., DiAugustine, R.P. and McLachlan, J.A. (1986) Purification and properties of an oestrogen-stimulated mouse uterine glycoprotein (approx 70kD). Biochem. J. 240, 413-422.

Tyrrell, D., James, P., Rao, N., Foxall, C., Abbas, S., Dasgupta, F., Nashed, M., Hasegawa, A., Kiso, M., Asa, D., Kidd, J. and Brandley, B.K. (1991) Structural requirements for the carbohydrate ligand of E-Selectin. Proc. Natl. Acad. Sci. USA 88, 10372-10376.

Valdizan, M.C. Julian, J. and Carson, D.D. (1992) WGA-binding mucin glycoproteins protect the apical cell surface of mouse uterine epithelial cells. J. Cell Physiol. 151, 451-465.

Wassarman, P.M. (1992) Mouse gamete adhesion molecules. Biol. Reprod. 46, 186-191.

Welsh, A. and Enders, A.C. (1989) Comparison of the ability of cells from rat and mouse blastocysts to alter complex extracellular matrix in vitro. In `Blastocyst implantation` Yoshinago K. ed., Adams publ. Co. Boston 55-74.

Wewer, U.A, Damjanov, A. Weiss, J. Liotta, L.L. and Damjanov, I. (1986) Mouse endometrial stromal cells produce basement-membrane components. Differentiation 32, 49-58.

White, S. and Kimber, S.J. (1993) Changes in α(1-2)fucosyltransferase activity in the murine endometrial epithelium during the estrous cycle, early pregnancy and after ovariectomy and hormone replacement. Biol. Reprod. in press.

Whyte, J. and Allen, W.R. (1985) Equine endometrium at preimplantation stages of pregnancy has specific glycosylated regions. Placenta 6, 537-542.

Willison, K.R., Karol, R.A., Suzuki, A., Kundu, S.U. and Marcus, D.M (1982) Neutral glycolipid antigens as developmental markers of mouse teratocarcinoma and early embryos: An immunological and chemical analysis. J. Immunol. 129, 603-609.

Winterhager, E., Brummer, F., Dermietzel, R., Hulser, D. and Denker, H.-W. (1988) Gap junction formation in rabbit uterine epithelium in response to embryo recognition. Dev. Biol. 126, 203-211.

Winterhager, E. Stutenkemper, R. Traub, O. Beyer, E. and Willecke, K. (1991) Expression of different connexin genes in rat uterus during decidualisation and at term. European J. Cell Biol. 55, 133-142.

Wood, M.J. Sjoblom, P., Lindenberg, S. and Kimber, S.J. (1992) Effect of slow and ultra-rapid freezing on cell surface antigens of 8-cell mouse embryos. J. Exp. Zool. 262, 330-339.

Wordinger, R.J., Brun-Zinkernagel, A.-M. and Jackson, T. (1991) An ultrastructural study of in-vitro interaction of guinea pig and mouse blastocysts with extracellular matrices. J. Reprod. Fert. 93, 585-597.

Yamagata, T. and Yamazaki, K. (1991) Implanting mouse embryos stain with LNF-1 bearing fluorescent probe at their mural trophectoderm side. Biochem. Res. Comm. 181, 1004-1009.

Zalik, S.E. (1991) On the possible role of endogenous lectins in early animal development. Anat. Embryol. 183, 521-536.

IN VITRO ANALYSIS OF EPITHELIAL SURFACE CHANGES DURING IMPLANTATION

John E. Morris, Sandra W. Potter, Georgeen Gaza

Department of Zoology
Oregon State University
Corvallis, OR 97331-2914

INTRODUCTION

Blastocyst implantation involves interaction between two independently controlled yet highly interdependent systems, the embryo and its maternal environment. In rodents the preimplantation blastocyst becomes closely surrounded by uterine epithelium in an "implantation chamber" (Enders, 1975), which precedes the first irreversible interaction between the embryo and uterus -- adhesion between the apical surfaces of the first embryonic epithelium (trophectoderm) and the uterine epithelium ("1" in Figure 1). In mice, adhesion occurs about 100 hours after the morning in which the coital vaginal plug is discovered (Potts, 1968). In rats this is brought about by a nidatory surge of estradiol, accompanying increasing levels of progesterone (Psychoyos, 1973). There is no clear separation in time between the appearance of apical adhesion and the initiation of basolateral changes, which occur between about 3.5 and 4.5 days after mating ("2" in Figure 1). These changes are characterized by a loosening of lateral cell associations, involving ultimately the junctional complexes ("3" in Figure 1) and a separation of the epithelium from its basal lamina (Schlafke et al., 1985), preparatory to its subsequent sloughing and cell death (Parr et al., 1987).

In this chapter we report our observations on these phenomena *in vitro*, using three different systems: hanging drop culture to study apical epithelial interactions, cell monolayer culture to study surface turnover, and rotation mediated suspension culture to examine junctional association. Although we have focused exclusively on the epithelium in the present work, it is important to emphasize that the exquisite sensitivity of the uterine epithelium to the maternal hormonal status throughout the estrous cycle and during the initiation of these changes may depend directly or indirectly on participation of the subadjacent stromal cells (Glasser and McCormack, 1981; Cunha et al., 1985; Astrahantseff and Morris, 1992).

CHANGES IN THE APICAL EPITHELIAL SURFACE

Because blastocysts *in vitro* are relatively non-adhesive toward living cells compared with their tenacious binding to and spreading on non-biological surfaces (Glass et al., 1980; Morris et al., 1983), the idea has developed that the pre-implantation uterine epithelial surface coat (the glycocalyx) blocks blastocyst adhesion (Enders and Schlafke, 1974; Morris and Potter, 1990). One group of candidate blocking molecules is the sialylated glycoconjugates, which are the dominant contributors to negative surface charge on these cells (Morris and Potter, 1984; Zhu et al., 1990). Net negative surface charge on both blastocysts and uterine epithelial cells decreases sharply as the time of blastocyst adhesion approaches, by the criteria of histological staining (Hewitt et al., 1979), electrophoretic mobility (Nilsson and Hjerten, 1982), and binding to DEAE beads (Morris and Potter, 1984). In contrast, there is an increase in specific cell adhesion molecules, notably lacto-N-fucopentaose-1 (see Kimber in the present volume and Kimber and Lindenberg, 1990; Lindenberg et al., 1990), which likely binds specific receptors on blastocysts (Yamagata and Yamazaki, 1991). Lacto-N-fucopentaose-1 may be one of the oligosaccharides comprising the heterogeneous galactosyltransferase-binding lactosaminoglycans, which have been identified as adhesive ligands between uterine epithelial cells (Dutt et al., 1987, 1988). There is evidence from studies of endothelial-leukocyte interaction that a transition from sialylated to fucosylated lactosaminoglycans may regulate cell adhesion (Lowe et al., 1990; Zhou et al., 1991), and it is reasonable to expect that there is a similar regulation of both adhesion inhibiting and promoting molecules also in blastocyst implantation.

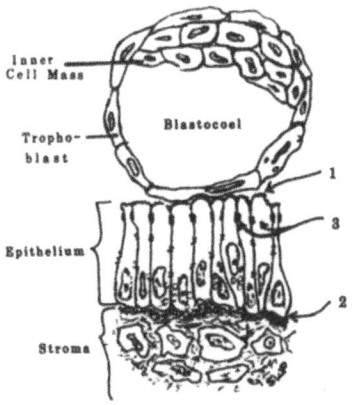

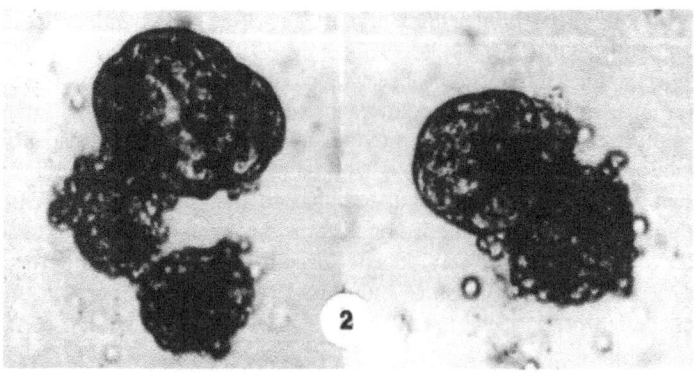

Figure 1. Cartoon of hatched mouse blastocyst prior to implantation, indicating the major epithelial events examined here using three different *in vitro* systems. Following this apposition stage, the blastocyst adheres to the apical surface of the uterine epithelium by specific adhesion sites (1). During the invasion stage the epithelium shows a loosening of basal and basolateral associations resulting in separation from the basal lamina (2) and a reorganization of lateral junctions (3).

Figure 2. Hanging drop cultures of mouse 3.5-day blastocysts with vesicles of uterine epithelium. Living blastocysts adhering to individual vesicles in drops after 2 days in culture. x560.

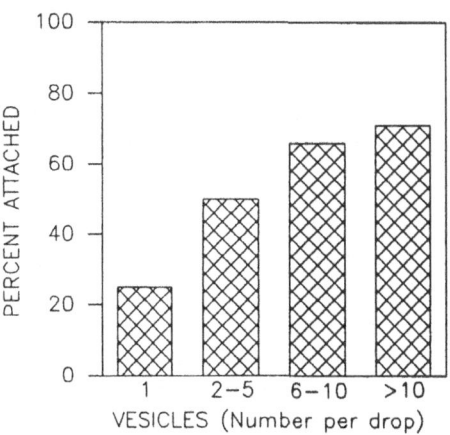

Figure 3. Influence of vesicle number on adhesion frequency. Hanging drop cultures of 3.5-day blastocysts were placed together with varying numbers of epithelial vesicles for 2 days in 15-20 μl drops of medium. Adhesion was determined by the failure of blastocyst and vesicle to separate when vibrated. Data from one experiment for 2-5 vesicles; others are averages of at least two experiments.

In order to examine various adhesion control mechanisms more critically, some years ago we extended the classical hanging drop culture system to permit the direct microscopic examination of adhesion between single blastocysts and single vesicles of uterine epithelium (Figure 2 and Morris et al., 1982; Morris and Potter, 1990). The major advantages of this system over systems in which blastocysts are placed on sheets of epithelial cells in culture (Salomon and Sherman, 1975; Enders et al., 1981) are that the sites of contact could be directly observed and that blastocysts interacted only with epithelial cells, not the culture substratum. A major disadvantage is that the frequency of adhesion is low and unpredictable, so that it is not possible to get statistically meaningful tests of inhibitors or stimulators.

A likely reason for the low adhesion frequency with the hanging drop system is the fact that within some drops the blastocyst and vesicle remain in intimate contact but in others they separate due to various factors ranging from vibrations in the incubator to interposition of cell secretions or debris. The need for sustained intimate contact is suggested by the apposition stage of implantation in which rodent blastocysts are tightly enclosed in the implantation chamber *in vivo* prior to attachment (Enders, 1975). The need for intimate contact *in vitro* in rabbits was shown by the requirement for bundling blastocysts and epithelium together in miniature dialysis bags, which gave a significantly higher frequency of adhesion than when they were in hanging drops (Hohn and Denker, 1990). Applying these ideas to our system, we have been able to greatly enhance adhesion frequency in hanging drops simply by packing enough uterine epithelial vesicles together with a single blastocyst into the drop to assure intimate contact (Figure 3). We cannot yet say whether adhesion frequency increased due to greater contact rather than, or in addition to, a higher concentration of a factor released from the epithelial vesicles.

CHANGES IN THE EPITHELIAL BASOLATERAL SURFACES AND BASAL LAMINA

Although events at the apical surface of uterine epithelial cells probably are the most important for initial adhesive interactions, changes at the more extensive basolateral surfaces and basal lamina may drive those at the apical surface. For example, uterine epithelial integrins may be retargeted from their more characteristic residence on the basal surface where they interact with the basal lamina (Lessey et al., 1992) to lateral or even the apical surface (see Denker in this volume), as in keratinocytes (Larjava et al., 1990) and endothelial cells (Conforti et al., 1992) in culture. Beyond this, associations between adjacent epithelial cells and between these cells and their matrix must undergo dramatic changes during the invasive stage of implantation, as the epithelium becomes penetrable by the invasive trophoblast (Damsky et al., 1992) and loosens its grip on its basal lamina. Although molecules moving laterally within the plasma membrane of most epithelia are restricted to their respective apical and basolateral compartments by the junctional complex that girdles and links the cells near the apical surface, there is some shuttling of plasma membrane proteins from the basal to apical surfaces of at least some epithelia by endocytosis and retargeting to the new surface (Nelson, 1992). Retargeting has not been directly demonstrated in uterine epithelium, but there is ultrastructural evidence for basal and basolateral to apical vesicular traffic (Parr, 1980). *In vitro* analysis of tissue from rat uteri cultured on membrane filters coated with a basal lamina preparation from Englebreth-Holmes-Swarm tumor (Matrigel) has revealed a distinct polarity of secretion (Glasser et al., 1988). Keratan sulfate and heparan sulfate are secreted into the medium predominantly from the apical surface (Carson et al., 1988) and prostaglandin$_{F2\alpha}$ from the basal surface (Jacobs et al., 1990).

In our work we have focused on the heparan sulfate proteoglycans (HSPGs) associated with the basolateral plasma membranes and in the basal lamina of mouse uterine epithelium as molecules of particular interest during implantation. In mammalian epithelia the dominant HSPGs exist either as syndecan, the 200-250 kDa HSPG of the basolateral plasma membrane (Saunders et al., 1989), or as perlecan, the 467 kDa HSPG of the basal lamina (Noonan et al., 1991). Syndecan anchors the uterine epithelium apparently by extending through the basal lamina to the underlying stromal matrix and binding collagen I, and perlecan acts within the basal lamina to maintain the integrity of this barrier between epithelial and stromal cells. In addition, both proteoglycans bind growth factors (Bernfield and Hooper, 1991; Ruoslahti and Yamaguchi, 1991; Yayon et al., 1991). As we show here, the two types of HSPG also appear to be sensitive to the uterine hormonal milieu in distinctly different ways.

Turnover of Cell Membrane HSPG

A cell membrane HSPG that is the major metabolically active proteoglycan in uterine epithelium *in vivo* and *in vitro* appears to be identical to the mouse mammary cell membrane HSPG, syndecan, or closely related to it, on the basis of its size and glycosaminoglycan composition (Morris et al., 1988a) and the reactivity of tissue sections to anti-syndecan antibody (Potter and Morris, 1992). The uterine epithelial HSPG, however, appears to have a much smaller ratio of chondroitin sulfate to heparan sulfate than that reported for syndecan (Saunders et al., 1989).

Despite the fact that in autoradiographs of ^{35}S-sulfate-labeled tissue sections the vast majority of exposed silver grains lie in the stromal matrix, extraction of the epithelial surface of the intact uterus with 1% non-ionic detergent, Nonidet P-40 (NP-40) for 1 minute removes label almost exclusively from the epithelium. Electron micrographs show that tissue extraction occurs down to the level of the basal lamina barrier but not below it. Nearly all of the epithelial labeling is on the basolateral and basal surfaces. We found label on the apical surface to be less predictable, always in lesser amounts when present, and completely removed by extraction with the NP-40 (Morris et al., 1988a). Following the intraperitoneal injection of 12.5 ng of estradiol into 3-week old sexually immature mice daily for 3 days (Figure 4) or into ovariectomized adult mice daily for 2 days (Morris et al., 1988a), there was little apparent change in the extractable cell membrane

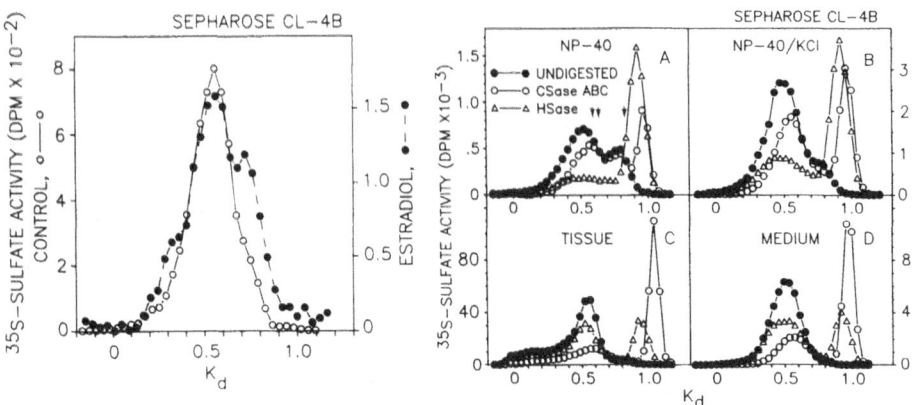

Figure 4. Proteoglycans from cultures of immature mice. Uteri were labeled in culture 1 hr, washed, and extracted with NP-40. The isolated proteoglycans were chromatographed on Sepharose CL-4B in 4 M guanidine-HCl and 0.5% Triton X-100.

Figure 5. Proteoglycans from cultures of uteri from pregnant mice. Uteri were labeled in culture 4 hr, removed from the culture medium (D), and sequentially separated into NP-40 extract (A), NP-40/KCl extract (B), and residual tissue (C). The isolated proteoglycans were chromatographed on Sepharose CL-4B as in Figure 4 following either no treatment (UNDIGESTED), digestion with chondroitinase ABC (Chase ABC) to reveal HSPGs, or digestion with heparitinase (HSase) to reveal DS/CSPGs. The columns in Figures 4 and 5 were calibrated with blue dextran (0 K_d), [^{3}H]glucosamine (1.0 K_d), and three protein standards (arrows above A, from the left): catalase (232K MW), aldolase (158K MW), and ribonuclease A (13.7K MW). (Figures 4 and 5 reprinted from *J. Biol. Chem.* 263:4712, 1988, by permission, The American Society for Biochemistry).

HSPG peak, but there was a sharp relative increase in the amount of label going into the lysosomally degraded HSPG (Morris et al., 1988b). An identical increase was seen in 4.5-day pregnant mice (Figure 5A). These observations, together with pulse-chase labeling and the use of chloroquine to block lysosomal activity provide evidence that the HSPG is endocytosed and degraded at the same time it is actively being synthesized and carried to the cell surface. The turnover in HSPG was not affected by prior injection with progesterone or its inclusion with estradiol, suggesting that the turnover during early pregnancy may also be an estradiol mediated event. It is significant that both during estradiol-induced mitotic activity in estrus (Finn and Publicover, 1981) and during establishment of new cell adhesive interactions in early pregnancy, the turnover of cell-surface macromolecules correlates with renewed biological activity in the cells.

Reorganization of HSPG within the Cell Membrane

If the metabolic turnover of membrane HSPG reflects a general remodeling of the cell membrane in uterine epithelium, a change in distribution of this HSPG might occur as it does in the branching of mouse mammary gland (Rapraeger et al., 1986) or its reorganization of adhesive junctions in culture (Rapraeger and Bernfield, 1985). Accordingly, we examined its localization during the estrous cycle and during the early stages of pregnancy (Potter and Morris, 1992). Using monoclonal antibody 281-2 (Rapraeger et al., 1986), we found that syndecan shifts from a basolateral to a predominantly basal localization, and into the basal lamina and subadjacent stromal matrix, as the estrous cycle progressed from metestrus toward estrus. Stromal localization of anti-syndecan-reactive HSPG has also been reported in estrous rats (Carson et al., 1993). This response could be induced by the injection of estradiol into intact immature or ovariectomized adult mice (Figure 6). Between 3.5 to 4.5 days of pregnancy the process was reversed, and both epithelial and stromal staining became much less intense. This distribution is quite

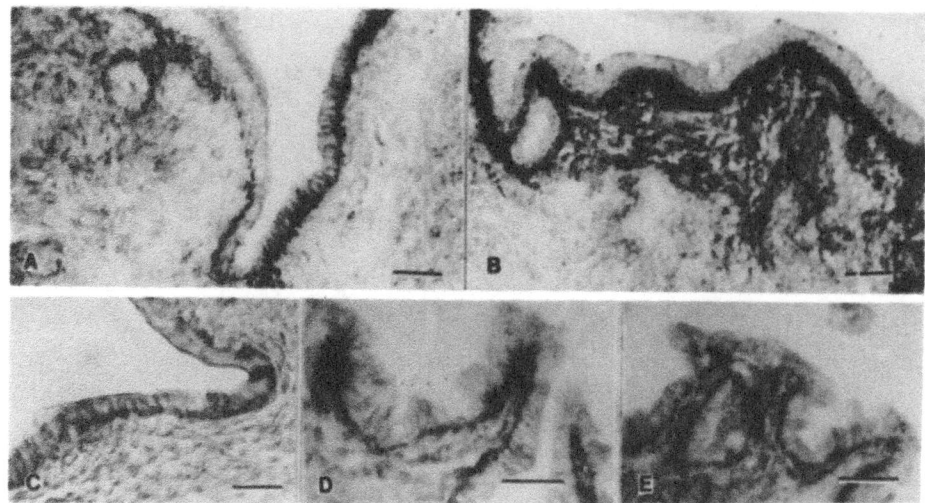

Figure 6. Comparison of anti-syndecan localization in uterine tissue from immature (A,B) and ovariectomized mature (C,D,E) mice injected with saline (A,C) or estradiol (B,D) or with simultaneous injections of estradiol and progesterone (E). Uteri of animals receiving estradiol showed more intense stromal reaction with anti-syndecan monoclonal antibody, 281-2 (B,D). Those receiving both steroids (E) had stromal staining equivalent to those receiving estradiol alone (D). Bars = 50 μm. (Reprinted from *Anat. Rec.* 284:383, 1992, with permission, John Wiley & Sons).

different from that of the cell adhesion molecule uvomorulin (E-cadherin), discussed in the last section.

Stability and Stimulation of Basal Lamina HSPG

On the other hand, the localization of perlecan, which is not inserted in the plasma membrane, appears unaffected by steroids in the uterus (Potter and Morris, 1992). Regardless of the stage of the estrous cycle or the biological changes up to at least 4.5 days of pregnancy a polyclonal antibody AF-2 (Hassell et al., 1985) or monoclonal antibody HK-84 or HK-102 (Kato et al., 1988) always localized to the basal lamina of the uterine epithelium and vascular endothelium (Figure 7 and Potter and Morris, 1992). The rigid localization only reflects the stability of basal lamina localization; it is not evidence of metabolic stability. Under conditions that clearly labeled the metabolically active cell membrane HSPG in uteri from pregnant mice, no peak of radioactivity was seen in a larger proteoglycan that could have represented basal lamina HSPG, even when the tissue was extracted with 1 M KCl, which should have released basal lamina proteoglycans (Figure 5B). The large proteoglycans that remained in the tissue after extraction (Figure 5C) and those that were released into the culture medium (Figure 5D) were predominantly chondroitin sulfate proteoglycans. In uteri of immature mice stimulated to grow in response to estradiol, however, a small but distinct enhancement of large HSPG synthesis was occasionally seen after only 1 hour of incubation with ^{35}S-sulfate (see peak at 0.2-0.3 K_d in Figure 4). These observations suggest that the basal lamina HSPG turns over very slowly and is under tight metabolic control.

To investigate the control of basal lamina HSPG turnover uterine epithelial cells from immature mice were isolated by digestion in pancreatin and trypsin and incubated with ^{35}S-sulfate in culture for 4 hours. During the first 6 hours of culture (4 hours of which were with ^{35}S-sulfate) the proteoglycans synthesized were indistinguishable from those synthesized in intact tissue, by the criteria of enzyme susceptibility and Sepharose

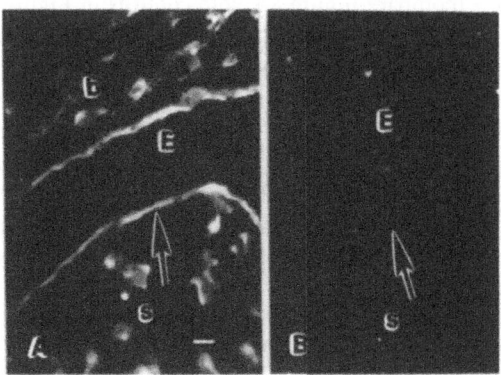

Figure 7. Immunohistochemical demonstration of basal lamina HSPG in mouse uterine epithelium. Sections of fresh-frozen tissue were fixed and permeabilized with methanol at -20 °C. Monoclonal antibody HK-84 against perlecan, the EHS tumor low-density basal lamina HSPG (A) and nonrelevant IgG-containing rat hybridoma supernatant (B) were stained with FITC-conjugated rabbit anti-rat IgG. Note that the basal lamina of blood vessels also is antigenic for HK-84. E, epithelium; b, blood vessels; *arrow*, basal lamina of epithelium.

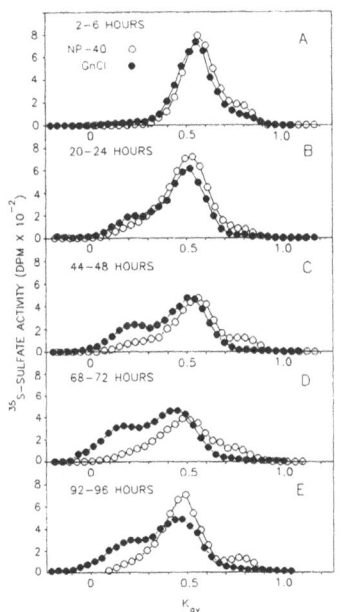
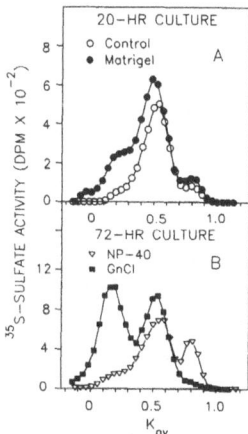

Figure 8. A large metabolically active and detergent resistant proteoglycan increased in isolated epithelium during culture. Flakes of epithelium from uteri of immature mice in culture were labeled with ^{35}S-sulfate for 4 hours during the interval indicated (A-E). A proteoglycan eluting at about 0.2 K_{av} from Sepharose CL-4B was labeled only after at least 6 hours of culture (B-E). It was not extracted with detergent (Nonidet-P40) but was extracted in the detergent-resistant fraction with guanidine-HCl (GnCl) and reacted with anti-perlecan antibody on western blots.

Figure 9. Relative increase of the 0.2 K_{av} peak during culture on Matrigel. Fragments were cultured for 20 hours (A) or 72 hours (B) on 1:1 dilution Matrigel. (A) Guanidine-HCl extracts of 20-hour cultures on Matrigel or on plastic (control) with ^{35}S-sulfate. They were sequentially extracted with NP-40 and guanidine-HCl (GnCl). Only the guanidine-HCl extracts are shown for A, and only the Matrigel culture is shown for B. (Figures 8 and 9 reprinted from *In Vitro Cell. Devel. Biol.*, in press by permission of The Tissue Culture Association of America).

CL-4B chromatography (compare Figure 8A with controls in Figure 4). In other words, the process of isolating the cells from the tissue had no significant influence on proteoglycan synthesis. After longer periods of culture, however, as the epithelial fragments had recovered and spread on the dish, a large HSPG gradually appeared in detergent and 4 M guanidine-HCl extracts (but not in detergent extracts without high ionic strength); small amounts appeared after 24 hours (Figure 8B), increasing to a maximum by 48 hours (Figure 8C), and maintaining a steady rate of accumulation up to at least 96 hours (Figure 8D,E). Susceptibility to heparitinase but not to chondroitinase and binding of monoclonal antibody HK-84 or HK-102 to transfer blots of the proteoglycans strongly suggest that the large HSPG is perlecan (Morris et al., 1993).

Culturing the cells on matrix obtained by detergent or EDTA treatment of stromal cell monolayers had no influence on the 20- or 72-hour pattern of HSPG synthesis (data not shown), but culture over Matrigel, a basal lamina extract from Engelbreth-Holm-Swarm (EHS) tumor, induced a dramatic increase in the accumulation of ^{35}S-sulfate in the large HSPG fraction (Figure 9). The maximal effect required culture with intact Matrigel, but we were able to obtain a partial stimulation from simple medium extracts of Matrigel obtained by either a 1-hour incubation with 5X concentrated DMEM (Figure 10C,D) or a 24-hour incubation with 1X DMEM. Because we observed less stimulation of large HSPG by Matrigel in the absence of serum (Figure 10B) than with it (Figure 10C), we asked whether the action of both Matrigel and serum could be due to growth factors. The examination of transforming growth factor-ß1 (TGF-ß1), basic fibroblast growth factor, epidermal growth factor, insulin-like growth factor-1, and interleukin-1 revealed that only

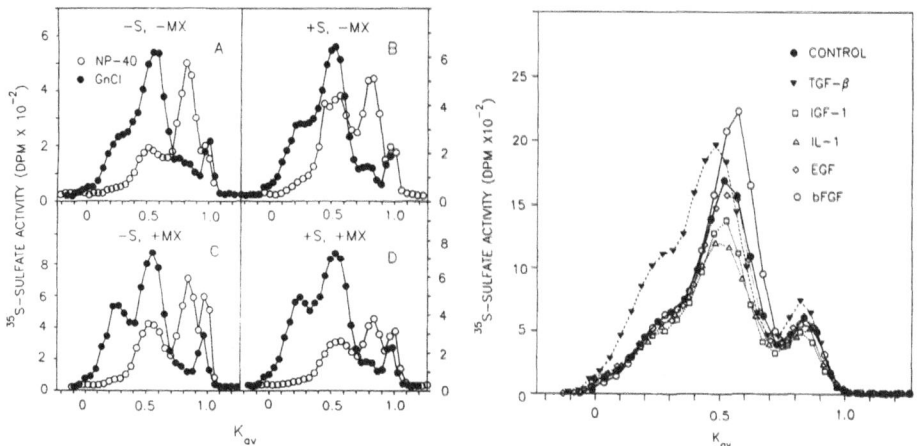

Figure 10. A saline extract of Matrigel is an effective inducer of large HSPG in cultured epithelial cells. Matrigel was extracted with 5X concentrated DMEM as described in the text. Cells were cultured for 2 days + or - serum (S) and + or - Matrigel extract (MX) and then labeled with ^{35}S-sulfate for a 3rd day. Compared to the control dish without serum or Matrigel (A), the radioactivity in the 0.2 K_{av} peak relative to the total radioactivity was 17% higher with serum added (B). Matrigel alone stimulated radioactivity 28% (C), and Matrigel with serum stimulated it 48% (D). (Reprinted from *In Vitro Cell. Devel. Biol.*, in press by permission of The Tissue Culture Association of America).

Figure 11. TGF-ß1 mimics the activity of Matrigel and Matrigel extract by the selective induction of large HSPG. Epithelial cells cultured 2 days in serum-free DMEM with 10 ng/ml each of several factors were given ^{35}S-sulfate during the 2nd day. Proteoglycans were isolated and chromatographed on Sepharose CL-4B. Only TGF-ß1 among those factors tested stimulated the 0.2 K_{av} peak. (Reprinted from *In Vitro Cell. Devel. Biol.*, in press by permission of The Tissue Culture Association of America).

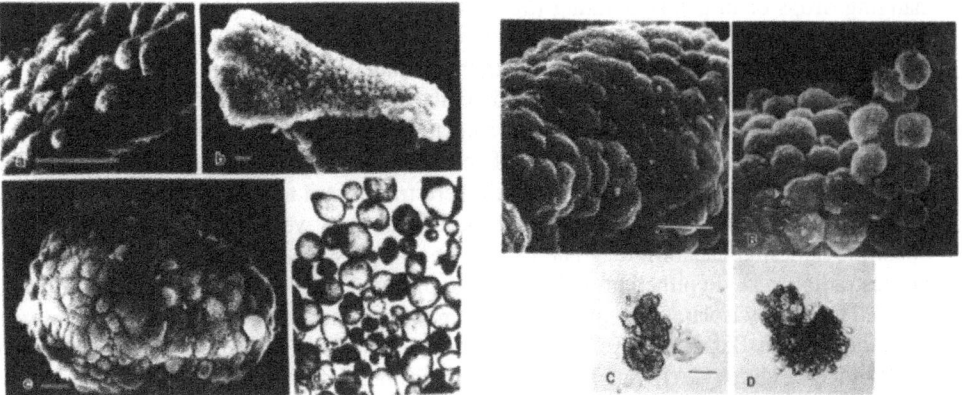

Figure 12. Development of vesicles from fragments of uterine epithelium. Epithelium was isolated from uteri of mice on day 4 of pregnancy (i.e., 3.5 days post-coitus) and cultured en masse on a shaker. Scanning electron micrographs were made initially (a), after 3 hours (b), and after 24 hours (c). Vesicles were selected for size and pooled (shown live in d). Note that the rolling of uterine fragments occurred with the apical surface facing outward and the basal surface inside the vesicle. Scale bar = 10 μm (a-c) 50 μm (d). (Reprinted from *Trophoblast Research* 4:51, 1990, with permission of Plenum Publishing Corp.).

Figure 13. Disruption of lateral cell associations by estradiol. Immature mice were injected for 3 days with 12.5 ng of estradiol daily prior to removal of uteri. Epithelial flakes from saline-injected (A,C) or estradiol-injected (B,D) mice were isolated (Morris and Potter, 1984) and immediately prepared for scanning electron microscopy (A,B). After 24 hours in rotation culture the vesicles were photographed (C,D). Epithelia from mice receiving estradiol were much more fragile and formed poorly organized and disintegrating aggregates. Such cells showed no abnormalities in their ability to attach to and spread on tissue culture plastic (not shown). Scale bar = 50 μm.

TGF-ß1 specifically stimulated the basal lamina HSPG (Figure 11). Based upon the binding of antiserum against TGF-ß1 to western blots, we estimate that Matrigel may contain 4-25 ng of TGF-ß1 per ml of Matrigel (range of two assays, Morris et al., 1993); on the basis of inhibition of cell division assays about 6.7-8.5 ng/ml appears to be present (Vukicevic et al., 1992).

These observations suggest that the basal lamina HSPG ordinarily turns over very slowly in uterine epithelium but can be stimulated by TGF-ß1, which is known to be synthesized by uterine epithelium, but not stroma, during the first 4 days of pregnancy in the mouse (Tamada et al., 1990). Because of the high affinity of collagen IV for TGF-ß1 (Paralkar et al., 1991), the basal lamina may act as a reservoir to release the factor to stimulate new lamina synthesis in surrounding intact epithelium (Silberstein et al., 1992), when it is degraded by the underlying stromal cells during the invasive stage of implantation (Schlafke et al., 1985) or when the cells are stimulated to divide by estradiol or by spreading on new surfaces.

CHANGES IN ASSOCIATION BETWEEN EPITHELIAL CELLS

Influence of Hormonal Status on Integrity of the Epithelium

There is a striking difference between the integrity of the epithelium in early pregnancy compared with other stages. Epithelium isolated from 3.5-day pregnant mice rolls up into transparent vesicles if maintained in suspension culture overnight, either in hanging drops or in gently gyrating (50 rpm) Erlenmeyer flasks (Figure 12). It is these vesicles that were used for the hanging drop experiments (Figures 2 and 3). Vesicles also were obtained from epithelium removed from immature mice, but these were not as symmetrical nor as large and tended to be collapsed and solid instead of hollow (Figure 13C). A striking difference was seen if immature mice had been treated with estradiol; the cells showed a nearly complete loss of cell-cell adhesion and formed loose fragmented aggregates (Figure 13D). If the aggregates were transferred to a dish, the cells were fully capable of spreading on the plastic, indicating that they were still viable and retained substrate adhesion capabilities. The difference in adhesiveness between cells of control and estradiol-treated epithelium did not develop during culture but was present from the start. The freshly isolated immature epithelium that had not been exposed to estradiol was in smooth intact flakes (Figure 13A), similar to the epithelium removed from pregnant adult mice (Figure 12A), but the epithelium from estradiol-treated mice consisted of rounded cells that appeared to be only loosely adhering to one another (Figure 13B). The implication is that adhesion molecules localized along the lateral surfaces of the cells either were removed or inactivated in response to estradiol.

As would be expected, naturally cycling and pregnant adult mice showed differences in their ability to form epithelial vesicles that correlated with their hormonal experience (S.W.P., unpublished data). The best vesicles were formed from epithelium in early estrus and the most fragmented in metestrus. Thus, the loss of cell-cell adhesion lagged somewhat behind the peak estradiol levels at estrus, but they correlate with the regression of the epithelial cell surface in post-estrous mice and may reflect the delay required between exposure to estradiol and translation into membrane-mediated adhesion changes. Immature mice injected with estradiol for only 1 day began to show a loss of epithelial integrity that was not fully developed until about 3 days. In pregnant mice, the attainment of optimum vesicles peaked at 3.5 days after mating, a time during which serum levels of estradiol in rats has peaked and progesterone is still increasing (Glasser and McCormack, 1981). Beyond 3.5 days the quality of vesicles had deteriorated (Morris and Potter, 1984), so that they appeared similar to vesicles produced by epithelium from estradiol-treated immature mice.

Influence of Hormonal Status on Uvomorulin (E-cadherin)

A clue to the target of estradiol inhibition of cell-cell adhesion came from the observation that in silver stained SDS-PAGE gels a 120 kDa band was present in extracts from the uteri of 2.5- through 4.5-day pregnant mice but was very weak in extracts from 0.5- and 1.5-day pregnant mice (Potter et al., 1991). The size of the protein represented by this band suggested that it may be the cell adhesion molecule uvomorulin (Hyafil et al., 1981) or E-cadherin (Yoshida and Takeichi, 1982), which are the mouse analogs of human cell-CAM 120/80 (Damsky et al., 1983). An 80 kDa fragment of cell-CAM 120/80 or of E-cadherin represents the extracellular adhesive domain, minus the plasma membrane-spanning and cytoplasmic domains. In our experiments anti-uvomorulin typically bound to both 120 kDa and 80 kDa bands, but only in western blots of detergent extracts that were from uteri of estrous mice. A more intensely reactive 120 kDa band, but typically no 80 kDa band, was seen in blots of extract from 4.5-day pregnant mice (Potter et al., in preparation).

E-cadherin is restricted to the lateral surface of rat uterine epithelial cells *in vitro* (Glasser et al., 1988). It is relatively uniformly distributed over the basolateral cell surface of mouse uterine epithelial cells *in vivo* during early pregnancy, but some localization appears at focal contact sites between the blastocyst and uterine epithelial surfaces after about 5 days (Kadokawa et al., 1989). We found that considerable change

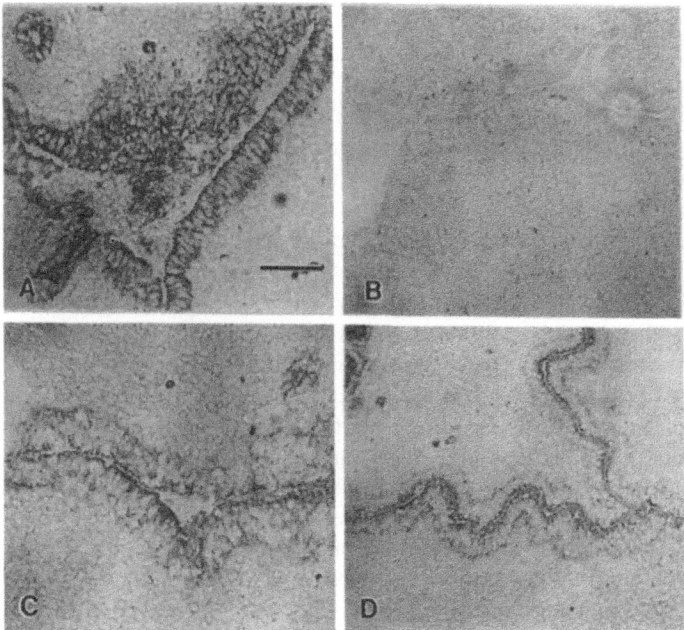

Figure 14. Changing distribution in E-cadherin during pregnancy. Uteri were frozen and permeabilized with methanol after sectioning. After incubation with monoclonal antibody against E-cadherin (uvomorulin) for 1 hr, the sections were washed and reacted with HRP-linked anti-rat IgG and stained with DAB. Shown are (A) 3.5-day pregnant uterus reacted with primary and secondary antibody or (B) with secondary antibody only as a control; (C) 4.5-day and (D) 5.5-day pregnant mice were reacted with both antibodies. During 4.5 to 5.5 days of pregnancy the antibody reaction product became progressively restricted and concentrated at the apical cell borders, in punctate regions, approximating the location of junctional complexes. Scale bar = 50 μm.

in localization occurs that correlates with the change in ability of the cells to form vesicles and with the change in reactivity of 120 kDa and 80 kDa E-cadherin on western bolts (Potter et al., in preparation). Epithelium between 3.5 and 5.5 days of pregnancy showed a progressive restriction of staining from general basolateral staining at 3.5 days to intense staining at the apical-basolateral boundary by 5.5 days (Figure 14). The latter staining was restricted to punctate regions at junctions between cells, probably within the junctional complexes. The failure of metestrous epithelium to form intact vesicles is evidence for an estradiol-triggered cleavage of cell adhesion molecules, as seen by the appearance of 80 kDa E-cadherin fragments at estrus. Restriction of E-cadherin to the junctional complex by 5.5 days correlates with the normal loosening of epithelium in preparation for invasion by the trophoblast and with the inability of epithelium beyond 3.5 days to form robust vesicles (Morris and Potter, 1984).

Other evidence suggests that the ability of uterine epithelium from 3.5-day pregnant mice to form hollow vesicles may be due at least in part to the presence of intact 120 kDa E-cadherin over the lateral cell surface. Medium containing 1:1,000, 1:500, and 1:250 dilutions of anti-uvomorulin monoclonal antibody showed a concentration-dependent inhibition of vesicle formation by epithelium from immature mice after 24 hours of culture (S.W.P., unpublished data). The affected epithelium resembled that from estradiol-treated immature mice (Figure 13).

SUMMARY

We have used *in vitro* methods to examine aspects of early implantation in mouse uterine epithelium. (1) The frequency of blastocyst adhesion to vesicles of uterine epithelium in hanging drop culture was shown to be directly related to the number of vesicles in the drop, indicating the importance of major epithelial contact and/or of epithelial factors for initiating attachment and implantation. (2) A cell-surface heparan sulfate proteoglycan (HSPG) was studied in detergent extracts of the epithelial surfaces of uteri. It turned over in the basolateral plasma membrane in response to increasing estradiol in uteri of sexually immature mice, ovariectomized adults, and in naturally pregnant mice. This turnover was associated with a change to a more basal localization of the antibodies against cell-membrane HSPG in response to estradiol in uterine epithelium of immature and ovariectomized mice and during estrus. In 3.5-day pregnant mice staining of the HSPG became very weak, particularly basally. (3) The synthesis of a large basal lamina HSPG was stimulated by cell outgrowth on plastic and even more strongly by culture on the EHS tumor basal lamina, Matrigel. A physiological salt extract of Matrigel or the growth factor TGF-ß1, which is tightly associated with Matrigel, also stimulated basal lamina HSPG. (4) Cell adhesion was maximal in vesicles prepared from isolated 3.5-day pregnant uterine epithelium, but adhesion was adversely affected by continued exposure to estradiol; only loosely associated cells were obtained from metestrous mice, immature females injected with estradiol, or mice beyond 3.5 days of pregnancy. In sections of 3.5-day pregnant uteri the distribution of E-cadherin was more or less uniform around the cell, but during 4.5 to 5.5 days of pregnancy it was selectively lost from around the basolateral and basal surfaces. Both 120 kDa E-cadherin and the unanchored 80 kDa extracellular fragment were present in estrous epithelium, but increased amounts of the 120 kDa E-cadherin and no 80 kDa fragment typically was present in 4.5-day pregnant epithelium. Epithelium from immature mice cultured with antibody against E-cadherin was as disrupted as that from mice that had been injected with estradiol.

Overall, the data suggest that during estrus, uterine epithelial cells may be tightly anchored to the underlying stromal matrix by basally localized syndecan, but they are relatively loosely associated with each other due to the presence of relatively large amounts of unanchored 80 kDa E-cadherin fragments. After about 3.5 days of pregnancy an adhesive apical surface appears for blastocyst attachment, driven by a reorganization of the cell surface, which is represented in the experiments presented here by enhanced turnover and relocalization of syndecan. The work of others discussed here suggests that there may also be an increase in lacto-N-fucopentaose-1 and a resorting of integrin. After about 4.5 days of pregnancy, when the blastocyst is attached, there is a loss of syndecan at the basal surface, resulting in a loss of stability of the basal lamina. This is followed at about 5.5 days by a loss of E-cadherin from all but the junctional complex and a loosening of the epithelial cell-cell associations, as the trophoblast begins to invade. Ultimate degradation of the basal lamina locally may result in the stimulation of new basal lamina synthesis in the surrounding epithelium or in the trophoblast by release of TGF-ß.

Acknowledgements. This research was supported by grant HD 19530 from the National Institutes of Health. For the generous provision of antibodies we thank Merton Bernfield for anti-syndecan and John Hassell and Koji Kimata, separately, for anti-perlecan.

REFERENCES

Astrahantseff, K.N., and Morris, J.E., 1992, Interactions between cultured mouse uterine epithelial and stromal cells during estrogen-induced proliferation, *Mol. Biol. Cell.* 3:143a.

Bernfield, M., and Hooper, K.C., 1991, Possible regulation of FGF activity by syndecan, an integral membrane heparan sulfate proteoglycan, *Ann. N.Y. Acad. Sci.* 638:182-194.

Carson, D.D., Tang, J.-P., Julian, J., and Glasser, S.R., 1988, Vectorial secretion of proteoglycans by polarized rat uterine epithelial cells, *J. Cell Biol.* 107:2425-2435.

Carson, D.D., Tang, Jy-P., and Julian, JoA., 1993, Heparan sulfate proteoglycan (Perlecan) expression by mouse embryos during acquisition of attachment competence, *Dev. Biol.* 155:97-106.

Conforti, G., Dominguez-Jimenez, C., Zanetti, A., Gimbrone, M.A., Jr., Cremona, O., Marchisio, P.C., and Dejana, E., 1992, Human endothelial cells express integrin receptors on the luminal aspect of their membrane, *Blood.* 80:437-446.

Cunha, G.R., Bigsby, R.M., Cooke, P.S., and Sugimura, Y., 1985, Stromal-epithelial interactions in adult organs, *Cell. Differen.* 17:137-148.

Damsky, C.H., Richa, J., Solter, D., Knudson, K., and Buck, C.A., 1983, Identification and purification of a cell surface glycoprotein mediating intercellular adhesion in embryonic and adult tissue, *Cell.* 34:455-466.

Damsky, C.H., Fitzgerald, M.L., and Fisher, S.J., 1992, Distribution patterns of extracellular matrix components and adhesion receptors are intricately modulated during first trimester cytotrophoblast differentiation along the invasive pathway, in vivo, *J. Clin. Invest.* 89:210-222.

Dutt, A., Tang, J.-P., and Carson, D.D., 1987, Lactosaminoglycans are involved in uterine epithelial cell adhesion in vitro, *Dev. Biol.* 119:27-37.

Dutt, A., Tang, J.P., and Carson, D.D., 1988, Estrogen preferentially stimulates lactosaminoglycan-containing oligosaccharide synthesis in mouse uteri, *J. Biol. Chem.* 263:2270-2279.

Enders, A.C., 1975, The implantation chamber, blastocyst and blastocyst imprint of the rat: A scanning electron microscope study, *Anat. Rec.* 182:137-150.

Enders, A.C., and Schlafke, S., 1974, Surface coats of the mouse blastocyst and uterus during the preimplantation period, *Anat. Rec.* 180:31-45.

Enders, A.C., Chàvez, D.J., and Schlafke, S., 1981, Comparison of implantation in utero and in vitro, *in* "Cellular and Molecular Aspects of Implantation", S.R. Glasser and D.W. Bullock, eds., Plenum Publ. Corp., New York. pp. 365-382.

Finn, C.A., and Publicover, M., 1981, Cell proliferation and cell death in the endometrium, *in* "Cellular and Molecular Aspects of Implantation", S.R. Glasser and D.W. Bullock, eds., Plenum Publishing Corp., New York. pp. 181-195.

Glass, R.H., Spindle, A.I., Maglio, M., and Pedersen, R.A., 1980, The free surface of mouse trophoblast in culture is non-adhesive for other cells, *J. Reprod. Fert.* 59:403-407.

Glasser, S.R., and McCormack, S.A., 1981, Separated cell types as analytical tools in the study of decidualization and implantation, *in* "Cellular and Molecular Aspects of Implantation", S.R. Glasser and D.W. Bullock, eds., Plenum Press, New York. pp. 217-239.

Glasser, S.R., Julian, J., Decker, G.L., Tang, J.-P., and Carson, D.D., 1988, Development of morphological and functional polarity in primary cultures of immature rat uterine epithelial cells, *J. Cell Biol.* 107:2409-2423.

Hassell, J.R., Leyshon, W.C., Ledbetter, S.R., Tyree, B., Suzuki, S., Kato, M., Kimata, K., and Kleinman, H.K., 1985, Isolation of two forms of basement membrane proteoglycans, *J. Biol. Chem.* 260:8098-8105.

Hewitt, K., Beer, A.E., and Grinnell, F., 1979, Disappearance of anionic sites from the surface of the rat endometrial epithelium at the time of blastocyst implantation, *Biol. Reprod.* 21:691- 707.

Hohn, H.-P., and Denker, H.-W., 1990, A three-dimensional organ culture model for the study of implantation of rabbit blastocysts in vitro, *Troph. Res.* 4:71-95.

Hyafil, F., Babinet, C., and Jacob, F., 1981, Cell-cell interaction in early embryogenesis: A molecular approach to the role of calcium, *Cell.* 26:447-454.

Jacobs, A.L., Decker, G.L., Glasser, S.R., Julian, J., and Carson, D.D., 1990, Vectorial secretion of prostaglandins by polarized rodent uterine epithelial cells, *Endocrinology.* 126:2125-2136.

Kadokawa, Y., Fuketa, I., Nose, A., Takeichi, M., and Nakatsji, N., 1989, Expression pattern of E- and P-cadherin in mouse embryos and uteri during the periimplantation period, *Develop. Growth & Differ.* 31:23-30.

Kato, M., Koiki, Y., Suzuki, S., and Kimata, K., 1988, Basement membrane proteoglycans in various tissues: Characterization using monoclonal antibodies to the Engelbreth-Holm-Swarm mouse tumor low density heparan sulfate proteoglycan, *J. Cell Biol.* 106:2203-2210.

Kimber, S.J., and Lindenberg, S., 1990, Hormonal control of a carbohydrate epitope involved in implantation in mice, *J. Reprod. Fert.* 89:13-21.

Larjava, H., Peltonen, J., Akiyama, S.K., Yamada, S.S., Gralnick, H.R., Uitto, J., and Yamada, K.M., 1990, Novel function for beta 1 integrins in keratinocyte cell-cell interactions, *J. Cell Biol.* 110:803-815.

Lessey, B.A., Damjanovich, L., Coutifaris, C., Castelbaum, A., Albelda, S.M., and Buck, C.A., 1992, Integrin adhesion molecules in the human endometrium. Correlation with the normal and abnormal menstrual cycle, *J. Clin. Invest.* 90:188-195.

Lindenberg, S., Kimber, S.J., and Kallin, E., 1990, Carbohydrate binding properties of mouse embryos, *J. Reprod. Fertil.* 89:431-439.

Lowe, J.B., Stoolman, L.M., Nair, R.P., Larsen, R.D., Berhend, T.L., and Marks, R.M., 1990, ELAM-1--dependent cell adhesion to vascular endothelium determined by a transfected human fucosyltransferase cDNA, *Cell.* 63:475-484.

Morris, J.E., and Potter, S.W., 1984, A comparison of developmental changes in surface charge in mouse blastocysts and uterine epithelium using DEAE beads and dextran sulfate in vitro, *Dev. Biol.* 103:190-199.

Morris, J.E., and Potter, S.W., 1990, An in vitro model for studying interactions between mouse trophoblast and uterine epithelial cells, *Troph. Res.* 4:51-69.

Morris, J.E., Potter, S.W., and Buckley, P.M., 1982, Mouse embryos and uterine epithelia show adhesive interactions in culture, *J. Exp. Zool.* 222:195-198.

Morris, J.E., Potter, S.W., Rynd, L.S., and Buckley, P.M., 1983, Adhesion of mouse blastocysts to uterine epithelium in culture: A requirement for mutual surface interactions, *J. Exp. Zool.* 225:467-479.

Morris, J.E., Potter, S.W., and Gaza-Bulseco, G., 1988a, Estradiol induces an accumulation of free heparan sulfate glycosaminoglycan chains in uterine epithelium, *Endocrinology.* 122:242-253.

Morris, J.E., Potter, S.W., and Gaza-Bulseco, G., 1988b, Estradiol-stimulated turnover of heparan sulfate proteoglycan in mouse uterine epithelium, *J. Biol. Chem.* 263:4712-4718.

Morris, J.E., Gaza, G., and Potter, S.W., 1993, Specific stimulation of basal lamina heparan sulfate proteoglycan in mouse uterine epithelium by Matrigel and by transforming growth factor-β1, *In Vitro Cell. Dev. Biol.*, in press.

Nelson, W.J., 1992, Regulation of cell surface polarity from bacteria to mammals, *Science*. 258:948-955.

Nilsson, B.O., and Hjerten, S., 1982, Electrophoretic quantification of the changes in the average net negative surface charge density of mouse blastocysts implanting in vivo and in vitro, *Biol. Reprod*. 27:485-493.

Noonan, D.M., Fulle, A., Valente, P., Cai, S., Horigan, E., Sasaki, M., Yamada, Y., and Hassell, J.R., 1991, The complete sequence of perlecan, a basement membrane heparan sulfate proteoglycan, reveals extensive similarity with laminin A chain, low density lipoprotein-receptor, and the neural cell adhesion molecule, *J. Biol. Chem*. 266:22939-22947.

Paralkar, V.M., Vukicevic, S., and Reddi, A.H., 1991, Transforming growth factor beta type 1 binds to collagen IV of basement membrane matrix: implications for development, *Dev. Biol*. 143:303-308.

Parr, E.L., Tung, H.N., and Parr, M.B., 1987, Apoptosis as the mode of uterine epithelial cell death during embryo implantation in mice and rats, *Biol. Reprod*. 36:211-225.

Parr, M.B., 1980, Endocytosis at the basal and lateral membrane of the rat uterine cells during early pregnancy, *J. Reprod. Fert*. 60:95-99.

Potter, S.W., and Morris, J.E., 1992, Changes in histochemical distribution of cell surface heparan sulfate proteoglycan in mouse uterus during the estrous cycle and early pregnancy, *Anat. Rec*. 234:383-390.

Potter, S.W., Gaza-Bulseco, G., and Morris, J.E., 1991, Changes in proteins in mouse uterine epithelial surface during early pregnancy, *J. Cell Biol*. 115:235a.

Potts, D.M., 1968, The ultrastructure of implantation in the mouse, *J. Anat*. 103:77-90.

Psychoyos, A., 1973, Hormonal control of ovimplantation, *Vit. Hormones*. 31:201-256.

Rapraeger, A., and Bernfield, M., 1985, Cell surface proteoglycan of mammary epithelial cells. Protease releases a heparan sulfate-rich ectodomain from a putative membrane-anchored domain, *J. Biol. Chem*. 260:4103-4109.

Rapraeger, A., Jalkanen, M., and Bernfield, M., 1986, Cell surface proteoglycan associates with the cytoskeleton at the basolateral cell surface of mouse mammary epithelial cells, *J. Cell Biol*. 103:2683-2696.

Ruoslahti, E., and Yamaguchi, Y., 1991, Proteoglycans as modulators of growth factor activities, *Cell*. 64:867-869.

Salomon, D.S., and Sherman, M.I., 1975, Implantation and invasiveness of mouse blastocysts on uterine monolayers, *Exp. Cell Res*. 90:261-268.

Saunders, S., Jalkanen, M., O'Farrell, S., and Bernfield, M., 1989, Molecular cloning of syndecan, an integral membrane proteoglycan, *J. Cell Biol*. 108:1547-1556.

Schlafke, S., Welsch, A.O., and Enders, A.C., 1985, Penetration of the basal lamina of the uterine luminal epithelium during implantation in the rat, *Anat. Rec*. 212:47-56.

Silberstein, G.B., Flanders, K.C., Roberts, A.B., and Daniel, C.W., 1992, Regulation of mammary morphogenesis: Evidence for extracellular matrix-mediated inhibition of ductal budding by transforming growth factor-β1, *Dev. Biol*. 152:354-362.

Tamada, H., McMaster, M.T., Flanders, K.C., Andrews, G.K., and Dey, S.K., 1990, Cell type-specific expression of transforming growth factor-beta 1 in the mouse uterus during the periimplantation period, *Mol. Endocrinol*. 4:965-972.

Vukicevic, S., Kleinman, H.K., Luyten, F.P., Roberts, A.B., Roche, N.S., and Reddi, A.H., 1992, Identification of multiple active growth factors in basement membrane Matrigel suggests caution in interpretation of cellular activity related to extracellular matrix components, *Exp. Cell Res.* 202:1-8.

Yamagata, T., and Yamazaki, K., 1991, Implanting mouse embryo stain with a LNF-I bearing fluorescent probe at their mural trophectodermal side, *Biochem. Biophys. Res. Commun.* 181:1004-1009.

Yayon, A., Klagsbrun, M., Esko, J.D., Leder, P., and Ornitz, D.M., 1991, Cell surface, heparin-like molecules are required for binding of basic fibroblast growth factor to its high affinity receptor, *Cell.* 64:841-848.

Yoshida, C., and Takeichi, M., 1982, Teratocarcinoma cell adhesion: Identification of a cell surface protein involved in calcium-dependent cell aggregation, *Cell.* 28:217-224.

Zhou, Q., Moore, K.L., Smith, D.F., Varki, A., McEver, R.P., and Cummings, R.D., 1991, The selectin GMP-140 binds to sialylated, fucosylated lactosaminoglycans on both myeloid and nonmyeloid cells, *J. Cell Biol.* 115:557-564.

Zhu, Z., Deng, H., Fenderson, B.A., Nudelman, E.D., and Tsui, Z., 1990, Glycosphingolipids of human myometrium and endometrium and their changes during the menstrual cycle, pregnancy and ageing, *J. Reprod. Fert.* 88:71-79.

STEROID REGULATION OF INFLAMMATORY RESPONSES IN THE REPRODUCTIVE TRACT

Maria C. Leiva, Lisa A. Hasty, and C. Richard Lyttle

Department of Obstetrics and Gynecology
Division of Reproductive Biology
University of Pennsylvania, School of Medicine
Philadelphia, PA 19104

INTRODUCTION

Steroid hormones induce profound changes in the function of the uterus, for example estradiol administration produces several responses that can be separated into early changes that occur within six hours of the administration of the hormone and among these are increased vascular permeability, and water imbibition; many of these early changes resemble an inflammatory response and may be regulated by uterine release of mediators of inflammation after estrogen administration. Late responses are considered genomic and include growth and differentiation of the glandular epithelial cells, infiltration by granulocytes of the stroma, and secretion and synthesis of specific proteins or messenger RNA's (mRNA) by one or more resident uterine cell types (Wheeler et al, 1987).

REGULATION OF INFLAMMATORY RESPONSES IN THE RAT UTERUS

Initial studies in our laboratory concentrated on the changes in the immature rat uterus after exposure to estradiol; specifically we looked at the increase in peroxidase activity (Lyttle and DeSombre, 1977), as well as the infiltration of the uterus by eosinophils; our results demonstrated that estradiol, acting via its nuclear receptor, stimulates the production of a uterine eosinophil chemotactic factor (ECF-U) (Lee et al, 1989). This factor is a newly produced or activated protein secreted by the stromal cells with an approximate molecular weight of 20-25 kilodaltons (Leiva et al, 1991). The role of infiltrating eosinophils after estrogen administration may be related to endometrial remodeling.

The function and regulation of different uterine secretory proteins are of major importance in understanding the endometrium's role in successful reproduction. One of these proteins which has been widely studied by our laboratory is the regulation of complement component C3 in the rat uterus. Estradiol administration stimulates the synthesis and secretion of C3 in the endometrium and this effect is reversed or delayed by the administration of progesterone (Sundstrom et al, 1989; Brown et al, 1990). This protein with a molecular weight of 180 kDa was identified by immunohistochemical studies and by RNA blots of epithelial versus stromal and myometrial cells to be of epithelial origin.   In the rat uterus we have also observed a physiologic pattern of hormonal regulation for the secretion of C3. C3 secretion is at its highest level during the estrogen dominated stages of the estrous cycle (proestrus and estrus) and decreased in the metaestrus and diestrus stages.

INFLAMMATORY RESPONSES IN ENDOMETRIOSIS

The presence of these multiple inflammatory type of responses in the rat focused our attention in the human normal reproductive tract. C3 in the female was first identified by Weed and Arquembourg as well as Bartosik; (Weed and Arquembourg, 1980; Bartosik, 1985).   In their studies, complement component 3 was present in the endometrial glands of patients with endometriosis and they attributed its presence to deposition from serum. Endometriosis is a pathological condition which has been classically associated with inflammation and infertility, the underlying mechanisms indicating an association between these conditions are not clearly understood. Investigators have looked at many factors such as luteal phase dysfunction, alteration on the sperm-egg interaction, phagocytosis of the sperm, among others. Thus, while many etiologies have been proposed, few have been substantiated. A theory which has been extensively studied recently, is the presence of an aseptic inflammation in the peritoneal cavity which will cause modifications in the peritoneal fluid and in the functioning of the pelvic organs. (Surrey and Halme, 1989).

Many of the known biological functions of C3 could play a role in the pathogenesis of the associated infertility in endometriosis, Isaacson incubated endometriotic tissue in minimum essential medium (MEM) containing $50\mu Ci/ml$ of ^{35}S methionine and samples were analyzed by SDS-PAGE and immunoprecipitated with a rabbit antihuman C3 antibody. These results indicated that endometriotic tissue is biologically active and is able to synthesize and secrete in vitro several proteins, one of which was further identified as being complement component 3 by immunoprecipitation with the specific antibody (Fig. 1), (Isaacson et al, 1989).   Another significant observation of these studies on the expression of C3, is that the endometrium of patients with endometriosis displays

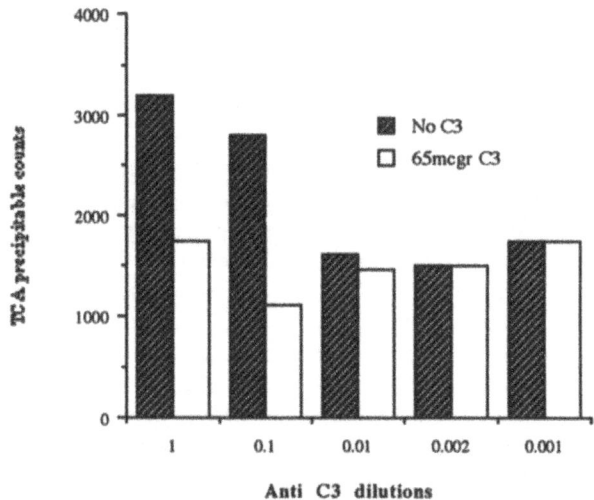

Figure 1. Competition of immunoprecipitated (^{35}S) C3 by unlabeled C3.   Aliquots containing 50,000 TCA precipitable counts were immunoprecipitated with different dilutions of rabbit antihuman C3 in the absence (line bars) or presence (open bars) of added C3 (65 µg).

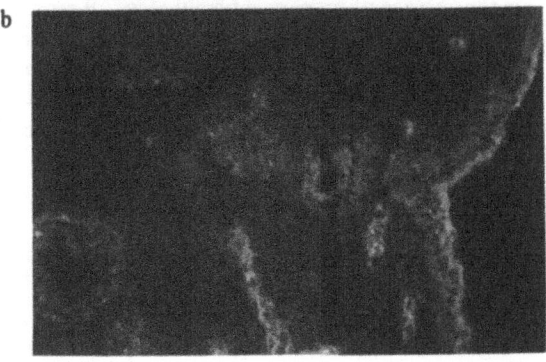

Figure 2. Immunohistochemical analysis of C3 in human endometrial tissue. 1μg of anti-human C3 antibody was diluted with 100 μl of PBS/BSA. Sections of late secretory phase endometrium from disease-free patients were stained for C3. A represents no primary antibody and B staining with anti-C3 antibody. (40x magnification).

significant C3 synthesis in contrast to the endometrium of normal patients which produces little if any C3 (Isaacson et al, 1990).

REGULATION OF COMPLEMENT COMPONENTS AND RECEPTORS THROUGHOUT THE MENSTRUAL CYCLE

Other complement proteins besides C3 have been detected in reproductive tissue: membrane cofactor protein (MCP) a regulator of complement activation is present in the acrosome-reacted human sperm (Anderson et al, 1989), and in trophoblast (Johnson et al, 1990).

Activation of complement by either classical or alternative pathways, leads to generation of products that mediate inflammation and tissue injury. The activation and regulation of each of these pathways requires the presence of other complement proteins or factors as well as receptors which are key in the activation of each of the pathways.

The classical pathway represents the adaptive immunity associated with memory in which the first stage leading to C3 fixation is the formation of the complex antigen-antibody; the initial step in the classical pathway requires the presence of C4 and C2. In contrast, the alternative pathway, does not require complexed antibody and activated alternative enzymes assemble in the target membrane and cleave C3b from C3 (Goldstein, 1988). The central controlling element of this pathway is a mechanism which prevents the

315

association between C3b and factor B. Factor B also enhances the macrophage-mediated cytotoxicity and activation of plasminogen; it is also necessary to have the presence of factor D and properdin. The activity of complement is regulated either through inactivation of C3b by membrane cofactor protein (MCP) or by the dissociation of the alternative pathway convertase by decay accelerating factor (DAF). (Kinoshita, 1991). Furthermore, MCP binds C3 and protects cells from lysis representing a self protected mechanism.

Recently we have examined the presence and the hormonal regulation of different complement components and receptors in the human endometrium at various phases of the menstrual cycle (Hasty et al, 1993). Tissues were obtained from women with no pelvic pathology; immunohistochemical analysis of endometrial sections was performed using monoclonal antibodies against C3, factor B, DAF, MCP, CR1, CR2 and CR3. Figure 2 demonstrates the presence of C3 in luteal endometrium: Fig. 2A represents luteal endometrium stained in the absence of primary antibody. Fig. 2B represents staining in the presence of antihuman C3 antibody. To further confirm our immunohistochemistry findings, the tissue was incubated with methionine-free MEM containing (^{35}S) methionine. Immunoprecipitations were performed with goat anti-human C3 antibody and the media proteins were analyzed by SDS-PAGE. C3 was found to be present in the glandular epithelial cells of luteal endometrium. Biosynthesis as analyzed by immunoprecipitation using the anti-C3 antibody was found to increase during the luteal phase of the cycle and to be minimal in the proliferative phase.

Like C3, Factor B and DAF were localized in the glandular epithelial cells of luteal endometrium (Fig. 3). In contrast, membrane cofactor protein (MCP) was found to be present throughout the menstrual cycle. CR1 was present only in the stromal compartment of luteal endometrium. CR3 was present only in the infiltrating leukocytes and CR2 was non detectable.

Local production of C3 was again demonstrated by immunoprecipitation of the incubated media with a goat anti-human antibody for C3. This confirmed the immunohistochemistry findings of C3 expression in the luteal endometrium. We have demonstrated a cycle specific appearance of complement and its binding proteins in humar endometrium. The synthesis and presence of C3 in the progesterone dominated phase of the menstrual cycle raises the possibility that in the human the expression of this particular protein may be regulated by progesterone. The underlying mechanisms for this regulation are not known; mediators such as interleukin 1 (IL-1) have been demonstrated to be regulated by progesterone and cellular levels of IL-1 mRNAs and serum levels of IL-1β

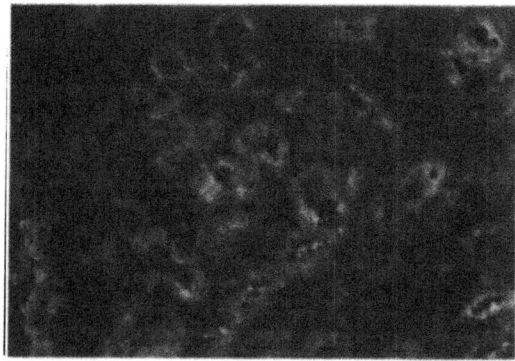

Figure 3. Immunohistochemical localization of DAF in disease-free human endometrial tissue in the secretory phase (day 24).

Table 1. Summary of the regulation of complement components throughout the menstrual cycle

Component	Proliferative	Secretory
C3	(-)	glands
DAF	(-)	glands
MCP	(-)	glands
factor B	(-)	glands
CR1	(-)	stroma

are increased during the secretory phase of the human endometrium, (Kauma et al, 1990; Cannon and Dinarello, 1985).

Table 1 presents a summary of the findings regarding the regulation of C3 and its regulatory proteins throughout the menstrual cycle. The function of C3 produced by cells of the female reproductive tract is unknown at present; the presence of Factor B, DAF, MCP and CR1 within the endometrial tissue suggest a functional complement system; The expression of C3, Factor B, DAF and MCP in the same cell type suggest that lytic functions of complement are directed to different targets. Whether the C3 produced by these cells acts in an autocrine fashion is unlikely since receptors for C3 have not been found on these glandular epithelial cells. The differential regulation of complement components suggests a tight control of whatever functions complement may play in the endometrium.

PERITONEAL FLUID CHANGES IN ENDOMETRIOSIS

Having demonstrated that endometriotic tissue actively secretes C3 we examined the peritoneal fluid, trying to explain the changes of the fluid in endometriosis and the secondary inflammation of the peritoneal cavity as another contributory mechanism. The peritoneal fluid arises from two different sources: the plasma as a transudate and the ovary as an exudate; it normally contains several types of blood cells with macrophages and lymphocytes being the most abundant although desquamated endometrial and mesothelial cells are also present (Haney, 1990). In the initial stages of the disease, endometriosis has been shown to increase the peritoneal fluid volume, the number of cells present and its concentration of lysosomic enzymes, probably a reflection of the increased number of cells and their state of activation (Halme et al, 1983).

Several studies have focused on the elevated numbers of macrophages in the peritoneal fluid of endometriosis patients which showed a 10 fold increase when compared to normal patients (Halme, 1989). Furthermore, the ratio of macrophages to lymphocytes is also higher when compared with fertile patients or with the ratio observed in peripheral blood samples (Hill et al, 1988). The majority of lymphocytes present consist of T-helper cells and natural killer cells with very few B lymphocytes being detected. This elevation of peritoneal leukocyte number is only observed in initial stages of the disease the reason for this is not known, neither are the mechanisms mediating this infiltration understood.

The study of macrophages and cytokines within the pelvic cavity has opened new possibilities in the understanding of endometriosis. The presence of activated macrophages in the fluid has been associated with an increase in the production of prostaglandins, tumor necrosis factor α and other compounds such as interleukin-1. Additionally, prostaglandins affect several functions associated with reproduction such as ovulation, tubal motility and implantation (Fakih et al, 1987; Rezai et al, 1987; Halme 1989). Recent reports have shown the production of MDGF (macrophage derived growth factor) by activated cells and this may be a new contributory factor in the associated-infertility of endometriosis patients (Halme et al, 1988). Further studies by Halme and others demonstrated that peritoneal macrophages in endometriosis exhibit higher levels of expression of acid phosphatase and low myeloperoxidase activity which would be in agreement with the view that endometriosis produces a more mature and activated population of macrophages when compared with normal patients (Halme et al, 1984).

CHEMOTACTIC ACTIVITY OF PERITONEAL FLUID IN ENDOMETRIOSIS

We have tested chemotactic activity of peritoneal fluid to macrophages and neutrophils derived from U937 and HL60 cell lines respectively (Fig.4). These cell types are leukemia cell lines which have proven chemotactic activity in the presence of chemoattractants such as histamine or fMLP (N-formyl-L-methionyl-L-leucyl-L-phenylalanine). Chemotaxis is performed utilizing a forty-eight well chemotactic chamber; samples in triplicate are applied to the lower chamber and incubated for 30 minutes at 37°C in humidified 5% carbon dioxide and 95% air atmosphere. After incubation, 50µl of the cell sample is applied to each well of the upper chamber which is incubated for another hour under the same conditions. Chemotaxis is assessed by counting the number of cells that migrate and attach to the bottom side of the filter (Leiva et al, 1991).

Peritoneal fluid obtained from patients with endometriosis has chemotactic activity for neutrophils and macrophages (Fig. 4). This activity is significantly different from the activity displayed by the fluid obtained from normal patients and it is comparable to the activity exhibited by fMLP. Another interesting observation from these studies was that treatment of the disease either with Medroxiprogesterone acetate or oral contraceptives reduces this activity to levels even smaller than the ones displayed by normal patients. (Leiva et al, 1993a)

To further resolve the characteristics of this chemoattractant peritoneal fluid was passed through a G-75 superfine Sephadex column, protein profiles of the eluant were monitored by absorbance at 280nm and major protein peaks analyzed for chemotactic activity (Fig. 5).

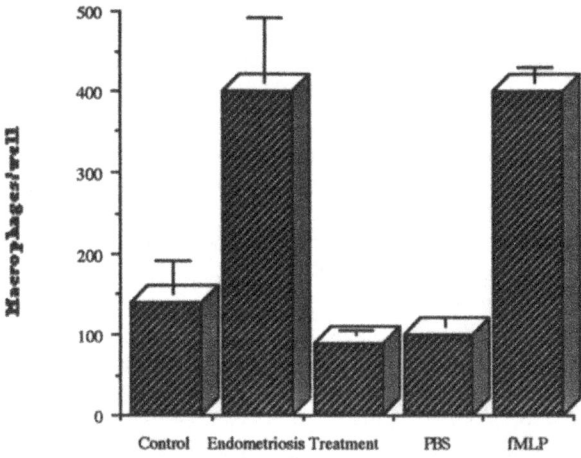

Figure 4. Chemotactic activity of peritoneal fluid. Macrophages were differentiated from U937 cells and tested for chemotaxis. Results are expressed as number of migrated cells ± SD. Controls were fMLP and phosphate buffered saline. Treatment group consisted of patients who had received either oral contraceptives or Medroxiprogesterone acetate.

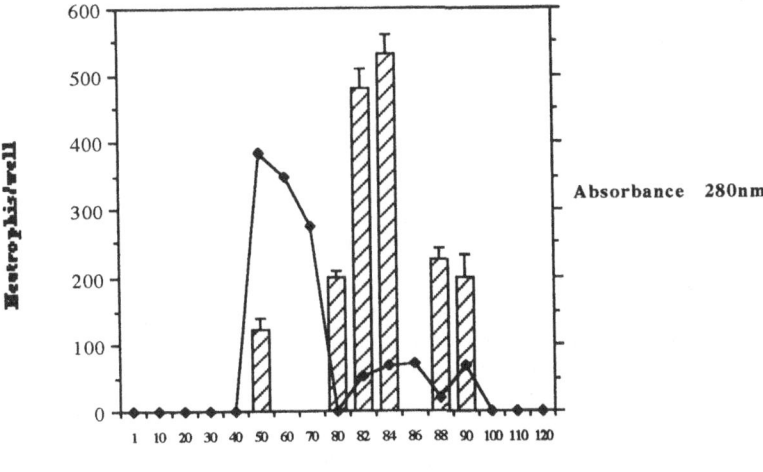

Figure 5. Protein profile and chemotactic activity of fractions from patients with endometriosis. Peritoneal fluid fractions from patients with endometriosis were applied to a G-75 superfine Sephadex column and the 280nm absorbance of each fraction was determined; The graph represents 280nm divided by 10. For evaluation of chemotaxis, 30 μl of each fraction was run in triplicate results are presented as number of cells/well ± SD.

Comparison of protein profiles from the control and endometriosis patients demonstrated a similar protein profile for both groups, with the majority of the proteins being of high molecular weight; however, the endometriosis group had an additional protein peak of low molecular weight, approximately 15 to 25 kDa where the majority of the chemotactic activity was present. The exact nature and sequence of this small protein peak is part of our current investigations.

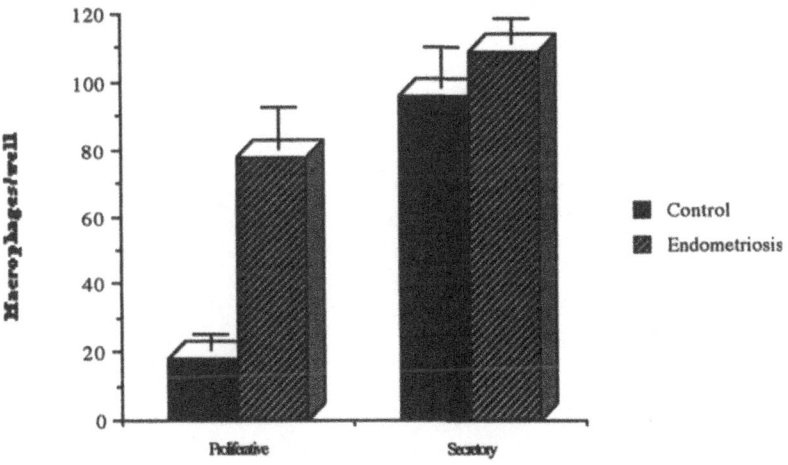

Figure 6. Chemotactic activity of endometrial biopsies from endometriosis and normal patients. Endometrial extracts obtained from proliferative and luteal phases of the cycle were tested for chemotactic activity towards macrophages differentiated from U937 cells. Results are expressed as number of cells/well ± SD.

ENDOMETRIAL CHANGES IN ENDOMETRIOSIS

Is the phenomena of aseptic inflammation in endometriosis localized to the peritoneal fluid and cavity or is it also present in other reproductive organs such as the endometrium? Previous investigations by Isaacson, (Isaacson et al, 1990) demonstrated that the proliferative endometrium of patients with endometriosis displayed significant C3 synthesis in contrast to the endometrium of normal patients which expresses little if any . These observations were further confirmed by Hasty, (Hasty et al.,1993) with immunohistochemical studies of endometrial biopsies stained for C3 throughout the menstrual cycle. To further expand on these observations endometrial biopsies from patients with mild to moderate endometriosis as well as normal patients were evaluated for chemotactic activity throughout the menstrual cycle. As shown in fig. 6, patients without the disease displayed a pattern of hormonally regulated chemotactic activity with increasing numbers of migrating cells in the luteal phase. In contrast, patients with minimal to moderate endometriosis exhibited high chemotactic activity for neutrophils and macrophages through out the menstrual cycle, (Leiva et al.,1993b).

After isolating the different uterine cell types from luteal biopsies (glandular epithelium, stroma and myometrium) to investigate the source of the chemotactic activity, the activity, the stromal compartment demonstrated higher chemotactic activity over the other two cell types (fig. 7). The immunohistochemical analysis of these sections confirmed the presence of greater amounts of infiltrating leukocytes in the endometriosis samples, the majority of which are monocytic in origin.

SUMMARY

All of our studies initially performed on rats and applied to humans have demonstrated a pattern of hormonal regulation for inflammatory responses in the reproductive organs. These responses are mostly physiological such as the regulation of the complement components throughout the menstrual cycle or the infiltration of blood marrow derived cells in the late luteal phase of the cycle or the peri-implantation period. However, our studies also indicate a modification of these responses in pathological conditions associated with infertility such as endometriosis. We do not know at this point

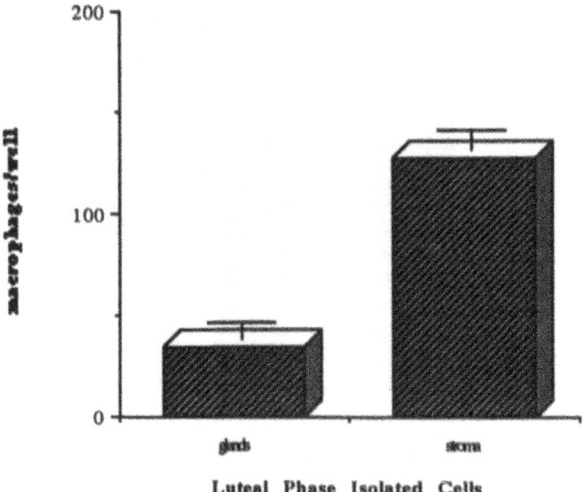

Figure 7. Chemotaxis of cellular extracts from luteal glandular epithelium and stroma from normal and endometriosis patients were evaluated with macrophages. Results are presented as in previous figures.

the exact nature of the chemoattractant which is present in the peritoneal fluid and the endometrium of patients with this disease. Neither do we know the exact role that it plays in the pathophysiology of endometriosis; however, based on these observations it may suggest that it is strongly associated with the aseptic inflammation of the pelvic cavity, and its almost complete disappearance after treatment may be a useful marker of therapy.

The endometrium of patients with endometriosis is able to synthesize and secrete inflammatory factors and proteins. Macrophages are routinely present in the connective tissue and stroma of several tissues including the endometrium; recently activation factors such as monocyte chemotactic protein-1 (MCP-1) have been shown to be modulated by endometrial stromal cells (Tabizadeh 1991). The regulation of this protein may be responsible for coordinating the recruitment of monocytes at specific times. In fact, one of the mechanisms proposed to explain early pregnancy loss is based on a suppresor immune-cell deficiency or in an increase in macrophage activation and function on the decidua of women with spontaneous abortions (Hill et al ., 1991).

Many inflammatory responses of the uterus appear to be regulated by steroidal hormones; abnormal functioning of these responses may be associated with reproductive failure and endometriosis.

REFERENCES

Anderson DJ, Michaelson JS and Johnson PM 1989. Trophoblast/leukocyte-common antigen is expressed by testicular germ cells and appears on the surface of acrosome-reacted sperm. Biol Reprod 41:285-293.

Bartosik D. 1985. Immunologic aspects of endometriosis. Semin Reprod Endocrinol 3:329-337.

Brown EO, Sundstrom SA, Kohm BS, Yi Z, Teuscher C, Lyttle CR. 1990 Progesterone regulation of estradiol-induced rat uterine secretory protein, complement C3. Biol Reprod 42:713-719.

Cannon JG, Dinarello CA. 1985. Increased plasma interleukin-1 activity in women after ovulation. Science 227:1247-1249.

Fakih H, Baggett B, Holtz G, Tsang KY, Lee JC, and Williamson HO. 1987. Interleukin-1: a possible role in the infertility associated with endometriosis. Fertil Steril 47:213-216.

Goldstein IM 1988: Complement: Biologically Active Products In: Inflammation Basic Principles and Clinical Correlates Gallin JI, Goldstein IM and Snyderman R. eds. Raven Press Ltd., New York 55-74.

Halme J, Becker S, Hammond MG, Raj MHG and Raj S. 1983 Increased activation of pelvic macrophages in infertile women with mild endometriosis. Am J Obstet Gynecol 145:333-337.

Halme J, Becker S, Wing R. 1984. Accentuated cyclic activation of peritoneal macrophages in patients with endometriosis. Am J. Obstet Gynecol 148:85-90.

Halme J, Becker SW, Haskill S. 1987 Altered maturation and function of peritoneal macrophages: Possible role in pathogenesis of endometriosis. Fertil Steril 50:216-222.

Halme J, White C, Kauma S, Estes J, Haskill S. 1988. Peritoneal macrophages from patients with endometriosis release growth factor activity in vitro. J. Clin Endocrinol Metab. 66:1044-1049.

Halme J. Release of tumor necrosis factor α by human peritoneal macrophages in vivo and in vitro. Am J Obstet Gynecol 161:1718-1723.

Haney AF 1990. The peritoneal environment and infertility . In: Evers JHL, Heinemann MJ, eds. From Ovulation to Implantation. Elsevier Science New York. 193-202.

Hasty LA, Lambris JD, Lessey BA, Pruksananonda K and Lyttle CR. 1993. Hormonal regulation of complement components and receptors throughout the menstrual cycle. Accepted for publication Am J Obstet and Gynecol.

Hill JA, Faris HM, Schiff I, Anderson DJ. 1988. Characterization of leukocyte sub populations in the peritoneal fluid of women with endometriosis. Fertil Steril 50:216-222.

Hill JA. 1991. Cellular immune mechanisms of early reproductive failure. Semin Perinatol. 15:225-229.

Isaacson K, Coutifaris C, Garcia CR and Lyttle CR. 1989. Production and secretion of complement component 3 by endometriotic tissue. J. Clin Endocrinol Metab 69:1003-1009.

Isaacson KB, Galman M, Coutifaris C and Lyttle CR 1990. Endometrial synthesis and secretion of complement component 3 by patients with and without endometriosis. Fertil Steril 53:836-841.

Johnson PM, Risk JM, Mwenda JM, Hart CA, Purcell DJF, Deacon NJ. 1990. Human trophoblast expression of retroviral-like activity and CD46 (membrane cofactor protein (MCP), Huly-m5 and H316 (TLX) antigen). In Mettler L, Billington WD eds. Reproductive Immunology. Elseviier Biomedical Press Amsterdam. 21-28.

Kauma S, Matt D, Strom S, Eierman D, Tuner T. 1990. Interleukin 1B (IL-1B), HLA-DRa and transforming growth factor B (TGF-B) expression in endometrium, placenta and placental membranes. Am J Obstet Gynecol 163:1430-1437.

Kinoshita T. 1991. Biology of complement: the overture. Immunol Today 12:291-295.

Lee YH, Howe RS, Sha SJ, Teuscher C, Sheehan D and Lyttle CR 1989. Estrogen regulation of an eosinophil chemotactic factor in the rat uterus. Endocrinology 125:3022-3028.

Leiva MC, Xu Q, Fishkoff SA and Lyttle CR 1991. Estrogen regulation of an eosinophil chemotactic factor by the uterine stromal cells. Abstract 88 Proceedings of the 73rd annual meeting of the Endocrine Society Washington June 1991.

Leiva MC, Hasty LA, Pfeifer S, Mastroianni L and Lyttle CR 1993a. Increased chemotactic activity of peritoneal fluid in patients with endometriosis. Am J Obstet Gynecol 168:592-598.

Leiva MC, Hasty LA, Lyttle CR. 1993b. Inflammatory changes in the endometrium of patients with minimal to moderate endometriosis. Submitted for publication.

Lyttle CR, De Sombre ER 1977. Uterine peroxidase as a marker for estrogen action. Proc Natl. Acad. Sci USA 74:3162.

Rezai N, Ghodgaonkar RB, Zacur HA, Rock JA, Dubin NH. 1987. Cul-de-sac fluid in women with endometriosis: fluid volume, protein and prostanoid concentration during the periovulatory period days 13-18. Fertil Steril 48:29-32.

Sundstrom SA, Kohm BS, Ponce-de-Leon H, Yi Z, Teuscher C and Lyttle CR 1989. Estrogen regulation of tissue-specific expression of complement C3. J. Biol Chem. 246:16941-16947.

Surrey ES, Halme J. 1989 Endometriosis as a cause of Infertility. Obstetrics and Gynecology Clinics of North America 16:79-89.

Tabibzadeh S. 1991. Human endometrium: An active site of cytokine production and action. Endocrine Reviews 12:272-290.

Weed JC, Arquembourg PC 1980. Endometriosis: can it produce an auto immune response resulting in infertility? Clin Obstet Gynecol 23:885-893.

Wheeler C, Kohm B, Lyttle CR. 1987. Estrogen regulation of protein synthesis in the immature rat uterus: the effects of progesterone on proteins released into the medium during in vitro incubations. Endocrinology 120:919-923

IN VITRO SYSTEMS TO STUDY EMBRYO-MATERNAL INTERACTIONS

J. K. Findlay

Prince Henry's Institute of Medical Research
P.O. Box 152
Clayton, Victoria 3168
Australia

INTRODUCTION

Is it possible to set up an *in vitro* culture system which will mimic implantation *in vivo*? Such a system would have to accommodate acquisition of receptivity of the endometrial cells, apposition and adhesion of the trophectoderm to the endometrial epithelial layer and, if appropriate to the species, penetration of the endometrium and the subsequent modifications to the subepithelial cells and the penetrating trophoblast. In 1981, Enders et al., critically compared the *in vitro* systems which were available at the time and said that there is no such thing as *in vitro* implantation, although "it is possible to make models, if not of normal implantation, at least of ectopic implantation". Significant advances in the techniques of cell culture have occurred in the past decade which, when applied to endometrium and blastocyst, have allowed us to study the interaction of the blastocyst and endometrium *in vitro* in a way which is better defined and more successful than was possible previously. Only time will tell whether or not these new *in vitro* models for studying embryo-maternal interactions are more relevant than earlier models. There have also been major advances in our understanding of the cytokines and other local regulatory substances acting between the trophectoderm and endometrium and within the endometrium, knowledge about the role of the extracellular matrix in cellular function, and information about mesenchymal-epithelial interactions.

IN VITRO METHODS

It is worth reminding ourselves about the limitations of *in vitro* models. Studies of whole organs or pieces of organs do not provide information about the functional differences and interactions between the individual cells that constitute that organ or the interactions between organs. This particularly applies to the endometrium which consists of two types of epithelial

Endocrinology of Embryo-Endometrium Interactions
Edited by S.R. Glasser *et al.*, Plenum Press, New York, 1994

cells (lumenal and glandular), endothelial cells and myoepithelial vascular cells, stromal fibroblasts (which may be several in type and decidualised) and a host of transient cells such as macrophages, lymphocytes, mast cells and leukocytes. It is only recently that we have had the techniques to identify most of these cells *in situ* and to isolate and study them *in vitro* (Findlay, Salamonsen and Cherny 1990; Clark 1993). However, these techniques have both advantages and disadvantages, some very pertinent to the study of embryo-maternal interactions *in vitro*.

The advantages of using isolated cells in culture are as follows:

(a) the individual functions and responses of one cell type can be identified and distinguished from those of other cells and from effects mediated by vascular components.

(b) cells can be studied *in vitro* in varying numbers for prolonged periods under the influence of specific stimuli or inhibitors, with relatively good reproducibility between replicates and batches.

(c) a study of isolated cells allows identification of autocrine and paracrine factors produced by those cells.

The disadvantages of using isolated cells are:

(a) An investigation of one population of cells *in vitro* does not define the activity of the organ as a whole.

(b) It is imperative that cross contamination of one cell type with another is known and preferably minimal and that cells are properly identified.

(c) A particular subpopulation of cells may be inadvertently selected for study, e.g., lumenal vs. glandular epithelium, making conclusions limited to that population.

(d) Isolated cells *in vitro* may not respond the same as they would *in situ*, either because they lack a basic requirement in the culture medium or were modified during isolation or because they are missing an essential regulatory influence which can only be provided by a surrounding matrix of ground substance and/ or other cells.

So, can we use these methods to isolate and identify individual cells from the endometrium, and then put them together again *in vitro* in such a way that a blastocyst will 'implant' *in vitro* in a manner similar to that *in vivo*? The answer is a cautious 'maybe'! Attempts to do these types of reconstruction experiments *in vitro* to study the function of the endometrium either alone or in the presence of a blastocyst, have relied on knowledge gained from other *in vitro* systems and have also produced some interesting results of their own. Some critical points are as follows:

(a) cells are influenced by their surrounding extracellular matrix, not only through physical support and determining their shape but also through mediation of transfer into cells, for determining their hormone responsiveness or as a source of mitogens.

(b) epithelial cells are normally polarized *in vivo*, with apical and basolateral domains of unique protein and lipid composition and tight inter-epithelial junctions, and they have preferential secretion of products apically and basally.

(c) mesenchymal (stromal) cells are known to exert an influence on the morphogenesis and cytodifferentiation of urogenital epithelial cells; they can also make the epithelial cells responsive to steroids *in vitro*.

There are now *in vitro* systems which can accommodate some or all of these requirements for epithelial endometrial cells (Findlay et al., 1990; Salamonsen and Nancarrow, 1994). Individual components of the extracellular matrix or a complex such as Matrigel, are now available commercially and can be added to cultures of endometrial cells with the result that epithelial cells, for example, become polarized and form tight junctions if grown on an appropriate support. Three dimensional culture systems such as the floating collagen gel or the millicell inserts can accommodate cocultures of epithelial and stromal cells in a matrix. Attempts have also been made to add blastocysts to these *in vitro* systems.

APPLICATION OF METHODS

With these issues in mind, we can return to the stages of implantation viz. apposition, adhesion and penetration of a receptive endometrium and ask whether or not the new *in vitro* systems can be used to contribute to our understanding of these processes.

The surfaces of the blastocyst and endometrium are epithelial sheets, and as opposing epithelial surfaces are generally non-adhesive to one another, inductive processes which allow apposition and adhesion must precede implantation. Hypotheses advanced to explain this phenomenon including production of a glycocalyx of carbohydrates by the trophoectoderm or by the epithelial cells of the endometrium (see Carson et al., 1992). The *in vitro* systems now available would allow critical testing of these hypotheses.

The acquisition of receptivity to the blastocyst by the progesterone dominated endometrial epithelial cell in response to nidatory estrogen is a key event in implantation but its mechanism is not understood (Glasser & Mulholland 1992). The advent of polarized endometrial epithelial cells in culture which are hormonally responsive should serve as an *in vitro* model for this aspect of implantation. For example, transient down-regulation of the expression of glycosylated mucin glycoproteins under steroid hormone control could facilitate access to the uterine surface (see Carson et al., 1992) and could be tested in this model.

Penetration of the endometrium will require a degree of extracellular remodelling which is known to involve a family of enzymes called matrix metalloproteinases (MMPs) (see Salamonsen & Nancarrow, 1994). The identities, source and activities of these MMPs and their regulation can now be studied using the *in vitro* models summarized above.

A combination of blastocyst (or trophectoderm) and polarized epithelial cells on a matrix containing stromal fibroblasts in a three dimensional system should be tested as a model for implantation. If successful, such a model would allow a study of the morphological and functional changes occurring before and during implantation and to examine the effects of cytokines, growth factors and potential inhibitors of implantation *in vitro*.

In those species such as rodents and primates in which the trophectoderm penetrates the subepithelial layer, some as yet unexplained mechanism limits the degree of penetration. Does this involve tissue inhibitors of MMPs? Is this a function of decidualized stromal cells or is it a function of the transient lymphomyeloid cells (ie. an immune response)? Perhaps this can also be examined in the 'reconstituted' *in vitro* systems.

The author is supported by the National Health & Medical Research Council of Australia.

REFERENCES

Carson, D.D., Jacobs, A.L., Julian, J., Rohde, L.H., and Valdizan, M.C., 1992, Glycoconjugates as positive and negative modulators of embryo implantation, *Reprod. Fert. Dev.* 4:271-274.

Clark, D.A., 1993, Cytokines, decidua and early pregnancy, *in* "Oxford Reviews of Reproductive Biology", S.R. Milligan, ed., Clarendon Press, Oxford. pp.83-111.

Enders, A.C., Chavez, D.J., and Schlafke, S., 1981, Comparison of implantation *in utero* and *in vitro*, *in* "Cellular and Molecular Aspects of Implantation", S.R. Glasser and D.W. Bullock, eds., Plenum Press, New York. pp. 365-382.

Findlay, J.K., Salamonsen, L.A., and Cherny, R.A. 1990, Endometrial function: studies using isolated cells *in vitro*, *in* "Oxford Reviews of Reproductive Biology" S.R. Milligan, ed., Clarendon Press, Oxford. pp. 181-223.

Glasser, S.R., and Mulholland, J., 1992, Receptivity is a polarity dependent special function of hormonally regulated uterine epithelial cells, *J. Electron Microscop. Tech* 25:.106-120.

Salamonsen, L.A., and Nancarrow, C.D., 1994, The cell biology of the oviduct and the endometrium, *in* "Molecular and Cellular Mechanisms in Female Reproduction", J.K. Findlay, ed., Academic Press, New York (in press).

THE INTERACTION OF TROPHOBLAST WITH ENDOMETRIAL STROMA

John D. Aplin[1] and Stanley R. Glasser[2]

[1]Department of Ob/Gyn and School of
   Biological Sciences
   University of Manchester, U.K.

[2]Department of Cell Biology
   Baylor College of Medicine
   Houston, Texas, U.S.A.

INTRODUCTION

Implantation involves initial interaction between the embryo and uterine epithelium. Where placentation is chorio-epithelial, this relationship continues throughout gestation. In other cases in which invasion occurs, the embryo interacts with mesenchymal elements of the uterus. In species showing interstitial implantation, the embryo-stromal inter-relationship is much more long-lived than the embryo-epithelial interaction and arguably more important in the success of the pregnancy since it is the means by which the embryo gains access to nutrients supplied through the maternal blood stream (Mossman, 1987). The intention of this short article is not to review decidualisation or placentation, but to comment on model systems that may offer promise for an improved understanding of morphogenesis in the trophoblastic interaction with maternal stroma in rodent and primate.

In establishing models to enable the interaction of embryo and maternal stroma to be studied in vivo or in vitro, the difficulty arises that two time-dependent processes are occurring: embryo (placental) development and differentiation (with or without overt decidualisation) of the maternal stroma. Selective retrieval of specific cell types, disturbances of the relative time course of differentiation of the cell populations present, and altered physical interrelationships between the cells are often the consequence of experimental manipulations in vitro. Thus it is essential to examine carefully the significance of any observations made to the processes of implantation in vivo in the relevant species. Inappropriate interpretations of data gathered from model systems permeate the literature.

THE DECIDUAL ENVIRONMENT

The precise timings and anatomical distributions of differentiating decidual cells differ between species; in human, implantation occurs into an initially undifferentiated stroma (Hertig et al, 1956; Enders, 1991), decidual cells appearing about 3 days after the first interaction between syncytiotrophoblast and the endometrial stroma. In rats, decidualisation of steroidally-sensitised stromal cells occurs as a result of a trans-epithelial stimulus from the pre-implantation embryo so that the implanting trophoblast encounters differentiated stromal cells from the first. Much evidence exists to show that provided (in rodents) the initial stimulus is given, decidual differentiation in both rodents and humans can occur independent of the presence of a conceptus (De Feo, 1967; Shelesnyak, 1986; Glasser, 1990). In addition to the cytological changes attendant on decidualisation, new secretory products are expressed by the cells (Bell, 1983; 1988; 1989; Jayatilak et al., 1989; Gu et al., 1992; Thomas, 1993) and extracellular matrix remodelling takes place (O'Shea et al., 1983; Zorn et al., 1986; Wewer et al., 1986; Glasser et al., 1987; Aplin and Jones, 1989; Aplin, 1989) with a reduction in abundance of banded collagen fibrils (Myers et al., 1990), loss of the microfibrillar collagen type VI (Aplin et al., 1988; Mulholland et al., 1993), loss of hyaluronate (Brown and Papaioannou, 1992) and the production of a pericellular basement membrane containing collagen type IV, laminin and heparan sulphate proteoglycan (Wewer et al., 1985; Faber et al., 1986; Glasser et al., 1987; Aplin et al., 1988; Kisalus and Herr, 1988).

In addition, altered cytokine networks are established (Mitchell et al., 1993; Wegmann et al., 1993) with the appearance of new populations of bone marrow-derived cells including large granulated lymphocytes and macrophages (Starkey, 1992).

In rat, regionally specific protein secretions are associated with the mesometrial and antimesometrial decidua; the latter produces a luteotropin and therefore functions as an endocrine tissue. The former tissue produces the protease inhibitor and growth factor binding protein α2-macroglobulin, which is more likely to function in the decidual-placental dialogue (Gu et al., 1992; Thomas, 1993). There is also evidence of regional variations in extracellular matrix glycoproteins (laminin isoforms in mouse: Farrar and Carson, 1992) and in pericellular glycosylation (specific subsets of decidual cells express terminal α2-3 sialic acid residues: Jones et al., 1993).

This regionally specific function may be related to the anatomy of placentation in rat and mouse, the yolk sac placenta forming adjacent to the antimesometrial decidua while the mesometrial decidua lies adjacent to the chorioallantoic placenta. However, there is a general problem of determining whether a product of decidual tissue is either an 'internal' requirement of differentiation, or involved in the interaction with trophoblast. Thus, for example, it is not clear whether the primary function of decidual laminin is in the physical integrity of decidual tissue or in providing a substrate for migrating trophoblast or bone marrow-derived cells (or all of these). Similarly, protease production is a requirement of the matrix remodelling observed to occur during decidualisation (Aplin, 1989), and decidua may regulate this process by producing protease inhibitors (α2-macroglobulin: Gu et al., 1992; Thomas, 1993; TIMP: Graham et al., 1993). The extent to which these components are involved directly in the interaction with trophoblast in vivo is not easy to determine. It is worth noting however that in species showing chorioepithelial placentation, matrix-directed proteolysis remains a feature of the endometrial stroma (Salamonsen, et al., 1991 and this volume).

The resident stromal cell population of the uterus can be isolated and maintained in culture (Glasser and Julian, 1986; Irwin et al., 1991; Fernandez-Shaw et al., 1992). Bone marrow-derived cells and vascular cells have also been recovered using enrichment procedures (Starkey et al., 1988). Given the likelihood of intercellular communication via soluble mediators (Wegmann, 1993), it is important to define and characterise the cells present in culture models. Their state of differentiation is also an important variable. In the case of human stromal cells, the addition of progesterone to the culture medium stimulates decidual differentiation as monitored morphologically and by the production of prolactin, IGFBP-1, laminin and other specific secretions (Irwin et al., 1991). In contrast, rat and rabbit cells differentiate spontaneously amd rapidly in vitro without the need for hormonal stimulation (Mani et al., 1992).

TROPHOBLAST DIFFERENTIATION

In both rat and human, trophoblast differentiation occurs from a stem cell population along several pathways giving rise to giant cells, spongiotrophoblast, glycogen cells, syncytiotrophoblast and endovascular trophoblast in the rat (Glasser and Davies, 1967; Soares et al., 1991), and extravillous interstitial cytotrophoblast, extravillous endovascular cytotrophoblast, villous syncytio- and cytotrophoblast and chorionic cytotrophoblast in the human (Pijnenborg et al., 1980; Benirschke and Kaufmann, 1990; Aplin, 1991).

The anatomical relationships between these subpopulations are very different in the two species. While a distant analogy might be drawn between the rodent ectoplacental cone and the human cytotrophoblastic shell (Pijnenborg et al., 1981), the interstitial infiltrative behaviour of extravillous cytotrophoblast observed in human has no parallel in rat or mouse. Human terminal chorionic villi occur in two forms: free floating and anchoring. Anchoring villi make contact with the decidua basalis. From these contact zones, rapid cytotrophoblast proliferation occurs into columns that are continuous with the cytotrophoblastic shell. From here mononuclear cytotrophoblast detaches to migrate into the maternal stroma (Pijnenborg et al., 1980). These interstitial cytotrophoblasts are more abundant in the vicinity of maternal spiral arteries than elsewhere in the stroma (Pijnenborg, 1990). Cytotrophoblast migration also occurs along the inner walls of maternal spiral arteries. The migrating populations of trophoblast are known as 'extravillous' and they produce a characteristic pericellular matrix (Feinberg et al., 1991; Fernandez et al., 1992; Blankenship et al., 1993). Eventually, presumably as a result of trophoblast-stimulated remodelling, the arterial walls are transformed with the loss of smooth muscle and replacement of elastic and collagenous extracellular matrix with a fibrinoid polymer. This allows an increased flow of blood to the placenta. This phenomenon is common to human, certain other primates (Enders and King, 1991; Blankenship et al., 1993) and rodents including rat, hamster and guinea pig (Pijnenborg et al., 1981). While in the macaque there appears to be a close correlation between arterial remodelling and the presence of trophoblast (Blankenship et al., 1993), in guinea pig and hamster arterial changes precede the appearance of trophoblast (Pijnenborg et al., 1981; Hees et al., 1987).

Significant changes in the cell surface expression of integrins occur during trophoblast differentiation into the extravillous pathway with the loss of integrin α6β4 and the appearance of integrins of the β1 family, specifically the fibronectin receptor α5β1 and the collagen/laminin receptor α1β1 (Korhonen et al., 1991; Damsky et al., 1992; Aplin, 1993; Damsky, this volume). Aplin et al. (1992) have also suggested that not only the level of surface expression, but also the activation state of the fibronectin receptor may affect the ability of trophoblast to migrate. Loss of molecules such as E-cadherin (Logan et al., 1992) that mediate cell-cell adhesion is also likely to be important in the ability of cytotrophoblasts to leave the columns.

There is much evidence to demonstrate the ability of trophoblast to produce proteases capable of the local degradation of extracellular matrix; these include plasminogen activator, interstitial collagenase, and 72kDA and 92kDa type IV collagenases (Strickland et al., 1976; Martin and Arias, 1982; Glass et al., 1983; Fisher et al., 1985; 1989; Puistola, 1989; Moll and Lane, 1990; Lala and Graham, 1990; Bischof et al., 1991; Librach et al, 1991; Autio-Harmainen et al., 1992; Behrendtsen et al., 1992; Fernandez et al., 1992; Logan et al., 1992; Graham et al., 1993). The nature of the enzymes produced is modulated in vitro by the substrate composition (Bischof et al., 1991). Proteolysis might be required during early implantation when the trophoblast penetrates the epithelial basement membrane (Turpeenniemi-Hujanen et al., 1992), or at later stages including the remodelling of maternal vessels by extravillous cytotrophoblast (Blankenship et al., 1993), but careful matching of data to species is required; for example, subepithelial decidual cells rather than trophoblast have been shown to effect local degradation of the basal lamina in rat (Schlafke et al., 1985; Welsh and Enders, 1987).

Cultured human trophoblast of first trimester requires secreted protease to infiltrate and cross a three-dimensional barrier of densely collagenous extracellular matrix (the amnion invasion assay; Yagel et al., 1988; Lala and Graham, 1990; Graham et al., 1993). The same cells also utilise proteases to penetrate filter wells coated with collagen gel or matrigel (Fisher et al., 1985; 1989; Librach et al., 1991).

A difficulty associated with the assignment of in vivo function to trophoblast proteases is that in situ localisation has not revealed major differences between subsets of trophoblast. Thus in later first trimester, both villous and extravillous trophoblast contain interstitial collagenase (Moll and Lane, 1990); the same applies to the 72kDa type IV collagenase (Autio-Harminen et al., 1992). One possible resolution of this issue is the existence of local environments in which an appropriate balance is achieved between protease activity and its inhibition. This may be achieved by a combination of autocrine and paracrine mechanisms. Extravillous cells produce specifically the inhibitor of plasminogen activator PAI-1 (Feinberg et al, 1989). Plasminogen activator (PA) is required to produce plasmin for the activation of latent collagenases (He et al., 1989). Lala and Graham (1990) have shown that TGFß produced by decidual cells acts to up-regulate production of the tissue inhibitor of metalloproteinase (TIMP) both in trophoblast and decidua.

EMBRYO TRANSPLANTATION EXPERIMENTS

Many authors have cited the work of Kirby (1963a,b) and Cowell (1969) who transferred blastocysts to ectopic soft tissue sites and non-progestational uteri respectively. Samuel and Perry (1972) transferred blastocysts from pig, in which epitheliochorial placentation occurs, from the uterine cavity to the stroma. All these authors reported a greater degree of 'invasive' behaviour by trophoblast than would be found in normal implantation sites. Aplin (1991) however pointed out the need to distinguish invasion (cell motility) from erosion (local degradation). It is now well known that trophoblast in all the species studied produces secretory proteases, and it is not clear to what extent the observations in these early papers arose from the failure of the host environment to regulate this activity or to what extent active migration of trophoblast may have occurred. Given the technical limitations and lack of markers available at the time they were performed and the relatively few published micrographs, it would be useful if these studies were repeated and extended. The data of Samuel and Perry (1972) indicating syncytialisation of trophoblast in eutopic stroma does indicate a possible role for uterine stimuli in trophoblast differentiation since syncytium does not form in the normal chorioepithelial placenta of pig.

Abrahamsohn and coworkers have more recently undertaken experiments in which mouse ectoplacental cones (EPCs; day 8) were transplanted to subcutaneous sites (Bevilacqua et al., 1991). It is interesting in the light of in vitro models in which EPC cells interact with two-dimensional substrates (see below), that although overall cell proliferation occurred resulting in growth of the subcutaneous grafts, no evidence for specific motile activity by trophoblast or infiltration of the surrounding tissues was reported. Instead, the enlarging grafts remained smooth-edged. Trophoblast differentiation occurred to give several cell types also found in eutopic placenta, with giant cells at the periphery (Soares, 1991). This evidence supports the notion of a programme of trophoblast differentiation much of the essential character of which is independent of tissue-specific signals from the maternal environment (Aplin, 1991).

Randall et al. (1987) studied the morphology of ectopic implantation sites in human Fallopian tubes and again found features including cell columns and anchoring villi that were very similar to those observed during placental development in utero.

EMBRYO OUTGROWTH ON EXTRACELLULAR MATRIX

A widely used approach that is often claimed to be relevant to the dissection of embryo-endometrial interactions involves plating blastocysts onto two-dimensional substrates containing extracellular matrix (ecm) components. Hatched mouse blastocysts attach, and trophoblast outgrowth can be observed, on substrata containing various matrix molecules including fibronectin, laminin, entactin and various collagen types (Armant et al., 1986; Carson et al., 1988; Yelian et al., 1993). This activity can be enhanced by exposure to growth factors including TGFα, EGF, PDGF, FGF and CSF-1 (Haimovici and Anderson, 1993), all of which are present in the uterus at the time of implantation (Tabibzadeh, 1991). There is also evidence that PDGF and FGF binding to ecm could enhance presentation to the blastocyst (Haimovici and Anderson, 1993). Trophoblast outgrowth can also occur from mouse day 7.5 ectoplacental cones (EPCs) onto laminin substrates (Romagnero and Babiarz, 1993).

Such assays demonstrate the potential for trophoblast to adhere to individual components of ECM and to use the interaction as a means to motile activity during the period immediately after hatching (i.e., when interaction with the antimesometrial decidua is occurring). However, relating the observations to the behaviour of trophoblast in vivo is less staightforward. In rats, the earliest trophoblast-uterine interactions occur at the anti-mesometrial pole where the primary decidual zone forms prior to direct heterotypic intercellular contact between trophoblast and decidual cells (Krehbiel, 1937; Welsh and Enders, 1985; 1987). The maternal blood supply to the yolk sac placenta is established following degenerative loss of first epithelial cells and later subepithelial decidual cells (Welsh and Enders, 1987), rather than as the result of trophoblast invasion. Later, beginning on day 10, the chorioallantoic placenta forms at the mesometrial pole after the disappearance of uterine epithelium (Welsh and Enders, 1991). There is evidence to suggest that the chorioallantoic placenta acquires its blood supply as the result of maternal angiogenesis rather than the penetration of trophoblast into the stroma (Welsh and Enders, 1991). In any case, this occurs at a developmental stage later than most of the above-mentioned assays were carried out.

The observation of trophoblast outgrowth from blastocysts or EPCs onto adhesive substrates is not in itself surprising -- most cells, plated in the form of aggregates onto an adhesive substrate, will behave in a similar fashion (Trinkaus, 1984). Lack of outgrowth in these circumstances may be a more useful observation, indicating either a non-permissive substrate composition or strong intercellular adhesion (Aplin and Foden, 1985). Chondroitin sulphate proteoglycan, which is produced by mouse stromal cells, can act as an inhibitor of outgrowth on fibronectin or collagen (Carson et al., 1992). Thus the local balance between permissive and inhibitory matrix components and cell-cell interactions needs to be determined.

Furthermore, loss of the antimesometrial-mesometrial polarity of the implantation site along with the spatial relationship between the embryo and inner cell mass is occurring during the process of trophoblast outgrowth in vitro. Outgrowth may occur as the result of proliferation, differentiation into flatter cells, migration and spreading of an existing trophoblast population, or a combination of these processes. It is difficult to determine whether the observed adhesive and migratory activities really reflect requirements for maintaining the integrity of the developing embryo, or for the trophoblast-maternal stroma interaction, or are artifacts of the culture system.

Cell shape and motility assays are probably more relevant to interactions that occur in vivo in the infiltration of human cytotrophoblast from columns into the maternal interstitium. Cultured normal human trophoblast of late first trimester (a time whem active infiltration is underway) has the ability to attach and spread on fibronectin, presumably as a result of the expression of integrin α5ß1 (Burrows et al., 1993). Since villous syncytiotrophoblast does not express significant quantities of α5ß1 (Korhonen et al., 1991; Damsky et al., 1992), this suggests that the cultures express certain characteristics of migrating extravillous cytotrophoblast. However, as pointed out by Aplin (1991) and Kliman and Feinberg (1992), a characteristic common to cultured trophoblast and choriocarcinoma cell lines is the loss of a regulated pattern of differentiation. These cells express numerous trophoblastic features, but cannot be identified with any one normal trophoblastic subpopulation. Immortalisation (Logan et al., 1992) or selection (Aplin et al., 1992) of cells with defined properties may help with characterisation of specific lineages, but control of differentiation in vitro remains a major challenge in trophoblast biology.

TROPHOBLAST OUTGROWTH ON DECIDUAL CELLS

In rats and mice the most superficial layers of antimesometrial decidual cells are closely packed, with prominent adherens and gap junctions (Finn and Porter, 1975; O'Shea et al., 1983; Parr and Parr, 1986; Winterhager, 1991 and this volume) between the cells and rather narrow residual intercellular spaces. Pericellular basement membrane components including laminin and collagen type IV may be detected immunocytochemically (Glasser et al., 1987), but ECM is not present in abundance. The mesometrial decidual cells are less close-packed (O'Shea et al., 1983). However, it seems likely that trophoblast-stromal interaction involves cell-cell as well as cell-matrix interactions. More extracellular matrix is usually visible between human decidual cells (Wynn, 1974; Aplin and Jones, 1989). Nevertheless, close apposition between infiltrating trophoblast and resident decidual cells is also evident.

Babiarz et al (1992) have shown that trophoblast of mouse EPCs (day 7) attaches to decidual cell monolayers and is capable of displacing the maternal cells to form outgrowths along the culture dish. There also appears near the interface a bilayer consisting of trophoblast giant cells overlaid with polygonal mononuclear trophoblast. The data suggest that less extensive outgrowth occurs in the presence of decidual cells than in control experiments on plastic. Outgrowth, but not attachment, could be inhibited using peptides of the RGD (fibronectin-derived, but also present in other ECM macromolecules including mouse laminin) or YIGSR (laminin-derived) families.

Although as pointed out by Mareel (1979), two dimensional assays are inadequate when used to model biological phenomena that occur in three dimensions, this is still a more realistic model of trophoblast-decidual interaction than outgrowth from newly hatched blastocysts. The data indicate that trophoblast might attach directly to decidual cells, and that an interaction with the maternal extracellular matrix could occur during formation of the chorioallantoic placenta (or alternatively that there is a role for laminin in the integrity of the EPC). Since outgrowth was inhibited by proteinase inhibitors, it may be that local trophoblast-derived proteolysis is required during placentation, but further investigation (especially the identification of physiological substrates) is required to determine the precise role of protease action. Although these data could support a role for the decidual cell in restricting invasion, either by inhibiting proteolysis or motility or by influencing the differentiation and multilayering of the trophoblast, it is not yet clear whether there is any cell type specificity associated with the restriction of outgrowth.

Aplin and Charlton (1990) have shown that choriocarcinoma cells are capable of displacing monolayers of endothelial cells in confrontation cultures; this may provide a useful model of the surface interactions that occur as trophoblast migrates within the lumena of maternal spiral arteries.

THREE DIMENSIONAL CELL CULTURE MODELS

Kliman et al. (1990) allowed aggregates of cultured term cytotrophoblast to attach to fragments of endometrium in a suspension culture system. Attachment to the open stromal surface took place and was followed by some limited infiltration of the tissue. An acellular or necrotic zone developed at or near contact sites, apparently not as a result of hypoxia or starvation.

Genbacev et al. (1992) cultured first trimester villi as explants on a Matrigel substrate and observed the outgrowth of cytotrophoblast with the features of extravillous cells, and these also showed matrix degrading activity. Vićovac et al. (1993) and Aplin et al. (1993) coincubated first trimester villous tissue explants with decidual tissue in a collagen gel support matrix. They observed the specific proliferation of cells at heterotypic contact zones into columns with the appearance of differentiation markers of the extravillous pathway such as integrin ß1 (Vićovac et al., submitted). Cells at the tips of the columns detached and infiltrated the adjacent decidua. Again, the collagen gel suffered degradation, but decidual tissue remained viable except for a thin zone of necrosis at the contact site. These observations however were not specific to decidual cocultures; villous tissue cocultured with peritoneum behaved in the same fashion (Vićovac et al., 1993). However, the formation of columns specifically at contact sites of villi either with matrix (Genbacev et al., 1992) or viable tissue (Vićovac et al., 1993; 1994; Aplin et al., 1993) does suggest a role for local signals in diverting villous stem cytotrophoblast into the extravillous differentiation pathway.

REFERENCES

Aplin, J.D., and Foden, L.J., 1985, Defective adhesion to
    extracellular matrix leads to altered social
    behaviour in cultured fibroblasts, *J. Cell Sci.* 76:199.

Aplin, J.D. Charlton, A.K., and Ayad, S., 1988, An
    immunohistochemical study of human endometrial
    extracellular matrix during the menstrual cycle and
    first trimester of pregnancy, *Cell Tissue Res.* 253:235.

Aplin, J.D., 1989, Cellular biochemistry of the endometrium,
    in: "Biology of the Uterus," R.M. Wynn, W.P. Jollie,
    eds., Plenum Press, New York.

Aplin, J.D., and Jones, C.J.P., 1989, Extracellular matrix in
    endometrium and decidua, *in*:"Placenta as a Model and
    Source," O. Genbacev, A. Klopper, R. Beaconsfield eds.,
    Plenum Press, New York.

Aplin, J.D., and Charlton, A.K., 1990, The role of matrix
    macromolecules in te invasion of deciua by trophoblast:
    model studies using BeWo cells, *in*:"Trophoblast Invasion
    and Endometrial Receptivity", H.-W. Denker, J.D.Aplin,
    eds. Trophoblast Research vol. 4, pp 139-158, Plenum
    Medical, New York.

Aplin, J.D., 1991, Implantation, trophoblast differentiation
    and haemochorial placentation: mechanistic evidence in
    vivo and in vitro, *J. Cell Sci.* 99:681.

Aplin, J.D., Sattar, A., and Mould, A.P., 1992, Variant
    choriocarcinoma (BeWo) cells with differing adhesion and
    migration to fibronectin display conserved patterns of
    integrin expression, *J. Cell Sci.* 103:435.

Aplin, J.D., 1993 Expression of integrin α6β4 in human
    trophoblast and its loss from extravillous cells,
    Placenta, 14:203.

Aplin, J.D., Sattar, A., and Vicovac, L., 1993, Cell
    interactions in trophoblast invasion, Serono symposium,
    in press.

Armant, D.R., Kaplan, H.A., and Lennarz, W.I., 1986,
    Fibronectin and laminin promote in vitro attachment and
    outgrowth of mouse blastocysts, *Dev. Biol.* 116:519.

Autio-Harmainen, H., Hurskainen, T., Niskasaari, K., Hoyhtya,
    M., and Tryggvason, K., 1992, Simultaneous expression of
    70kDa type IV collagenase and type IV collagen α1(IV)
    chain genes by cells of early human placenta and
    gestational endometrium, *Lab. Invest.* 67:191.

Babiarz, B.S., Romagnano, L.C., and Kurilla, G.M., 1992,
    Interaction of mouse ectoplacental cone trophoblast and
    uterine decidua in vitro, *In Vitro Cell Dev.Biol.*28A:500.

Behrendtsen, O., Alexander, C.M., and Werb, Z., 1992,
    Metalloproteinases mediate extracellular matrix
    degradation by cells from mouse blastocyst outgrowths.
    Development, 114:447.

Bell, S.C., 1983, Decidualisation: regional differentiation
    and associated function, *in*:"Oxford Reviews of
    Reproductive Biology", 5:220.

Bell, S.C., 1988, Synthesis and secretion of proteins by the
    endometrium and decidua. *in*:"Implantation. Biological
    and Clinical Aspects", M. Chapman, G. Gruzinskas and T.
    Chard, eds., Springer-Verlag, London.

Bell, S.C., 1989 Decidualisation and IGF-BP: implications for its role in stromal cell differentiation and the decidual cell in haemochorial placentation. *Human Reprod*. 4:125.

Benirschke, K., and Kaufman, P., 1990, Pathology of the Human Placenta (2nd ed.) Springer-Verlag, New York.

Bevilacqua, E.M.A.F., Faria, M.R., and Abrahamsohn, P.A., 1991, Growth of mouse ectoplacental cone cells in subcutaneous tissues. Development of placental-like cells, *Am. J. Anat*. 192:382.

Bischof, P., Friedli, E., Martelli, M., and Campana, A., 1991, Expression of extracellular matrix-degrading metalloproteinases by cultured human cytotrophoblast cells: effects of cell adhesion and immunopurification. *Am. J. Obstet. Gynecol*. 165:1791.

Blankenship, T.N., Enders, A.C., and King, B.F., 1993, Trophoblastic invasion and the development of uteroplacental arteries in the macaque, *Cell Tissue Res*. 272:227.

Brown, J.J.G., and Papaioannou, V.E., 1992, Ontogeny of hyaluronan expression during preimplantation differentiation of the mouse endometrium. *Differentiation* 52:61.

Burrows, T.D., King, A., and Loke, Y.W., 1993, Expression of integrins by human trophoblast and differential adhesion to laminin or fibronectin, *Human Reprod*. 8:475.

Carson, D.D., Tang, J.-P., and Gay, S., 1988, Collagens support embryo attachment and outgrowth in vitro: effects of the RGD sequence. *Dev. Biol*. 127:368.

Carson, D.D., Julian, J., and Jacobs, A.L., 1992, Uterine stromal cell chondroitin sulphate proteoglycans bind to collagen type I and inhibit outgrowth in vitro, *Dev. Biol*. 149:307.

Cowell, T.P., 1969, Implantation and development of mouse eggs transferred to the uteri of non-progestational mice. *J. Reprod. Fert*. 19:239.

Damsky, C.H., Fitzgerald, M.L, and Fisher, S.J., 1992, Distribution patterns of extracellular matrix components and adhesion receptors are intricately modulated during first trimester cytotophoblast differentiation along the invasive pathway in vivo, *J. Clin. Invest*. 89:210.

Davies, J., and Glasser, S.R., 1967, Histological and fine structural observations on the placenta of the rat. *Acta Anat*. 69:542.

De Feo, V.J., 1967, Decidualisation, *in*:"Cellular Biology of the Uterus", R.M.Wynn, ed. North-Holland, Amsterdam.

Enders, A.C., 1991, Current topic: structural responses of the primate endometrium to implantation, *Placenta* 12:309.

Enders, A.C., and King, B.F., 1991, Early stages of trophoblastic invasion of the maternal vascular system during implantation in the macaque and baboon, *Am. J. Anat*. 192:329.

Faber, M., Wewer, U.M., Berthelsen, J.G., Liotta, L.A., and Albrechtsen, R., 1986, Laminin production by human endometrial stromal cells relates to the cyclic and pathological state of the endometrium, *Am. J. Pathol*. 124:384.

Farrar, J.D., and Carson, D.D., 1992, Differential temporal and spatial expression of mRNA encoding extracellular matrix components in decidua during the peri-implantation period, *Biol. Reprod.* 46:1095.

Feinberg, R.F., Kao, L.-C., Haimowitz, J.E., Queenan, J.T., Wun, T-C., Strauss, J.F., and Kliman, H.J., 1989, Plasminogen activator inhibitor types 1 and 2 in human trophoblasts. PAI-1 is an immunocytochemical marker of invading trophoblasts, *Lab. Invest.* 61:20.

Feinberg, R.F., Kliman, H.J., and Lockwood, C.J., 1991, Is oncofetal fibronectin a trophoblast glue for human implantation? *Am. J. Pathol.* 138:537.

Fernandez, P.L., Merino, M.J., Nogales, F.F., Charonis, A.S., Stetler-Stevenson, W., and Liotta, L., 1992, Immunohistochemical profile of basement membrane proteins and 72kDa type IV collagenase in the implantation placental site, *Lab. Invest.* 66:572.

Fernandez-Shaw, S., Shorter, S.C., Naish, C.E., Barlow, D.H., and Starkey, P.M., 1992, Isolation and purification of human endometrial stromal and glandular cells using immunomagnetic microspheres, *Human Reprod.* 7:156.

Finn, C.A., and Porter, D.G., 1975, "The Uterus", Elek Science, London.

Fisher, S.J., Leitch, M.S., Kantor, M.S., Basbaum, C.B., and Kramer, R.H., 1985, Degradation of extracellular matrix by the trophoblastic cells of first-trimester human placentas, *J. Cell. Biochem.* 27:31.

Fisher, S.J., Cui, T.-y., Zhang, L., Hartman, L., Grahl, K., Guo-Yang, Z., Tarpey, J., and Damsky, C.H., 1989, Adhesive and degradative properties of human placental cytotrophoblast cells in vitro, *J. Cell Biol.* 109:891.

Genbacev, O., Sshubach, S., and Miller, R.K., 1992, Villous culture of first trimester human placenta -- model to study extravillous trophoblast differentiation, *Placenta* 13:439.

Glass, R.H., Aggeler, J., Spindle, A.I., Pederson, R.A., and Werb, Z., 1983, Degradation of extracellular matrix by mouse trophoblast outgrowths: a model for implantation. *J. Cell Biol.* 96:1108.

Glasser, S.R., 1990, Biochemical and structural changes in uterine endometrial cell types following natural or artificial deciduogenic stimuli, *Trophoblast Res.* 4:377.

Glasser, S.R., and Julian, J., 1986, Intermediate filament protein as a marker of uterine stromal cell decidualisation, *Biol. Reprod.* 35:463.

Glasser, S.R., Lampelo, S., Munir, M.I. and Julian, J.A., 1987, Expression of desmin, laminin and fibronectin during in situ differentiation (decidualisation) of rat uterine stromal cells, *Differentiation* 35:132.

Graham, C.H., McCrae, K.R., and Lala, P.K., 1993, Molecular mechanisms controlling trophoblast invasion of the uterus, *Trophoblast Res.* 7:237.

Gu, Y., Jayatilak, P.G., Parmer, T.G., Gauldie, J., Fey, G.H., and Gibori, G., 1992, α2-macroglobulin expression in the mesometrial decidua and its regulation by decidual luteotropin and prolactin, *Endocrinology* 131:1321.

Haimovici, F, and Anderson, D.J., 1993, Effects of growth factors and growth factor-extracellular matrix interactions on mouse trophoblast outgrowth in vitro. *Biol. Reprod.* 49:124.

He, C., Wilhelm, S.M., Pentland, A.P., Marker, B.L., Grant, G.A., Eisen, A.Z., and Goldberg, G.I., 1989, Tissue cooperation in a proteolytic cascade activating human interstitial collagenase, *Proc. Nat. Acad. Sci. USA* 86: 2632.

Hees, H., Moll, W., Wrobel, K.-H., and Hees, I., 1987, Pregnancy- induced structural changes and trophoblastic invasion in the segmental mesometrial arteries of the guinea pig, *Placenta* 8:609.

Hertig, A.T., Rock, J., and Adams, J., 1956, A description of 34 human ova within the first 17 days of development, *Am. J. Anat.* 98:435.

Herz, J., Clouthier, D.E., and Hammer, R.E., 1992, LDL receptor-related protein internalises and degrades uPA-PAI-1 complexes and is essential for embryo implantation, *Cell* 71:411.

Irwin, J.C., Utian, W.H., and Eckert, R.L., 1991, Sex steroids and growth factors differentially regulate the growth and differentiation of cultured human endometrial stromal cells, *Endocrinology* 129:2385.

Jayatilak, P.G., Puryear, T.K., Herz, Z., Fazleabas, A., and Gibori, G., 1989, Protein secretion by mesometrial and antimesometrial rat decidual tissue: evidence for differential gene expression, *Endocrinology* 125:659.

Jones, C.J.P., Aplin, J.D., Mulholland, J., and Glasser, R., 1993, Patterns of sialylation in differentiating rat decidual cells as revealed by lectin histochemistry, *J. Reprod. Fert.*, in press.

Kisalus, L., and Herr, J.C., 1988, Immunocytochemical localisation of heparan sulphate proteoglycan in human decidual cell secretory bodies and placental fibrinoid. *Biol. Reprod.* 39:419.

Kirby, D.R.S., 1963a, The development of mouse blastocysts tranferred to the spleen, *J. Reprod. Fert.* 5:1.

Kirby, D.R.S., 1963b, The development of the mouse blastocyst transferred to the cryptorchid and scrotal testis, *J. Anat.* 97:119.

Kliman, H.J., Feinberg, R.F., and Haimowitz, J.E., 1990, Human trophoblast-endometrial interactions in an in vitro suspension culture system, *Placenta* 11:349.

Kliman, H.J., and Feinberg, R.F., 1992, Differentiation of the trophoblast, *in:*"The First Twelve Weeks of Pregnancy", E.R. Barnea, J. Hustin, and E. Jauniaux, eds., Springer-Verlag, Berlin.

Korhonen, M., Ylanne, J., Laitinen, L., Cooper, H.M., Quaranta, V., and Virtanen, I., 1991, Distribution of the α_1-α_6 integrin subunits in human developing and term placenta, *Lab. Invest.* 65:347.

Krehbiel, R.H., 1937, Cytological studies of the decidual reaction in the rat during early pregnancy and the production of deciduomata, *Physiol. Zool.* 10:212.

Lala, P.K., and Graham, C.H., 1990, Mechanisms of trophoblast invasiveness and their control: the role of proteases and protease inhibitors, *Cancer Metast. Rev.* 9:369.

Librach, C.L., Werb, Z., Fitzgerald, M.L., Chiu, K., Corwin, N.M., Esteves, R.A., Grobelny, D., Galardy, R., Damsky, C.H., and Fisher, S.J., 1991, 92-kD type IV collagenase mediates invasion of human cytotrophoblasts, *J. Cell Biol.* 113:437.

Logan, S.K., Fisher, S.J., and Damsky, C.H., 1992, Human placental cells with temperature-sensitive simian virus 40 are immortalised and mimic the phenotype of invasive cytotrophoblasts at both permissive and nonpermissive temperatures, *Cancer Res.* 52:6001.

Mani, S.K., Julian, J., Lampelo, S., and Glasser, S.R., 1992, Initiation and maintenance of in vitro decidualisation are independent of hormonla sensitization in vivo, *Biol. Reprod.* 47:785.

Mareel, M.M.K., 1979, Is invasiveness in vitro characteristic of malignant cells? *Cell Biol. Int. Rep.* 3:627.

Martin, O., and Aarias, F., 1982, Plasminogen activator production by trophoblast cells in vitro: effect of steroid hormones and protein synthesis inhibitors, *Am. J. Obstet. Gynecol.* 142:402.

Mitchell, M.D., Trautmann, M.S., and Dudley, D.J., 1993, Cytokine networking in the placenta, *Placenta* 14:249.

Moll, U.M., and Lane, B.L., 1990, Proteolytic activity of first trimester human placenta: localization of interstitial collagenase in villous and extravillous trophoblast, *Histochemistry* 94:555.

Mossman, H.W., 1987, Vertebrate Fetal Membranes. MacMillan, London.

Mulholland, J., Aplin, J.D., Ayad, S., Hong, S., and Glasser S.R., 1992, Loss of type VI collagen from rat endometrium during decidualisation, *Biol. Reprod.* 46:1136.

Myers, D.B., Clark, D.E., and Hurst, P.R., 1990, Decreased collagen concentration in rat uterine implantation sites compared with non-implantation tissue at days 6-11 of pregnancy, *Reprod. Fert. Dev.* 2:607.

O'Shea, J.D., Kleinfeld, R.G., and Morrow, H.A., 1983, Ultrastructure of decidualisation in the pseudopregnant rat, *Am. J. Anat.* 166:271.

Parr, M.B., Tung, H.N., and Parr, E.L., 1986, The ultrastructure of the rat primary decidual zone, *Am. J. Anat.* 176:423.

Pijnenborg, R., 1990, Trophoblast invasion and placentation in the human: morphological aspects, *in:* Trophoblast Invasion and Endometrial Receptivity, H.-W. Denker and J.D. Aplin, eds, Trophoblast Research Vol. 4, pp 33-47, Plenum Medical, New York, London.

Pijnenborg, R., Dixon, G., Robertson, W.B., and Brosens, I., 1980, Trophoblastic invasion of human decidua from 8 to 18 weeks of pregnancy, *Placenta* 1:3.

Pijnenborg, R., Robertson, W.B., Brosens, I., and Dixon, G. 1981, Review article: trophoblast invasion and the establishment of haemochorial placentation in man and laboratory animals, *Placenta* 2:71.

Puistola, U., Ronnberg, L. Martikainen, H., and Turpeenniemi-Hujanen, T. (1989) The human embryo produces basement membrane collagen (type IV collagen)-degrading protease activity, *Human Reprod.* 4:309.

Randall, S., Buckley, C.H., and Fox, H., 1987, Placentation in the Fallopian tube, *Int. J. Gynaecol. Pathol.* 6:132.

Romagnano, L., and Babiarz, B., 1993, Mechanisms of murine
    trophoblast interaction with laminin, *Biol.Reprod*.49:374.
Salamonsen, L.A., Nagase, H., and Wooley, D.E., 1991,
    Production of matrix metalloproteinase 3 (stromelysin) by
    cultured ovine endometrial cells, *J. Cell Sci.* 100:381.
Samuel, C.A., and Perry, J.S., 1972, The ultrastructure of pig
    trophoblast transplanted to an ectopic site in the
    uterine wall, *J. Anat.* 113:139.
Schlafke, S., Welsh, A.O., and Enders, A.C., 1985,
    Penetration of the basal lamina of the uterine luminal
    epithelium during implantation in the rat, *Anat. Rec.*
    212:47.
Shelesnyak, M.C., 1986, A history of research on nidation,
    Ann. N.Y. Acad. Sci. 476:5.
Soares, M.J., Faria, T.N., Roby, K.F., and Deb, S., 1991,
    Pregnancy and the prolactin family of hormones:
    coordination of anterior pituitary, uterine and placental
    expression, *Endocrine Rev.* 12:402.
Starkey, P.M., 1993, The decidua and factors controlling
    placentation, *in*:"The Human Placenta", C.W.G.Redman,
    I.L.Sargent, P.M.Starkey, eds, Blackwell Scientific
    Publications, Oxford.
Starkey, P.M., Sargent, I.L., and Redman, C.W.G., 1988, Cell
    populations in human early pregnancy decidua:
    characterisation and isolation of large granular
    lymphocytes by flow cytometry, *Immunology* 65:129.
Strickland, S., Reich, E., and Sherman, M.I., 1976,
    Plasminogen activator in early embryogenesis: enzyme
    production by trophoblast and parietal endoderm, *Cell* 9:
    231.
Tabibzadeh, S., 1991, Human endometrium: an active site of
    cytokine production and action, *Endocrine Rev.* 12:272.
Thomas, T., 1993, Distribution of α2-macroglobulin and α1-acid
    glycoprotein mRNA show regional specialisation in rat
    decidua, *Placenta* 14:417.
Trinkaus, J.P., 1984, Cells into Organs: the Forces that Shape
    the Embryo, 2nd edition, Prentice-Hall, New York.
Turpeenniemi-Hujanen, T., Ronnberg, L., Kauppila, A., and
    Puistola, U., 1992, Lamininin the human embryo
    implantation: analogy to invasion by malignant cells,
    Fertil Steril. 58:105.
Vicovac, Lj., Papic, N., and Aplin, J.D., 1993, Tissue
    interactions in first trimester trophoblast-decidua
    cocultures, *Trophoblast Res.* 7:223.
Vicovac, Lj., Jones, C.J.P., and Aplin, J.D., 1994,
    Trophoblast differentiation during anchoring villus
    formation in a coculture model of the human implantation
    site in vitro, Submitted.
Wegmann, T.G., Lin, H., Guilbert, L., and Mossmann, T.R.,
    1993, Bidirectional cytokine interactions in the
    maternal-fetal relationship: is successful pregnancy a
    TH2 phenomenon? *Immunology Today* 14:353.
Welsh, A.O., and Enders, A.C., 1985, Light and electron
    microscopic examination of mature decidual cells of the
    rat with emphasis on the antimesometrial decidua and its
    degeneration, *Am. J. Anat.* 172:1.
Welsh, A.O., and Enders, A.C., 1987, Trophoblast-decidual
    cell interactions and establishment of maternal blood
    circulation in the parietal yolk sac placenta of the rat,
    Anat. Rec. 217:203.

Welsh, A.O., and Enders, A.C., 1991, Chorioallantoic placenta
    formation in the rat: II.   Angiogenesis and maternal
    blood circulation in the mesometrial region of the
    implantation chamber prior to placenta formation, *Am. J.
    Anat.* 192:347.
Wewer, U.M., Faber, M., Liotta, L.A., and Albrechtsen, R.,
    1985, Immunochemical and ultrastructural assessment of
    the nature of the pericellular basement membrane of human
    decidual cells, *Lab. Invest.* 53:624.
Winterhager, E., Stutenkemper, R., Traub, O., Beyer, E., and
    Willecke, K., 1991, Expression of different connexin
    genes in rat uterus during decidualisation and at term,
    Eur. J. Cell Biol. 55:133.
Wynn, R.M., 1974, Ultrastructural development of the human
    decidua, *Am. J. Obstet. Gynecol.* 118:652.
Yagel, S., Parhar, R.S., Jeffrey, J.J., and Lala, P.K., 1988,
    Normal nonmetastatic human trophoblast cells share in
    vitro invasive properties of malignant cells, *J. Cell
    Physiol.* 136:455.
Yelian, F.D., Edgeworth, N.A., Dong, L.-J., Chung, A.E., and
    Armant, D.R., 1993, Recombinant entactin promotes mouse
    primary trophoblast cell adhesion and migration through
    the RGD recognition sequence, *J. Cell Biol.* 121:923.
Zorn, T.M.T., Bevilacqua, E.M.A.F., and Abrahamsohn, P.A.,
    1986, Collagen remodelling during decidualisation in the
    mouse, *Cell Tissue Res.* 244:443.

INDUCTION OF DIFFERENT GAP JUNCTION CONNEXINS
IN RESPONSE TO EMBRYO IMPLANTATION

E. Winterhager[1], R. Grümmer[1], O. Traub[2]

[1]Institute of Anatomy, University Essen, 4300 Essen, Germany
[2]Institute of Genetics, University Bonn, 5300 Bonn, Germany

INTRODUCTION

Implantation requires an interaction of the blastocyst with the uterine endometrium which initiates an orderly morphological, physiological and biochemical alteration of the endometrial cell populations. This successful interaction develops only on a synchronous program of the maternal side due to estrogen and progesterone which transforms the uterus into its ovoreceptive stage (Psychoyos 1976). Though the modus of implantation is quite different among different species the preconditioning of the uterus by steroid hormones and, in addition, a local interaction between the blastocyst and the endometrium seem to be common in mammalian implantation. In recent years it became more evident that some of the cell biological differentiation programs are similar between all invasive types of implantation including apposition and adhesion of the trophoblast before it penetrates the uterine epithelium (Schlafke and Enders 1975).

The local rearrangement of the endometrium and the formation of a decidua in rodents seem to be needed for permission and control of trophoblast invasion (for review see Glasser 1990).

A high amount of embryonic signals and their effects on the maternal endometrium, especially decidualization, are discussed in literature (for review see DeFeo 1967, Kennedy 1983a and Glasser 1990) but little is known how they can regulate the maternal differentiation program and what kind of expression pattern is really changed. Recently we found evidence that the induction of cell-cell communication via gap junctions is one of the impressive maternal program changes in response to an embryonic signal (Winterhager et al. 1988, 1991, 1993).

Each connexon of a gap junctional plaque is composed of six connexins (cx) which belong to a multigene family with more than 10 members (Beyer et al. 1990, Willecke et al. 1991). These connexins are cloned and their sequences show very high sequence-identity between the species. Cell-cell communication is involved in ion transfer like in the myometrium and heart (Garfield et al. 1980, Beyer et al. 1987), and in regulation of tissue homeostasis (Yamasaki 1991), in regulation of cell differentiation as well as cell proliferation (Mehta et al. 1989). In addition, gap junctions normally treated more or less as a housekeeping gene in an adult organism demonstrate a dynamic structure with modulations and a high turnover in the uterus throughout pregnancy. The purpose of this review is to give evidence that the induction of gap junctions during embryo implantation seems to be an important step in regulating the invasion process.

RESULTS

Gap Junction Formation during Embryo Implantation in Rabbits

Implantation in rabbits occurs by adhesion of the trophoblastic knobs followed by membrane fusion of the apical trophoblast membrane with the apical membrane of the uterine epithelial cells thereby penetrating this barrier (Enders and Schlafke 1971). During nonpregnancy as well as during the preimplantation phase up to day 6 post coitum (p.c.) the uterine epithelium is noncoupled as evidenced by lucifer yellow injection (Fig. 2a), freeze fracture and immunocytochemistry. At implantation (day 6 p.c.), however, cx32 is detected in the uterine epithelium of the implantation chamber surrounding the blastocyst (Fig. 1a/b). At this stage the blastocyst is still separated from the uterine epithelium by the blastocyst covering (Fig. 1a). One day later a fully coupled epithelium restricted to the implantation chamber (Fig. 2b) is evidenced by lucifer yellow injection. This local induction of cell-cell communication spreads out on day 8 of pregnancy leading to connexin expression in the epithelium of the whole uterus. In contrast, uterine epithelium of pseudopregnant animals in comparable phases demonstrates a very low or no expression of gap junctions (Fig. 3); tubal ligation experiments which permit comparison of the blastocyst bearing uterus with the blastocyst-free uterus within the same animal give evidence that the induction of cx32 depends on the presence of a blastocyst when examined on day 7 of pregnancy. These experiments and the comparison with pseudopregnant animals support the idea that in a

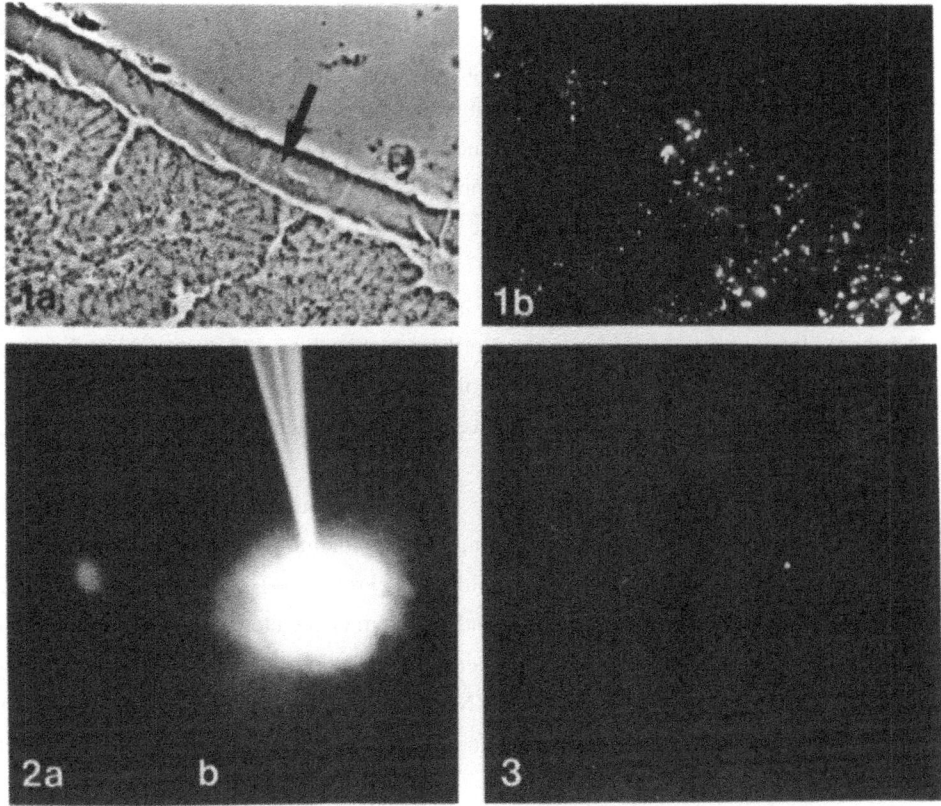

Figure 1a,b. Implantation chamber (6 d p.c.) of a rabbit uterus. The phase contrast micrograph reveals that the blastocyst is still surrounded by the coverings (arrow) which stick to the uterine epithelium (a). The uterine epithelium adjacent to the trophoblast exhibits an intense immunostaining to cx32 (b).

Figure 2. Dye coupling proved by lucifer yellow injection. The dye remains in one injected uterine epithelial cell of a nonpregnant rabbit (a). Uterine epithelial cells in the implantation chamber (7 d p.c.) demonstrate a high dye coupling (b).

Figure 3. Endometrium of a pseudopregnant rabbit (7 d p.c.). The uterine epithelium of a pseudopregnant animal lacks the reaction to cx32.

preconditioned uterus embryonic signals seem to trigger this differentiation step prior to trophoblast invasion (Winterhager et al. 1988).

Gap Junction Formation during Embryo Implantation in Rats

If gap junction induction at implantation is an essential step to trophoblast invasion it should be verified in other species with a different implantation modus. Thus we have investigated the role of gap junctional communication in rats. Here the trophoblast invades the endometrium by destruction of the uterine epithelium (Enders and Schlafke 1967) and, in contrast to the rabbit which ovulates spontaneously after mating, rats are already governed by steroid hormones during the different cyclic phases of nonpregnancy. We have investigated four different connexins - cx26, cx32, cx37 and cx43 - during all cyclic phases as well as during early pregnancy up to day 9 p.c. using immunohistochemistry, freeze fracture and Northern blot analysis. As evidenced by Northern blot analysis and immunohistochemistry, cx32 and cx37 do not play that important role during the cyclic phases of nonpregnancy as well as during early pregnancy. Therefore we focused our interest on the expression of cx26 and cx43.

In all cyclic phases of nonpregnancy cx43 is not found in the endometrium using immunolabeling; cx26 exhibits some staining of the epithelium in the deeper crypts (Fig. 4) and a very weak staining of the luminal epithelium. From day 1 p.c. onwards neither cx26 nor cx43 are found throughout the entire endometrium of the uterus up to the beginning of implantation. Northern blot analysis, however, reveals that during the different cyclic phases of nonpregnancy cx26- as well as cx43-mRNA are expressed in the rat endometrium (Figs. 9, 10) whereas - as described above - the antigen is not (cx43) or only faintly (cx26) detected.

It was of our main interest to investigate if this suppression of the connexin synthesis during the preimplantation phase is abolished by an embryonic signal at implantation. In rats the implantation reaction starts by invagination of the uterine epithelium thereby catching the blastocyst and forming an interstitial implantation chamber (Fig. 5) (Enders and Schlafke 1967). Staining with anti-cx26 reveals an intense reaction in the uterine epithelium of this implantation chamber surrounding the trophoblast on day 5 p.c. (Fig. 6). Interestingly cx26 labeling decreases with a gradient towards the epithelial cells of the uterine lumen (data not shown). The developing decidual cells arround this implantation chamber exhibit a weak but clear expression of cx43 (Fig. 7). Thus prior to trophoblast invasion induction of cx26 is evidenced locally in the surrounding epithelium and cx43 in the developing decidua. In the rat the invasion process proceeds by destruction of the uterine epithelium of the implantation chamber which leads to a confrontation of the trophoblast with the compact decidual tissue. At this stage in addition to cx43 immunostaining of cx26 is detected in the decidual cell

Figure 4. Endometrium of a nonpregnant rat during estrus. The epithelium of the uterine glands reveals a weak immunoreaction to cx26.

population in the vicinity of the blastocyst, the so-called primary decidual zone. This phenomenon is enhanced one day later (7d p.c.): cx26 antigen is strongly expressed in the specialized decidual cells surrounding the embryo showing decreasing intensity with increasing distance from the trophoblast (Fig. 8). At the same time cx43 has spread out through the whole decidual tissue up to the myometrial layer but is still enhanced in the vicinity of the implantation chamber. This situation does not change up to day 9 of pregnancy: Cx26 expression stays in the decidual cell population around the growing embryo whereas cx43 is the connexin of all decidual cells.

The corresponding Northern blot analysis reveals a clear cut relationship to the results obtained with immunolabeling (Figs. 9, 10). Cx43 mRNA is 13-fold increased at day 6 p.c. compared to nonpregnancy and stays nearly constant up to day 9 (Fig. 9). Cx26 transcripts increases from day 5 p.c. onwards up to day 9 p.c. (Fig. 10). In contrast to the constant increasing of the cx43 mRNA levels cx26 transcript is slightly enhanced on day 4 p.c. compared to day 5 p.c. The result obtained on day 4 p.c. could be correlated with the high expression of cx26 one day later in the uterine epithelium. The decrease of the cx26 mRNA level on day 5 p.c. could be due to the shift of the expression from the epithelial to the decidual tissue.

The connexin expression results in numerous gap junctional plaques as we could evidence by freeze fracture from day 7 of pregnancy onwards (Fig. 11). Double immunolabeling (Fig. 12) of tissue from the primary decidual zone with anti-cx26 and anti-cx43 demonstrate that most of the gap junctional plaques coexpress both connexins.

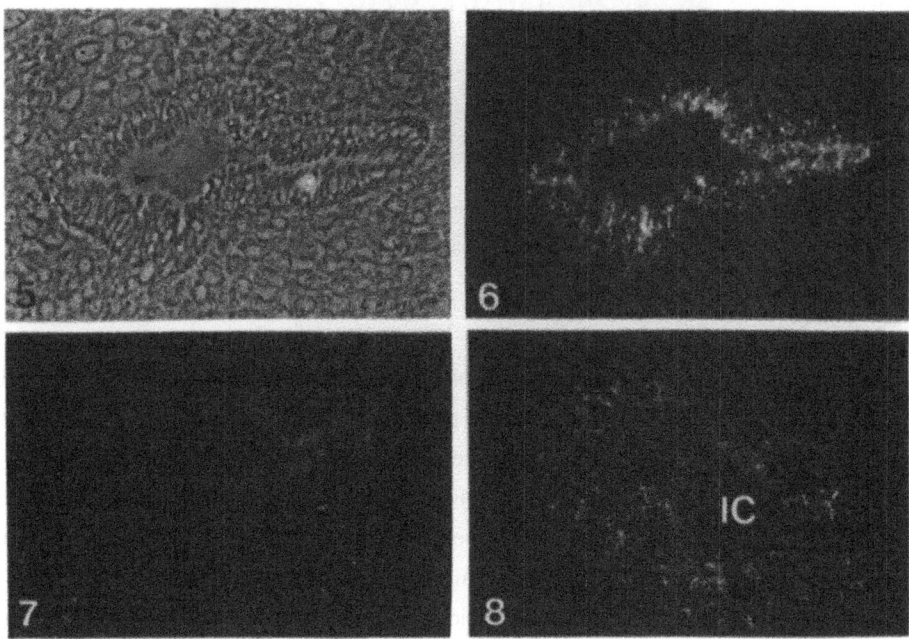

Figure 5. Implantation chamber in the rat uterus (5 d p.c.). The phase contrast micrograph shows the interstitial implantation chamber surrounded by the uterine epithelium enclosing the blastocyst (arrow head). The stroma demonstrates developing decidual cells.

Figure 6. The same implantation chamber described in figure 5. The uterine epithelium surrounding the blastocyst is strongly labeled with cx26 antigen.

Figure 7. The same implantation chamber described in figure 5. The reaction to cx43 is found in the developing decidua surrounding the implantation chamber.

Figure 8. Implantation chamber in the rat uterus (7 d p.c.). The primary decidual cell population surrounding the implantation chamber exhibits cx26 with a decreasing gradient from the implantation chamber (IC).

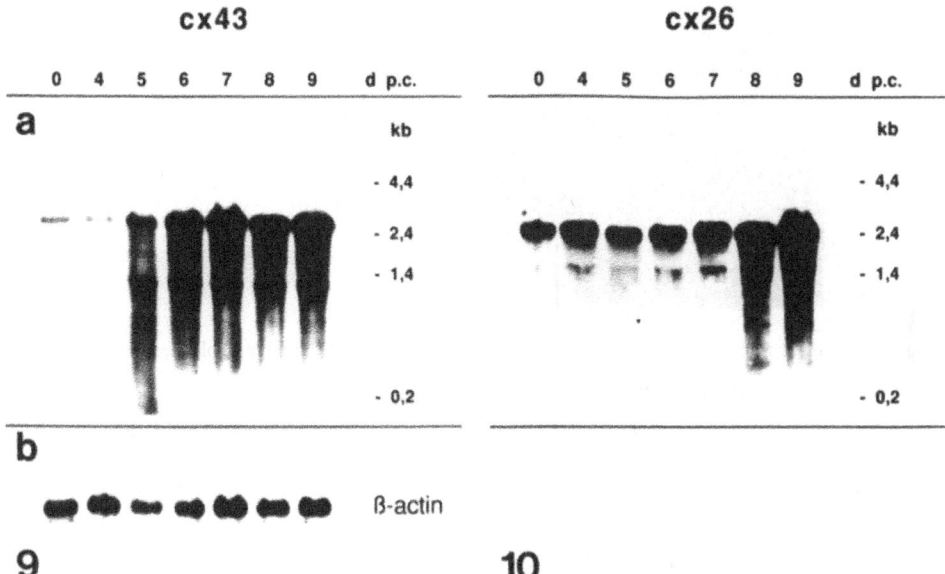

cx43

0 4 5 6 7 8 9 d p.c.

a

kb

- 4,4
- 2,4
- 1,4

- 0,2

b

β-actin

9

cx26

0 4 5 6 7 8 9 d p.c.

kb

- 4,4
- 2,4
- 1,4

- 0,2

10

Figure 9. Northern blot analysis of cx43 mRNA in the endometrium of nonpregnant and pregnant rats (4-9 d p.c.). Hybridization was performed with a rat cDNA probe (1.4 kb) specific for cx43. The level of cx43 mRNA increases rapidly from 4 to 6 d p.c.; from 6d p.c. onwards all probes demonstrate about the same mRNA levels (a). The same Northern blot was rehybridized to a ß-actin probe. No apparent differences in actin mRNA levels could be seen.

Figure 10. Northern blot analysis of cx26 mRNA in the endometrium of nonpregnant and pregnant rats (4-9 d p.c.). Hybridization was performed with a cDNA probe (1.1 kb) corresponding to the coding region of the rat cx26 gene. In pregnant rats the level of cx26 mRNA increases constantly from day 5 to day 6 p.c. compared to nonpregnant rats. The cx26 transcript is elevated on day 4 p.c. just before implantation starts.

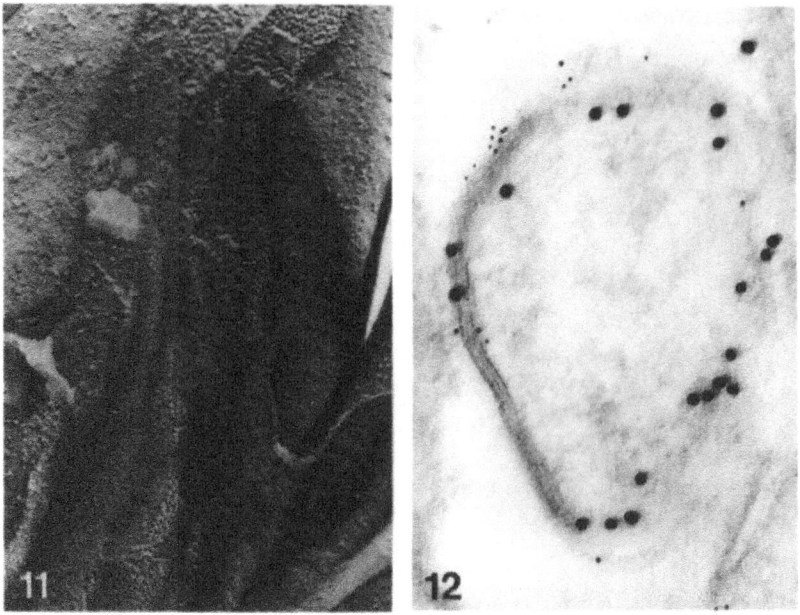

Figure 11. Freeze fracture replica of decidual cells (8 d p.c.). Large gap junctional plaques could be found in high amounts between the decidual cells.

Figure 12. Double immunocytochemical labeling demonstrates that both connexins are integrated in the same gap junctional plaque. Big beats represent cx43, small beats cx26.

SUMMARY AND CONCLUSIONS

Cell-cell coupling via gap junctional channels seems to be induced in the uterine epithelium in rats as well as in rabbits as a response to embryo recognition. The expression of cx26 in the uterine epithelium of the rat is a cell marker of differentiation corresponding to cx32 expression in rabbit uterine epithelium. Preconditioning of the endometrium by maternal steroid hormones seems to suppress cell-cell communication during the preimplantation phase. In addition, the rabbit lacks cell-cell communication in nonpregnancy probably because of noncycling. The program of connexin expression seems to be triggered locally by the embryo only in a hormonally preconditioned endometrium. Although the implantation modus of rat and rabbit are quite different (Schlafke and Enders 1975) the response of the maternal tissue to this process is similar which may indicate the importance of this cell biological differentiation step in regulating implantation. In rat due to the implantation reaction (especially decidualization) it becomes evident that the event of connexin induction is not only restricted to the uterine epithelium but accompanies the route of invasion in the surrounding tissue penetrated by the trophoblast. Spatial and temporal expression of the two different connexins cx26 and cx43 demonstrates that cx43 becomes distributed all over the decidua, while the induction of cx26 stays restricted to the trophoblast penetration pathway into the endometrium (Winterhager et al. 1993). The local and limited expression of cx32 in rabbit uterine epithelium and cx26 in rat endometrium suggests an influence of the embryo via unknown signals as postulated by several authors (Hoffman et al. 1977, DeFeo 1967, Kennedy 1983a, b, 1985). The embryonic signals which could be related to such a process remain to be identified. Signals discussed include a high number of different compounds which probably could act like a cascade including different steroid hormones, second messengers, growth factors, prostaglandins and even mechanical manipulations. Several of these stimuli induce decidualization in a hormonally preconditioned rat (Kennedy 1983b) or rabbit (Hoffman 1977) uterus.

The results described give evidence that the decidual cell population alters its program to a different extent in response to these stimuli and is not only regulated by ovarian hormones but in a more sophisticated way. It has already been shown that connexin channels exhibit different unit conductivities as proved by investigations in stable transfections (Eghbali et al. 1990, Fishman et al. 1991). However, nothing is known in regard to the differences in metabolic coupling so we could only speculate about the role of cx26 induction in the rat. Cx26 accompanies the route of trophoblast invasion in the surrounding tissue probably to maintain a coordinate differentiation program which may be regulated in a synchronous way by the induction of such specific gap junctional channels.

ACKNOWLEDGEMENTS

This work was supported by the Deutsche Forschungsgemeinschaft (DFG) through SFB 354, project 12, Universität GHS Essen and SFB 284, project 2, Universität Bonn. We want to thank Georgia Rauter for performing immunohistochemistry as well as freeze fracturing, Detlef Kittel for photographic work and Gudrun Hellmund-Risse for typing the manuscript.

REFERENCES

Beyer, E.C., Paul, D.L., and Goodenough, D.A., 1987, Connexin43: A protein from rat heart homologous to a gap junction protein from liver. *J. Cell Biol.* 105: 2621-2629.

Beyer, E.C., Paul, D.L., and Goodenough, D.A., 1990, Connexin family of gap junctions proteins. *J. Membr. Biol.* 116: 187-194.

DeFeo, V.J., 1967, Decidualization. *In* "Biology of the Uterus" (R.M. Wynn, Ed.), pp. 192-290. Appleton-Century-Crofts, New York.

Eghbali, B., Kessler, J.A., and Spray, D.C., 1990, Expression of gap junction channels in communication-incompetent cells after stable transfection with cDNA encoding connexin32. *Proc. Natl. Acad. Sci. USA* 87: 1328-1331.

Enders, A.E. and Schlafke, S., 1967, A morphological analysis of the early implantation stages in the rat. *Am. J. Anat.* 120: 185-226.

Enders, A.E. and Schlafke, S., 1971, Penetration of the uterine epithelium during implantation in the rabbit. *Am. J. Anat.* 132: 219-240.

Fishman, G.I., Moreno, A.P., Spray, D.C., and Leinwand, L.A., 1991, In vitro studies on the control of human myometrial gap junctions. *Int. J. Gynaecol. Obstet.* 25: 241-248.

Garfield, R.E., Kannan, M.S., and Daniel, E.E., 1980, Gap junction formation in myometrium: Control by oestrogens, progesterone and prostaglandins. *Am. Physiol. Soc.* 238: C81-89.

Glasser, S.R., 1990, Biochemical and structural changes in uterine endometrial cell types following natural or artificial deciduogenic stimuli. *In*: "Trophoblast Research," Vol. 4, "Trophoblast Invasion and Endometrial Receptivity" (H.-W. Denker and J. Aplin, Eds.), pp. 377-416. Plenum Medical Book Co., New York/London.

Hoffman, L.H., Strong, G.B., Davenport, G.R., and Frölich, J.C., 1977, Deciduogenic effect of prostaglandins in the pseudopregnant rabbit. *J. Reprod. Fert.* 50: 231-237.

Kennedy, T.G., 1983a, Embryonic signals and the initiation of blastocyst implantation. *Aust. J. Biol. Sci.* 36: 531-543.

Kennedy, T.G., 1983b, Prostaglandin E2, adenosin 3´5´-cyclic monophosphate and changes in endometrial vascular permeability in rat uteri sensitized for the decidual cell reaction. *Biol. Reprod.* 29: 1069-1076.

Kennedy, T.G., 1985, Evidence for the involvement of prostaglandins throughout the decidual reaction in the rat. *Biol. Reprod.* 29: 140-146.

Mehta, P.P., Bertram, J.S., and Loewenstein, W.R., 1989, The actions of retinoids on cellular growth correlate with their actions on gap junctional communication. *J. Cell Biol.* 108 (3): 1053-1065.

Psychoyos, A., 1976, Hormonal control of uterine receptivity for nidation. *J. Reprod. Fertil. Suppl.* 25: 17-28.

Schlafke, S. and Enders, A.C., 1975, Cellular basis of interaction between trophoblast and uterus at implantation. *Biol. Reprod.* 12: 41-65.

Willecke, K., Hennemann, H., Dahl, E., Jungbluth, S., and Heynkes, R., 1991, The diversity of connexin genes encoding gap junctional proteins. *Eur. J. Cell Biol.* 56: 1-7.

Winterhager, E., Brümmer, F., Dermietzel, R., Hülser, D.F. and Denker, H.-W., 1988, Gap junction formation in rabbit uterine epithelium in response to embryo recognition. *Dev. Biol.* 126: 203-211.

Winterhager, E., Grümmer, R., Jahn, E., Willecke, K., and Traub, O., 1993, Spatial and temporal expression of connexin26 and connexin 43 in rat endometrium during trophoblast invasion. *Dev. Biol.* 157: 399-409.

Winterhager, E., Stutenkemper, R., Traub, O., Beyer, E., and Willecke, K., 1991, Expression of different connexin genes in rat uterus during decidualization and at term. *Eur. J. Cell Biol.* 55: 133-142.

Yamasaki, H., 1991, Abberant expression and function of different gap junctions during carcinogenesis. *Environ. Health Perspect.* 93: 191-197.

CHANGES IN INTEGRIN EXPRESSION REFLECT A STEPWISE DIFFERENTIATION PROGRAM IN MOUSE AND HUMAN TROPHOBLAST

Caroline H. Damsky[1,2] and Susan J. Fisher[1,2,3,4]

Departments of [1]Stomatology; [2]Anatomy; [3]Obstetrics, Gynecology and
Reproductive Sciences; and [4]Pharmaceutical Chemistry
University of California, San Francisco, San Francisco, CA 94143, USA

INTRODUCTION

The differentiation program for trophoblast is an intricate one that begins at the 8-16 cell stage and continues throughout the first third of gestation (reviewed in Damsky, et al., 1993). Along with the extraembryonic endoderm it is the first terminal differentiation program to be executed in the mammalian embryo, a reality necessitated by the nutritional needs of the growing embryo. It therefore has lessons for students of differentiation mechanisms in other tissues. Trophoblast differentiation is of particular interest to tumor biologists as well as to developmental biologists because trophoblasts acquire an invasive phenotype as part of their normal differentiation program. Although as vigorous in their invasive capacity as metastatic tumor cells, the invasiveness of trophoblasts in vivo is closely regulated both spatially and temporally. The mechanisms by which the invasive phenotype of trophoblasts is developed and controlled are therefore of widespread interest.

The acquisition of an invasive phenotype by trophoblasts is a complex process involving the appropriate regulation of several sets of molecules including cell-cell adhesion molecules (e.g. the cadherins), extracellular matrix molecules and their cell surface receptors (primarily the integrins), matrix-degrading proteinases and their inhibitors (primarily metalloproteinases–MMPs, and tissue inhibitors of metalloproteinases–TIMPs). According to this scenario, invasion takes place when the net effect of the functions of these molecules tips the balance in favor of cell migration and matrix degradation (Alexander and Werb,

1991). The spatial and temporal regulation of this balance is therefore likely to be critical for regulating both the depth of trophoblast invasion and the time period over which it takes place. Growth factors and cytokines surely play a critical role in regulating the expression of adhesion molecules, ECM components and proteinases and their inhibitors. However, it is likely that signals triggered through cell-cell and cell-ECM adhesion receptors themselves, also supply regulatory information that influences expression of other classes of molecules that determine the net invasiveness of trophoblast (e.g. growth factors, proteinases). This has been demonstrated in other systems (Damsky and Werb, 1992; Juliano and Haskill, 1993). Thus, net invasiveness likely involves a network of inductive and feedback regulatory loops.

We have studied the early stages of trophoblast differentiation and invasion in the mouse (Sutherland et al., 1993) and later stages of trophoblast differentiation and placentation in the human (Fisher, et al., 1989; Librach, et al., 1991a; Damsky et al., 1992; Zhou et al., 1993; Damsky et al., in preparation). The architecture of the placentas in these two species is very different (Aplin, 1991; Rossant and Croy, 1985), although both are characterized as hemochorial. In the mouse, the initial stages of penetration and invasion of the uterine wall, and later events in placentation appear to be carried out by different populations of trophoblast. The mural trophectoderm (precursor of primary giant cell population) of the mouse embryo initiates the process of blastocyst-uterine epithelial contact and trophoblast penetration of the epithelial basement membrane, and contributes to the yolk sac placenta. The polar trophectoderm forms the ectoplacental cone (EPC), which is the stem cell population for additional invasive (secondary) giant cells, and also contributes to the chorioallantoic placenta. The situation is less clear in the human because of the difficulty in obtaining data on the very early stages of implantation. The histology of early human implantation sites suggested that the early trophoblast syncytium might initiate implantation. However, evidence is not firm (discussed in Enders and Schlafke, 1979). Clearly this issue requires further careful work. After further morphogenesis, the cytotrophoblasts (CTB) in the human, like the polar trophectoderm in the mouse, form a proliferative stem cell population and specialized structures (anchoring villi, as compared to the EPC in the mouse) which supply the bulk of the cells (mononuclear CTB) that invade the uterine wall (Table 1; see Damsky et al., 1992; Sutherland et al., 1993). Although invasion is much more extensive, and there is much greater intermingling of fetal and maternal cells in the human than in the mouse, in both cases, direct contact between fetal trophoblast and maternal blood results.

As we will describe below, there are striking similarities in the patterns of expression of integrins (Sutherland, et al., 1993; Damsky, et al., 1992), proteinases (Behrendtsen, et al 1992; Librach et al., 1991a) and proteinase inhibitors (Nomura et al., 1989; Fisher, unpublished experiments) in differentiating mouse and human trophoblast. Therefore, insights from early trophoblast differentiation gained from the mouse system should be relevant to the human system. That said, however, the unique features of human placentation make it imperative to study the biology of human trophoblast as extensively as possible, to provide insight into the etiology of disorders of implantation that have a profound impact on fertility and pregnancy outcomes in humans. The utility of this advice is demonstrated by our finding that the regulation of integrin expression by differentiating human CTB in placentas from preeclamptic pregnancies is abnormal.

INTEGRINS ARE INTRICATELY REGULATED IN MOUSE AND HUMAN TROPHOBLAST

Table 1 shows the switching of the integrin repertoire that takes place temporally and spatially during differentiation of mouse and human trophoblast. We do not have complete information for all integrins in both systems. In particular, determining the spatial distribution of several integrins in the mouse has been hampered by the lack of antibodies that both recognize mouse integrin α subunits and stain tissue. Thus, a combination of immunocytochemistry, embryo dissection followed by immunoprecipitation, and RT-PCR have provided the information in the mouse (Sutherland et al., 1993), whereas immunocytochemistry has been the primary tool in the human system. In addition, the information in the mouse is at an earlier stage than we have been able to examine in the human. Despite these caveats, two striking similarities stand out (Damsky et al., 1992; Sutherland et al., 1993). Firstly, α6-containing integrins are expressed in early stages of primary trophoblast differentiation and in the stem cell trophoblast populations present in the chorionic villi in human and EPC in the mouse. These are down regulated in trophoblasts that are differentiating to an invasive phenotype in both species. Secondly, expression of the α1 and α7 integrin subunits is upregulated in differentiating trophoblast. In the mouse, α7 is detected at the time the hatched blastocyst becomes attachment-competent, while α1 is detected only after trophoblast outgrowth has started. In the human the expression pattern of α7 is unknown, but the α1 integrin is present only in CTB that have invaded the uterine wall.

The α6 and α7 integrins, receptors for several forms of laminin, and the α1ß1 integrin, a receptor for laminin and collagen IV, recognize basement membrane components. Thus, the switching from one integrin subset to another that takes place during trophoblast differentiation suggests that trophoblast stem cells and invasive trophoblast interact in qualitatively different ways with basement membrane components in their environments. This, in turn should elicit distinctive behavioral responses. Trophoblast interactions with basement membrane components are likely to be important during implantation. During the decidual response, individual uterine stromal fibroblasts enlarge and surround themselves with a basement membrane containing laminin, entactin and collagen IV (Wewer et al., 1986). In contrast, interstitial matrix components such as fibronectin and fibrillar collagens are downregulated. The α6ß1 and α7ß1 integrins interact with the long arm of the cross-shaped laminin heterotrimer (Hall et al., 1990; Kramer et al., 1991). Comparative mapping of the two interaction sites within the long arm has not been carried out. However, it is reasonable to suggest that binding via the two different integrins have distinct consequences. The α1β1 integrin interacts with the cross region of laminin (Hall et al., 1990). Interestingly, in some systems, contact with this region of laminin promotes motility rather than adhesion (Calof and Lander, 1991), supporting the suggestion that distinct integrin repertoires promote distinct cellular behaviors. Determining what downstream signals are activated following binding via these integrin complexes may shed light on why stem cell TB and invasive TB behave so differently. The availability of in vitro model systems for trophoblast differentiation in both the mouse and human should enable us to address such questions.

INTEGRIN SWITCHING HAS FUNCTIONAL CONSEQUENCES FOR TROPHOBLAST INVASIVENESS

Human cytotrophoblasts upregulate counterbalancing adhesion mechanisms during their differentiation

To determine if the extensive changes in the integrin phenotype observed during CTB differentiation have important functional consequences, we employed a model for human trophoblast differentiation originally developed to study proteinase function (Librach et al., 1991a). In that model, CTB stem cells, isolated from floating chorionic villi of placentas of different gestational age, are put into culture on surfaces coated with the reconstituted basement membrane, Matrigel. CTB from first trimester placentas adhere, aggregate and invade the Matrigel over 18-36 h. The invasive capacity of cells from second trimester placentas is reduced, while CTB from term placentas do not invade at all in this assay. Using this model, Librach et al. (1991a), determined that the 92 kD MMP is rate-limiting for CTB invasion and that its production by CTB declines over the course of gestation. Thus, this proteinase is likely to be critical in determining invasive capacity of CTB in vivo. The 92 kD MMP is also rate-limiting for mouse trophoblast invasion (Behrendtsen et al., 1992).

To use this model for determining the functional consequence of integrin switching for CTB function, we first determined that CTB invading Matrigel displayed the same integrin and ECM phenotype as CTB invading the uterus in situ. As shown in Table 2, this criterion was fulfilled, thereby validating the model for testing the role in CTB invasiveness of those integrins that are modulated during CTB differentiation in situ (Librach et al., 1991b; Damsky et al., in preparation). We then used an antibody-perturbation approach to determine which CTB-ECM interactions were critical for invasiveness.

Table 1. Integrin switching during trophoblast differentiation in mouse and human

Mouse Trophoblast		Human Trophoblast	
Stage	TE/TB Integrins	Zone of AV	1st TM CTB Integrins
Early Blastocyst	$\alpha V\beta 3$ $\alpha 6\beta 1$*		
		I	$\alpha 6\beta 4$
		II	
Hatched Blastocyst	$\alpha V\beta 3$ $\alpha 6\beta 1$* $\alpha 7\beta 1$	III	$\alpha 6\beta 4\downarrow \alpha 6\beta 1\downarrow$ $\alpha 5\beta 1\uparrow$
		IV	$\alpha 5\beta 1$ $\alpha 1\beta 1\uparrow$
Embyro Outgrowth	$\alpha V\beta 3$ $\alpha 7\beta 1$ $\alpha 1\beta 1$		
Ectoplacental cone	$\alpha V\beta 3$ $\alpha 6\beta 1$¶ $\alpha 7\beta 1$ $\alpha 1\beta 1$ $\alpha 5\beta 1$		

TE (labels on diagrams)

B (blood vessel labels)

E
P
C

c o r e

G C

E E Ec

T B

*polarized to basal surface of TE/TB
¶restricted to stem cell core of EPC

TE, trophectoderm; TB, trophoblast; 2° GC, secondary giant cells; EEEc, extraembryonic ectoderm; EPC, ectoplacental cone; CTB, cytotrophoblast; STB, syncytiotrophoblast; TM, trimester; AV, anchoring villus; BV, blood vessel

Table 2. Integrins and ECM ligands expressed by first trimester human CTB invading uterine wall *in vivo* and isolated CTB invading Matrigel *in vitro*

	CTB: Zone I	II	III	IV	CTB in Matrigel
ECM components					
Fibronectin	-	-	+	+/-	+
Collagen IV	-	-	+	+/-	+
Laminin A	+	+	+	+	+
Laminin B1	+	+	+	+	+
Laminin B2	+	+	+	+	+
S-laminin	+	-	-	-	-
Merosin	+	-	-	-	-
Integrins					
$\alpha 1\beta 1$	-	-	-	+	+
$\alpha 5\beta 1$	-	-	+	+	+
$\alpha 6\beta 4$	++	+	+/-	-	-
$\alpha 6\beta 1$	-	-	?*	-	+/-

*since $\beta 4$ and $\beta 1$ are both present in this zone, and $\alpha 6$ pairs preferentially with $\beta 4$, it is not clear whether $\alpha 6\beta 1$ is also present.

Table 3. Effects on CTB invasion of Matrigel of antibodies that interfere with CTB-laminin interactions or CTB-fibronectin interactions

	Reagent added to invasion assay	Invasiveness of 1st TM CTB (control=1.0)
CTB Interactions with Laminin and Collagen	Anti-laminin	0.1
	Anti-collagen IV	0.1
	Anti-$\alpha 1$ integrin	0.4
	Anti-$\alpha 6$ integrin	0.9
	Anti-$\alpha 1$+Anti-$\alpha 6$	0.2
CTB Interactions with Fibronectin	Anti-fibronectin	1.2
	Anti-$\alpha 5$ integrin	2.4
	Exogenous fibronectin in Matrigel	0.2
	Anti-$\alpha 5$+exogenous fibronectin	1.4

As indicated in Table 3, CTB interactions with laminin or collagen are critical for invasion, as antibodies against them blocked CTB invasion. Antibodies against the α1ß1 integrin collagen/laminin receptor, which is upregulated during CTB invasion in situ, also blocked CTB invasion significantly, suggesting that α1ß1 is critical for mediating CTB interactions with basement membrane components during invasion. The GoH3 antibody against the α6 subunit, which affects interactions of α6ß1, but not α6ß4 with laminin, had a small effect on invasion that was apparent only when it was added together with the anti-α1 antibody. Thus, a small amount of α6ß1 remaining after differentiation of CTB, may make a small contribution to CTB interactions with laminin. However, the α1ß1 receptor is clearly the major receptor affecting invasiveness of first trimester CTB. We obtained further evidence for the significance of the α1ß1 integrin by examining the onset of expression of the α1 integrin subunit in CTB plated on a monolayer of Matrigel. In first trimester CTB, this integrin was initially expressed on only 15% of the cells. A plateau level of 60% α1-positive cells was reached by 18 h. In second trimester CTB, α1 upregulation was slower, but the maximum levels obtained after 36 hr were the same as for first trimester CTB. In term placenta, CTB did not upregulate the α1 subunit even after 72 hr in culture. Thus, there is a gestational delay in the ability of CTB to express the α1 subunit (Lim, et al., 1993). This parallels the decline of invasive capacity and MMP production in CTB isolated from different trimesters of gestation (Fisher, et al., 1989; Librach, et al., 1991a).

In contrast to the results for CTB interactions with laminin and collagen, CTB-Fn interactions appeared to restrain invasion. Antibody to the integrin fibronectin receptor, α5ß1, stimulated CTB invasion, while addition of exogenous Fn to the Matrigel reduced CTB invasion. This was at first surprising. However, this is consistent with results from tumor cells. In those studies, CHO cell lines with low levels of the α5ß1 were more tumorigenic than those with higher levels and restoration of higher expression of α5ß1 in the tumorigenic lines, reduced their tumorgenicity (Giancotti and Ruoslahti, 1990). Taken together, these results suggest that, in the course of differentiation to an invasive phenotype, human CTB upregulate counterbalancing adhesion mechanisms that are invasion-enhancing (laminin and collagen interactions with α1ß1) and invasion-restraining (fibronectin interactions with α5ß1) adhesion mechanisms (Librach, et al., 1991b; Damsky, et al., in preparation). We hypothesize that careful regulation of these mechanisms plays a significant role in regulating the depth of CTB invasion during implantation. If this is the case, we would expect abnormal regulation of integrin expression in implantation disorders in which CTB invasion is excessively deep or excessively shallow.

Abnormal regulation of CTB integrin switching is a striking feature of preeclampsia

Preeclampsia is a serious disorder of pregnancy that is characterized by high blood pressure and proteinuria in the mother, and by intrauterine growth retardation of the fetus.

Examination of the placenta at the histological level reveals that trophoblast invasion is abnormally shallow and blood vessel invasion virtually absent. Given our analysis of integrin expression and function, it is reasonable to suggest that the invasion-promoting arm of the adhesion mechanism proposed above is not upregulated during CTB differentiation in preeclamptic pregnancies. This prediction was borne out when patterns of integrin expression in placental bed biopsies of placentas from age-matched normal and preeclamptic pregnancies were examined by immunocytochemistry (Table 4; Zhou, et al., 1993).

Table 4. Comparison of the pattern of integrin expression on CTB in the placental bed tissue from age-matched late second trimester placentas (22-27 weeks) from control and preeclamptic pregnancies

	Integrin	Control*	Preeclampsia
CTB in chorionic villi	α6ß4	+	+
	α3ß1	+	+
CTB in Placental Bed	α6ß4	-	+
	α6ß1	+¶	?¶
	α3ß1	+	+
	α5ß1	+	+
	α1ß1	+	-
CTB Present in Maternal Blood Vessels		CTB Present	CTB Not present
	α1ß1≠	+	
	α5ß1	+	

*The differences in integrin profile in the control CTB at this stage when compared to the first trimester CTB (Tables 1, 2 and Damsky, et al 1992) reflect normal gestational changes of the integrin profile in normal tissue.

¶since ß4 and ß1 are both abundant in the preeclamptic placental bed, and α6 pairs preferentially with ß4, it is not clear whether α6ß1 is also present. Since ß4 is not present in control placental bed, α6ß1 is unambiguously the α6-containing integrin present.

≠CTB lining blood vessel walls in normal second trimester placental bed stained particularly strongly for a1.

Firstly, the α6ß4 integrin characteristic of undifferentiated villus CTB, is not down regulated on preeclamptic CTB in the placental bed. Secondly, the α1ß1 integrin characteristic of CTB within the uterine wall in normal placenta, is not upregulated on CTB from preeclamptic placental bed tissue. Thirdly, the α5ß1 integrin is upregulated normally in CTB from preeclamptic tissue. Finally, CTB from the preeclamptic placenta are confined to the superficial region of the placental bed, and have not invaded and modified maternal blood vessels. These data support the idea that in preeclampsia, CTB differentiate abnormally: they retain a major stem cell adhesion molecule, and upregulate the integrin suggested to restrain invasion (α5ß1), but do not upregulate the integrin (α1ß1) suggested to accelerate invasion. Thus, CTB in preeclampsia do not assume the optimum adhesion phenotype for invasion to the appropriate depth during placentation.

Disregulation of integrins during CTB differentiation is not likely to be the primary defect causing preeclampsia. However, since preeclampsia is a disease of the placenta (recovery results from delivery of the placenta) it is reasonable to propose that defective differentiation of the cell type responsible for invasion of the uterine wall during placentation, is a proximal consequence of the initial defect(s), and has significance for the further development of systemic disease. The fact that preeclampsia is primarily a disease of primagravidas suggests that the maternal immunologic response to the conceptus at an early stage of implantation could be abnormal. Over 50% of the maternal cells in first trimester decidua are bone marrow-derived (Starkey, et al., 1988), and such cells are active producers of cytokines and growth factors. Therefore, an altered immune response could result in abnormal production of signals that modulate trophoblast-uterine interactions.

How are integrins regulated during trophoblast differentiation ?

Data described above make understanding how integrin expression is modulated during normal and defective placentation a high priority. The fact that the differentiation of villus (zone I) CTB to an invasive phenotype can take place in vitro suggests that the normal program for integrin switching is regulated in an autocrine manner. This is supported by our work in the mouse, in which the downregulation of α6ß1 on trophoblast and the sequential upregulation of α7ß1 and α1ß1 at the time of blastocyst attachment and outgrowth take place in embryos cultured in vitro from the 2 cell stage onward, precluding a significant maternal (paracrine) effect on the normal integrin expression pattern from the uterine wall (Sutherland, et al., 1993). Growth factors and cytokines are prime candidates for regulating integrin expression. Trophoblasts have been shown to express several growth factors and growth factor and cytokine receptors (reviewed in Bass, et al., 1993; Wegmann and Guilbert, 1992), some of which are known to regulate integrin expression in other cell types. These are candidates for regulating normal CTB integrin expression in an autocrine manner. This does not mean, however, that signals from the uterine wall could not modulate the depth of invasion or provide cues for trophoblast targeting to maternal blood vessels. Autocrine regulation by trophoblast of their normal pattern of integrin expression also does not preclude aberrant signals from the endometrium interfering with the normal program of integrin regulation in the case of preeclampsia. We are currently examining which factors normally made by CTB, or by the endometrium, are able to modulate normal CTB invasion.

These will be tested for their ability to accelerate or depress the onset of expression of the $\alpha 1$ integrin subunit in CTB of different gestational ages in vitro.

FUTURE QUESTIONS

Many important questions remain concerning the consequences of altered integrin regulation for the etiology of preeclampsia. The most obvious link is that CTB do not migrate normally and therefore do not reach the appropriate depth within the uterine wall. More specifically, they may not have the appropriate integrin repertoire to interact properly with maternal blood vessels, resulting in the inability of the trophoblasts to invade and modify them. This in turn could lead to poor perfusion of blood to the developing fetus. It is also possible that the altered integrin repertoire has effects on the expression by trophoblasts of other components that are required for invasion. For example, integrins have been shown to regulate the expression of MMP in fibroblasts (Werb, et al., 1989) and in melanoma cells (Seftor, et al., 1992). If inappropriate regulation of integrins leads to an altered proteinase repertoire, this could seriously impair matrix degradation and therefore invasiveness. We are currently investigating this possibility in our in vitro model.

Integrins are likely not the only adhesion molecules affected in preeclampsia. Previous studies have shown that the cell-cell adhesion molecule E-cadherin, which is strongly expressed by the polarized villus CTB monolayer, is downregulated in invasive CTB in vitro (Fisher, et al., 1989) and in vivo (Damsky and Fisher, unpublished experiments). E-cadherin expression and/or function are also downregulated in invasive carcinomas, leading to the idea that E-cadherin and its cytoplasmic binding proteins constitute a tumor suppressor system (Behrens, et al., 1993). E-cadherin is a member of a large family of calcium-dependent cell adhesion molecules. Cadherin switching is a feature of the differentiation of most tissues (Takeichi, 1991), and is likely to occur during human CTB differentiation. If cadherin regulation is abnormal in preeclampsia, this could affect the ability of CTB to undergo the step-wise epithelial-mesenchymal transformation necessary for it to differentiate from a polarized monolayer sitting on the trophoblast basement membrane, to a large aggregate of nonpolarized cells (the cell column), and finally, to the largely single cell population that actively invades the uterine wall.

Finally, understanding the biology of uterine blood vessels and their endothelium could provide important links between the specific defects in trophoblast adhesion we have observed and the systemic nature of preeclampsia. If one consequence of preeclampsia is aberrant activation of endothelium (either insufficient or excessive), this could alter immune cell trafficking and cytokine production, thereby modifying the normal trophoblast program for regulating adhesion receptors. Clearly, much fascinating, but difficult work is needed to understand what role trophoblast adhesion molecules play in the etiology of preeclampsia.

REFERENCES

Alexander , C., and Werb, Z., 1991, Extracellular matrix degradation, *in* : "Cell Biology of the Extracellular Matrix", E.D. Hay, ed., Plenum Press, New York. p. 255-304.

Aplin, J.D., 1991, Implantation, trophoblast differentiation and hemochorial placentation: mechanistic evidence in vivo and in vitro, *J. Cell Sci.* 99:681-692.

Bass, K., Roth, B., Damsky, C., and Fisher, S., 1993, The regulation of human cytotrophoblast invasion, *in*:: "Nonhuman Primate IVF", D. Shomberg, ed., Springer Verlag, N. Y., in press.

Behrendsten, O., Alexander, C. M. and Werb, Z., 1992, Metalloproteinases mediate extracellular matrix degradation by cells from mouse blastocyst outgrowths. *Development* 114:447-456.

Behrens, J., Vaeket, L., Friis, R, Winterhager, E., Van Roy, F., Mareel, M., and Birchmeier, W., 1993, Loss of epithelial differentiation and gain of invasiveness correlates with tyrosine phosphorylation of the E-cadherin/ß catenin complex in cells transformed with a temperature sensitive v-SRC gene, *J. Cell Biol.* 120:757-766.

Calof, A. L. and Lander, A. D. (1991) Relationship between neuronal migration and cell-substratum adhesion: laminin and merosin promote olfactory neuronal migration but are anti-adhesive. *J. Cell Biol.* 115:779-794.

Damsky, C.H., Fitzgerald, M.L., and Fisher, S.J., 1992, Distribution patterns of extra-cellular matrix components and adhesion receptors are intricately modulated during first trimester differentiation along the invasive pathway, in vivo. *J. Clin. Invest.* 89:210-222.

Damsky, C., and Werb, Z., 1992, Signal transduction by integrin receptors for extra-cellular matrix: cooperative processing of extracellular information. *Curr. Opin. Cell Biol.*. 4:772-781.

Damsky, C.H., Sutherland, A.E., and Fisher, S.J., 1993, Adhesive interactions in early mammalian embryogenesis, implantation and placentation, *FASEB J.*, in press.

Enders, A.C., and Schlafke, S., 1979, Comparative aspects of blastocyst-endometrial interactions at implantation, *in*: "Maternal Recognition of Pregnancy", Ciba Foundation Series, 64. Excerpta Medica.

Fisher, S., Cui, T.-Y., Zhang, L., Hartman, L., Grahl, K., Zhang, G.-Y., Tarpey, J., and Damsky, C., 1989, Adhesive and invasive interactions of human cells in vitro, *J. Cell Biol.* 109:891-902.

Giancotti, F.G., and Ruoslahti, E., 1990, Elevated levels of the α5ß1 fibronectin receptor suppress the transformed phenotype of CHO cells, *Cell* 60:849.

Hall, D., Reichardt, L., Crowley, E., Moezzi, H., Holley, B., Sonnenberg, A., and Damsky, C.H., 1990, The α1/ß1 and α6/ß1 integrin heterodimers mediate cell attachment to distinct sites on Ln. *J. Cell Biol.* 110:2175-2184.

Juliano, R. and Haskill, S., 1993, Signal transduction from extracellular matrix. *J. Cell Biol.* 120:577-585.

Kramer, R. H., Vu, M. P., Cheng, Y-F., Ramos, D. M., Timpl, R. and Waheh, N., 1991, Laminin-binding integrin α7ß1: Functional characterization and expression in normal and malignant melanocytes. *Cell Reg.* 2: 805-817.

Librach, C.L., Werb, Z., Fitzgerald, M.L., Chiu, K., Corwin, N.M., Esteves, R.A., Grobelny, D., Galardy, R., Damsky, C.H., and Fisher, S.J., 1991a, 92-kD type IV collagenase mediates invasion of human s, *J. Cell Biol.* 113:437-449.

Librach, C., Fisher, S.J., Fitzgerald, M.L., and Damsky, C.H., 1991b, Cytotrophoblast-fibronectin and cytotrophoblast-laminin interactions have distinct roles in cytotrophoblast invasion, *J. Cell Biol.* 115:6a

Lim, K-H, Bass, K., Kosten, K., Damsky, C. and Fisher, S., 1993, Developmental delay of cytotrophoblast differentiation along the invasive pathway. Soc. Gynecolog. Invest. S166.

Normura, S., Hogan, B. L. M., Wills, A. J., Heath, J. and Edwards, D. R. 1989 Developmental expression of tissue inhibitor of metalloproteinase (TIMP) RNA. *Development* 105: 575-583.

Rossant, J., and Croy, B.A., 1985, Genetic identification of tissue of origin of cellular populations within the mouse placenta, *J. Embryol. Exp. Morph.* 86:177-189.

Seftor, R.E.B, Seftor, G.A., Gehlsen., K., Stetler-Stevenson, W.G., Brown, P.D., Ruoslahti, E., and Hendrix, M., 1992, The role of αVß3 integrin in human melanoma cell invasion, *Proc. Natl. Acad. Sci USA* 89:1557-1561.

Starkey, P.M., Sargent, I.L and Redman, C.W.G. (1988) Cell populations in human early pregnancy decidua: characterization and isolation of large granular lymphocytes by flow cytometry. *Immunology* 65, 129-134.

Sutherland, A., Calarco, P., and Damsky, C., 1993, Developmental regulation of integrin expression at the time of implantation in the mouse embryo, *Development*, 119, in press.

Takeichi, M., 1991, Cadherin cell adhesion receptors as a morphogenetic regulator, *Science* 251:1451-1455.

Wegmann, T.G., and Guilbert, L.J., 1992, Immune signalling at the maternal-fetal interface and trophoblast differentiation, *Dev. Comp. Immunol.* 16:435-440.

Werb, Z., Tremble, P.M., Behrendtsen, O., Crowley, E., and Damsky, C.H., 1989, Signal transduction through the fibronectin receptor induces collagenase stromolysin gene expression, *J. Cell Biol.* 109:877-889.

Wewer, U., Damjanov, A., Weiss, J., Liotta, L., and Damjanov, I., 1986, Mouse endometrial stromal cells produce basement-membrane components. *Different.* 32:49-58.

Zhou, Y., Damsky, C., Chiu, K., Roberts, J., and Fisher, S., 1993, Preeclampsia is associated with abnormal expression of adhesion molecules by invasive cytotrophoblasts, *J. Clin. Invest.* 91: 950-960.

MATRIX METALLOPROTEINASES: A ROLE IN IMPLANTATION?

Lois A. Salamonsen[1], Rikako Suzuki[2], Hideaki Nagase[2]
and David E. Woolley[3]

[1]Prince Henry's Institute of Medical Research
P.O. Box 152, Clayton, Victoria 3168, Australia
[2]Department of Biochemistry and Molecular Biology
University of Kansas Medical Center, Kansas City
Kansas 66103, USA
[3]University Department of Medicine, University
Hospital of South Manchester, West Didsbury
Manchester M20 8LR, UK

INTRODUCTION

Implantation of the conceptus is a progressive process but can conveniently be divided into three phases, blastocyst adhesion, trophoblast invasion and placentation. In all mammalian species invasion of the trophoblast is accompanied by degradation and remodelling of uterine tissue although there is a line of increasing invasiveness from those species where there is simple attachment of the trophoblast to the luminal uterine epithelium to those in which implantation involves deep embedding of the blastocyst into the endometrium. The extent of tissue degradation is also related to the form of placentation that ultimately results. The human (hemochorial placentation) and the sheep (epitheliochorial placentation) represent two extremes (Renfree, 1982; Finn, 1990). In the human and other primates, implantation is highly invasive and commences immediately after the hatched blastocyst (in the region of the polar trophectoderm) adheres to a receptive endometrial epithelium. Subsequently the trophoblast cells begin to intrude between the epithelial cells and eventually breach the basal lamina to reach the endometrial stroma. Further penetration of the extracellular matrix (ECM) occurs with erosion of local blood vessels, bringing the trophoblast into direct contact with maternal blood (Pijnenborg et al., 1981). By contrast, in the sheep, relatively little invasion by the trophoblast occurs. The ovine blastocyst attaches at clearly-defined sites, the caruncles, by fusion of binucleate cells of trophoblast origin with caruncular epithelial cells (Wooding & Morgan 1989). Some penetration of the basal lamina is seen at defined sites near areas of formation of minisyncytia. In addition, degenerative changes in the intercaruncular regions (the glandular areas between the caruncles) accompany attachment and implantation with columnar surface epithelium being lost, the trophoblast apparently making use of the cellular debris by ingestion (Boshier, 1969). Little is known of events in the endometrial stroma at this time but reorganization of the vasculature and associated structural changes in the interstitial matrix would be required to bring a sufficient blood supply into the area.

Extracellular matrix (ECM) degradation occurs at a number of sites during the implantation process. The basal lamina (associated with both the uterine epithelium and the endometrial blood vessels) must be degraded to allow passage of the trophoblast and access to maternal blood. Remodelling of the interstitial matrix is necessary to allow for new vessel growth and to accommodate the invading cells. Because the tissue barriers are made up of variety of ECM macromolecules, proteolytic enzymes capable of degrading each component are required. Therefore, it would seem important to identify the proteolytic enzymes responsible for the different phases of successful implantation, their cellular sources and the temporal and spatial relationship of their expression to the nidatory events.

Matrix Metalloproteinases

Matrix metalloproteinases (MMPs) are a family of enzymes which together are capable of degrading most of the components of the ECM. Along with tissue inhibitors of MMPs (TIMPs) the enzymes may be synthesized and secreted locally by the resident cells of a tissue or by recruited inflammatory cells such as neutrophils and macrophages. They have been studied extensively in inflammatory connective tissue disorders and in situations in which tissues are being invaded (Woessner, 1991; Matrisian, 1990; Mignatti and Rifkin, 1993) but their potential role in the remodelling of endometrial tissue or in implantation has not been widely explored. However, the inflammatory-type responses to implantation (Finn 1986) and the invasive nature of trophoblast tissue (Fisher et al., 1989) suggests that similar mechanisms will exist.

MMPs are calcium-dependent zinc proteinases with optimal activity at neutral pH. A number of the human MMP family members have now been cloned and their products sequenced (Birkedal-Hansen et al., 1993). The MMPs can be classified into major groups on the basis of preferential activity towards specific ECM macromolecular components, such as the collagens, proteoglycans, fibronectin and laminin (Woessner et al., 1991; Birkedal-Hansen et al., 1993). Those most likely to be important in the endometrium are: MMP-1 (interstitial collagenase), which is capable of initiating the degradation of the interstitial collagens; MMP-2 (gelatinase A; 72kDa gelatinase), which further degrades the products of MMP-1 in the interstitium as well as having activity towards some basement membrane components; MMP-3 (stromelysin 1), which digests proteoglycans, fibronectin and collagen IV and also degrades other components of the basal lamina; and MMP-9 (gelatinase B; 92kDa gelatinase) which appears to be important in tumor invasion degrading many of the same substrates as MMP-2 but also collagen XI (Table 1). Unique features of MMPs which distinguish them from other metalloendopeptidases are that each is secreted as a latent zymogen or precursor which can be activated by treatment with mercurial compounds or proteinases *in vitro* and by certain proteinases *in vivo* (Table 2). In this context, MMP-3 is likely to be a pivotal enzyme in the generation of collagenolytic activity since it is able to activate both MMP-1 and MMP-9. Thus the activation of MMP-3 may be critical (Nagase et al., 1991) and may initiate a cascade of proteolytic activity if co-expressed with other MMPs. All MMPs are inhibited by tissue inhibitors of metalloproteinases (TIMPs) by the formation of 1:1 complexes (Woessner, 1991) and by α_2-macroglobulin.

Table 1. Substrate specificities of matrix metalloproteinases of likely importance in implantation

MMP#	Alternative names	Extracellular matrix substrates
MMP-1	interstitial collagenase	collagens I, II, III, VII, X gelatin, proteoglycans
MMP-2	gelatinase A 72kDa gelatinase	gelatin collagens IV, V, VII, X laminin, elastin, fibronectin proteoglycans
MMP-3	stromelysin 1	proteoglycans collagens IV, V, IX, X laminin, fibronectin procollagen peptides activates proMMP-1 and proMMP-9
MMP-7	matrilysin	fibronectin, laminin, gelatin collagen IV proteoglycans elastin activates proMMP-1
MMP-9	gelatina ,e B 92 kDa gelatinase	gelatin collagens IV, V laminin, elastin

Adapted from Nagase et al. (1991)

Table 2. Natural activators of matrix metalloproteinases

proMMP-1	proMMP-2	proMMP-3	proMMP-9
trypsin plasma kallikrein plasmin MMP-3	not known	trypsin chymotrypsin plasmin plasma kallikrein neutrophil elastase cathepsin G thermolysin	trypsin MMP-3

Induction of genes for MMPs and their activators by phorbol myristate acetate (PMA) and by cytokines and growth factors such as tumor necrosis factor α (TNF), epidermal growth factor (EGF), basic fibroblast growth factor (bFGF) and interleukin-1 (IL-1) and of TIMP-1 and matrix proteins by transforming growth factor β (TGFβ), platelet-derived growth factor (PDGF), and epidermal growth factor (EGF) have been described in a range of tissues and cell cultures (Woessner, 1991). Thus, control of the activity of each MMP is multifactorial and controlled at a number of different levels (Figure 1). Many of these regulatory molecules (growth factors and cytokines) are locally produced in the uterus (Tabibzadeh, 1991; Salamonsen 1992) though whether or not their release coincides temporally or spatially with MMP expression has not been explored.

Matrix Metalloproteinases in the Embryo-Uterine Axis

There have been a number of reports of MMPs or other matrix-degrading enzymes in the embryo-uterine axis. Human embryos produce collagen-degrading activity (Puistola et al.,

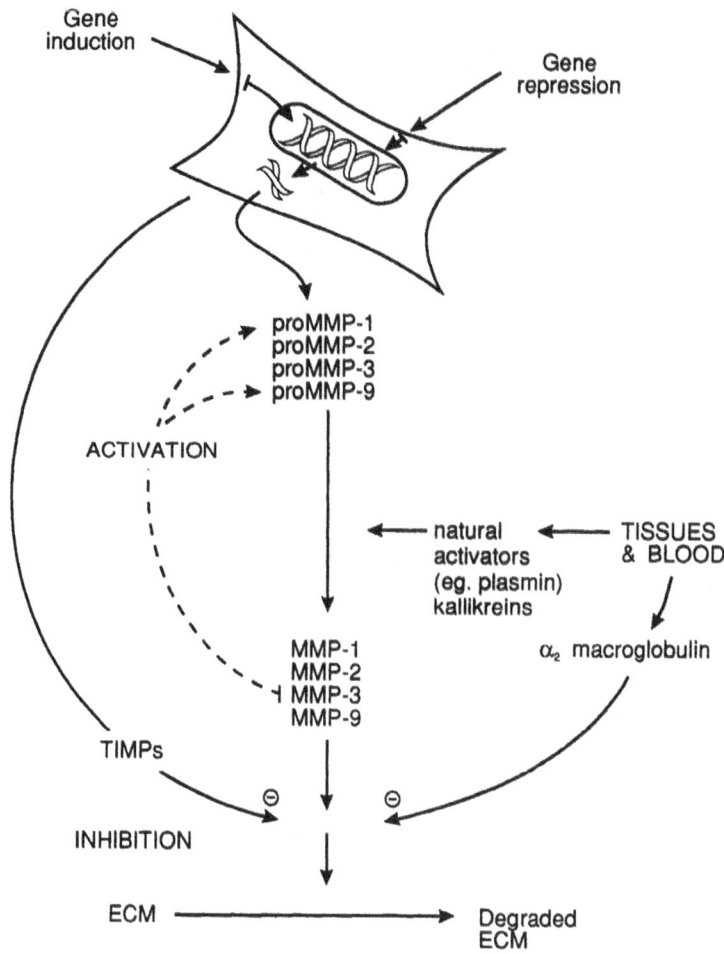

Figure 1. Regulation of matrix metalloproteinases.

1989) and mRNA transcripts for collagenase, stromelysin and TIMP, have been demonstrated in mouse embryos using the technique of polymerase chain reaction; as maternal transcripts in the unfertilized egg, in the zygote and cleavage stages but increasing at the blastocyst stage (Brenner et al., 1989). In addition, MMPs, particularly gelatinases are produced by human cytotrophoblast cells derived from first trimester pregnancies (Fisher et al., 1989; Bischof et al., 1991, this volume; Librach et al., 1991; Fernandez et al., 1992). *In vitro*, these cytotrophoblasts invade cultured endometrial glands, but not epithelial cells from other sources, suggesting that not only are the trophoblast cells invasive but also that the uterine epithelial cells must have a specific response to the trophoblast factors (Librach et al., 1991). Furthermore, decidual cells are the first cells to penetrate the uterine luminal basal lamina of rats and also invade the basal lamina of maternal blood vessels (Schlafke et al., 1985). Rat uterine cells have a greater capacity to degrade a variety of complex extracellular matrices *in vitro* than either rat or mouse trophoblast cells (Welsh and Enders, 1989). These studies support an active role for the uterine stroma in implantation and placentation in the rat and suggest that at least in this species, trophoblast invasiveness is not all important for implantation. Thus, it would seem important to determine the expression and secretion of MMPs from both endometrial as well as trophoblast cells in a number of animal species with different forms of placentation.

MATRIX METALLOPROTEINASES IN THE EMBRYO-MATERNAL UNIT OF THE SHEEP

Endometrium

Ovine endometrial cells in primary culture have the potential to express a range of MMPs (Salamonsen et al., 1991, 1993). Endometrial tissue was derived from estrogen- and progesterone-treated ovariectomized Corriedale ewes and dissociated cell suspensions were obtained following digestion with bacterial collagenase. Cultures of either mixed endometrial cells (Salamonsen et al., 1985) or highly purified epithelial and stromal cells (Cherny & Findlay, 1990) were plated in medium 199 with 10% charcoal-stripped fetal calf serum and antibiotics for 24-48 hours and the medium then changed to serum-free with or without PMA (100nM) for a further 48 hours. Conditioned medium was collected for analysis and the cells were either retained for DNA assay or fixed for immunocytochemistry following treatment with monensin (which inhibits transport of proteins through the Golgi apparatus) for 4 hours. The identity of MMPs in the culture medium was established by the use of specific enzyme assays for collagenase (MMP-1), gelatinases (MMP-2 + MMP-9) and stromelysin (MMP-3), by zymography and by Western blot analysis. Northern blot analysis using cDNA probes for human MMP-1 and human MMP-3 identified specific mRNA as a component of the total RNA extracted from the PMA-treated cells.

Collagenase, gelatinase and stromelysin activity was detected in the culture medium from mixed endometrial cells by specific enzyme assays. Most of the enzymes were present in their latent form as evidenced by the increased activity measured following treatment of the medium with 4-(aminophenyl)mercuric acetate (APMA). Both collagenase and stromelysin activities were increased following treatment of the cells with PMA but gelatinase activity was not significantly affected (Figure 2). These effects of PMA were both time- and dose-responsive (Salamonsen et al., 1991 1993).

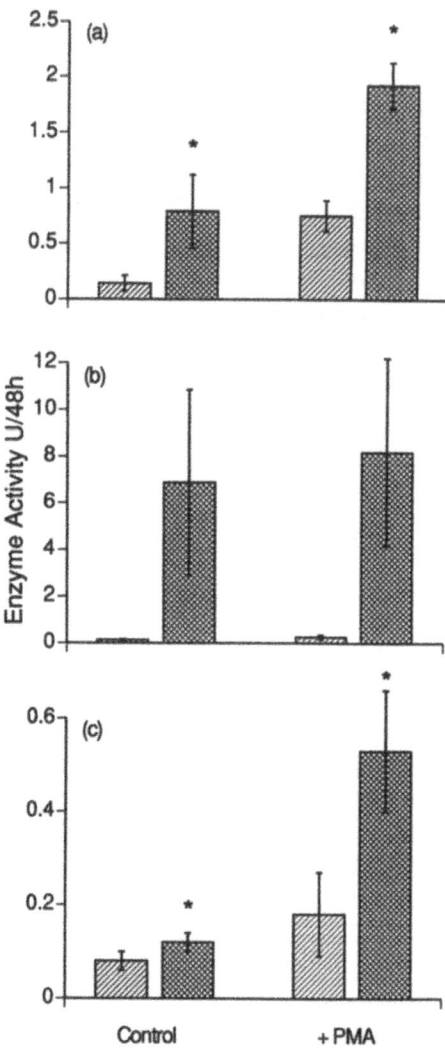

Figure 2. Relative production of (a) collagenase, (b) gelatinase and (c) stromelysin by ovine endometrial cells cultured without (control) or with PMA. Enzyme activities are expressed as units released into culture medium in 48h. Hatched bars = without activation; cross-hatched bars = following treatment with APMA in vitro. * significantly different (P<0.05). (Compiled from data in Salamonsen et al 1991, 1993.)

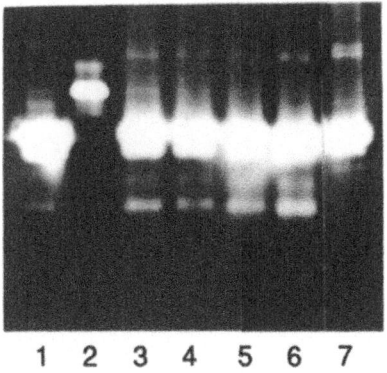

1 2 3 4 5 6 7

Figure 3. Gelatin zymography of culture medium from mixed cultures of ovine endometrial cells. Lane 1: pro MMP-2 (human standard), lane 2: proMMP-9 (human standard), lanes 3 & 4: culture medium from ovine endometrial cells, unstimulated, lanes 5 & 6: culture medium from ovine endometrial cells stimulated with PMA (100nM, 48 hours), lane 7: human synovial cell culture medium. (reprinted from Salamonsen et al., 1993, with permission).

Gelatin zymography demonstrated that gelatinases of molecular weights corresponding to those of human proMMP-2, but not proMMP-9, were present in the culture medium together with an additional gelatinolytic band of higher molecular weight (approximately 110kDa) (Figure 3). Both these gelatinases were also present in culture medium from human synovial cells. The higher molecular weight species was less prevalent than proMMP-2. Each gelatinase was secreted primarily in its latent form as demonstrated both by molecular weight determination and by its activation by APMA. Further evidence for the identification of these gelatinases as MMPs was demonstrated by their complete inhibition using EDTA and o-phenanthroline. Both gelatinases were produced by the cells without stimulation *in vitro* and PMA did not significantly increase their production.

Expression of specific mRNA for MMP-1 and MMP-3 in these cell preparations was confirmed by Northern blot analysis of total RNA derived from the cultured cells following stimulation with PMA for 48 hours (Figure 4).

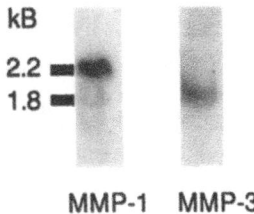

MMP-1 MMP-3

Figure 4. Northern blot analysis of total RNA (20µg) derived from cultured ovine endometrial cells treated with PMA (48 hours, 100nM), and hybridized with ^{32}P-cDNA probes specific for human MMP-1 and human MMP-3.

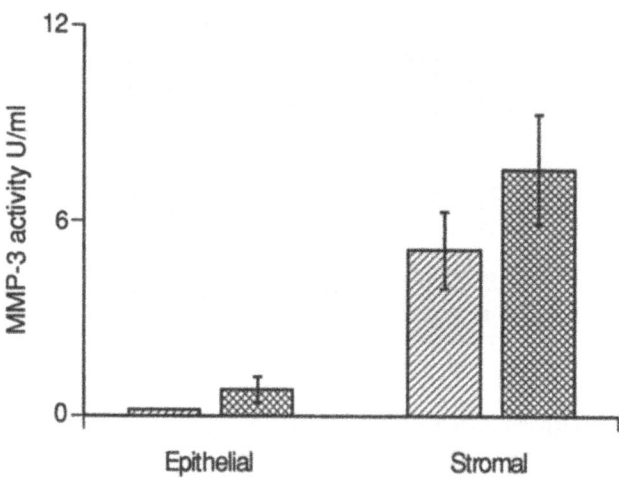

Figure 5. MMP-3 production by purified ovine epithelial and stromal cells in primary culture, either (single hatched bars) untreated or (cross-hatched bars) treated with PMA (100nM, 48 hours). Culture medium was treated with APMA prior to assay.

Matrix metalloproteinases are produced predominantly by appropriately stimulated connective tissue cells (Matrisian and Hogan, 1990) and in the sheep endometrium it was the stromal and not the epithelial cells which produced MMPs 1, 2 & 3 into the culture medium (Salamonsen et al., 1993; Figure 5), observations confirmed by immunocytochemical staining of the cells (Figure 6).

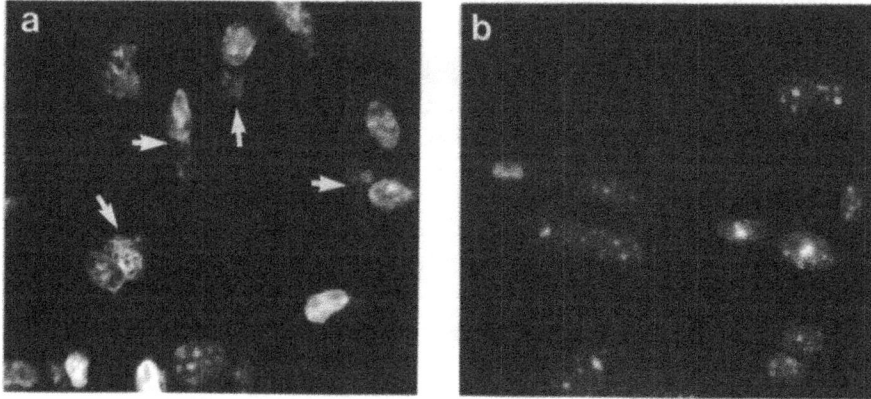

Figure 6. Immunocytochemical staining for MMP-3. (a) in ovine endometrial stromal cell cultures following stimulation with PMA for 48 hours followed by treatment with monensin (3µM) for 4 hours. (b) in ovine endometrial epithelial cells treated similarly. Antiserum was raised in rabbits against human MMP-3 and detected using FITC-labelled anti-rabbit IgG. Sections were counterstained with ethidium bromide. Arrows highlight positive staining for MMP-3 in the Golgi apparatus of stromal but not epithelial cells.

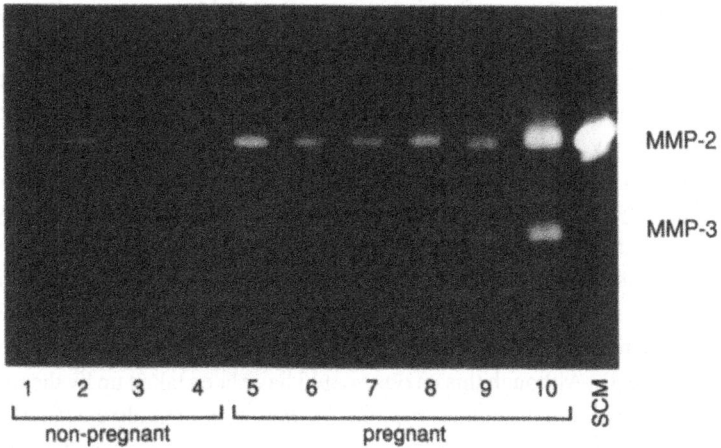

Figure 7. Gelatin zymography of uterine flushings from day 16 non-pregnant (lanes 1-4) and day 16 pregnant (lanes 5-10) ewes and synovial cell culture medium (contains proMMP-2, lane 11).

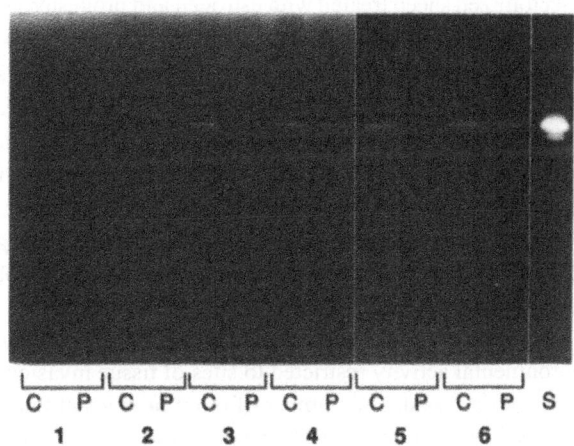

Figure 8. Gelatin zymography of medium following culture of trophoblast tissue from 6 individual sheep blastocysts (1-6); C = cultured without PMA, P = cultured in the presence of PMA, S = synovial cell culture medium (proMMP-2).

Blastocyst

The uteri of anaesthetized pregnant and non-pregnant ewes were flushed with saline on day 16 of the cycle (estrus, day 0; non-pregnant) or day 16 of pregnancy (day of attachment) (Salamonsen et al., 1986). Uterine flushings were collected, blastocysts retrieved and cellular debris removed by centrifugation. Pieces of trophoblast tissue (approximately 1cm) were thoroughly washed and maintained in serum-free culture for 48 hours with and without PMA (100nM) (Salamonsen et al., 1984). Both flushings and culture medium were subjected to gelatin zymography. Whereas uterine flushings from four day 16 non-pregnant ewes contained little or no gelatinase activity, flushings from six pregnant ewes contained at least two bands of activity, one being identified as proMMP-2 and the other as probably proMMP-3 by molecular weight estimation (Figure 7, Salamonsen et al., 1992). Furthermore, following culture of trophoblast tissue, proMMP-2 activity was identified by zymography in the culture medium from 4 of 6 separate blastocysts. In each case there was no more proMMP-2 released in the presence than in the absence of PMA suggesting that its secretion by trophoblast is constitutive (Figure 8). Although this enzyme could have been taken up by the trophoblast *in utero* and then released into the culture medium *in vitro*, a more likely explanation is that the preimplantation trophoblast is synthesizing proMMP-2 and secreting it into the culture medium. This requires confirmation.

In summary, sheep endometrial stromal but not epithelial cells in primary culture secrete proMMP-1, proMMP-2 and proMMP-3. The production of proMMP-1 and proMMP-3 can be further stimulated *in vitro* by PMA, but synthesis of proMMP-2 appears to be either constitutive or to have been previously stimulated *in vivo* by some bioregulator present in the uterus of the ovariectomized sheep treated with estrogen and progesterone to mimic the luteal phase of the estrous cycle. The nature of the natural stimuli of gene expression within the endometrium, the extent and mediators of activation of endometrial MMPs *in vivo,* and to what extent the MMPs are bound by TIMPs or α_2-macroglobulin following secretion, remains to be determined. It is possible that the co-production of proMMP-1 and proMMP-3 from endometrial cells favours activation of proMMP-1 *in vivo*. Preliminary evidence also suggests that proMMP-2 is produced by sheep blastocysts at the time of attachment. The developmental stage of the sheep blastocyst when gene transcription and translation begins, the duration of release of MMP-2 and whether the enzyme is present at the site of trophoblast invasion of the basal lamina of the uterine epithelium, remain to be established. It is to be expected that, as in disorders such as rheumatoid arthritis and tumor invasion, MMP production and activation reflects microenvironmental activity restricted to sites of tissue invasion.

The sheep blastocyst produces a number of other proteins at the time of implantation (Godkin et al., 1982, Salamonsen et al., 1984) the most dominant of these being a trophoblast interferon (Roberts et al., 1991) which modifies both prostaglandin and protein release from the endometrium (Salamonsen et al., 1988). Whether or not any of these proteins act in a paracrine manner to modify MMP production by the endometrial cells or whether any of them are capable of activating MMPs of either endometrial or blastocyst origin are other important questions which require resolution.

MATRIX METALLOPROTEINASES IN THE HUMAN ENDOMETRIUM

Relatively subtle changes in the interstitial extracellular matrix components of the human endometrium can be observed throughout the menstrual cycle with the collagen type VI network that interconnects the major structural collagens being less prominent around the time of implantation. There are also cyclical changes in the relative amounts of the components of the basal lamina (laminin, entactin/nidogen, collagen type IV, heparan sulphate proteoglycan) (Aplin et al., 1988, Aplin 1989). In particular, decidualization of the stromal fibroblasts is associated with development of a basal lamina around the resultant decidual cells and hence there is a relatively large total increase in the associated component molecules. As the human embryo invades through the luminal epithelium into the decidualized stroma, it seems likely that both interstitial ECM and basal lamina will be broached by the local production and activation of MMPs. Identification of the MMPs and their endometrial or embryonic origin should improve our understanding of the implantation process.

Production of MMPs *in vitro* by human endometrium has recently been described. Collagenase (MMP-1), and gelatinases A (MMP-2) and B (MMP-9) are secreted into culture medium by human endometrial explants (Marbaix et al., 1992) and are stimulated upon withdrawal of progesterone from the explants. The same three MMPs plus MMP-3 have been detected by enzyme assay and by gelatin and casein zymography in culture medium from human endometrial stromal cells in both primary and secondary culture (Rawdanowicz et al., 1992). Following treatment of these cells with PMA (100nM) the secretion of MMP-2 is not altered whereas that of the other three MMPs is enhanced, suggesting differential regulation as seen for MMPs in the ovine endometrium. Messenger RNAs for both MMP-1 and MMP-3 have been detected by Northern analysis following extraction from the human endometrial stromal cell cultures (Hampton and Salamonsen, unpublished observations). In another study (Rodgers et al., 1992) both the enzyme matrilysin (MMP-7) and its mRNA have been localized to epithelial cells of the human endometrium being particularly abundant in the perimenstrual period. This epithelial origin of an MMP is unusual, most MMPs being produced from cells of connective tissues or from tumor or trophoblast cells. The time of the appearance of MMP-7 and the upregulation of other endometrial MMPs by progesterone withdrawal suggests that in women, MMPs of endometrial origin may have a particularly important role in menstruation. For ethical reasons it will be difficult to establish the role of MMPs in the early stages of blastocyst invasion in the human.

CONCLUSIONS

Although only limited data are available it is apparent from our studies and those of others that MMPs are likely to have a pivotal role in implantation. It will be important to establish the temporal and spatial location of the MMP production in relation to the invasion of trophoblast, the factors regulating transcription of the genes for MMPs in the endometrium and the trophoblast, the identification of specific enzyme activators and the control of inhibitor production. Ultimately, an improved understanding of these mechanisms may offer new avenues for the regulation of fertility and infertility not only in man, but also in domestic and endangered species.

ACKNOWLEDGMENTS

The authors wish to thank Anne Hampton and Tom Rawdanowicz for expert laboratory assistance, Sue Panckridge for the graphical presentations and Claudette Thiedeman for assistance with presentation of the manuscript. The authors' work was supported by the NH&MRC of Australia (LAS) and by NIH grant AR39189 (HN).

REFERENCES

Aplin, J.D., 1989, Cellular biochemistry of the human endometrium, in "Biology of the Uterus," R.M. Wynn and W.P. Jollie, ed., Plenum Medical Book Company, New York p89-108.

Aplin, J.D., Charlton, A. K., and Ayad, S., 1988, An immunohistochemical study of human endometrial extracellular matrix during the menstrual cycle and first trimester of pregnancy. *Cell Tissue Res.* 253:231-240.

Birkedal-Hansen, H., Moore, W.G.I., Bodden, M.K., Windsor, L.J., Birkedal-Hansen, B., De Carlo, A. and Engler, J.A., 1993, Matrix metalloproteinases : A review. *Critical Rev. Oral Biol. Med.* 4:197-250.

Bischof, P., Friedli, E., Martelli, M., and Campana., A., 1991, Expression of extracellular matrix-degrading metalloproteinases by cultured human cytotrophoblast cells: effects of cell adhesion and immunopurification. *Am. J. Obstet. Gynecol.* 165:1791-1801.

Boshier, D.P., 1969, A histological and histochemical examination of implantation and early placentome formation in sheep. *J. Reprod. Fertil.*, 19, 51-61.

Brenner, C.A., Adder, R.A.., Rappolee, D.A., Pedersen, R.A., and Werb, Z., 1989, Genes for extracellular matrix degrading metalloprotcinases and their inhibitor, TIMP, are expressed during early mammalian development, *Genes and Deve.* 3:848-859.

Cherny, R.A., and Findlay, J.K., 1990, Separation and cultures of ovine endometrial and stromal cells: evidence of morphological and functional polarity, *Biol. Reprod.* 43:241-250.

Fernandez, P.L., Merino, M.J., Nogales, F.F., Charonois, A.S., Stetler-Stevenson, W., and Liotts, L., 1992, Immunohistochemical profile of basement membrane proteins and 72 kilodalton type IV collagenase in the implantation placental site. *Lab. Invest.* 66:572-579.

Finn, C.A., 1986, Implantation, menstruation and inflammation, *Biol. Rev.* 61:313-328.

Finn, C.A., 1990, Species variation, location and attachment of blastocysts, in "Blastocyst Implantation", K. Yoshinaga, ed., Adams Publishing Group, Boston. p47-54.

Fisher, S.J., Cui, T., Zhang, L., Hartman, L., Grahl., K., Guo-Zhang, Z., Tarpey, J. and Damsky, C.H., 1989, Adhesive and degradative properties of human placental cytotrophoblast cells *in vitro*, *J. Cell Biol.* 109:891-902.

Godkin, J.D., Bazer, F.W., Moffatt, J., Sessions, F. and Roberts, R.M., 1982, Purification and properties of a low molecular weight protein released by the trophoblast of sheep blastocysts at day 13-21, *J. Reprod. Fertil.*, 65:141-150.

Librach, C.L., Werb, Z., Fitzgerald, M.L., Chiu, K., Corwin, N.M., Esteves, R.A., Grobelny, D., Galardy, R., Damsky, C.H. and Fisher, S.J., 1991, 92-kD type IV collagenase mediates invasion of human cytotrophoblasts. *J. Cell Biol.*, 113:437-449.

Marbaix, E., Donnez, J., Courtoy, P.J., and Eeckhout, Y., 1992, Progesterone regulates the activity of collagenase and related gelatinases A and B in human endometrial explants, *Proc. Natl. Acad. Sci. USA.* 89:11789-11793.

Matrisian, L.M., 1990, Metalloproteinases and their inhibitors in matrix remodelling. *TIG.* 6:121-125.

Matrisian, L.M., and Hogan, B.L.M., 1990, Growth factor-regulated proteases and extracellular matrix remodelling during mammalian development, *Curr. Topics. Dev. Biol.* 24:219-259.

Mignatti, P. and Rifkin, D.B., 1993, Biology and biochemistry of proteinases in tumor invasion. *Physiol. Rev.* 73, 161-195.

Nagase, H., Ogata, Y., Suzuki, K., Enghild, J.J. and Salvesen, G., 1991, Substrate specificity and activation mechanisms of matrix matalloproteinases. *Biochem. Soc. Trans.* 19:715-718.

Pijnenborg R., Robertson, W.B., Brosens, I. and Dixon G., 1981, Review article: Trophoblast invasion and the establishment of haemochorial placentation in man and laboratory animals. *Placenta*, 2, 71-92.

Puistola, U., Ronnberg, L., Martikainen, H., and Turpeenniemi-Hujanen, T., 1989, The human embryo produces basement membrane collagen (type IV collagen)-degrading activity, *Human Reprod.* 4:309-311.

Rawdanowicz, T.J., Hampton, A.L., Nagase, H., Woolley, D.E., and Salamonsen, L.A., 1992, Human endometrial stromal cells in culture secrete matrix metalloproteinases: identification of MMP-1, MMP-2 and MMP-3, *Proc. Aust. Soc. Med. Res.*, P18.

Renfree, M.B., 1982, Implantation and placentation, *in* "Reproduction in Mammals, Book 2." Austin, C.R. and Short, R.V. eds, 2nd edition, Cambridge University Press, Cambridge. p26-69.

Roberts, R. M., Klemann, S.W., Leaman, D. W., Bixby, J.A., Cross, J.C., Farin, C.E., Imakawa, K., and Hansen, T.R., 1991, The polypeptides and genes for ovine and bovine trophoblast protein-1, *J. Reprod. Fertil. Suppl.*, 43:3-12.

Rodgers, W.H., Osteen, K.G., Matrisian, L.M., Navre, M., Guidice, L.C. and Gorstein, F., 1992, Expression and localization of matrilysin, a matrix metalloproteinase in human endometrium during the reproductive cycle. *Am. J. Obstet. Gynecol.*, 168:253-260.

Salamonsen, L.A., 1992, Local regulators and the establishment of pregnancy, *Reprod. Fertil. Dev.* 4:125-134.

Salamonsen, L.A., Doughton, B. and Findlay, J.K., 1984, Protein synthesis by preimplantation sheep blastocysts, *in* "Reproduction in Sheep," D.A.Lindsay and D.T.Pearce, eds, Aust. Acad. Sci., Canberra, p115-117.

Salamonsen, L.A., O, W.S., Doughton, B. and Findlay, J.K., 1985, The effects of estrogen and progesterone *in vivo* on protein synthesis and secretion by cultured epithelial cells from sheep endometrium, *Endocrinol.* 117:2148-2159.

Salamonsen, L.A., Doughton, B. and Findlay, J.K., 1986, The effects of the preimplantation blastocyst in vivo and in vitro on protein synthesis and secretion by cultured epithelial cells from sheep endometrium, *Endocrinol.* 119:622-628.

Salamonsen, L.A., Stuchbery, S.J., O'Grady, C.M., Godkin, J.D., and Findlay, J.K., 1988, Interferon α mimics the effects of ovine trophoblast protein-1 on prostaglandin and protein secretion by ovine endometrial cells *in vitro*, *J. Endo.* 117:R1-R4.

Salamonsen, L.A., Nagase, H., and Woolley, D.E., 1991, Production of matrix metalloproteinase 3 (stromelysin) by cultured ovine endometrial cells, *J. Cell. Scie.* 100:381-385.

Salamonsen, L.A., Nagase, H., Suzuki, R. and Woolley, D.E., 1992, Differential expression of matrix metalloproteinases by constituents of the embryo-maternal axis of the sheep prior to implantation. *Proc. Aust. Soc. Reprod. Biol.* 24, 77.

Salamonsen, L.A., Suzuki, R., Nagase, H. and Woolley, D.E., 1993, Production of matrix metalloproteinase 1 (interstitial collagenase) and matrix metalloproteinase 2 (gelatinase A: 72kDa gelatinase) by ovine endometrial cells *in vitro*: different regulation and preferential expression by stromal fibroblasts, *J. Reprod. Fertil.*, 98 (1) (in press).

Schlafke, S., Welsh, A.O., and Enders, A.C., 1985, Penetration of the basal lamina of the uterine luminal epithelium during implantation in the rat, *Anat. Rec.*, 212:47-56.

Tabibzadeh, S.S., 1991, Human endometrium: an active site of cytokine production and action. *Endocrine Rev.*, 12, 272-290.

Welsh, A.O., and Enders, A.C., 1989, Comparison of the ability of cells from rat and mouse blastocysts and rat uterus to alter complex extracellular matrix *in vitro*, *in* "Blastocyst Implantation," K. Yoshinaga, ed., Adams Publishing Group, Boston. p55-74.

Woessner, J.F.Jr., 1991, Matrix metalloproteinases and their inhibitors in connective tissue remodeling, *FASEB J.* 5:2145-2154.

Wooding, F.B.P., and Morgan, G., 1989, Fetomaternal cell fusion at ruminant implantation, *in* "Blastocyst Implantation", K. Yoshinaga, ed., Adams Publishing Group, Boston. p117-123.

CONTROLLED EXTRACELLULAR-MATRIX DEGRADATION: A FUNDAMENTAL MECHANISM IN THE IMPLANTATION PROCESS

Paul Bischof, Marzia Martelli and Aldo Campana

Infertility and Gynecologic Endocrinology Clinic
Department of Obstetrics and Gynecology
University of Geneva
Geneva, Switzerland

TROPHOBLAST INVASION AND PROTEOLYSIS

Implantation in humans occurs in a non decidualized endometrium. It is believed to begin 5 days after ovulation. After an initial apposition and attachment phase, penetration starts by thin folds of trophoblast progressing between adjacent endometrial epithelial cells. Once the blastocyst is completely embedded, the embryonic disc is surrounded by a cytotrophoblast layer and more distally by a highly proliferative syncytium (formed by the fusion of cytotrophoblastic cells). Radially distributed lacunae appear in the syncytium and cytotrophoblast cells proliferate between the lacunae (cytotrophoblastic columns, Boyd and Hamilton, 1970). This process leads to the formation of trophoblastic villi (lacunar stage). The mature villi, consist of an outer layer of syncytium, an inner layer of cytotrophoblast (villous trophoblast) surrounding a fetal stroma. Trophoblastic invasion does not end with the formation of the villi. At the tips of certain villi, cytotrophoblast cells break through the syncytium and begin to invade the endometrial stroma (anchored villi). These trophoblastic cells can either fuse secondarily to form syncytial islets or invade the maternal tissues (including the endometrial capillaries) as single cells. This trophoblastic tissue is known as the intermediate trophoblast (Yeh and Kurman, 1989). Thus, implantation is characterized by two distinctive waves of invasion, the first is due to syncytial trophoblast (prelacunar and lacunar stages) and the second to cytotrophoblast (lacunar and villous stages). It must be remembered that at the villous stage, syncytium is not invasive anymore. These two waves of trophoblastic invasion occur in two different types of endometria: the "syncytial wave" occurs in a non decidualized endometrium whereas the "cytotrophoblastic wave" occurs in a decidualized endometrium. To what extent decidualization triggers cytotrophoblastic invasion remains to be explored.

As reviewed recently (Bischof and Martelli, 1992), trophoblastic invasion is not due to passive growth pressure but to an active biochemical process. Trophoblast is invasive partly by virtue of its ability to secrete proteinases capable of degrading its surrounding extracellular matrix (Mullins and Rohrlich, 1983, Aplin 1991, Bischof and Martelli 1992). Several proteolytic enzymes associated with a phenotypically invasive trophoblast have been described. Moll and Lane (1990) described a 55 kDa interstitial collagenase produced by villous and extravillous trophoblast cells. Several distinct Matrix-Metalloproteinases (including the 72 kDa and 92 kDa gelatinases) where shown to be secreted by first trimester cytotrophoblast (Fisher et al 1989, Bischof et al 1991). Plasminogen activator (uPA, Martin and Arias 1982, Queenan et al 1987) and plasminogen activator inhibitor (PAI-I and PAI-2, Feinberg et al 1989) were found in villous and extravillous trophoblast. That these enzymes are involved in trophoblast invasion is evidenced by the following observations: Collagenase (55 kDa) and gelatinases (72 and 92 kDa) are the only enzymes capable of digesting collagen type I and type IV respectively. These collagens are the major constituents of the basement membrane (Type IV) and extracellular matrix (type I) of the endometrium. uPA and PAI-I contribute indirectly to matrix breakdown by controlling the generation of plasmin from its precursor plasminogen; plasmin activating procollagenase into collagenase(He et al 1989). As convincingly shown by Yagel et al (1988), trophoblast cells invade denuded amnion in vitro (basement membrane) and this invasion is abrogated by inhibitors of collagenase or plasmin. Furthermore, Librach et al (1991) showed that antibodies to uPA blocked only partially the invasion of cytotrophoblast cells into a basement membrane analogue Matrigel, whereas antibodies to the 92 kDa gelatinase inhibited this invasion completely.

Taken together these observations support the concept that trophoblast invasion into the endometrium depends on the activity of trophoblastic proteases capable of digesting the endometrial basement membrane and extracellular matrix.

REGULATION OF METALLOPROTEASE ACTIVITY

By using the classical technique of zymography on gelatin containing polyacrylamide gels (Bischof et al 1991) we analysed the culture medium of different cells types isolated from trophoblast, endometrium, fallopian tubes as well as granulosa cells, cumulus cells and human triploid 8 cells embryos (Tab. I). We were surprised to observe that all cells, except first trimester syncytium, expressed several gelatinases. Stromal and epithelial cells from the endometrium as well as lymphomyloid cells (CD45 positive cells) from the same tissue expressed several similar gelatinases. (Martelli et al 1993, Tab. I). The 89 kDa gelatinase

Table 1. Comparison of gelatinases secreted in vitro by different cell types of human reproductive tissues

Cell type	Molecular weight (kDa)							
	248	197	113	105	89	82	64-58	53
Trophoblast								
cytotrophoblast	-	(+)	-	-	+	+	+	-
syncytiotrophoblast	-	-	-	-	-	-	-	-
Endometrium								
Stroma cells	-	-	+	+	+	-	+	(+)
Epithelial cells	-	-	-	-	+	-	+	-
Lymphomyloid cells from[1]								
endometrial stroma	(+)	-	-	-	+	-	(+)	-
endometrial epithelium	-	-	-	-	+	-	(+)	-
trophoblast	+	+	-	-	+	(+)	+	(+)
fallopian tube	+	+	-	-	+	+	-	(+)
Embryo								
8 cell with 3 pronuclei	-	-	-	-	+	-	-	-
Other tissues								
Cumulus cells	-	-	-	-	+	-	-	-
Granulosa cells	-	+	-	-	+	-	-	-

[1]lymphomyloid cells were immunopurified from different tissues after enzymatic disruption and cultured in-vitro according to Bischof et al (1991).

(corresponding to the 92 kDa procollagenase type IV) was expressed by all cell types studied (Tab. I) whereas the 64-58 kDa gelatinase (corresponding to the 72 kDa procollagenase type IV and its activated form) was expressed by most cell types with the exception of 8 cells triploid human embryos, cumulus and granulosa cells (Tab. I). These observations imply that the secretion of metalloproteinases (gelatinases and collagenases) is not specific to cytotrophoblast cells and thus one can no longer view trophoblast invasion as the simple penetration of proteolytically active trophoblastic cells into a proteolytically inactive endometrium. Since convincing in vitro evidence (reviewed above) shows the necessity of active protease secretion for trophoblast invasion one must admit that during the invasion process, endometrial and trophoblastic metalloproteinases are both topographically and temporally precisely regulated.

Inhibitors

Many factors have been shown to regulate the activity of metalloproteinases. Inhibitors of metalloproteinases activity are present both in the circulation and the extracellular matrix: alpha 2 macroglobulin and alpha 1 antitrypsin are the major circulating inhibitors whereas Tissue Inhibitors of Metalloproteinases (TIMP-I and TIMP-2) are present in the extracellular matrix. TIMP-1 a 28 kDa glycoprotein forms a 1:1 stoichiometric complex with the 55 kDa interstitial collagenase and with the 92 kDa collagenase type IV (Cawston 1986). TIMP-2 (also called metalloproteinase inhibitor MI) is a 20 kDa glycoprotein, which has a 38 % sequence homology with TIMP-1, forms a complex with the 72 kDa collagenase type IV (Boone et al 1990). It is unknown if these inhibitors are produced by human endometrium and decidua. A TIMP like molecule is however produced by decidual cells in mice (Nomura et al 1989). None of these inhibitors appear highly selective in their inhibition and hence their precise role in modulating the activity of collagenases remains obscure. However, as shown by Librach et al (1991), TIMP-I inhibits both enzyme activity and trophoblast invasion into a basement membrane.

Endocrine factors

The endocrine regulation of collagenases and their inhibitors is not well understood and conflicting results have been published. Oestradiol and progesterone were shown to stimulate collagenase production in human cervical fibroblasts (Saito et al 1981). In contrast, in rat cervical fibroblasts, oestradiol and progesterone decreased the levels of procollagenase mRNA while increasing TIMP mRNA (Saito et al 1991). In our hands, oestradiol (30 nM), progesterone (300 nM) or oestradiol and progesterone decreased endometrial stromal cell collagenase type IV activity by 80, 90 and 90% respectively (Bischof, unpublished results) a result which was also observed by others (Dr. F. Schatz personal communication).

Para-autocrine factors

Increases in collagenase expression are induced by many growth factors and cytokines including Interleukin 1 (IL-I), Tumor Necrosis Factor (TNF), Epidermal Growth Factor (EGF), Platelet Derived Growth Factor (PDGF) and Fibroblast Growth Factor (FGF, see Tryggvason et al 1987 for review). Transforming Growth Factor ß (TGF ß) is a well recognized inhibitor of collagenase expression in synovial and lung fibroblasts (Edwards et al 1987) and inhibits trophoblast invasion in-vitro (Lala and Graham 1990). The inhibitory effect that conditioned medium from decidual cell cultures exerts on trophoblast invasiveness seems to be due to TGF ß since an antibody to this growth factor abolishes the inhibitory effect of decidual cell culture supernatant on cytotrophoblast invasiveness. Furthermore, an antibody to TIMP tested in the same conditions has the same effect as anti-TGF ß indicating that TGF ß inhibits collagenase activity and trophoblast invasiveness by inducing the release of TIMP. (Lala and Graham 1990). It is not known however if human decidua produces TGF ß.

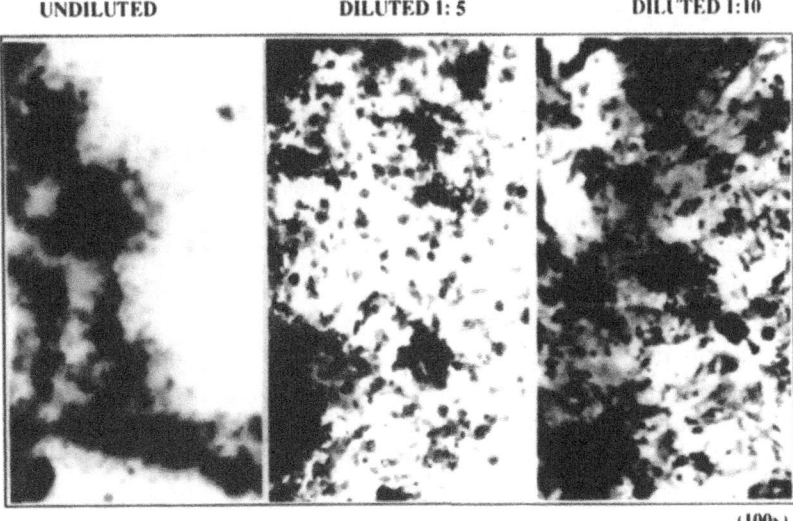

UNDILUTED   DILUTED 1: 5   DILUTED 1:10

(100x)

Figure 1. Effect of matrigel dilution on the proteolytic activity of first trimester cytotrophoblastic cells. Cells were plated at the same density (510^5 cells/well) on the same volume of matrigel (200 ul). Matrigel was prediluted with culture medium and allowed to gel before the cells were added.

In addition to the effects of cytokines and growth factors we made some interesting observations on other potential regulators. When cytotrophoblast cells are cultured on an analogue of basement membrane, Matrigel, they gradually digest the matrigel within a few days (Bischof et al 1991). However, the digestion areas become smaller when Matrigel is diluted before coating the plates (Fig. 1) and the activity of the 92 and 72 kDa type IV collagenases is reduced when analysed on a zymogram (result not shown). Matrigel is mainly composed of collagen type IV and laminin (an extracellular matrix glycoprotein). Thus by decreasing the concentration of these factors the activity of type IV collagenase decreases, indicating that factors from the extracellular matrix in the cellular micro-environment can also modulate the collagenolytic activity of the cytotrophoblast cells. This observation prompted us to analyse the effects of soluble laminin and fibronectin given to cytotrophoblast cells grown on plastic. As shown in Fig. 2, both matrix glycoproteins, laminin and fibronectin, increased the activity of the 92 and 72 kDa collagenase type IV. A similar observation had been published earlier by Emonard et al (1990). They showed that laminin increased the activity of trophoblastic interstitial collagenase (55 kDa). These observations give credit to the hypothesis that the cellular environment (matrix glycoprotein) modulates the proteolytic behaviour and probably the invasive potential of trophoblast cells.

The role of integrins

Cells recognize their environment (other cells, basement membrane or extracellular matrix) through membrane associated adhesion molecules. Adhesion molecules are grouped into five superfamilies: the integrins, the cadherins, the selectins, the immunoglobulins and the EGF family (Bock 1991). Integrins are a family of cell-surface receptors involved in cell-matrix adhesion (Buck and Horwitz 1987) and in cell-cell interaction (Hynes 1987). These membrane receptors are heterodimeric glycoproteins composed of an alpha and a ß subunit. The specificity of the integrin towards a ligand (matrix glycoprotein) depends on the type of alpha and ß subunit combination for example: alpha5 ß1 is the major fibronectin receptor, alpha2 ß1 is the major collagen type I receptor whereas alpha6 ß4 functions as a laminin receptor. (Albelda and Buck 1990). The alpha and ß subunits have a large amino-terminal extracellular domain, a hydrophobic transmembrane domain and a short carboxy-terminal cytoplasmic domain. The extracellular domain binds certain glycoprotein of the surrounding matrix and the intracellular domain is associated with the actin filaments of the cytoskeleton (Albelda and Buck 1990). Integrins are thus transducers of signals from the extracellular environment to the interior of the cell. This signal turns on genes the product of which will then modify the cellular microenvironment. Werb et al (1989) convincingly demonstrated in rabbit synovial fibroblasts that a monoclonal antibody to the fibronectin receptor which interferes with the binding of cells to

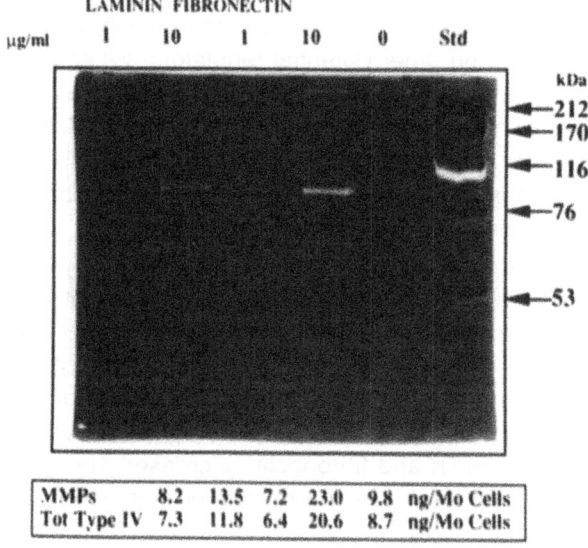

		LAMININ		FIBRONECTIN			
µg/ml		1	10	1	10	0	Std

						kDa
						◄—212
						◄—170
						◄—116
						◄—76
						◄—53

MMPs	8.2	13.5	7.2	23.0	9.8	ng/Mo Cells
Tot Type IV	7.3	11.8	6.4	20.6	8.7	ng/Mo Cells

Figure 2. Zymogram of supernatants of purified first trimester cytotrophoblast cells cultured on plastic (510[5] cells/well) in presence or absence of laminin or fibronectin. Proteolytic activity was also measured in the supernatants using denatured (60°C 30 min) ^{3}H collagen type IV. Digested ^{3}H collagen type IV was measured in the supernatant in presence or absence of 20mM Ethylene diamino tetraacetate (EDTA an inhibitor of metalloproteases MMP) Results are given by comparison to a standard of bacterial collagenases (Clostridium histolyticum) and expressed in ng/10[6] cells. <u>MMPs</u>, matrix metalloproteases, activity inhibitable by EDTA, <u>Tot-type IV</u>, Total type IV activity , total activity .

fibronectin (thus recognizes the antibody as it would be fibronectin) induces the expression of genes encoding interstitial collagenase and stromelysin (another metalloproteinase). An observation which fits very well with our observation of fibronectin stimulating type IV collagenases in cytotrophoblast cells. The mechanism of transduction from the binding of fibronectin to its membrane receptor to the induction of the genes coding for the collagenases is unknown, however it does require aggregation and conformational changes of the receptor (Werb et al 1989).

Cytotrophoblast cells do also express integrins. Immunolocalization studies (Korhonen et al 1991, Damsky et al 1992, Bischof et al 1993) showed that villous cytotrophoblast cells express an alpha6 ß4 integrin (laminin receptor) in a highly polarized pattern: alpha6 ß4 is only distributed at the zones of contact between the cytotrophoblast cells and the villous basement membrane. Intermediate cytotrophoblast cells (extravillous) expressed both the alpha6 ß4 and the alpha4 ß1 integrin (laminin and fibronectin receptors), but the alpha6 ß4 distribution was not polarized. Deeper in the endometrium, the cytotrophoblast cell expressed only the alpha5 ß1 integrin. Thus during trophoblastic invasion of the endometrium, cytotrophoblast cells change their integrin repertoire from being alpha6 ß4 positive and alpha5 ß1 negative to alpha6 ß4 negative and alpha5 ß1 positive. Note that villous cytotrophoblast cells (alpha6 ß4 positive) are non invasive whereas the intermediate cytotrophoblast cells (alpha5 ß1 positive) are highly invasive cells. One can speculate that these two types of cytotrophoblastic cells must secrete quite different amounts of metalloproteinases. Experiments are underway in our laboratory to investigate this possibility.

In order to see if collagenase expression in cytotrophoblast cells was also modulated by binding of matrix glycoproteins to their receptors, we incubated purified (Bischof et al 1991) first trimester cytotrophoblast cells with monoclonal antibodies to the alpha 2, alpha 5, and alpha 6 subunits of integrin. As shown in Fig. 3, in three different experiments run in duplicates, all antibodies used inhibited the secretion of the 72 and 92 kDa collagenase type IV as measured by zymography or by a quantitative assay. Note that in contrast to the study of Werb et al (1989) reported above, the antibodies used here did not inhibit the adhesion of the cells to the culture dish and thus the inhibition of collagenases secretion is most probably due to an inhibition of receptor aggregation and consequently to an inhibition of the integrin transduction mechanism. We thus conclude that metalloprotease secretion by cytotrophoblast cells is under the control of the cellular microenvironment through, at least, the fibronectin, the laminin and the collagen type I receptors.

CONCLUSION

Based on the summarized herein observations but fully aware that the results obtained in-vitro with first trimester cytotrophoblast cells might not accurately reflect the in-vivo behaviour of blastocyst's trophectodermal cells, we

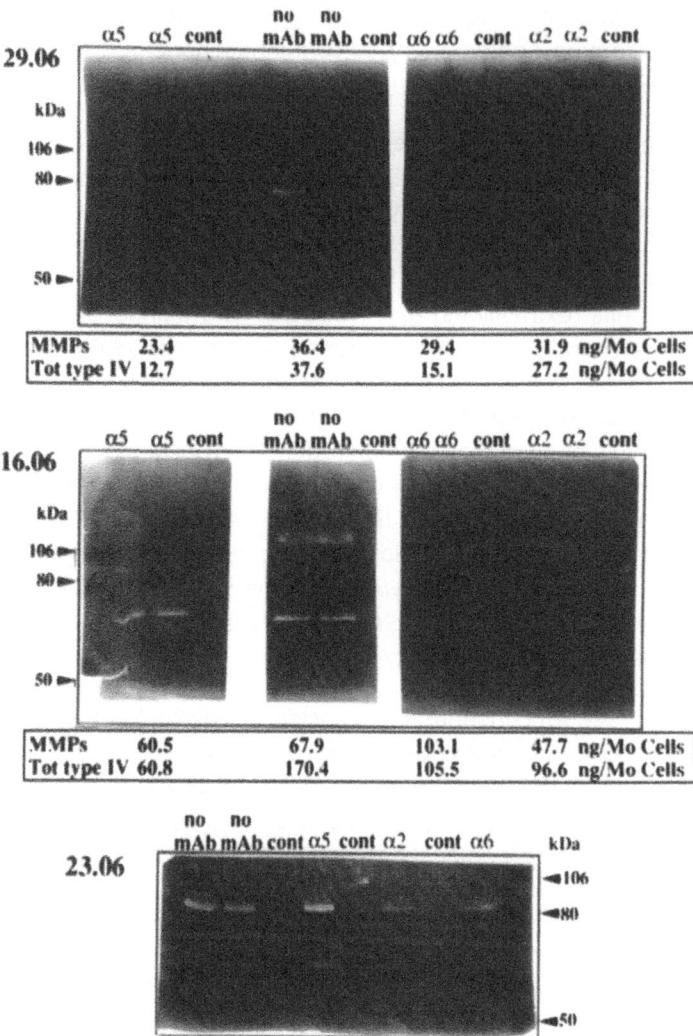

Figure 3. Effect of monoclonal antibodies (diluted 1:20) to alpha 2, alpha 5 and alpha 6 integrin subunits on the expression of collagenases by purified first trimester cytotrophoblast cells (510^5 cells/well). Zymograms and quantitative measurements (see legend of fig. 2) of 3 different experiments run in duplicates.

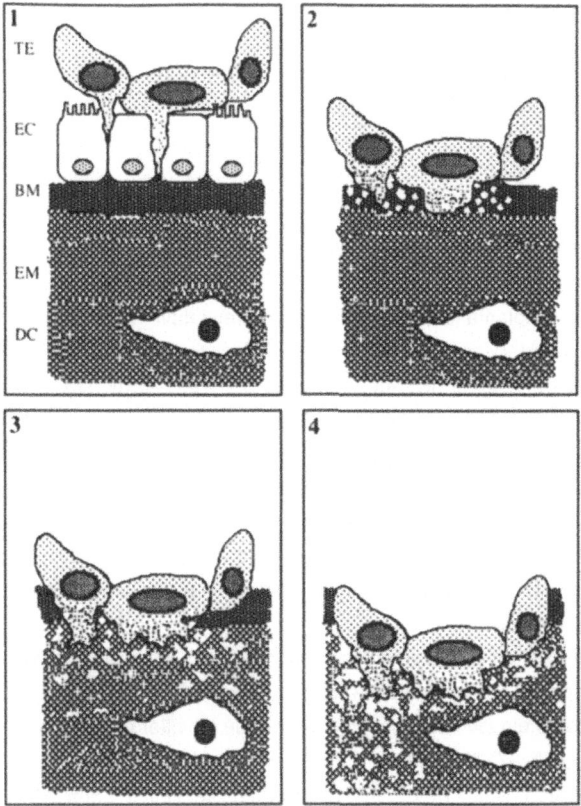

Figure 4. A molecular model for blastocyst implantation (detailed description see conclusions).

would like to propose a molecular model for the mechanism of trophoblast invasion (Fig. 4). After an initial attachment to the endometrial lining, the blastocyst's trophectodermal cells progress towards the endometrial basement membrane in an intrusive way by expanding invadopodes. These invadopodes bind to the basement membrane through their laminin (alpha 6, ß 4 integrins) and collagen type IV (alpha 1, ß 1 integrins) receptors (Fig. 4.1). This binding between the basement membrane glycoproteins and their receptors on the trophectodermal cells induces the secretion of the 72 and 92 kDa collagenases type IV which digest the basement membrane underneath the cells (Fig. 4.2). This allows the trophectodermal cells to penetrate through the basement membrane and to come in contact with the underlying extracellular matrix. The trophectodermal cells bind to the extracellular matrix through their fibronectin receptors (alpha 5, ß 1 integrin). This binding induces the secretion of interstitial collagenase which digests the surrounding extracellular matrix allowing the cells to progress further (Fig. 4.3). During their progression, the collagenolytic activity of the trophectodermal cells is modulated by inhibitory factors (TIMP and/or TGF ß) secreted by the endometrial cells (Fig. 4). Although this model is most probably an over simplification of the many mechanisms controlling trophoblast cell mobility and proteases secretion, we found it helpful to formulate our hypothesis.

ACKNOWLEDGEMENTS

The authors express their gratitude to the Swiss National Fund for Scientific Research which funded our research.

REFERENCES

Albelda, S.M. and Buck, C.A., 1990, Integrins and other adhesion molecules, *FASEB J.* 4: 2868-2880.

Aplin, J.D., 1991, Implantation, trophoblast differentiation and haemochorial placentation: mechanistic evidence in vivo and in vitro, *J. Cell. Sci.* 99: 681-692.

Bischof, P., Friedli, E., Martelli, M. and Campana, A., 1991, Expression of extracellular matrix-degrading metalloproteinases by cultured human cytotrophoblast cells: effect of cell adhesion and immunopurification, *Am. J. Obstet. Gynecol.* 165: 1791-1801.

Bischof, P. and Martelli, M., 1992, Proteolysis in the penetration phase of the implantation process, *Placenta* 13: 17-24.

Bischof, P., Redard, M., Gindre, P., Vassilakos, P. and Campana, A., 1993, Localization of alpha 2, alpha 5 and alpha 6 integrin subunits in human endometrium, decidua and trophoblast, *Placenta Submitted.*

Bock, E., 1991, Cell-cell adhesion molecules. *Biochem. Soc. Transactions,* 19: 1076-1081.

Boone, T.C., Johnson, M.J., De Clerk, Y.A. and Langley, K.E., 1990, cDNA cloning and expression of a metalloproteinase inhibitor related to tissue inhibitor of metalloproteinases, *Proc. Natl. Acad. Sci. USA.* 87: 2800-2804.

Boyd, J.D. and Hamilton, W.J., 1970, The Human Placenta, *Hefer, Cambridge.*

Buck, C. and Howitz, A., 1987, Cell-surface receptors for extracellular matrix molecules, *Annu. Rev. Cell. Biol.* 3: 179-205.

Cawston, T.E., 1986, Protein inhibitors of metalloproteinases, In Proteinase Inhibitors, *Barret and Salvesen (ed), Elsevier Amsterdam* 589-610.

Damsky, C.H., Fitzgerald, M.L. and Fisher, S.J., 1992, Distribution patterns of extracellular matrix components and adhesion receptors are intricately modulated during first trimester cytotrophoblast differentiation along the invasive pathway, in-vivo, *J. Clin. Invest.* 89: 210-222.

Edwards, D.R., Murphy, G., Reynolds, J.J., Whitham, S.E., Docherty, A.J.P., Angel, P. and Heath, J.K., 1987, Transforming growth factor ß modulates the expression of collagenase and metalloprotease inhibitor, *EMBO. J.* 6: 1899-1904.

Emonard, H., Christiane, Y., Smet, M., Grimaud, J.A. and Foidart, J.M., 1990, Type IV and interstitial collagenolytic activities in normal and malignant trophoblast cells are specifically regulated by extracellular matrix, *Invasion & Metas.* 10: 170-177.

Feinberg, R.F., Kao, L.C., Haimowitz, J.E., Queenan, J.T., Wun, T.C., Strauss III, J.F. and Kliman, H.J., 1989, Plasminogen activator inhibitor type 1 and 2 in human trophoblast, PAI-I is an immunocytochemical marker of invading trophoblast, *Lab. Invest.* 61: 20-26.

Fisher, S.J., Li Zhang, T.C., Hartmann, L., Grahl, K., Guo-Yang, Z., Tarpey, J. and Damsky, C., 1989, Adhesive and degradative properties of human placental cytotrophoblast cells in vitro, *J. Cell. Biol.* 109: 891-902.

He, C., Wilhelm, S.M., Pentland, A.P., Marker, B.L., Grant, G.A., Eisen, A.Z. and Goldberg, G.I., 1989, Tissue cooperation in a proteolytic cascade activating human interstitial collagenase, *Proc. Nat. Acad. Sci. USA* 86: 2632-2636.

Hynes, R.O., 1987, Integrins: a family of cell surface receptors, *Cell* 48: 549-555.

Korhonen, M., Ylänne, J., Laitinen, L., Cooper, M.M., Quaranta, V. and Virtanen, I., 1991, Distribution of alpha 1, alpha 6 integrin subunits in human developing and term placenta, *Lab. Invest.* 65: 347-356.

Lala, P.K. and Graham, C.H., 1990, Mechanisms of trophoblast invasiveness and their control: the role of proteases and protease inhibitors, *Cancer Metas. Review* 9: 369-379.

Librach, C.L., Werb, Z., Fitzgerald, M.L., Chin, K., Couvin, N.M., Esteves, R.A., Grobelny, D., Galardy, R., Damsky, C.H. and Fisher, S.J., 1991, 92 kDa Type IV collagenase mediates invasion of human cytotrophoblasts, *J. Cell. Biol.* 113: 437-449.

Martelli, M., Campana, A. and Bischof, P., 1993, Secretion of matrix-metalloproteinases by human endometrial cells in vitro, *J. Reprod. Fertil. in Press.*

Martin, O. and Arias, F., 1982, Plasminogen activator production by trophoblast cells in vitro: effect of steroid hormones and protein synthesis inhibitors, *Am. J. Obstet. Gynecol.* 142: 402-409.

Moll, U.M. and Lane, B.L., 1990, Proteolytic activity of first trimester human placenta: localization of interstitial collagenase in villous and extravillous trophoblast, *Histochem.* 94: 555-560.

Mullins, D.E. and Rohrlich, S.T., 1983, The role of proteinases in cellular invasiveness, *Biochim. Biophys.* 695: 177-214.

Nomura, S., Hogan, B.L.M., Wills, A.J., Heath, J.K. and Edwards, D.R., 1989, Developmental expression of TIMP RNA, *Develop.* 105: 575-583.

Queenan, J.T., Kas, L.C., Arboleda, C.A., Ulloa-Aguirre, A., Golos, T.G., Cines, D.B. and Strauss III, J.F, 1987, Regulation of urokinase-type plasminogen activator production by cultured human cytotrophoblasts, *J. Biol. Chem.* 262: 10903-10906.

Saito, Y., Takahashi, S. and Meki, M., 1981, In vitro effect of some free estrogen or estrogen precursors on collagenase activity of uterine cervix, *Nippon Samka Fujinka Gakkai Zasshi* 32: 827-832.

Sato, T., Ito, A., Mori, Y., Yamashita, K., Hayakawa, T. and Nagase, H., 1991, Hormonal regulation of collagenolysis in uterine cervical fibroblasts, *Biochem. J.* 275: 645-650.

Tryggvason, K., Höyhtyä, M. and Sato, T., 1987, Proteolytic degradation of extracellular matrix in tumor invasion, *Biochem. Biophys. Acta.* 907: 191-217.

Werb, Z., Tremble, P.M., Behrendtsen, O., Crowley, E. and Damsky, C., 1989, Signal transduction through the fibronectin receptor induces collagenase and stromelysin gene expression, *J. Cell. Biol.* 109: 877-889.

Yagel, S., Parhar, R.S., Jeffrey, J.J. and Lala, P.K., 1988, Normal nonmetastatic human trophoblast cells share in-vitro invasive properties of malignant cells, *J. Cell. Physiol.* 136: 455-462.

Yeh, I.T. and Kurman, R.J., 1989, Functional and morphological expression of trophoblast, *Lab. Invest.* 61: 1-14.

FUTURE PROSPECT OF RESEARCH ON ENDOCRINOLOGY OF EMBRYO-ENDOMETRIAL INTERACTIONS

Koji Yoshinaga

Reproductive Sciences Branch,
Center for Population Research
NICHD, NIH, Bethesda, MD 20892

INTRODUCTION

Although there are many interesting maternal-embryo interactions that deserve careful attention, this presentation will focus on the following four topics that represent, in my personal opinion, areas of investigation that attract research interest in the near future: (1) endocrine regulation in early pregnancy, (2) uterine receptivity for blastocyst implantation, (3) *in vitro* culture systems, and (4) step-by-step consideration of blastocyst-endometrial interactions.

ENDOCRINE REGULATION OF EARLY PREGNANCY

It is well established that progesterone plays an important role in preparation of the endometrium for the implanting blastocyst and in the subsequent maintenance of pregnancy. We know that there are species variations in the luteotropic complex during early pregnancy. In the human the pituitary gonadotropins, particularly LH, are responsible for the initial maintenance of luteal function, while in small laboratory rodents it is essential to have prolactin secretion to activate the non-functional corpora lutea of the estrous cycle. After fertilization the peri-implantation embryo in the genital tract plays an important role to assure an adequate supply of progesterone.

There are at least three types of embryonic participation in luteal maintenance; the species of mammals that belong to these types are (1) primates, (2) rodents, and (3) sheep and cows. The first two types use luteotropic hormones of the trophoblastic origin to stimulate the luteal cells directly, i.e., by CG (chorionic gonadotropin, an LH-like hormone) in primates, or by placental luteotropin (a prolactin-like hormone) in rodents (Yoshinaga, 1977). In the third type, trophoblastic proteins include interferons and these proteins act on the uterine epithelium to dampen the pulsatile secretion of prostaglandin $F_{2\alpha}$ (PGF $_{2\alpha}$) which is a potent luteolytic factor, prior to the expected onset of luteolysis so that progesterone secretion by the corpus luteum is maintained (Roberts, 1989).

According to a recent report (Wiltbank et al., 1992), the sheep conceptus also appears to exert a direct effect on the corpus luteum. These workers found that $PGF_{2\alpha}$ inhibited progesterone production by lipoprotein-stimulated large luteal cells in vitro. This anti-steroidogenic action was blocked in a dose-dependent manner by conceptus proteins (28kd) secreted from day 15 embryos. Purified ovine trophoblast protein-1 (oTP-1) did not exhibit the anti-PGF $_{2\alpha}$ activity. On the other hand, conceptus-derived proteins devoid of oTP-1 did prevent the antisteroidogenic effects of PGF $_{2\alpha}$ This recent finding adds a new dimension to the luteotropic complex during early pregnancy in the sheep.

The involvement of embryonic interferons in the luteotropic complex in the sheep and cows makes one think whether embryonic interferons are also involved in the luteotropic complex, or in maternal recognition of pregnancy in other species of animals. What we know about the endocrine regulatory mechanism is limited to only a few mammalian species and the complexity of the luteotropic mechanisms in these known species appears to warrant further investigation in other species of animals. Interferons have antiviral and immunological activities. Recent findings of the involvement of lymphokines and cytokines in endocrine functions make it necessary to investigate the mode of interaction between the endocrine system and the immune system. If there is cross-talk between the immunological system and the endocrine system in terms of establishment and maintenance of early pregnancy, the communication between the two systems may not only exist between the uterus and the ovary, but may also involve intracellular signal transduction systems within the target cells that contain receptors for not only endocrine/paracrine regulators, but for immune ligands as well.

Although the corpus luteum of early pregnancy is maintained mainly by the hormones of pituitary and trophoblastic origin, there is, at least in the rat, a luteotropin secreted from decidual tissue (Gibori et al., 1987). Human decidual cells secrete prolactin, but the luteotropic activity of human decidual tissue has not been substantiated.

UTERINE RECEPTIVITY FOR BLASTOCYST IMPLANTATION

Using the rat model, Psychoyos (1973) demonstrated that the uterus can accept the blastocyst to implant only for a brief period of time. He defined the state of the uterine sensitivity (the neutral, receptive and refractory phases) for blastocyst implantation. Electronmicroscopic studies revealed that the endometrial luminal epithelial cells exhibit numerous large protrusions (pinopods) into the lumen during the receptive period. The stage-specific appearance of pinopods and glycoproteins on the surface of uterine epithelial cells (Martel et al., 1989) and reduction in the negative charge of the luminal epithelium have been reported to be characteristics of the receptive endometrium. However, these indices have not been used in clinical studies because they are not practically applicable. From the clinical point of view, there is an urgent need for some measurable markers that indicate the uterus is indeed receptive or going to become receptive within a certain period of time, so that one can perform a timely transfer of embryos fertilized *in vitro*. But these markers unfortunately are not available at the present time.

As one of the steps to solve this problem, the National Cooperative Program on Markers of Uterine Receptivity for Blastocyst Implantation has recently been initiated. This program is, by a collaborative effort of multiple research sites, to identify and characterize some useful markers of the receptive uterus for non-human blastocyst implantation. It is hoped that this program will expand in the future to include human studies.

IN VITRO CULTURE SYSTEMS

In research on implantation, one of the difficulties is accessibility to the site of implantation. One way to resolve this problem is to establish *in vitro* models. Appropriate usage of extracellular matrix components appears to be important to establish polarized uterine epithelial cell culture systems which are hormone responsive (Glasser et al., 1991). The article by Kimber in this volume provides the information that attachment of the blastocyst is influenced by manipulating the ratio of progesterone to estrogen. Studies of this kind are considered to be especially useful to focus one's attention on some of the specific stages of the implantation process. It is a difficult task to develop one culture system which has the same structure and function of the endometrium *in vivo*. Instead, it would be more practical to develop separate culture systems that may be appropriate for studies on only one step of the implantation process. For example, one system may be useful for a study of attachment by cell adhesion, another culture system may be suitable for a study of penetration of the basement membrane, and still another system may be useful for a study of further invasion into the stromal tissue. Much effort and time will be needed to establish *in vitro* systems for studies of sophisticated local embryo-endometrial structures (e.g. epithelium + basal lamina; trophoblast + epithelium; etc.). However, *in vitro* systems will become essential when specific molecules, e.g. "receptive uterus marker genes", are found and one tries to do gene expression studies *in vitro* using the DNA transfected in a cell line which is unrelated to the reproductive system.

STEP-BY-STEP CONSIDERATION OF TROPHOBLAST-UTERINE INTERACTIONS

The concept of involvement of ligand-receptor interactions during the implantation process was developed by early researchers who found mouse blastocysts attach selectively to extracellular matrix components. These workers confirmed the presence of the receptors on the trophoblast by pretreatment of blastocysts with an excess amount of ligands to prevent their attachment to the respective component. Since this concept was reviewed by this author at a previous symposium (Yoshinaga, 1989), it has been gaining ground with increasing evidence that the uterine epithelial membrane contains ligands and the trophectoderm membrane contains their receptors, and that the appearance of some of these molecules is hormone dependent (Dey *et al.*, 1991). Identification of some molecules in the trophectoderm such as EGF receptors, and localization of their ligands on the apical surface of the receptive uterine epithelial cells give the impression that a ligand-receptor interaction is very likely to be taking place at the time of attachment.

Cell to Cell Attachment - Cell Surface Glycoproteins

Analysis of the stage-specific and hormone-dependent glycoproteins on the apical surface of the receptive uterine luminal epithelial cells is expected to reveal what kind of glycoproteins, integrins and their ligands, are involved in the attachment of trophoblast cells to the uterine epithelial cells. Particularly it has been suggested that the structure of oligosaccharide is important for attachment of blastocysts to the uterine surface (Kimber and Lindenberg, 1990). A recent report suggests that apical mucin glycoproteins of uterine epithelial cells provide an enzymatically resistant barrier, and drastic reduction of mucin expression is likely to be required to permit cellular access to the apical surface of uterine epithelial cells. The importance of heparan sulfate proteoglycan has also been implicated in embryo attachment to the uterine surface (Carson *et al.*, 1993). Further analysis of the cell surface molecules appears to warrant new information useful for elucidation of the initial attachment mechanism. One thing we have to bear in our mind is that the cell surface changes caused by estrogen alone should be distinguished from those changes caused by the estrogen-progesterone regimen during the preimplantation period. Since pinopods are said to be one of the markers of receptivity, functional studies of this stage specific structure are needed.

Trophoblastic Intrusion of the Epithelium

In guinea pigs, rhesus monkeys and humans where implantation is the intrusive type and, even in the rat where implantation is the displacement type, the initial attachment of trophoblastic membrane on the apical surface of the endometrial epithelial cells is followed by intrusion of the trophoblast processes between the epithelial cells. The epithelial cells are tightly bound together near their apical regions with tight junctions. How, then, can trophoblast cells insert themselves between the two cells? Since the epithelial cells at this early stage of implantation do not appear to be degenerative, it is reasonable to assume that the tight junctions are either "relaxed" or dissociated to give way to the intruding trophoblastic processes.

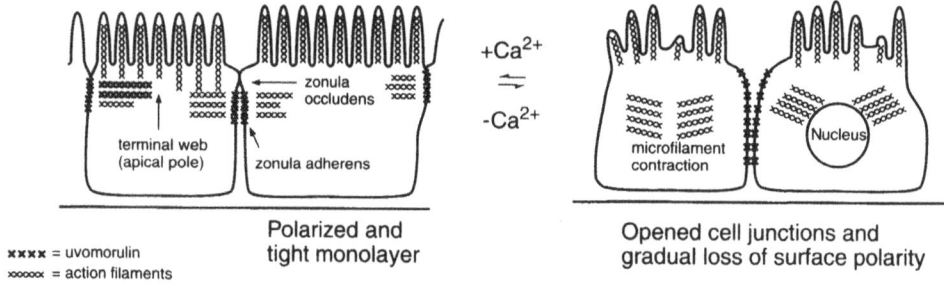

Figure 1. Effect of Ca^{++} on cell junctions. (Modified from Gumbiner & Simons, 1986)

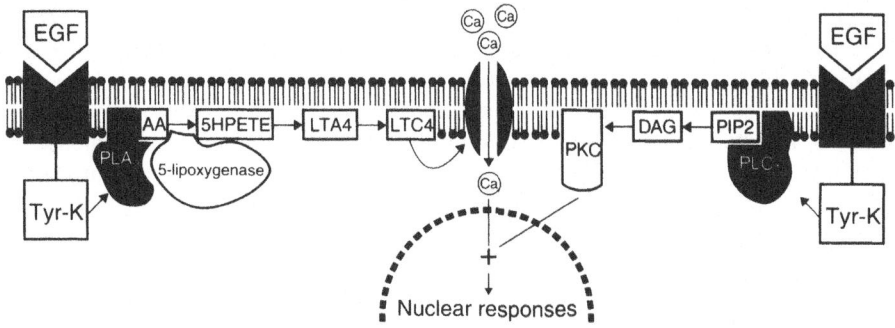

Figure 2. Epidermal growth factor activates calcium channels. (from M.P. Peppelenbosch et al.,1992)

If the tight junctions are dissolved or digested by the intruding trophoblast processes, it may be possible that the trophoblast processes may secrete specific enzymes to digest proteins specific to tight junction such as ZO-1 or cingulin (Citi, 1993). Another possibility is that the tight junctions are dissociated by the trophoblast processes. If the latter is the case, how does it happen? The following is a hypothetical mechanism: EGF on the uterine epithelium binds the EGF receptors on the trophoblast cell membrane; this activates the Ca^{++} channel of the trophoblast cell and increases Ca^{++} influx into the trophoblast cell, thus resulting in a reduction in Ca^{++} concentration in the uterine fluid trapped between the epithelial cells and the trophoblast cell; the reduction of Ca^{++} in the uterine fluid locally dissociates the tight junctions of the uterine epithelial cells. Figure 1 illustrates the relationship between the establishment of tight junctions and Ca^{++} in the culture medium of established polarized epithelial cells (Gumbiner and Simons, 1986).

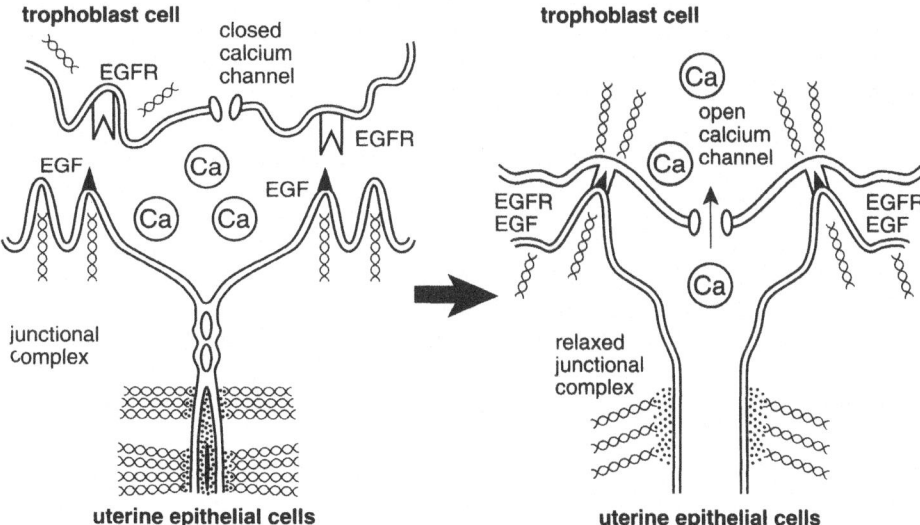

Figure 3. A hypothetical mechanism of trophoblast intrusion between the uterine epithelial cells at the initial stage of implantation (see text for explanation).

When EGF binds to EGF receptors, EGF induces a Ca^{++} influx in many cell types (Figure 2, Peppelenbosch et al., 1992). Estrogen enhances levels of transcripts for HB-EGF (heparin binding EGF-like growth factor) in the rat uterine luminal epithelium (Zhang et al., 1994). Ligands for EGF receptors (EGF/TGF/HB-EGF) increases in the apical portion of the uterine epithelial cells during the receptive period of the mouse (Das et al., 1993). EGF receptors appear on the trophoblast membrane prior to the time of implantation (Dey et al., 1991). Therefore, this hypothetical mechanism (Figure 3) appears theoretically possible, but needs verification.

Junctional Complex Formation

The electronmicroscopic studies of cell to cell attachment at the early stages of implantation clearly indicate the establishment of junctional complexes between the trophoblast cell and the uterine epithelial cell. These junctional complexes established between the trophoblastic cell membrane and that of the maternal cell are tight junctions and desmosomes. The desmosome is a complex disk-shape structure at the surface of one cell that is matched with an identical structure at the surface of another cell. The cell membranes are very flat and some material is often present in the intercellular space. Inside the membrane of each cell is a circular plaque and groups of intermediate filaments of the cytokeratin family are inserted into the plaque or make hairpin turns to the cytoplasm. The function of the desmosome is to provide a firm attachment between the cells. Since the formation of junctions is considered to require a prior cell adhesion molecule-mediated event, it is quite likely that the initial attachment of the trophoblast cell membrane and the uterine epithelial cell membrane involves cell adhesion proteins.

Although the process of the establishment of junctions between the trophoblastic cells and the uterine luminal epithelial cells has not been clarified, it may be postulated that trophoblast and epithelial cells bind each other by means of a carbohydrate linkage and a ligand-receptor binding mechanism. Many membrane receptors have been shown to be coupled to G proteins, and the binding of a ligand to its receptor results in activation of G protein as a means of signal transduction. The assembly of cytoskeleton components such as actin and tubulin has been shown to be controlled by G proteins, and the ligand binding results in rearrangement of cytoskeletal elements. The ligand-receptor binding also causes lateral movement of the bound receptors in the plasma membrane resulting in clustering, patching and capping of the receptors. This lateral movement of bound receptors in the membrane is regulated by cytoskeletal tubules and microfilaments. Thus, it may be considered that the formation of desmosomes is a result of ligand-receptor binding. The desmosome formation between the trophoblast and the uterine cells is a very important process for firm attachment and further penetration of the entire blastocyst into the stroma and for establishment of placental blood circulation.

Movement of Blastocyst after Desmosome Formation

The formation of desmosomes between the trophoblast and uterine epithelial cells appears to be important for secure attachment of the blastocyst to the uterine wall. However, how does the attached blastocyst insert itself into the epithelium? The reader is referred to Preston et al. (1990) for comprehensive writings on the cytoskeleton and cell motility. The mechanism for the crawling movement of tissue cells such as fibroblasts appears to be useful for our investigation of the trophoblast cell movement. Fibroblasts advance *in vitro* across the substratum by the forward protrusion of the anterior of the cell. In this connection it is interesting to note that the point of attachment of a cell to the substratum by means of the ligand-receptor binding mechanism is often associated with actin filaments, thus attachment and mobility appear to be coordinated at the subcellular local level. The movement of membrane and cytoskeletal fiber elements in the advancing pseudopod of an Ascaris (pin worm) sperm, for example, is done by continuous assembly and disassembly of proteins and lipids. It is fascinating to imagine that the fibrous complex, called a villipodium in this species, may be considered an equivalent of the desmosome in trophoblast cells of the implanting blastocyst. Clarification of the cellular and molecular mechanisms involved in the blastocyst movement in implantation is an area wide open for research.

Penetration of the Basement Membrane

Trophoblast cells invading through the uterine luminal epithelium stop at the basement membrane when they reach it before breaching the membrane (Schlafke *et al.*, 1985). It takes some time to penetrate the membrane. As has been suggested by Liotta *et al.*, (1990) for the cellular mechanism of cancer cell invasion of the basal lamina, attachment of the trophoblast cells to the basal lamina appears to be mediated by the receptors on the invading cell surface to their ligands in the basal lamina. *In vitro* mouse trophoblast cells attach to various components of extracellular matrix including laminin, type IV collagen and fibronectin by ligand-receptor binding mechanism. A recent study (Yelian *et al.*, 1993) indicates that entactin, another component of the basement membrane, plays an important role in blastocyst penetration through this membrane. Using a mouse trophoblast outgrowth model system, these workers showed that the outgrowth is promoted by a mechanism mediated by the amino acid sequence Arg-Gly-Asp (RGD), the integrin recognition site. If we apply the cancer cell invasion hypothesis to the blastocyst invasion of the basal lamina, the trophoblast on the basal lamina attaches itself by means of receptors such as laminin receptors and anchors itself to the membrane. The trophoblast, next, secretes proteases to break down type IV collagen that is one of the major components of the basal lamina. The trophoblast cell then penetrates the basal lamina. As trophoblast cells invade the uterus, it is ordinarily considered that the invading trophoblast cells breach the basement membrane. However, it is decidual cells, the cells on the other side of the basement membrane, that penetrate the basement membrane in the case of the rat (Schlafke *et al.*, 1985). This may be explained by a hypothesis that prostaglandins produced by trophoblast cells penetrate the basement membrane and act on decidual cells to stimulate the activity of type IV collagenase. Type IV collagenase breaks the membrane from the stromal side. Since prostaglandin is produced by blastocysts and prostaglandin has been shown to activate type IV collagenase in different cell types, localization of the collagenase activity will solve this problem.

Although I have reviewed only some areas of the research I am interested in, the research on the maternal-embryo interactions during the periimplantation period needs to be carried out in a step-by-step manner as emphasized by Schlafke and Enders (1975). Further elucidation of the mechanisms applicable to each step will bring us useful information to construct a complete picture of blastocyst implantation.

SUMMARY

In summary, we need more information on endocrine regulation of early pregnancy. Mechanisms by which the pituitary-ovarian feedback system is altered from the cyclic pattern to that of pregnancy is not well elucidated. Particular attention is needed to the local mechanisms: secretion of molecules by the conceptus, response of the uterus to the conceptus molecules and secretion of uterine specific molecules that influence the fetus and the ovarian tissues, and alteration of the ovarian function by the presence of conceptus in the uterus. Simple extrapolation from one species to another may not be possible because of well-known species variations in the mode of blastocyst implantation.

We need to clarify what is the receptivity of the uterus that allows the blastocyst to implant. How the uterine surface that is usually indifferent to blastocyst is made responsive to the presence of a mature blastocyst? What markers that indicate the uterus is receptive are available? Can we use the markers, when and if they are available, to predict the uterus is going to become receptive so that we can improve the pregnancy rate after embryo transfer?

We need to establish useful *in vitro* culture systems to study blastocyst - uterine interactions. However, in establishing a system, we have to aim at clarification of the mechanisms in volved in a specific stage of implantation or a specific cell-to-cell interaction.

We need to start sorting out all of the hormones and factors thus far found to be involved in implantation, and to construct a sequence of events in considering the morphological aspect of the cell-to-cell interactions. This practice will assist finding more new molecules and to assign the positions and roles of essential molecules in the "cascade" of implantation phenomenon.

REFERENCES

Carson, D.D., Tang, J.-P. and Julian, J., 1993, Heparan sulfate proteoglycan (perlecan) expression by mouse embryos during acquisition of attachment competence. Dev. Biol., 155: 97-106.

Citi, S., 1993, The molecular organization of tight junctions, J. Cell Biol., 121: 485-489.

Das, S.K., Wang, X.N., Klagsbrun, M., Abraham, J. and Dey, S.K., 1993, Heparin-binding EGF-like growth factor (HB-EGF) gene is expressed in the periimplantation mouse uterus and regulated by progesterone and estrogen, Abstr. #69, Annual Meeting of the Society for the Study of Reproduction, Fort Collins, CO.

Dey, S.K., Paria, B.C. and Andrews, G.K., 1991, Uterine EGF ligand-receptor signalling and its role in embryo-uterine interactions during implantation in the mouse, in: "Uterine and Embryonic Factors in Early Pregnancy", J.F. Strauss, III and C.R. Lyttle, eds., Plenum Press, New York, pp 51-69.

Gibori, G., Jayatilak, P.G., Kahn, I., Rigby, B., T. Puryear, T., Nelson, S. and Herz, Z., 1987, Decidual luteotropin secretion and action: its role in pregnancy maintenance in the rat. in: "Regulation of Ovarian and Testicular Function," V.B. Mahesh et al., eds. Plenum Press, New York pp 379-397.

Glasser, S.R., Mani, S.K., and Mulholland, J., 1991, In vitro models of implantation, *in*: "Uterine and Embryonic Factors in Early Pregnancy," J.F. Strauss, III and C.R. Lyttle eds., Plenum Press, New York pp 33-50.

Gumbiner, B. and Simons, K., 1986, The role of uvomorulin in the formation of epithelial occluding junctions. in: "Junctional Complex of Epithelial Cells" (Ciba Foundation Symposium 125) Wiley, Chichester, pp 168-186.

Martel, D., R. Frydman, R., Sarantis, L., Roche, D. and Psychoyos, A., 1989, Scanning electron microscopy of the uterine luminal epithelium as a marker of the implantation window, *in*: "Blastocyst Implantation", K. Yoshinaga, ed., Adams Publishing Group, Boston (Serono Symposia, USA) pp 225-230.

Kimber, S.J. and Lindenberg, S., 1990, Hormonal control of a carbohydrate epitope involved in implantation in mice, J. Reprod. Fertil. 89:13-21.

Liotta, L.A., Steeg, P.S. and Stetler-Stevenson, W.G., 1990, Cancer metastasis and angiogenesis: an imbalance of positive and negative regulation, Cell 64: 327.

Peppelenbosch, P., Tertoolen, L.G.J., den Hertog, J. and de Laat, S.W., 1992, Epidermal growth factor activates calcium channels by phospholipase A_2/5-lipoxydase-mediated leukotriene C_4 production, Cell, 69:295-303.

Preston, T.M., King, C.A. and Hyams, J.S., 1990, The Cytoskeleton and Cell Motility, Blackie and Son Ltd., Glasgow.

Psychoyos, A., 1973, Hormonal control of ovoimplantation, Vitam. Hormon. 31:205-255.

Roberts, R.M., 1989, Minireview: Conceptus interferons and maternal recognition of pregnancy, Biol. Reprod. 40: 449-452.

Schlafke, S. and Enders, A.C., 1975, Cellular basis of interaction between trophoblast and uterus at implantation, Biol. Reprod. 12:41-65.

Schlafke, S., Welsh, A.O. and Enders, A.C., 1985, Penetration of the basal lamina of the uterine luminal epithelium during implantation in the rat. Anat. Rec. 212:47.

Wiltbank, M.C., Wiepz, G.J., Knickerbocker, J.J., C.J. Belfiore, C.J. and Niswender, G.D., 1992, Proteins secreted from the early ovine conceptus block the action of prostaglandin $F_{2\alpha}$ in large luteal cells. Biol. Reprod. 46:475-482.

Yelian, F.D., Edgeworth, N.A., Dong, L.-J., Chung, A.E. and Armant, D.R., 1993, Recombinant entactin promotes mouse primary trophoblast cell adhesion and migration through the Arg-Gly-Asp (RGD) recognition sequence. J.Cell Biol. 121:923-929.

Yoshinaga, K., 1977, Hormonal interplay in the establishment of pregnancy, *in*: "International Review of Physiology, Reproductive Physiology II," R.O. Greep, ed., University Park Press, Baltimore pp 201-223.

Yoshinaga, K., 1989, Receptor concept in implantation research, *in*: "Development of Preimplantation Embryos and Their Environment", K. Yoshinaga and T. Mori, eds., Alan R. Liss, Inc., New York pp 379-387.

Zhang, Z., Funk, C., Roy, D., Glasser, S.R. and Mulholland, J., 1994, Differential regulation of HB-EGF expression in the rat uterus by steroid hormones. Endocrinology 134: 1089–1094.

INDEX

The manufacturer's authorised representative in the EU is Springer
Nature Customer Service Centre GmbH, Europaplatz 3, 69115 Heidelberg,
Germany. If you have any concerns regarding our products, please
contact ProductSafety@springernature.com

Printed and bound by CPI Group (UK) Ltd, Croydon, CR0 4YY
23/04/2026
02095628-0014